PROLACTIN

ENDOCRINE UPDATES

Shlomo Melmed, M.D., Series Editor

---

1. E.R. Levin and J.L. Nadler (eds.): Endocrinology of Cardiovascular Function. 1998. ISBN: 0-7923-8217-X
2. J.A. Fagin (ed.): Thyroid Cancer. 1998. ISBN: 0-7923-8326-5
3. J.S. Adams and B.P. Lukert (eds.): Osteoporosis: Genetics, Prevention and Treatment. 1998. ISBN: 0-7923-8366-4.
4. B.-Å. Bengtsson (ed.): Growth Hormone. 1999. ISBN: 0-7923-8478-4
5. C. Wang (ed.): Male Reproductive Function. 1999. ISBN 0-7923-8520-9
6. B. Rapoport and S.M. McLachlan (eds.): Graves' Disease: Pathogenesis and Treatment. 2000. ISBN: 0-7923-7790-7.
7. W. W. de Herder (ed.): Functional and Morphological Imaging of the Endocrine System. 2000. ISBN 0-7923-7923-9
8. H.G. Burger (ed.): Sex Hormone Replacement Therapy. 2001. ISBN 0-7923-7965-9
9. A. Giustina (ed.): Growth Hormone and the Heart. 2001. ISBN 0-7923-7212-3
10. W.L. Lowe, Jr. (ed.): Genetics of Diabetes Mellitus. 2001. ISBN 0-7923-7252-2
11. J.F. Habener and M.A. Hussain (eds.): Molecular Basis of Pancreas Development and Function. 2001. ISBN 0-7923-7271-9

PROLACTIN

edited by

Nelson D. Horseman
University of Cincinnati

KLUWER ACADEMIC PUBLISHERS
Boston / Dordrecht / London

Distributors for North, Central and South America:
Kluwer Academic Publishers
101 Philip Drive
Assinippi Park
Norwell, Massachusetts 02061 USA
Telephone (781) 871-6600
Fax (781) 871-6528
E-Mail <kluwer@wkap.com>

Distributors for all other countries:
Kluwer Academic Publishers Group
Distribution Centre
Post Office Box 322
3300 AH Dordrecht, THE NETHERLANDS
Telephone 31 78 6392 392
Fax 31 78 6546 474
E-Mail <orderdept@wkap.nl>

Electronic Services <http://www.wkap.nl>

Library of Congress Cataloging-in-Publication Data

A C.I.P. Catalogue record for this book is available
from the Library of Congress.

Printed on acid-free paper.

Printed in the United States of America

Contents

Preface

It is an authentic privilege to have the opportunity to assemble and edit a new volume on "Prolactin," the first in several decades to be devoted to this fascinating hormone in all its aspects. The obvious clinical rationale for understanding prolactin (PRL) is the frequent occurrence of prolactinomas, the most common type of pituitary tumor. Fortunately, medical management of prolactinomas can be based on our understanding of the physiology of hypothalamic control of the lactotroph. Armed with this knowledge, therapies for prolactinomas are highly successful and well tolerated. Because of the historical and practical importance of knowledge regarding the hypothalamus-lactotroph axis, the first chapters of this volume are dedicated to reviewing the physiology, development, and cell biology of lactotroph regulation. Chapters focusing on prolactinomas and related clinical issues follow these.

PRL is the primary hormone that is responsible for "parental care" in many vertebrate species. This reproductive strategy is not unique to mammals, but it has developed through evolution to be the central distinguishing feature of the mammalian life cycle. Among the mammals, mice have become the most effective research species in recent years. This can be traced to the development of a wide range of methods for manipulating mouse genetics, and thereby influencing development, physiology and behavior. Mice also provide a profound illustration of the physiological challenges faced in maternity. Female mice undergo a post-partum estrous, and often are both pregnant and lactating simultaneously. When they give birth, they are generally in the process of weaning a litter that has a total weight at least three times greater than that of the mother herself, and facing a new litter of constantly hungry neonates. Despite all we have learned about the fundamental molecular mechanisms of hormone action in general, and the signal transduction pathways for PRL in particular, our understanding of the roles of PRL in the integration of maternal physiology is still very incomplete. The merger of genetics, developmental biology, and physiology has offered new avenues to study these essential functions. Several chapters in this book review our current understandings of behavioral, reproductive, metabolic and developmental functions of PRL that relate to maternal care.

Biochemistry and molecular biology transformed endocrinology in recent decades by providing the means to discover and analyze the intracellular signal transducing pathways for hormones. The fundamental PRL signaling pathways, including relevant kinases, transcription factors, and comodulatory proteins were elucidated during the decade of the '90s. Appropriately, several chapters review these findings from the receptor level to the nucleus.

The authors of this volume were all aware of the importance of establishing this book as a guidepost for the students and developing scientists that will join the effort to understand PRL in coming years. In recognition of this, the book is dedicated to the students and postdoctoral fellows who bring joy to my career: past, present and future.

Nelson D. Horseman, Ph.D.

Chapter 1

Hypothalamic Control of Prolactin Synthesis and Secretion

Nira Ben-Jonathan

University of Cincinnati, Cincinnati, OH

INTRODUCTION

The synthesis and release of prolactin (PRL) by the lactotrophs is subjected to multiple regulators. These can be classified into four broad categories: endocrine, paracrine, juxtacrine and autocrine. Endocrine agents originate from the hypothalamus, gonads and the neural lobe of the pituitary and reach the lactotrophs via a humoral route. Paracrine factors are produced by cells of the intermediate and anterior lobes and reach their target cells by diffusion. Juxtacrine interactions emanate from extracellular matrix of cells adjacent to the lactotrophs. Autocrine agents are synthesized by the lactotrophs themselves. Consequently, the overall secretory activity of the lactotrophs reflects a balance between local and distant releasing and inhibiting factors (Figure 1-1).

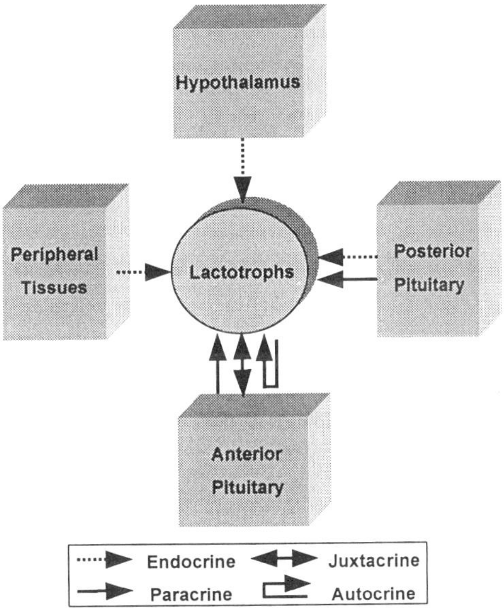

Figure 1-1. Schematic presentation of the four types of regulation of the lactotrophs

The foregoing discussion on PRL regulation is not comprehensive, but it highlights recent developments in the field. Emphasis has been placed on the distribution, regulation and mechanism of action of dopamine. Dopamine, acting via dopamine type-2 receptors (D2R) expressed by the lactotrophs, is the primary inhibitor of PRL. On the other hand, numerous peptides and growth factors can stimulate PRL either by altering dopamine input to the lactotrophs, or by directly affecting the lactotrophs.

The direct prolactin regulating factors (PRFs) can be further subdivided into two categories: those that rapidly stimulate release of stored PRL and those that primarily induce the PRL geneleading to a secondary increase in PRL secretion.

Another section of this chapter focuses on transgenic animals. Animals with PRL deficiency have been instrumental for delineating the full spectrum of its functions. Among animals with natural mutations that reduce PRL are the Snell and Ames dwarf mice and the hypoprolactinemic nude rat. Recently, targeted mutagenesis has generated several lines of transgenic mice with altered genes that affect the neuroendocrine axis the regulates PRL. Absence of a hormone or its receptor often reveal unexpected functions or previously unknown interactions with other regulators. A refinement of this technology to generate transgenic animals whose genes can be turned on/off at specific times in selective sites would be of particular benefit in future studies.

THE HYPOTHALAMIC DOPAMINERGIC SYSTEMS

ANATOMY AND RELATIVE CONTRIBUTION TO PRL RELEASE

The brain contains several well-defined dopaminergic systems. Three of these originate in the midbrain and comprise the nigrostriatal, mesolimbic and mesocortical systems which are involved in the control of locomotion, emotion and mentation, with no direct effect on pituitary function. Within the hypothalamus, dopamine perikarya are located in the posterior hypothalamus (A11), arcuate nucleus (A12), zona incerta (A13) and lateral and ventral hypothalamus (A15) (1). Based on anatomical projections and functional studies, the two arcuate (A_{12}) systems that regulate PRL are the TIDA (tubero-infundibular) and THDA (tubero-hypophysial), while the incerto-hypothalamic neurons project diffusely within the hypothalamus and play a role in the control of GnRH and LH release. Because of lower abundance and heterogenous distribution, the hypothalamic DA neurons are not as well characterized as their midbrain counterparts.

The TIDA neurons provide the major dopaminergic input to the anterior pituitary. Most of their perikarya are located in the arcuate nucleus with a smaller population arising from the periventricular nucleus. Their relatively short axons terminate in the external zone of the median eminence near the primary capillary loops of the hypophysial portal vessels (Figure 1-2). As the median eminence contains a negligible number of cell bodies, there are no classical synapses. Instead, the TIDA system is composed of true neurosecretory neurons, whose product is released into perivascular spaces and carried by the portal blood to the anterior pituitary. Evidence based on measurement of dopamine in hypophysial portal blood and determination of dopamine concentration and turnover in the arcuate nucleus has established that the activity of

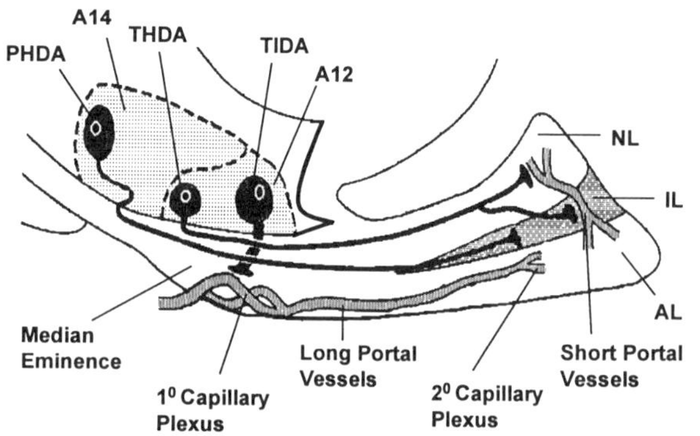

Figure 1-2. The three hypothathalamic dopaminergic systems that regulate PRL secretion (redrawn and modified from Ref. 5).

these neurons is altered under many physiological conditions known to affect PRL release (2). However, the early view that the TIDA neuronal activity must be negatively correlated with PRL release is too simplistic because the underlying reciprocal relationship can be masked by other factors that stimulate PRL release.

The THDA neurons originate from the rostral arcuate nucleus and send axons that course through the pituitary stalk and terminate in the neural (NL) and intermediate (IL) lobes of the pituitary. They are joined by another system, termed PHDA (periventricular hypothalamic DA), with perikarya in the periventricular nucleus and terminals in the IL (3). Whereas the THDA terminals in the NL are neurosecretory, those in the IL form true synapses with melanotrophs, where they suppress their proliferation (4) and inhibit β-endorphin and α-MSA release (5). Posterior pituitary lobectomy provided the first evidence that dopamine from the neurointermediate lobe suppresses PRL release in both males and cycling females (6,7). This was recently supported by the rise in plasma levels of PRL and α-MSH following denervation of the neurointermediate lobe (8).

The short portal vessels connecting the neural and anterior lobes provide vascular channels for delivery of dopamine to the lactotrophs, but as they are inaccessible for sampling, there is no direct evidence for this delivery route. The means by which dopamine from the avascular IL reaches the lactotrophs remains enigmatic, and may involve diffusion to the lactotrophs that line the pituitary cleft. Another possibility is that dopamine from the IL is indirectly involved in PRL regulation by suppressing the release of a local PRF (9). Although early studies failed to show correlation between the activity of the THDA neurons and PRL release (10), recent studies reported that the PHDA/THDA neurons are activated during the afternoon of proestrous (11) and in response to exogenous PRL (12). Still, of the three hypothalamic dopaminergic systems, the TIDA plays the predominant role.

REGULATION OF HYPOTHALAMIC DOPAMINE BY PRL AND ESTROGEN

In the absence of target gland hormones to provide feedback control over the lactotrophs, PRL itself serves as the major regulator of the hypothalamic dopaminergic systems. Many studies have established that an increase in either endogenous or exogenous PRL results in higher activity of the TIDA neurons while a decrease in circulating PRL levels, either by hypophysec-tomy, immunoneutralization or dopaminergic agonists, lowers TIDA neuronal activity (reviewed in [2]). The TIDA neurons respond to both acute and chronic changes in PRL. Under some conditions, e.g., pregnancy, lactation and prolactinomas, the dopaminergic neurons become refractory to the elevated PRL levels, thus enabling the maintenance of physiological or pathological hyperprolactinemia.

The existence of a short loop feedback arrangement between PRL and dopamine raises the following questions: a) are the effects of PRL direct or indirect, and if indirect, what are the mediators? b) are both the TIDA and THDA neurons regulated by PRL? c) how does PRL gain access to the TIDA neurons? d) what is the cellular distribution of hypothalamic PRL receptors and how are they regulated? and e) which components of the dopaminergic neurons are affected by PRL? At present, there are only partial answers to these questions.

The issue of direct vs. indirect effects of PRL on the TIDA neurons is difficult to resolve, given the scarcity of suitable *in vitro* systems. Cultured fetal hypothalamic neurons incubated with PRL had increased TH activity (13). Although most of the TH-expressing neurons were immunopositive for PRL receptors, they constituted less than 10% of the total cell population and the PRL receptors were also expressed by 20-25% of other neurons. Also, the stimulatory effect of PRL on DOPA accumulation was modest, required a serum factor, and took several days. In another study, dopamine concentration in TIDA neurons increased within 1hr after injecting ovine PRL to ovariectomized rats (12), implying a rapid and direct activation. A delayed increase in DA turnover in both the IL and NL indicates that all three populations of dopaminergic neurons were activated by PRL. Neither study, however, ruled out the possibility that the action of PRL was mediated by substances such as neurotensin, neuropeptide Y and opioids.

Circulating proteins are normally excluded from the brain except for the circumventricular regions such as the median eminence and NL, which are outside the blood brain barrier. It is unclear how PRL reaches the TIDA perikaya, which are located within the medial basal hypothalamus. PRL, which is detectable in many hypothalamic and extra-hypothalamic sites, can come from two sources: transport from the circulation and local synthesis. Transport of PRL to the CSF occurs by a receptor-mediated uptake at the choroid plexus (14), which contains the highest density of PRL receptors within the brain. The mechanism by which this receptor functions as a transporter is unknown. Once PRL gains access to the CSF, it can be distributed to various sites such as the arcuate nuclei, located in close proximity to the third ventricle. The hypothalamus is also capable of *de novo* synthesis of PRL, which is regulated by estrogen (15). Whether locally-produced PRL participates in the short loop feedback on the TIDA neurons is unknown.

The PRL receptor, a member of the cytokine/GH/PRL receptor super-family, is a protein with a single-pass transmembrane domain. Alternative splicing can generate several isoforms that differ in the length of the cytoplasmic domain and there is evidence for the existence of a soluble isoform (16). The rat brain expresses two main isoforms of the PRL-R, short and long. Second to the choroid plexus, the hypothalamus contains the highest concentration of PRL-R, with a K_d of 0.2-0.3 nM and B_{max} of 5-10 fmol/mg protein (17). Many hypothalamic nuclei, including the arcuate nucleus, express PRL-R mRNA, and contain immunoreactive PRL receptors. Co-localization of PRL-R in some, but not all, TH-positive neurons in the arcuate nucleus has been reported (13,18). The long isoform is predominant, but there is some evidence that only the short form is regulatable.

Alterations in circulating PRL levels during lactation (19) and after hypophysectomy (17) increase and decrease, respectively, the hypothalamic PRL receptors, suggesting that PRL regulates its receptors. Tyrosine hydroxylase, the rate limiting enzyme in dopamine synthesis, is the major target of PRL action. Both short term enzyme activation by phosphorylation and long term induction by increased enzyme expression, have been shown. Assuming that hypothalamic PRL receptors function similarly to their peripheral counterparts, the Jak2/STAT signaling pathway is likely to be involved (16).

Estrogens affect PRL homeostasis at several levels, i.e., the hypothalamus, posterior pituitary and anterior pituitary, and it is often difficult to distinguish between these functionally related sites. Estrogen action is mediated by two receptors, ERα and ERβ, that are the translation products of different genes. Both receptors have a similar modular structure of five functional regions, with high homology in the DNA- and ligand-binding domains but distinct regulatory domains. They bind estradiol with a similar affinity, and after homo- or hetero-dimerization they activate the same estrogen response element (ERE) located in the regulatory region of target genes. Only a fraction of TH-positive neurons in the arcuate nucleus express estrogen receptors with a predominance of ERα (20). It is debatable whether, similar to GnRH neurons, the action of estrogen on the TIDA neurons is indirect, involving other neurotransmitters or neuropeptides.

The importance of estrogen in the overall regulation of the TIDA neurons is underscored by the higher basal activity of TIDA neurons in females, the suppression of this activity by ovariectomy and its reversal by estrogen. Yet, length of exposure to estrogens, dosages, physiological state and species of the experimental animals often yield conflicting results. Unlike rats, estrogen alone does not alter TH expression in the arcuate nucleus of monkeys, but when followed by progesterone, there is a significant decrease (21). A non-genomic, rapid stimulation of dopamine release *in vitro* from the posterior pituitary, but not the median eminence, has also been reported (22). Since the latter was abolished by co-incubation with the opioid inhibitor naloxone, the effect of estradiol is apparently mediated by by local opioid peptide(s).

INTERACTIONS OF DOPAMINE WITH NEUROPEPTIDES AND NEUROTRANSMITTERS

Several neuropeptides stimulate PRL release by inhibiting the TIDA neurons. The best established are the opioid peptides. Systemic and central administration of morphine, the opioid agonist, lowers TIDA neuronal activity and suppresses DA concentration in hypophysial portal blood. Conversely, central administration of either μ or κ opioid antagonists increases TIDA neuronal activity (reviewed in [23]). The hypothalamus contains perikarya of the three opioid classes, i.e. POMC, endorphins and enkephalins, all of which participate in the regulation of PRL release under some conditions. Although the arcuate nucleus contains POMC-positive cells, only a minority of the TIDA perikarya receive contacts from β-endorphin terminals and most interactions appear to occur at their median eminence terminals (24). On the other hand, most of the TIDA neurons have synaptic contacts with dynorphin terminals and opioid peptides belonging to the proenkephalin and prodynorphin families may act as paracrine regulators of the TIDA neurons. Two lines of evidence indicate direct action of opioid peptides on the TIDA neurons. One is activation of potassium conductance and hyperprolarization of dopaminergic neurons by a μ-opioid agonist (25), and the other is inhibition of dopamine release from incubated median eminences by β-endorphin (26).

Interactions between opioid peptides and TH-positive neurons are especially extensive during the PRL surge on the afternoon of proestrus. The nocturnal PRL rise in pregnant rats and the suckling-induced PRL surge. Most studies documenting these effects utilized naloxone, a broad-spectrum opioid receptor antagonist. For example, infusion of naloxone during suckling resulted in marked suppression of plasma PRL levels and a concomitant increase in TIDA activity and tyrosine hydroxylase expression. Such alterations occurred in both the arcuate nucleus and median eminence, but not in the zona incerta (27). In addition, pups in the naloxone-treated group failed to gain weight, indicating less milk availability and inhibition of oxytocin release. These studies, and similar studies conducted with cycling and pregnant rats, revealed that the opioid peptides comprise an integral component of the system that generates PRL rises, but they do not rule out opiodergic interactions with PRF-producing neurons. Other neuropeptides, including neuropeptide Y, bombesin-like peptides, neurotensin and cholecystokinin have been implicated in the indirect stimulation of PRL release via suppression the hypothalamic dopaminergic systems [reviewed in (23)].

Among the neurotransmitters, GABA has a dual effect on PRL release: a direct inhibitory action at the level of the anterior pituitary and a central stimulation. However, the exact relationship between GABA and the TIDA neurons is unclear. Similarly, the role played by the cholinergic system in the regulation of PRL release is controversial, and the observed effects vary with the site of administration, the dose and the physiological conditions of the animals. There is an endogenous rhythm of cholinergic neurons which may contribute to the circadian changes that occur in both the TIDA neuronal activity and PRL release, but whether cholinergic neurons synapse with opiodergic or TIDA neurons remains to be determined (23). Histamine can also be added to the list of centrally and peripherally acting neurotransmitters with an unclear mechanism of action.

The stimulatory effect of serotonin on PRL release has long been established. Serotonergic neurons originating from the raphe nucleus, and they terminate in many hypothalamic nuclei including the superchiasmatic and arcuate nuclei. Drugs that enhance serotonergic function such as precursors, receptor agonists and uptake inhibitors, induce transient rises in PRL release in rodents but are less effective in humans. Conversely, drugs that impair serotonergic neurotransmission or lesions of the raphe nucleus prevent rises in PRL release in response to stress and suckling (reviewed in [2]). Although serotonin administration reduces dopamine levels in hypophysial portal blood and suppresses tyrosine hydroxylase activity, a consistent effect of serotonin on TIDA neuronal activity has not been observed and the current dogma is that it affects PRL secretion primarily by stimulating PRF release. The latter is supported by the increased PRL release from anterior pituitaries co-cultured with posterior pituitaries exposed to serotonin (28).

PROLACTIN REGULATING FACTORS

TRH AND VIP

In contrast to the many substances that affect PRL release by altering the hypothalamic dopaminergic system, several peptides affect both PRL gene expression and release by acting directly on the lactotrophs. Following the discovery of the hypothalamic releasing/inhibiting factors during the 1960s and 1970s, strict criteria were established for this class of compounds. These include their localization in the hypophysiotropic area of the hypothalamus, presence in portal blood, binding to specific receptors on hormone-producing pituitary cells and alterations in their expression/release that reflect changes in PRL secretion. As most of these compounds were later found to be produced by the pituitary itself, and the complexity of paracrine/autocrine regulation of the lactotrophs became apparent, it necessitated broadening of the criteria for PRL regulating factors.

TRH, initially defined by its ability to stimulate TSH release, is also an effective stimulator of PRL secretion. TRH is a tripeptide with a C-terminal amide and a pyro-Glu residue at the N-terminus. ProTRH is cleaved by prohormone convertases at paired basic residues to yield five copies of TRH (29). TRH neurons are located in the paraventricular nuclei of the hypothalamus. They have terminals in the external layer of the median eminence and secrete their product into portal blood. TRH binds to G protein-coupled receptors membrane receptors that are present on both thyrotrophs and lactotrophs (reviewed in [30]). A recently identified second TRH receptor, named TRH-R2, is mainly distributed within the brain, with only weak expression in the pituitary (31). TRH stimulates PRL release *in vivo* and *in vitro*, especially when the dopaminergic input is absent or reduced. Its primary signaling is via the phosphoinositide-calcium transduction pathway, and it induces a rapid, biphasic rise in intracellular free calcium concentrations. TRH also induces the PRL gene and recent evidence suggests that the signaling pathway utilizes by TRH also involves PKC- and calcium-dependent activation of MAP kinase (32).

In spite of the accumulating knowledge on the intracellular mechanisms mediating the action of TRH, its role as a primary physiological PRF has been challenged. First, little correlation exists between TRH concentrations in portal blood and PRL secretion under several physiological conditions. Second, a temporal dissociation often occurs between the release of TSH and PRL. Third, the data on the effect of immunoneutralization against TRH on PRL release are conflicting. Therefore, the current notion is that TRH may be an important modulator of PRL release, but is unlikely to be the primary PRF. Nonetheless, a TRH stimulation test is widely used by clinicians to distinguish prolactinomas from other causes of hyperprolactinemia (33).

VIP is a 28 amino acid peptide originally isolated from porcine intestines, and subsequently found in many sites within the central and peripheral nervous system. High concentrations of VIP are present in the superchiasmatic and paraventricular nuclei and in portal blood. VIP has been proposed as a physiological regulator of PRL based on its ability to stimulate PRL release *in vivo* and *in vitro*. This was supported by passive immunization studies, and the increases in hypothalamic VIP mRNA levels during lactation and following estradiol treatment (reviewed in [34]). Additionally, hyperprolactinemia lowers VIP concentrations in portal blood, indicating negative feedback by PRL. VIP is also produced by the lactotrophs (35), where its expression is stimulated by estrogens. The latter suggests that VIP acts as a paracrine/autocrine agent that maintains elevated basal PRL release. There is strong evidence that VIP functions as the primary PRF during the reproductive cycle of avian species (36).

VIP binds to two subclasses of G protein-coupled receptors named VIP-R1 and VIP-R2. Other members of the VIP family, e.g., PACAP, also bind to such receptors with different affinities (37). Despite the considerable homology between the receptors, each subtype is coupled to specific signaling pathways, and its expression is tissue and cell specific. Unlike TRH, VIP acts by increasing the production of cyclic AMP, followed by activation of protein kinase A. When compared on a molar basis, VIP is more sluggish and less potent than TRH as a PRL secretagogue in rat pituitary cells. Like TRH, VIP stimulates PRL gene expression, and studies conducted with turkey pituitaries showed increased transcription of the PRL gene and enhanced mRNA stability in response to VIP (38).

PROLACTIN-RELEASING PEPTIDES (PRRP)

Two peptides with potential PRL-regulating activity were recently cloned from bovine hypothalami. Based on their ability to increase PRL release *in vitro* they were named PRL releasing peptides (PrRP20 and PrRP31). Although it is presently controversial whether they truly deserve their assigned name, the strategy used for their identification and isolation was most interesting. Given that a multitude of hormones transmit their signals by binding to seven-transmembrane domain receptors, Hinuma and colleagues cloned a G-protein-coupled orphan receptor from the human pituitary, which they named hGR3 (39). Knowing that mRNA for this receptor was most abundant in the pituitary, they screened cells transfected with the receptor for an easily quantifiable response to hypothalamic extracts. They found that arachidonic acid, a lipid metabolite reported to mediate the release of PRL, was released in response to the extract. Combining the functional assay with chromatography, they isolated two peptides with overlapping sequences. The peptides were produced by proteolytic cleavage from a larger precursor molecule and have amidated carboxy termini.

When tested with CHO cells transfected with the receptor, both peptides stimulated arachidonic acid release with an ED_{50} of $5x10^{-10}M$, and bound to the cell membranes with a Kd of $2.5x10^{-11}M$. Upon incubation with rat pituitary cells, only PRL release was increased while other pituitary hormones were unchanged. However, the peptides stimulated PRL release from a rat pituitary adenoma-derived cell line at a 100 fold higher concentration than that required for arachidonic acid release. They were even less potent when tested with primary pituitary cells (39). Another group reported that when tested with female but not male pituitary cells, the peptides were effective only at a micromolar dose compared with the 1 to 10 nM effective dose of TRH (40). Recent work demonstrated that the PrRP peptides activate the PRL promoter in a time-dependent manner and their action is mediated via G proteins, followed by induction of both ERK and JNK signaling pathways in an Ets-dependent manner (41).

Since multiple factors can stimulate PRL release *in vitro,* it was important to demonstrate that the newly identified PRL-releasing peptides are active *in vivo* and are localized within the hypophysiotropic area of the hypothalamus. Intravenous injections of PrRP31 increased plasma PRL in a dose-dependent manner in cycling females but required significantly higher doses than TRH to elicit a similar stimulation (42). In addition, peptide expression was detected in many areas throughout the brain, with the highest levels observed in the nucleus of the solitary tract and the adrenal medulla. However, PrRPs were undetectable in the ventral hypothalamus or median eminence (43), raising the question of how they might reach the pituitary gland. As both the peptides and their receptors are abundant in rat (43) and human (44) pituitaries, one possibility is that they act as autocrine/paracrine factors. If so, they do not differ from other locally-produced peptides/growth factors capable of stimulating PRL secretion, but cannot be categorized as hypothalamic releasing factors. The wide extra-hypothalamic distribution of the peptides and their receptor suggest additional roles unrelated to pituitary function. The use of selective agonists and antagonists should help define the spectrum of action of these novel peptides.

PRFS FROM THE POSTERIOR PITUITARY

The posterior pituitary in most species is composed of two adjacent tissues, the neural and intermediate lobes, which differ in embryonic origin, innervation, vascularization, cell types and hormonal products. The NL extends from the median eminence and is composed of nerve endings and pituicytes, while the IL shares its embryonic origin and many features with the glandular anterior lobe. The main cell type within the IL is the POMC-producing melanotroph, with additional cells classified as interstitial, endothelial, fibroblasts and squamous cleft cells. The short portal vessels that connect the neural and anterior lobes do not penetrate the IL and this layer of cells is practically avascular (see Figure 1-2). Two compounds, oxytocin from the neural lobe, and as yet unidentified substances from the IL act as PRFs.

Oxytocin is a nonapeptide produced by the magnocellular neurons of the paraventricular and supraoptic nuclei. Posttranslational processing of prooxytocin takes place inside secretory granules during axonal transport to the NL. Both oxytocin and its neurophysin I carrier protein are stored in nerve terminals in the NL and released in equimolar amounts in response to action potentials from the hypothalamic perikarya. Presumably, oxytocin can reach the anterior lobe either from the median eminence or from the NL and bind to specific receptors on the lactotrophs. Oxytocin is released at times of elevated PRL release such as during suckling, in the afternoon of proestrus, following estradiol treatment and during immobilization stress (reviewed in [45]). Since oxytocin and PRL have different thresholds of activation and dissimilar kinetics, co-release could be coincidental or functionally related. The notion that oxytocin is a physiological PRF is controversial, however. Although oxytocin can stimulate PRL release, its potency is much lower than that of TRH (46,47). Further, immuno-neutralization of oxytocin and administration of oxytocin antagonists, only partially suppressed stimulated PRL release. Thus, oxytocin is unlikely a primary PRF, but could serve as a modulator of PRL release under some conditions.

Another PRF was serendipitously discovered when posterior pituitary lobectomized (LOBEX) rats were used to investigate the role of the THDA neurons in the suckling-induced rise in PRL (48). Basal PRL release was elevated in LOBEX lactating rats, confirming that dopamine from the THDA neurons contributes to the tonic inhibition of PRL release. However, suckling, with or without oxytocin administration, failed to increase plasma PRL levels, revealing the presence of PRF. Subsequent *in vivo* studies have shown that this PRF participates in the generation of the proestrus PRL surge, mediates the acute estradiol-induced rise in PRL, contributes to the nocturnal rise in PRL during early pregnancy, but is not involved in the stimulation of PRL release caused by ether or serotonin (reviewed in [9]).

Additional studies revealed that PRF activity is localized in the IL, with little or no activity present in the NL, hypothalamus or cerebral cortex (49). Posterior pituitary extracts stimulated PRL release in a rapid, dose-dependent and hormone-specific manner both *in vivo* and *in vitro* (reviewed in (9). Co-culturing posterior pituitary and GH_3 cells resulted in a dramatic increase in PRL mRNA levels, suggesting that this PRF also affects PRL gene expression (50). Attempts to isolate this PRF from bovine posterior pituitaries were unsuccessful, however, resulting in the sequencing of an unrelated peptide, named Peptide V (51).

Following the recent cloning of a PRF-producing cell line from a mouse intermediate lobe tumor (52), another substance with PRF activity was discovered. Only cells of unknown lineage, designated as mIL5, were capable of PRF synthesis and release whereas those mIL39 cells that and classified as melanotrophs, were devoid of PRF activity (Figure 1-3). Although this PRF does not rapidly stimulate PRL secretion, it is a very potent inducer of the PRL gene. Unexpectedly, this PRF also binds to heparin, and in this respect it differs from classical neuropeptide releasing factors but resembles several members of the class of heparin-binding growth factors that are strong stimulators of PRL gene

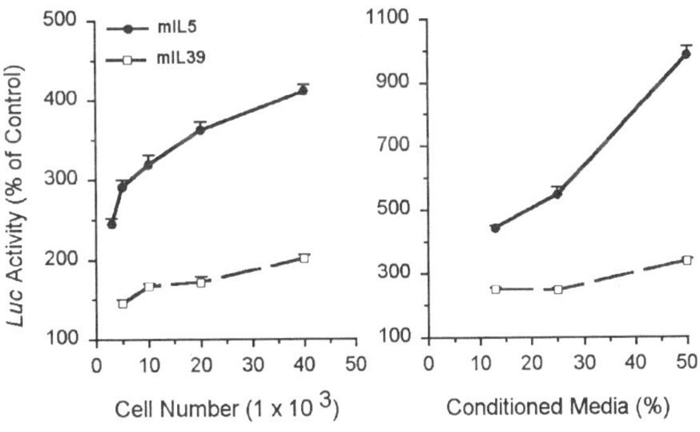

Figure 1-3. Stimulation of PRL gene expression in GH3 cells incubated with mIL5 cells (left panel) or their conditioned media (right panel) and lack of effect of mIL39 cells (Modified from Ref. 52).

expression. Based on three complimentary approaches including heparin-affinity chromatography, Western blotting and immunoneutralization, this PRF differs from FGF-2 and HB-EGF (heparin binding EGF-like growth factor), and its identity is presently under investigation.

MECHANISM OF ACTION OF DOPAMINE AND PRFS ON THE LACTOTROPHS

RAPID VS SLOW EFFECTS OF DOPAMINE

Although dopamine has been recognized as the primary inhibitor of the lactotrophs for three decades, its exact mechanism of action is not as well delineated as that of other hypothalamic factors. There are several reasons for this uncertainty. First, it is more difficult to study inhibition than stimulation. Second, lactotrophs are highly heterogeneous with respect to morphology, membrane potential, basal release and responsiveness to releasing factors. Third, there is a paucity of lactotroph cell lines with functional dopamine receptors. Finally, at extremely low concentrations, dopamine has been reported to stimulate, rather than inhibit, PRL secretion.

Dopamine exerts its action on the lactotrophs by binding to dopamine D_2 receptors. High affinity dopamine binding sites with a dissociation constant (Kd) of 4.4×10^{-10}M have been identified in the pituitary, and were later localized to both lactotrophs (53) and melanotrophs (54). The dopamine receptor is a seven transmembrane domain protein belonging to the G-protein-coupled receptor family. The family includes five members, subdivided into two groups based on amino acid sequence and structural similarities. The D1-like receptors (D1 and D5) increase adenylyl cyclase activity in response to dopamine while the D2-like receptors (D2, D3 and D4) inhibit adenylyl cyclase and regulate intracellular calcium. Alternative splicing of exon 6 of the D2 receptor gene yields two isoforms, long (D_2L) and short (D_2L). The short isoform has a 29 amino acid deletion in the third intracytoplasmic domain, where interactions with G_0/Gi proteins (in particular $G_{i/0}$, $G_{i3}\alpha$ and $G_0\alpha$) occur (55). The long isoform is predominant in the pituitary and its relative abundance, i.e., the utilization of a specific splice site, is regulated by sex steroids (56). Although both isoforms exhibit similar binding affinity to dopamine, they may differ in their coupling to second messengers, accounting for the different responsiveness of lactotroph sub-populations to dopamine under various physiological conditions.

Lactotrophs *in vivo* are continuously exposed to dopamine. When cultured without dopamine, their cellular properties can be altered, complicating interpretation. In spite of these restrictions, the diverse effects of dopamine on the lactotrophs is best characterized by observing temporal events. Within seconds, dopamine increases potassium conductance and inactivates voltage-sensitive calcium channels, resulting in membrane hyper-polarization, reduced intracellular free calcium and inhibition of PRL release. Within minutes to hours, dopamine suppresses adenylyl cyclase activity as well as inositol phosphate metabolism and arachidonic acid release, resulting in the downregulation of the PRL gene. Within days, dopamine suppresses lactotroph proliferation. It remains to be determined whether all these events occur within the same cell or in different cell subpopulations and to what extent they proceed sequentially or in parallel.

Application of whole cell recording and patch clamp techniques revealed that lactotrophs are electrically excitable. Many, though not all, exhibit spontaneous, voltage-dependent action potentials at frequencies of 0.2 to 0.5 Hz and a resting membrane potential of -40 to -60 mV (55). An interplay between tetrodotoxin-sensitive sodium current, T- and L-type calcium (K) channels and Ca-dependent potassium channels underlie this electrical activity. Individual lactotrophs exhibit different patterns of cytosolic calcium: repetitive rhythmic oscillations, irregular small fluctuations and quiescent stable levels (57). Dopamine induces membrane hyperpolarization by activating a nonvoltage-dependent K current (58) together with a differential inhibition of the T- and L-type voltage-dependent calcium channels (59). This leads to a rapid decrease in intracellular calcium and cessation of spontaneous calcium fluctuations. Since PRL release from storage granules is coupled to calcium availability, these processes explain the high basal release of PRL characteristic of lactotrophs *in vitro* in the absence of dopamine and the immediate suppression of hormone release upon exposure to dopamine.

Several laboratories have reported that dopamine at extremely low concentrations (below 10^{-10}M) stimulates PRL release (60,61). The stimulation is quite modest and has not been confirmed by all investigators. The proponents of this notion considered several mechanisms to explain this phenomenon, including predominance of the short isoform of the dopamine receptor (62), mediation by D1 and D5 dopamine receptors (63), involvement of $G_i\alpha_3$ protein (64), and dopamine-induced increases in intracellular calcium by lactotroph subpopulations (61). Opponents argue that the stimulatory activity can be easily explained by dopamine withdrawal, resulting in increased cytosolic free calcium concentration (57) and a rebound of PRL secretion that exceeds baseline levels. When rebound occurs in the presence of TRH or VIP, it results in a pronounced potentiation of their PRL-releasing actions (65). Evidently, the issue of the stimulatory effect of dopamine remains controversial and needs further assessment.

INTRACELLULAR SIGNALING AND PRL GENE EXPRESSION

Many neuropeptides and neurotransmitters bind to membrane receptors whose action is mediated by G-proteins. The G-proteins are divided into several subfamilies, each of which interacts with a different signal transduction pathway. One is the adenylyl cyclase-cAMP-protein kinase A generation system. Of the peptides that stimulate PRL secretion, VIP utilizes the cAMP signaling pathway. As VIP is also synthesized within the anterior pituitary and may act as a paracrine/autocrine factor (35), suppression of elevated cAMP in lactotrophs exposed to VIP is carried out by dopamine. Indeed, earlier studies reported a dopaminergic-induced decrease in cAMP content and an increase in intracellular cAMP in response to dopamine withdrawal (65). However, the cAMP pathway is now considered secondary in importance to the effects of dopamine on intracellular calcium via membrane channels (reviewed in [66]).

Another signal transduction pathway that affect the lactotrophs is phosphoinositide metabolism. This is the main mechanism by which TRH and angiotensin II (AII) increase PRL release and gene expression. TRH activates phospholipase C, leading to inositol phosphate hydrolysis and generation of inositol triphosphate and diacylglycerol (67). Together they mobilize calcium from intracellular stores and activate calcium-dependent protein kinase C (PKC). PRL release in response to TRH has two components, a rapid phase that results from redistribution of intracellular calcium and a delayed, prolonged phase that depends on calcium entry from extracellular sources (68). Several studies showed partial inhibition of TRH-induced phosphoinositide metabolism by dopamine (65), and redistribution of PKC to the particulate cellular fraction upon dopamine withdrawal (69). TRH and AII also liberate arachidonate from cellular lipids at a dynamics that parallels PRL secretion (70). Dopamine attenuates the secretagogue-induced release of arachidonate, perhaps secondary to its ability to suppress calcium fluxes. Again, the importance of this pathway in the regulation of PRL release is uncertain.

The PRL promoter is the best characterized of all hormone regulatory sequences. Multiple *cis*-elements that mediate tissue- and hormone-specific regulation of the PRL gene have been mapped in both the proximal and distal regions of the promoter. Many advances in our understanding of promoter function were made possible by the availability of GH cells, a somatolactotroph cell line that has remained stable for 30 years (reviewed in [71]). However, GH cells lack high affinity dopamine receptors and do not respond to dopamine. Attempts to overcome this problem by transfecting GH3 cells with dopamine receptors were only partially successful. Whereas treatment with EGF (72) or NGF (73) induced expression of functional dopamine receptors in GH3 cells, it is unclear whether this phenotype remained stable. Another lactotroph cell line, MMQ, expresses functional D2R. Although these cells have a low receptor density and express only the long D2R isoform, the inhibitory effect of dopamine on adenyl cyclase activity, inositol phosphate production and intracellular free calcium concentration were clearly demonstrated (74). These cells should be more widely used to study the dopaminergic regulation of PRL gene expression.

An imperfect Estrogen Responsive Elemant (ERE) site that mediates the direct stimulatory action of estrogen on PRL gene expression is localized to the distal promoter region of the PRL gene, adjacent to a Pit-1 site (75). Both ERα and ERβ are expressed by the majority of lactotrophs (76). In addition to these, pituitary-specific truncated estrogen receptor products, named TERP1 and TERP2, have been detected in the rat pituitary (Figure 1-4) and are upregulated by estrogen (77). TERP could act as a dominant negative regulator, or conversely, as a recruiter of adpator proteins, and either of these possibilities remain to be investigated. Overexpression of ERβ in GH3 cells results in increased estrogen-induced PRL gene expression, suggesting that this receptor is involved in mediating the action of estradiol on PRL synthesis (78). Whether this is accomplished by homodimeriza-tion of ERβ or hetero-dimerization with ERα remains to be determined. In addition to endogenous estrogens, PRL synthesis and release increase in response to a variety of substances, termed xenoestrogens, that are abundant in the environment, bind at low affinity to the estrogen receptor and act as agonists/partial agonists of estrogen (79).

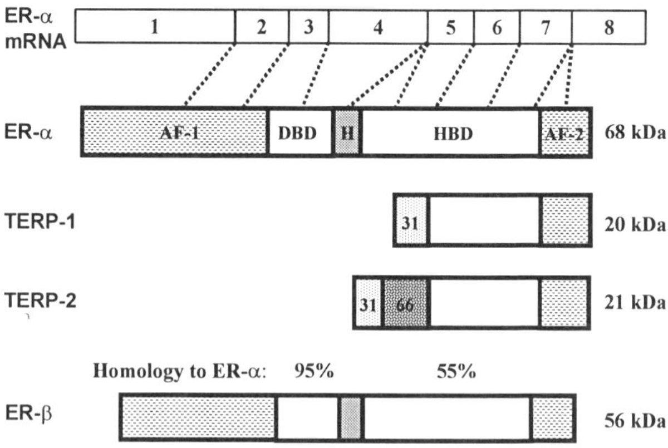

Figure 1-4. Diagram of the different estrogen receptors expressed by the pituitary gland. DBD and HBD - DNA and Hormone Binding Domains; AF-1 & AF-2 - transactivation domains; H - hinge. Numbers within TERPs show unique amino acids.

LACTOTROPH PROLIFERATION

Most anterior pituitary cells are quiescent, with the exception of lactotrophs that proliferate during pregnancy lactation and the estrous phase of the reproductive cycle. Estrogen is the major mitogenic hormone for the lactotrophs while dopamine has anti-mitogenic properties. The action of estrogen is likely to be mediated by paracrine interactions via estrogen-responsive substances such as VIP, galanin and TGFβ. The anti-proliferative action of the dopaminergic agonist bromocriptine has been valuable for treating patients with prolactinomas. In addition to suppressing serum PRL level 3, bromocriptine reduces tumor mass by decreasing cell volume, suppressing cell proliferation and increasing apoptosis (80). Because of its high potency, stability and prolonged effects, bromocriptine is the most widely used dopamine agonist. However, not all of its actions are mediated by dopaminergic receptors, since bromocriptine also suppresses proliferation of immune and smooth muscle cells lacking dopamine receptors (81).

Several pathways have been implicated in the anti-mitogenic activity of dopamine. One involves calcium-dependent PKC, which increases in response to estrogen and is reversed by dopamine. The identity of the PKC isoenzyme involved and the coupling of the dopamine receptor to pertussin-toxin-sensitive Gi protein is controversial, however (82,83). Other studies suggest that suppression of cAMP/PKA signaling is responsible, at least in part, for the inhibition of lactotroph proliferation by dopamine (84). Studies that specifically examine interaction of dopamine with mediators of the cell cycle and apoptosis are clearly needed.

LESSONS LEARNED FROM DEFICIENT MUTANT MICE

DELETIONS OF DOPAMINE D2 RECEPTOR (D2R) AND TRANSPORTER (DAT)

Two types of transgenic mice with alterations in dopamine have been generated: those with a deletion of the D2R that prevents dopamine action, and those with a deletion of the dopamine transporter (DAT), which caused increased dopamine availability. Two laboratories (85,86) have described the consequences of loss of D2R on the neuroendocrine axis. The major phenotype is chronic hyperprolactinemia and lactotroph hyperplasia, which develops into adenoma in aged females. Pituitary PRL expression is robust in D2R null mice with a slight decrease in GH and no major change in other hormones. Null mice of either sex have a 3-4 fold higher basal PRL levels and, as expected, do not respond to the dopamine receptor antagonist haloperidol. While D2R mutant males display normal spermatogenesis, young females have irregular estrous cycles and lower serum estrogen levels, but with little impact on fertility. D2R null females have higher serum PRL levels than males, underscoring the importance of estrogen in the regulation of PRL production. The lack of D2R does not affect expression of hypothalamic regulatory factors such as TRH and VIP, although this issue has not been examined in depth. Notably, pituitary PRL receptor expression was also higher in the D2R deficient mice, perhaps secondary to the pituitary hyperplasia discussed below.

Lactotroph hyperplasia, followed by adenoma formation, differ in onset and magnitude between the two sexes. At 3 months of age, there is no discernible difference in pituitary size or appearance between null and normal mice. By 9 to 12 months, females develop extensive lactotroph hyperplasia that is barely seen in males. At this time, many dilated blood-filled spaces are seen, but there are no signs of neoplastic transformation. Older D2R-null females (17 to 20 months old) develop lactotroph adenomas, with some pituitaries increasing in size as much as 50-fold above controls. There was some invasion into the brain but no evidence of metastasis (87). The pituitary in age-matched males doubles in size with only microscopic foci of lactotroph adenomas. The mechanisms responsible for tumor formation in females only and its late onset are not understood. Whereas PRL itself has been considered, there is only weak evidence that it acts as mitogenic hormone on the lactotrophs. In contrast, involvement of local growth factors such as galanin, normally subjected to dopaminergic inhibition (88), should be considered. Also, angiogenic/anti-angiogenic factors must underlie the observed perturbation in the local vasculature.

Impairment of dopamine transmission in the nigra-striatal system results as expected, in deficits in locomotion, while linear growth, onset of puberty and fecundity are not altered. The observed effects of D2R deletion on the IL differ between the two laboratories, presumably because of a different genetic background of the mice. Kelly *et al* (86) found no difference in the thickness of the IL and its β-endorphin content between D2R-deficient and normal mice while saiardi *et al* (85) reported an enlargement of this tissue and increased POMC expression in null mice.

Disruption of DAT results in an almost opposite phenotype compared with D2-R disruption (89). The DAT is a protein that spans the pre-synaptic membrane of dopaminergic neurons 12 times and is responsible for terminating the action of dopamine by a Na^+/K^+ ATPase-dependent reuptake mechanism. Reuptake is a critical process for a rapid removal of dopamine from synapses, but its relative importance to the TIDA or THDA neurons that terminate in vascular spaces and do not form typical synapses was controversial (90,91) until the generation of the DAT-null mice. Deletion of DAT results in increased dopaminergic tone, anterior pituitary hypoplasia, dwarfism and inability to lactate. The reduced pituitary size is due to a decreased number of lactotrophs and somatotrophs without a change in thyrotrophs or gonadotrophs.

With more dopamine presumably reaching the pituitary, one would expect a moderate reduction in pituitary PRL content, due to downregulation of the PRL gene, and a marked suppression of serum PRL levels. However, the DAT-deficient mice presented three unexpected phenotypes: a) a 70-80% reduction in pituitary PRL content, b) unchanged basal serum PRL levels, and c) unresponsiveness to the dopamine receptor antagonist sulpiride. Perhaps the major effect of increased dopaminergic input at an early age is on lactotroph development or survival, resulting in a lower number of functional lactotrophs and accounting for the reduced tissue PRL concentration. Homeostatic mechanisms, resulting in rapid turnover of PRL within the remaining lactotrophs, or an increase in their sensitivity to PRFs, may explain the maintenance of near normal basal serum PRL levels. Compensatory mechanisms acting to diminish the effects of enhanced dopamine input may underlie lactotroph unresponsiveness to blockade of D2R. These include down-regulation of the dopamine receptor or suppression of tyrosine hydroxylase and lowering of dopamine synthesis. The inability to lactate in spite of normal circulating PRL levels could be due to several factors, including pituitary insensitivity to PRL secretagogues or insufficient pituitary PRL reserves that are needed for the development/maturation of the mammary gland and milk production.

The mechanism underlying dwarfism in DAT-deficient mice was also examined. Similar to PRL, they have a marked decrease in number of somatotrophs, reduced pituitary GH content, but normal circulating GH levels. The expression of Pit-1 at embryonic day 16.5 is similar in DAT-null and controls, but was decreased by 50% in adult

DAT-deficient mice. The decrease in Pit-1 expression in adults could be the consequence of a lower number of pituitary cells expressing Pit-1 or due to a direct suppression of Pit-1 expression by the enhanced dopamine input. Increased dopamine availability also results in suppression of hypothalamic GHRH expression with no change in somatostatin. As the GHRH neurons are localized in the arcuate nucleus next to TIDA neurons, GHRH suppression by elevated dopamine could be responsible for decreased GHRH supply to the anterior pituitary, failure of somatotroph proliferation and the resultant dwarfism.

PRL DEFICIENCY AND DELETION OF THE PRL RECEPTOR

The consequences of a complete absence of PRL and inactivation of its receptor have been examined in naturally-occurring as well as mutated animals. The phenotype of Snell and Ames dwarf mice is an absence of lactotrophs, somatotrophs and thyrotrophs and their respective hormones. The Snell mouse has a mutation in the Pit-1 gene (92) a pituitary-specific transcription factor that is essential for differentiation and maintenance of the three pituitary cell types. The Ames mouse has a mutation in the Prop-1 gene which activates Pit-1, with residual lactotrophs and somatotrophs (93). Homozygous animals are growth-retarded, sterile and unable to lactate. Dopamine levels are severely depressed in both the TIDA and THDA systems, but not in other brain regions, with no change in norepinephrine (94). Notably, the TIDA neurons are similar in dwarfs and normal siblings until day 21 of age, when dopamine levels and TH-immunoreactive neurons in normal mice continue to increase while those in dwarfs remain unchanged or decline.

The dopamine deficit in dwarf mice can be reversed by PRL replacement, but only if initiated at an early postnatal period; adults are refractory to PRL administration (95). This indicates that PRL acts as a neurotropic factor for the TIDA neurons during a specific developmental window of time (see Chapter 2; Phelps and Hurley). An unresolved issue is whether early postnatal TIDA development is independent of PRL or is supported by PRL that is provided via the maternal milk. Finally, hypothalamic PRL releasing factors such as TRH, galanin and VIP are unaffected by the absence of PRL in dwarf mice, while anterior pituitary galanin is undetectable (96).

Two laboratories have generated mice with PRL signaling deficiencies: one resulting from deletion of the PRL receptor (97) and another resulting from targeted mutation of the PRL gene (98). Unlike humans who have a single PRL-R transcript, mice have one long and three short PRL-R isoforms. Deletion of exon 5 in the mouse PRL-R gene results in undetectable levels of the receptor protein in the null mice and no specific binding of PRL to liver microsomes, confirming the absence of functional PRL receptors. In contrast, the PRL gene, composed of 5 exons, was mutated by an insertion of a neomycin resistance gene downstream of Ser 117 (exon 4), resulting in the production of an 11 kDa immunoreactive PRL fragment by the pituitary. This truncated PRL, however, does not possess bioactivity in the Nb2 proliferation bioassay. Although the truncated product is similar in size to 16 kDa PRL with anti-angiogenic properties (99), there is no evidence that they share similar bioactivities.

Among the many characteristics of PRLR- and PRL-null animals, we will discuss only those pertaining to the neuroendocrine axis and reproduction. The features shared by both animals are irregular estrous cyclicity, absence of pseudopregnancy, abnormal mammary development and female sterility. In spite of repeated mating and presence of oviductal ova, preimplantation embryo development in the PRL-R-null mice is arrested.

Since their fertilized eggs can develop into normal offspring when implanted in foster intact dams, a deficiency in oviductal environment is indicated rather than an abnormal oocyte development. The failure of the uterus to support blastocyst implantation is secondary to a lack of progesterone production by the PRL-supported corpus luteum and renders these animals unable to maintain pregnancy. Males with the disrupted PRL gene are fully fertile (98). As expected, mammary gland development is abnormal in both types of mice, with unchanged ductal branching but an absence of terminal or lateral lobulation (see Chapter 11; Ormandy).

The ligand- and receptor-deficient mice differ in several respects. One is mammary gland development in heterozygotes. Whereas PRL heterozygous mice produce normal litter sizes and can nurse their young, the PRL-R-heterozygotes have impaired lactation in their first pregnancy which is restored with successive pregnancies. This suggests that two functional alleles of the PRL-R are required for efficient mammary gland development, with hormones other than PRL (gonadal steroids, GH and placental lactogens) providing sufficient stimulation leading to delayed development of lactation-competent mammary glands. On the other hand, normal levels of PRL can be synthesized from a single allele in PRL-heterozygotes. Maternal behavior is severely impaired in PRL-R-heterozygotes (100) but is normal in either heterozygote or homozygote-PRL deficient mice (98). This difference may involve PRL-like hormones (e.g. placental lactogens) which act via the PRL-R to modify maternal behavior in mice.

The hypothalamo-pituitary axis has not been extensively studied in either type of PRL-deficient mice. Plasma PRL in PRL-R-null mice is increased 30-fold in males and 100-fold in females and their pituitary size is somewhat larger with increased blood supply. Treatment of PRL-R-deficient mice with bromocriptine normalizes PRL levels, indicating retention of lactotrophs responsiveness to dopamine (PA Kelly; personal communication). While hypothalamic dopamine has not been examined in these animals,

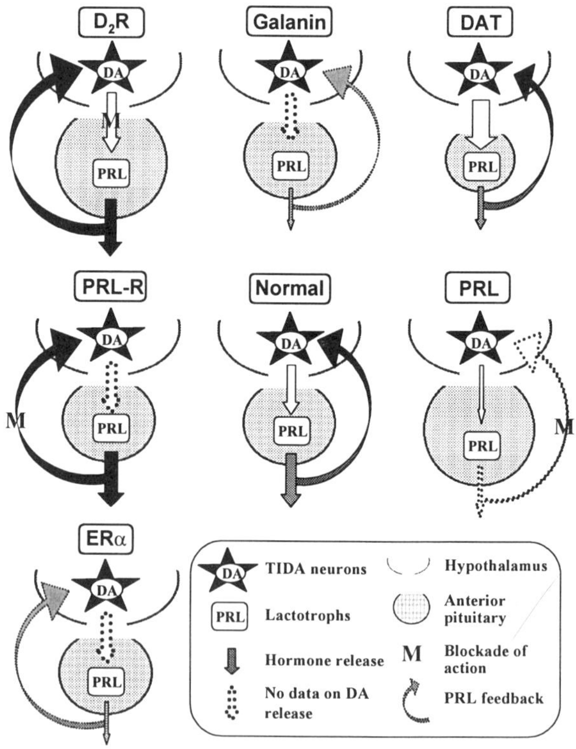

Figure 1-5. Schematic presentation of the effects of gene deletions on PRL and dopamine release and pituitary size in transgenic mice. The size and color of the arrows and the pituitary designate increases/decreases from normal.

one would predict that the absence of PRL-R results in reduced synthe-sis/production of DA as a compensatory response for a perceived lack of PRL. In PRL-deficient females, dopamine concentration and TH immunoreactivity are reduced in A12 neurons, but the number of TIDA neurons is unchanged (101). In PRL-deficient males, dopamine content in the median eminence, but not the medial basal hypothalamus, is reduced by 50%, with no change in norepinephrine (102). The most striking outcome of PRL deficiency is pituitary hyperplasia, seen already in young males and females and developing into large and hemorrhagic adenomas in aged females. This is reminiscent of the late onset adenoma that develops in D2R knockout mice which have impaired dopaminergic inhibition but elevated PRL levels. The common feature to the two types of transgenic mice is loss of negative growth control by dopamine which might involve mediation by local factors such as galanin.

ESTROGEN RECEPTORS AND GALANIN

The recent generation of ERα- and ERβ-deficient mice, named αERKO and βERKO respectively, opened up opportunities for delineating the diverse actions of estrogens. Most attention has been focused on the lower reproductive tract with only limited information available on the neuroendocrine axis that regulates PRL. Surprisingly, ERβ transcript is very low or undetectable in the mouse pituitary, while both receptors are expressed in the hypothalamus, with a predominance of ERα (20). The αERKO males and females exhibit a 10- and 20-fold reduction in PRL mRNA levels compared with wild-type animals. The change in circulating PRL levels in females is less dramatic, showing only a 5-fold decrease suggestive of compensatory mechanisms. The reduction in pituitary PRL expression in αERKO females is more severe than that seen in ovariectomized wild type mice, raising the possibility of a decreased number of lactotrophs (103). Neither TIDA/THDA neurons nor pituitary D2R expression have been evaluated in these animals. Since ERβ is not expressed in the mouse pituitary and βERKO females have normal pregnancy and lactation, ERβ does not appear to play an important role in the regulation of PRL in the mouse.

Galanin is a 29 amino acid peptide that is highly expressed in both the hypothalamus and pituitary gland. Galanin is synthesized primarily by the lactotrophs where it is co-localized with PRL in the same secretory granules (104). Like PRL, galanin is extremely sensitive to estrogens. In estrogen-treated F344 rats, its expression increases over 1000-fold above that in ovariectomized animals. In addition, TRH and dopamine stimulate and inhibit, respectively, galanin release from cultured anterior pituitary cells (88). Galanin-containing neurons are also found in the arcuate nucleus and like TIDA/THDA neurons, where they project to both the median eminence and posterior pituitary (105). Galanin fulfills a number of functions, including feeding behavior, memory and cognition and regulation of pituitary function. It exerts it actions by binding to one of several receptor subtypes, all of which are members of the large family of G-protein coupled receptors, but with little homology to other neuropeptide receptors.

Recently, two types of transgenic mice were generated, one with galanin deficiency and another with galanin overexpression in the pituitary, which exhibit opposite effects on PRL production and lactotroph proliferation. The galanin-deficient mice were produced by deleting the first five exons containing the entire coding region for galanin and had undetectable levels of galanin (105). These mice were viable, grew normally, underwent puberty at the appropriate time and were fully fertile. Pituitary PRL mRNA levels and contents in the galanin-deficient mice were reduced 30-40% but basal serum PRL levels are unchanged. Mutant females, however, were unable to lactate, at least after the first two pregnancies, owing to markedly reduced plasma PRL levels. Their response to a stimulatory dose of estrogen was also impaired. Interestingly, similar to the situation in PRL receptor heterozygotes, subsequent pregnancies restored lactation capability, suggesting delayed maturation of the mammary glands. Pituitary expression of other hormones, including GH and LH, were unchanged in the mutants, as were hypothalamic VIP and TRH levels. There is no information on the status of hypothalamic dopamine in these animals.

Transgenic mice overexpressing galanin in the pituitary gland carry a transgene composed of the mouse galanin gene ligated downstream of a 2.5-kb fragment of the rat PRL promoter (106). Both male and female transgenic mice had 100 to 1000-fold increases in galanin mRNA and peptide levels in the anterior pituitary with no effects on the posterior pituitary or hypothalamus. The effect of galanin overexpression on hypothalamic hormones has not been examined. In 2-4 month old transgenic females, pituitary PRL content was increased but plasma PRL levels did not rise until 6 months of age; other pituitary hormones, including GH, TSH and LH, showed little or no change. By 11 months, the female pituitaries are hyperplastic with increased percentages of lactotrophs and higher PRL gene expression, while none of these parameters were altered in males. Estrogen is likely to be the driving force for these changes, as judged by its ability to increase plasma PRL levels in transgenic males and the suppression of PRL release following ovariectomy of transgenic females. About 20%0 of the hyperplastic pituitaries in one year old females developed into tumors, and this percentage may increase with age, as was seen in both PRL- and D2R-deficient animals. Again, the mechanism for increased lactotroph proliferation in the presence of estrogen is unclear, and may involve a reduction in dopamine input, a direct effect of galanin, or any other local mitogenic factor that is regulated by estrogen, dopamine or both.

SUMMARY AND PERSPECTIVES

It should be remembered that PRL is an adaptive, rather than a survival, hormone; and its functions vary among species. For instance, PRL is important for osmoregulation in fish and amphibians but this function in mammals is rudimentary. PRL plays a critical role in corpus luteum maintenance and progesterone production in rodents, but not in other mammals, while synthesis of PRL by the decidua is unique to humans. With such diversity of actions, it is not surprising that both the dynamics of PRL release and the regulatory factors involved are not uniform across species. The rat has served as the primary animal model for neuroendocrine interactions and the cumulative knowledge on PRL in the rat exceeds that of all other species combined. PRL secretion in the rat increases in a biphasic manner during proestrus, rises over 100 fold in response to suckling and is stimulated by stress. In contrast, serum PRL levels in mice exhibit only minor changes during the estrous cycle, are only slightly reduced after ovariectomy, and respond only modestly to estrogen or bromocriptine administration. Because of the inability to take sequential blood samples, the dynamics of PRL release in the mouse under most physiological conditions is unknown. Thus, it is important to take in account these uncertainties when extrapolating results obtained with transgenic mice to other species, expecially humans. Future development of transgenic rats or other species will help in resolving this issue.

REFERENCES

1. Moore RY, Bloom FE. Central catecholamine neuron systems: anatomy and physiology of the dopamine systems. Ann Rev Neurosci 1978; 1:129-169
2. Ben-Jonathan N. Dopamine: A prolactin inhibiting hormone. Endocr Rev 1985; 6:564-589
3. Lerant A, Herman ME, Freeman ME. Dopaminergic neurons of periventricular and arcuate nuclei of pseudopregnant rats: Semi-carcadian rhythm in fos-related antigens (FRAs) immunoreactivities and in dopamine concentration. Endocrinology 1996; 137:3621-3628
4. Chronwall BM, Millington WR, Griffin WST, Understall JR, O'Donohue TL. Histological evaluation of the dopaminergic regulation of proopiomelanocortin gene expression in the intermediate lobe of the rat pituitary, involving in situ hybridization and ^{3}H thymidine uptake measurement. Endocrinology 1987; 120:1201-1211
5. Pardy K, Carter D, Murphy D. Dopaminergic mediation of physiological changes in proopiomelanocortin messenger ribonucleic acid expression in the neurointermediate lobe of the rat pituitary. Endocrinology 1990; 126:2960-2964
6. Peters LL, Hoefer MT, Ben-Jonathan N. The posterior pituitary: Regulation of anterior pituitary prolactin secretion. Science 1981; 213: 659-661
7. Froehlich JC, Ben-Jonathan N. Posterior pituitary involvement in the control of luteinizing hormone and prolactin secretion during the estrous cycle. Endocrinology 1984; 114:1059-1064
8. DeMaria JA, Zelena D, Vecsernyes M, Nagy GM, Freeman ME. The effect of neurointermediate lobe denervation on hypothalamic neuroendocrine dopaminergic neurons. Brain Res 1998; 806:98-94
9. Ben-Jonathan N, Laudon M, Garris PA. Novel aspects of posterior pituitary function: Regulation of prolactin secretion. Front Neuroendocrinol 1991; 12:231-277
10. Lookingland KJ, Farah JM, Lovell KL, Moore KE. Differential regulation of tuberohypophysial dopaminergic neurons terminating in the intermediate lobe and in the neural lobe of the rat pituitary gland. Neuroendocrinology 1985; 40:145-151
11. DeMaria JA, Livingstone JD, Freeman ME. Characterization of the dopaminergic input to the pituitary gland throughout the estrous cycle of the rat. Neuroendocrinology 1998; 67:377-383
12. DeMaria JA, Lerant A, Freeman ME. Prolactin activates all three populations of hypothalamic neuroendocrine dopaminergic neurons in ovariectomized rats. Brain Res 1999; 837:236-241
13. Arbogast LA, Voogt JL. Prolactin (PRL) receptors are co-localized in dopaminergic neurons in fetal hypothalamic cell cultures: effect of PRL on tyrosine hydroxylase activity. Endocrinology 1997; 138: 3016-3023
14. Mangurian LP, Walsh RJ, Posner BI 1992 Prolactin enhancement of its own uptake at the choroid plexus. Endocrinology 131:698-702
15. DeVito WJ, Avakian C, Stone S, Ace C. Estradiol increases prolactin synthesis and prolactin messenger ribonucleic acid in selected brain regions in the hypophysectomized female rat. Endocrinology 1992; 131:2154-2160
16. Bole-Feysot C, Goffin V, Edery M, Binart N, Kelly PA. Prolactin and its receptor: actions, signal transduction pathways and phenotypes observed in PRL receptor knockout mice. Endocr Rev 1998; 19:225-268
17. Muccioli G, Ghe C, Di Carlo R. Distribution and characterization of prolactin binding sites in the male and female rat brain: effects of hypophysectomy and ovariectomy. Neuroendocrinology 1991; 53:47-53
18. Lerant A, Freeman MC. Ovarian steroids differentially regulate the expression of PRL-R in neuroendocrine dopaminergic neuron populations: a double label confocal microscopic study. Brain Res 1998; 802:141-154

19. Pi XJ, Grattan DR Increased prolactin receptor immunoreactivity in the hypothalamus of lactating rats. J Neuro-endocrinol 1999; 11:693-705
20. Couse JF, Korach KS Estrogen receptor null mice: What have we learned and where will they lead us? Endocr Rev 1999; 20:358-417
21. Kohama SG, Bethea CL Steroid regulation of tyrosine hydroxylase messenger ribonucleic acid in dopaminergic sub-populations of monkey hypothalamus. Endocrinology 1995; 136:1790-1800
22. Garris PA, Ben-Jonathan N Estradiol rapidly stimulates dopamine release from the posterior pituitary *in vitro* via an opioidergic mechanism. Neuroendocrinology 1991; 53:601-607
23. Pan J-T Neuroendocrine functions of dopamine. In: Trevor WS (ed) CNS neurotransmitters and neuromodulators: Dopamine, 1996; CRC Press, Boca Raton, FA, p 213-231
24. Horvath TL, Naftolin F, Leranth C B-endorphin innervation of dopamine neurons in the rat hypothalamus: a light and electron microscopic double immunostaining study. Endocrinology 1992; 131: 1547- 1555
25. Loose MD, Ronnekleiv OK, Kelly MJ Membrane properties and response to opioids in identified dopmaine neurons in the guinea pig hypothalamus. J Neurosci 1990; 10:3627-3634
26. Garris PA, Ben-Jonathan N Regulation of dopamine release *in vitro* from the posterior pituitary by opioid peptides. Neuroendo-crinology 1990; 52:399-404
27. Arbogast LA, Voogt JL Endogenous opioid peptides contribute to suckling-induced prolactin release by suppressing tyrosine hydro-xylase activity and messenger ribonucleic acid levels in tubero-infundibular dopaminergic neurons. Endocrinology 1998; 139: 2857-2862
28. Balsa JA, Sanchez-Franco F, Pazos F, Lara JI, Lorenzo MJ, Maldonado G, Cacicedo L Direct action of serotonin on prolactin, growth hormone, corticotropin and luteinizing hormone release in co-cultures of anterior and posterior pituitary lobes: autocrine and/or paracrine action of vasoactive intestinal peptide. Neuroendocrinology 1998; 68:326-333
29. Pu L-P, Ma W, Barker JL, Loh YP Differential co-expression of genes encoding prothyrotropin-releasing hormone (Pro-TRH) and prohormone convertases (PC1 and PC2) in rat brain neurons: implications for differential processing of Pro-TRH. Endocrinology 1996; 137:1233-1241
30. Gershengorn MC, Osmar R Molecular and cellular biology of thyrotropin-releasing hormone receptors. Physiol Rev 1996; 76:175-191
31. Wang W, Gershengorn MC Rat TRH receptor type 2 exhibits higher basal signaling activity than TRH receptor type 1. Endocrinology 1999 ;140:4916-4919
32. Wang Y-H, Maurer RA A role for the mitogen-activated protein kinase in mediating the ability of thyrotropin-releasing hormone to stimulate the prolactin promoter. Mol Endocrinol 1999; 13: 1094-1104
33. Faglia G The clinical impact of the thyrotropin-releasing hormone test. Thyroid 1998; 8:903-908
34. Lam KSL Vasoactive intestinal peptide in the hypothalamus and pituitary. Neuroendocrinology 1991; 53:45-51
35. Nagy G, Mulchahey JJ, Neill JD Autocrine control of prolactin secretion by vasoactive intestinal peptide. Endocrinology 1988; 122:364-366
36. Chaiseha Y, El Halawani ME Expression of vasoactive intestinal peptide/peptide histidine isoleucine in several hypothalamic areas during the turkey reproductive cycle: relationship to prolactin secretion. Neuroendocrinology 1999 ; 70:402-412
37. Gourlet P, Vandermeers A, Van Rampelbergh J, De Neef P, Cnudde J, Waelbroeck M, Robberecht P Analogues of VIP, helodermin and PACAP discriminate between rat and human VIP1 and VIP2 receptors. Anal NY Acad Sci 1998; 865:247-252

38. Tong Z, Pitts GR, You S, Foster DN, El Halawani ME Vasoactive intestinal peptide stimulates turkey prolactin gene expression by increasing transcription rate and enhancing mRNA stability. J Mol Endocrinol 1998 ; 21:259-266
39. Hinuma S, Habata Y, Fujii R, Kawamata Y, Hosoya M, Fukusumi S, Kitada C, Masuo Y, Asano T, Matsumoto H, Sekiguchi M, Kurokawa T, Nishimura O, Onda H, Fujino M A prolactin-releasing peptide in the brain. Nature 1998 ; 393:272-276
40. Samson WK, Resch ZT, Murphy TC, Chang JK Gender-biased activity of the novel prolactin releasing peptides: comparison with thyrotropin releasing hormone levels reveals only pharmacological effects. Endocrine 1998; 9:289-291
41. Kimura A, Ohmichi M, Tasaka K, Kanda Y, Ikegami H, Hayakawa J, Hisamoto K, Morishige K, Hinuma S, Kurachi H, Murata Y Prolactin-releasing peptide activation of the prolactin promoter in differentially mediated by extracellular signal-regulated protein kinase and c-Jun N-terminal protein kinase. J Biol Chem 2000; 275:3667-3674
42. Matsumoto H, Noguchi J, Horikoshi Y, Kitada C, Hinuma S, Onda H, Nishimura O, Fujino M Stimulation of prolactin release by prolactin-releasing peptide in rats. Biochem Biophys Res Commun 1999; 259:321-324
43. Sutton SW, Wilson SJ, Luo L, Pyati J, Huvar R, Erlander MG, Lovenberg TW Anatomical distribution of prolactin-releasing peptide and its receptor suggests additional functions in the central nervous system and periphery. Endocrinology 1999; 140:5736-5745
44. Zhang X, Danila DC, Katai M, Swearingen B, Klibansky A Expression of prolactin-releasing peptide and its receptor messenger ribonucleic acid in normal human pituitary and pituitary adenomas. J Clin Endocrinol Metab 1999; 84:4652-4655
45. Ben-Jonathan N Regulation of prolactin secretion. In: Imura H (ed) The Pituitary Gland, 2ed, 1994; Raven Press, New York, p 261-283
46. Samson WK, Lumpkin MD, McCann SM Evidence for a physiological role for oxytocin in the control of prolactin secretion. Endocrinology 1986; 119:554-560
47. Liu J-W, Ben-Jonathan N Prolactin-releasing activity of neuro-hypophysial hormones: structure-function relationship. Endocrinology 1994; 134:114-118
48. Murai I, Ben-Jonathan N Posterior pituitary lobectomy abolishes the suckling-induced rise in prolactin: evidence for a prolactin-releasing factor in posterior pituitary. 1987; Endocrinology 121:205-211
49. Laudon M, Grossman DA, Ben-Jonathan N Prolactin-releasing factor: cellular origin in the intermediate lobe of the pituitary. Endocrinology 1990; 126:3185-3192
50. Steinmetz R, Gutierrez-Hartmann A, Bigsby RM, Ben-Jonathan N Activation of the prolactin promoter in transfected GH_3 cells by posterior pituitary cells. Endocrinology 1994; 135:2737-2741
51. Liu J-W, Andrews PC, Mershon J, Yan C, Allen DL, Ben-Jonathan N Peptide V: a VGF-derived neuropeptide purified from bovine posterior pituitary. Endocrinology 1994; 135:2742-2748
52. Hnasko RM, Khurana S, Shackleford N, Steinmetz R, Low MJ, Ben-Jonathan N Two distinct pituitary cell lines from mouse intermediate lobe tumors: A cell that produces prolactin-regulating factor and a melanotroph. Endocrinology 1997; 138:5589-5596
53. Kukstas LA, Domec C, Bascles L, Bonnet J, Verrier D, Israel JM, Vincent JD Different expression of the two dopaminergic D2 receptors, $D2_{415}$ and $D2_{444}$, in two types of lactotrophs each characterized by their response to dopamine, and modification of expression by sex steroids. Endocrinology 1991; 129:1101-1103
54. Chronwall BM, Beatty DM, Sharma P, Morris SJ Dopamine D_2 receptors regulate *in vitro* melanotroph L-type Ca^{2+} Channel activity via c-fos. Endocrinology 1995; 136:614-621
55. Corrette BJ, Bauer CK, Schwarz JR Electrophysiology of anterior pituitary cells. In: Scherubl H, Hescheler J (eds) The electrophysiology of neuroendocrine cells, 1995; CRC Press, London, p 101-143

56. Guivarc'h D, Vincent J-D, Vernier P Alternate splicing of the D_2 dopamine receptor messenger ribonucleic acid is modulated by activated sex steroid receptors in the MMQ prolactin cell line. Endocrinology 1998; 139:4213-4221
57. Ho M-Y, Kao JPY, Gregerson KA Dopamine withdrawal elicits prolonged calcium rise to support prolactin rebound release. Endocrinology 1996; 137:3513-3521
58. Einhorn LC, Gregerson KA, Oxford GS D2 dopamine receptor activation of potassium channels in identified rat lactotrophs: whole-cell and single-channel recording. J Neurosci 1991; 11:3727-3737
59. Lledo P-M, Legendre P, Israel J-M, Vincent J-D Dopamine inhibits two characterized voltage-dependent calcium currents in identified rat lactotroph cells. Endocrinology 1990; 127:990-1001
60. Denef C, Manet D, Dewals R Dopaminergic stimulation of prolactin release. Nature 1980 ; 285:243-246
61. Burris TP, Freeman ME Low concentrations of dopamine increase cytosolic calcium in lactotrophs. Endocrinology 1993; 133:63-68
62. Shin SH Dopamine agonists both stimulate and inhibit prolactin release in GH4ZR7 cells. Eur J Endocrinol 1999; 141:387-395
63. Porter TE, Grandy D, Bunzow J, Wiles CD, Civelli O, Frawley LS Evidence that stimulatory dopamine receptors may be involved in the regulation of prolactin secretion. Endocrinology 1994; 134:1263-1268
64. Kineman RD, Gettys TW, Frawley LS Paradoxical effects of dopamine (DA): G_{ia3} mediates DA inhibition of PRL release while masking its PRL-releasing activity. Endocrinology 1994; 135:790-793
65. De La Escalera GM, Weiner RI Dissociation of dopamine from its receptor as a signal in the pleiotropic hypothalamic regulation of prolactin secretion. Endocr Rev 1992 ; 13:241-255
66. Lamberts SWJ, MacLeod RM Regulation of prolactin secretion at the level of the lactotroph. Physiol Rev 1990; 70:279-318
67. Martin TFJ, Kowalchyk JA Evidence for the role of calcium and diacylglycerol as dual second messengers in thyrotropin-releasing hormone action: involvement of calcium. Endocrinology 1984; 115:1527-1536
68. Gershengorn MC, Thaw C Thyrotropin-releasing hormone (TRH) stimulates biphasic elevation of cytoplasmic free calcium in GH3 cells. Further evidence that TRH mobilizes cellular and extracellular calcium. Endocrinology 1985; 116:591-596
69. Mau SE Effects of withdrawal of dopamine on translocation of protein kinase C isozymes and prolactin secretion in rat lactotroph-enriched pituitary cells. modulation by substance P-mediated responses. J Mol Endocrinol 1997; 18:181-191
70. Judd AM, MacLeod RM Thyrotropin-releasing hormone and lysine-bradykinin stimulate arachidonate liberation from rat anterior pituitary cells through different mechanisms. Endocrinology 1992; 131: 1251-1260
71. Gourdji D, Laverriere JN The rat prolactin gene: a target for tissue-specific and hormone-dependent transcription factors. Mol Cell Endocrinol 1994; 100:133-142
72. Gardette R, Rasolonjanahary R, Kordon C, Enjalbert A Epidermal growth factor treatment induces D2 dopamine receptors functionally coupled to delayed outward potassium current (Ik) in GH4C1 clonal anterior pituitary cells. Neuroendocrinology 1994 ; 59:10-19
73. Missale C, Boroni F, Sigala S, Zanellato A, Dal Toso R, Balsari A, Spano P Nerve growth factor directs differentiation of the bipotential cell lineGH-3 into the mammotroph phenotype. Endocrinology 1994; 135:290-298

74. Ventra C, Meucci O, Grimaldi M, Scorziello A, Porcellini A, Schettini G Absence of D2S dopamine receptor in prolactin-secreting MMQ pituitary clone: characterization of a wild D2L receptor coupled to native transduction mechanisms. J Mol Endocrinol 1995; 14:375-389
75. Maurer RA, Notides AC Identification of an estrogen-responsive element from the 5'-flanking region of the rat prolactin gene. Mol Cell Biol 1987 ; 7:4247-4254
76. Mitchner NA, Garlick C, Ben-Jonathan N Cellular distribution and gene regulation of estrogen receptors a and b in the rat pituitary gland. Endocrinology 1998 ;139:3976-3983
77. Schreihofer DA, Resnick EM, Soh A.Y, Shupnik MA Transcriptional regulation by a naturally occurring truncated rat estrogen receptor (ER), truncated ER Product-1 (TERP-1). Mol Endocrinol 1999; 126:2960-2964
78. Mitchner NA, Garlick C, Steinmetz RW, Ben-Jonathan N Differential regulation and action of estrogen receptors a and b in GH3 cells. Endocrinology 1999; 140:2651-2658
79. Ben-Jonathan N, Steinmetz R Xenoestrogens: The emerging story of bisphenol A. Trends Endocrinol Metab 1998; 9:124-128
80. Asa SL, Ezzat S The cytogenesis and patogenesis of pituitary adenomas. Endocr Rev 1998 ; 19:798-827
81. Zhang Y, Cincotta AH Inhibitory effects of bromocriptine on vascular smooth muscle cell proliferation. Atherosclerosis 1997; 133:37-44
82. Senogles SE The D2 dopamine receptor mediates inhibition of growth in GH4ZR7 cells: involvement of protein kinase-CE. Endocrinology 1994; 134:783-789
83. Levy L, Alvaro V, Dubray C, Joubert D Ca^{++}-dependent protein kinase C isoforms in rat pituitary hyperplasia: effect of *in vivo* treatment with quingolide. Eur J Pharmacol 1994 ;16:327-334
84. Suzuki S, Yamamoto I, Arita J Mitogen-activated protein kinase-dependent stimulation of proliferation of rat lactotrophs in culture by 3',5'-cyclic adenosine monophosphate. Endocrinology 1999; 140: 2850-2858
85. Saiardi A, Bozzi Y, Baik J-H, Borrelli E Anti-proliferative role of dopamine: loss of D_2 receptors causes hormonal dysfunction and pituitary hyperplasia. Neuron 1997; 19:115-126
86. Kelly MA, Rubinstein M, Asa SL, Zheng G, Saez C, Bunzow JR, Allen R, Hnasko RM, Ben-Jonathan N, Grandy DK, Low MJ Pituitary lactotroph hyperplasia and chronic hyper-prolactinemia in dopamine D2 receptor-deficient mice. Neuron 1997;19:103-113
87. Asa SL, Kelly MA, Grandy D, Low MJ Pituitary lactotroph adenomas develop after prolonged lactotroph hyperplasia in dopamine D2 receptor-deficient mice. Endocrinology 1999 ; 140:5348-5355
88. Hyde JF, Keller BK Galanin secretion from anterior pituitary cells *in vitro* is regulated by dopamine, somatostatin, and thyrotropin-releasing hormone. Endocrinology 1991; 128:917-922
89. Bosse R, Fumagalli F, Jaber M, Giros B, Gainetdinov RR, Wetsel WC, Missale C, Caron MG Anterior pituitary hypoplasia and dwarfism in mice lacking the dopamine transporter. Neuron 1997; 19:127-138
90. Garris PA, Ben-Jonathan N Effects of reuptake inhibitors on dopamine release from the stalk-median eminence and posterior pituitary *in vitro*. Brain Res 1991; 556:123-129
91. Meister B, Elde R Dopamine transporter mRNA in neurons of the rat hypothalamus. Neuroendocrinology 1993; 58:388-395
92. Li S, Crenshaw III EB, Rawson EJ, Simmons DM, Swanson LW, Rosenfeld MG Dwarf locus mutants lacking three pituitary cell types result from mutations in the POU-domain gene pit-1. Nature 1999; 347:528-533

93. Sornson MW, Wu W, Dasen JS, Flynn SE, Norman DJ, O'Conell SM, Gukovsky I, Carriere C, Ryan AK, Miller AP, Zuo L, Gleiberman AS, Anderson B, Beamer WG, Rosenfeld MG Pituitary lineage of the *prophete of pit-1* homeodomain factor defective in Ames dwarfism. Nature 1996; 384:327-333
94. Morgan WW, Bartke A, Pfeil K Deficiency of dopamine in the median eminence of Snell dwarf mice. Endocrinology 1981; 109:2069-2075
95. Phelps C Pituitary hormones as neurotropic signals: Anomalous hypophysiotrophic neuron differentiation in hypopituitary dwarf mice. Proc Soc Exp Biol Med 1994; 206:6-23
96. Hyde JF, Bartke A, Davis BM Galanin gene expression in the hypothalamopituitary axis of the Ames Dwarf mouse. Mol Cell Neurosci 1993 ; 4:298-303
97. Ormandy CJ, Camus A, Barra J, Damotte D, Lucas B, Buteau H, Edery M, Brousse N, Babinet C, Binart N, Kelly PA Null mutation of the prolactin receptor gene produces multiple reproductive defects in the mouse. Genes Dev 1997; 11:167-178
98. Horseman ND, Zhao W, Montecino-Rodriguez E, Tanake M, Nakashima K, Engle SJ, Smith F, Markoff E, Dorshkind K Defective mammopoiesis, but normal hematopoiesis, in mice with a targeted disruption of the prolactin gene. EMBO J 1997; 16:6926-6935
99. Ferrara N, Clapp C, Weiner R The 16K fragment of prolactin specifically inhibits basal or fibroblast growth factor stimulated growth of capillary endothelial cells. Endocrinology 1991; 129:896-900
100. Lucas BK, Ormandy CJ, Bridges RS, Kelly PA Null mutation of the prolactin receptor gene produces a defect in maternal behavior. Endocrinology 1998; 139:4102-4107
101. Phelps C, Horseman ND Hypophysiotropic dopamine neurons in prolactin-knockout mice. Annual Meeting of the Endocrine Society, 1999; San Diego:P2-492 (abstract)
102. Steger RW, Chandrashekar V, Zhao W, Bartke A, Horseman ND Neuroendocrine and reproductive functions in male mice with targeted disruption of the prolactin gene. Endocrinology 1998; 139:3691-3695
103. Scully KM, Gleiberman AS, Lindzey J, Lubahn DB, Korach KS, Rosenfeld MG Role of estrogen receptor-a in the anterior pituitary gland. Mol Endocrinol 1997; 11:674-681
104. Hyde JF, Engle MG, Maley BE Co-localization of galanin and prolactin within secretory granules of anterior pituitary cells in estrogen-treated Fischer 344 rats. Endocrinology 1991; 129:270-276
105. Wynick D, Small CJ, Bloom SR Targeted disruption of the murine galanin gene. Anal NY Acad Sci 1998; 863:22-47
106. Cai A, Hayes JD, Patel N, Hyde JF Targeted overexpression of galanin in lactotrophs of transgenic mice induces hyperprolactinemia and pituitary hyperplasia. Endocrinology 1999; 140: 4955-4964

Chapter 2

Role of Prolactin in Developmental Differentiation of Hypothalamic Dopaminergic Neurons

Carol J. Phelps and David L. Hurley

Tulane University, New Orleans, LA

INTRODUCTION

Among the myriad biological actions of prolactin (PRL) that have been described across vertebrate classes (see [1] and other chapters in this volume) is a stimulatory effect in the brain that is part of the principle of endocrine feedback. In neuroendocrine terminology, "short-loop" feedback refers to the effects of anterior pituitary hormones on the hypophysiotropic neurons in the hypothalamus that regulate pituitary secretion. Because the majority of hypophysiotropic hormones are stimulatory ("releasing"), short-loop feedback is classically negative. However, in the case of neurons that produce inhibitory factors, pituitary feedback is positive.

The predominant regulation of PRL secretion is inhibitory, by dopamine (DA; for review see Ben-Jonathan [2]) produced by neurons in the medial basal hypothalamus (MBH) and delivered to the anterior pituitary *via* axons terminating in the external median eminence (ME) on hypophysial portal capillaries (3). Although DA is a transmitter in extrahypothalamic regions, such as the nigrostriatal pathway, neurons that regulate pituitary function are limited to the tuberoinfundibular (thus "TIDA" neurons) arcuate nucleus and, possibly, to anterior periventricular DA neurons (4); these areas are designated A12 and A14, respectively, according to the catecholaminergic (CA) nomenclature of Dahlstrom and Fuxe (5). The feedback effect of PRL on TIDA/A12 neurons is to increase expression of tyrosine hydroxylase (TH), the rate-limiting enzyme in DA synthesis (6) and activate DA synthesis (2). The mechanisms involved in physiological feedback of PRL on TIDA neurons is an important and current field of study, and readers are referred to numerous recent publications by Arbogast, Bartke, and Voogt.

It is the purpose of this review to provide evidence that the effect of PRL on hypothalamic DA neurons extends beyond dynamic feedback, such that PRL exerts neurotrophic effects on A12 phenotype differentiation and maintenance, perhaps including cell survival and axonal guidance. Evidence for this function has been gathered by study of several genetic models, both spontaneous and transgenic, of lifelong PRL deficiency or excess. In nearly all cases, the PRL abnormality is, or is accompanied by, alteration in growth hormone (GH), and the differentiation of GH-regulating neurons also is affected; for those effects, the reader is referred to other recent reviews (7-9). Following a description of these genetic models, the text will cover the adult status of TIDA neurons in the animals, developmental studies in PRL-deficient mice, and direct evidence for PRL neurotrophic effect from hormone replacement experiments.

ANIMAL MODELS

Two types of dwarf mouse, Snell (*dw*, [10]; dw^j [11]) and Ames (*df*, [12]), with absent GH and PRL (13) and thyrotropin (TSH; 14-16), were used to elucidate physiological roles of GH and PRL long before the nature of the spontaneous mutations was known. Both mutations are simple recessives, and dwarfism is manifested in the homozygous condition. The Snell dwarf mutations, *dw* and dw^j, were characterized in 1990 (17) as, respectively, a base substitution and a sequence rearrangement in the pituitary-specific transcription factor Pit-1 (18, 19), which is necessary for the expression of GH, PRL and TSH. Ames dwarf mice were characterized as having a T to C mutation at codon 83 in the transcription factor Prophet of Pit-1, which is necessary for Pit-1 activation (20). The genetic nomenclature for the mouse dwarf mutations has been changed officially to reflect characterization of the genes in which they occur, from *dw* to Pit1<dw>, dw^j to Pit1<dwJ>, and *df* to Prop1<df> (21). However, in this review, the original allele designations have been used in text and figures, including *DW/?* or *DF/?* for normal heterozygous or homozygous dominant siblings of Snell and Ames dwarfs. Transgenic dwarf mice were produced by ablating somatotrophs through expression of the diphtheria toxin gene driven by the GH promoter, the "GH-DT" (tg, hGRF, MT-GH, DTA, Bri78) dwarf (22). Transgenic mice engineered to produce heterologous bovine (b) or human (h) GH (23) show high circulating levels by production in non-pituitary sites, with circulating levels depending on gene variant type, copy number, and promoter construct (reviewed by Bartke *et al.* [24]).

ADULT STATUS

The trophic influence of a factor upon differentiation of specific cell types may be deduced initially by assessing the adult condition in an animal that genetically lacks or overproduces that factor, because the condition is lifelong. Assessment of the adult condition of PRL-regulating hypothalamic neurons in rodent models that either fail to produce or overexpress PRL has been accomplished for a number of spontaneous or transgenically engineered animals. An overview of these assessments is given in Table 2-1.

Table 2-1. TIDA neurons in models of lifelong altered PRL feedback

Model	TIDA neurons	reference
n.d. mGH + n.d. mPRL Ames dwarf mouse (*df/df*)	↓ DA	25 Morgan *et al.*, 1981; 26 Phelps *et al.*, 1993; 27 Phelps *et al.*, 1985
	↓ cell number	28 Morgan and Besch., 1990; 29 Phelps *et al.*, 1994
n.d. mGH + n.d. mPRL Snell dwarf mouse (*dw/dw*)	↓ DA	25 Morgan *et al.*, 1981; 27 Phelps *et al.*, 1985
	↓ cell number	30 Phelps, 1987; 31 Phelps *et al.*, 1997
↓ rGH dwarf rat (*dw/dw*)	DA: no effect ↑ cell number	32 Thomas *et al.*, 1997
n.d. mGH + ↓ mPRL GH-DT tg dwarf mouse	↓ DA ↓ cell number TH mRNA: no effect	33 Phelps *et al.*, 1991
+bGH tg giant mouse	DA: no effect cell number: no effect	34 Phelps and Bartke, 1997
+hGH ("mPRL") tg giant mouse	↑ DA ↑ cell number	34 Phelps and Bartke, 1997
↓ rGH +hGH ("rPRL") Tgr growth retarded rat	↑ DA, cell number: no effect	32 Thomas *et al.*, 1997

Legend: n.d. = not detectable; ↓ = decreased; ↑ = increased;
m = mouse; h = human; r = rat;
tg = transgenic; Tgr = transgenic growth retarded.

PROLACTIN DEFICIENCY: SPONTANEOUS AND TRANSGENIC DWARF MICE

Prolactin is not produced at any time in development by Snell (*dw/dw* or *dw^j/dw^j*) or Ames dwarf (*df/df*) mice (35, 36); pituitary PRL is reduced severely in transgenic dwarfs (22, 33). Morgan and colleagues (27) reported the first investigation of hypophysiotropic DA in PRL-deficient dwarf mice, in which levels were measured separately in ME (A12 terminals) and hypothalamus (including A12 perikarya but excluding ME) in both Snell (*dw/dw*) and Ames (*df/df*) dwarfs; ME DA was severely reduced (to ≅ 20%) in both dwarf types, but hypothalamic DA was not affected. That study isolated the hypothalamic deficit to DA, because ME norepinephrine (NE) levels were comparable to those in normal mice of each strain. A subsequent morphologic assessment of the DA deficit used a paraformaldehyde-induced CA fluorescence technique (37) to visualize endogenous transmitter (25). As shown in Figure 2-1, CA histofluorescence was decreased in both cells and terminals of dwarf A12. Fluorescence of DA in cell bodies in *substantia nigra* (SN), as well as of NE terminals in paraventricular nucleus (PVN) and supraoptic nucleus (SON) were comparable to those of normal mice in both dwarf types (Figure 2-2). Morgan and King reported subsequently (38) that DA synthetic rate was decreased markedly in the ME of Ames dwarfs.

Those studies prompted the question of whether the number of A12 DA neurons is reduced in the dwarf hypothalamus. In order to quantify neuronal cell number, A12 perikarya may be marked by immunocytochemistry (ICC) of TH. Although TH is the

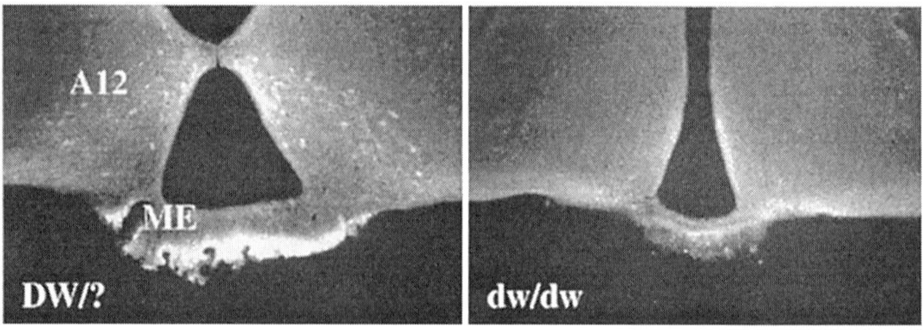

Figure 2-1. Endogenous paraformaldehyde-induced CA histofluorescence in TIDA neuronal perikarya (A12) and terminals (ME) of adult normal (*DW/?*) and Snell dwarf (*dw/dw*) mouse. Original objective magnification = 10x for both.

Figure 2-2. CA fluorescence of noradrenergic terminals in paraventricular nucleus (PVN) and supraoptic nucleus (SON) and in dopaminergic cell bodies in midbrain *substantia nigra* (SN) of dwarf mice (*dw/dw* and *df/df*). Original objective magnification 25x for all. From Phelps *et al.*, Cell Tiss Res 1985; 240:19. © Springer-Verlag.

synthetic enzyme for both DA and NE synthesis, previous studies of DA *vs.* NE systems by microspectrophotometry and CNS lesions (4) allow differentiation of DA and NE neuronal groups by location. Immunostaining provides a permanent record for CA, as opposed to formaldehyde-induced fluorescence, which fades rapidly upon exposure to intense blue-violet light microscopically. Figure 2-3 illustrates TH immunoreactivity in three dopaminergic areas in the diencephalon of normal (*DF/?*) and dwarf (*df/df*) mice: A13, in medial *zona incerta*, is most dorsal; A14, located from the anterior commissure rostrally through the level of ME caudally, is intermediate; and area A12 is the ventral ME-afferent and PRL-inhibitory group, which is shown at a middle ME level. A distinct qualitative decrease in TH immunostaining intensity is shown for the dwarfs, in both

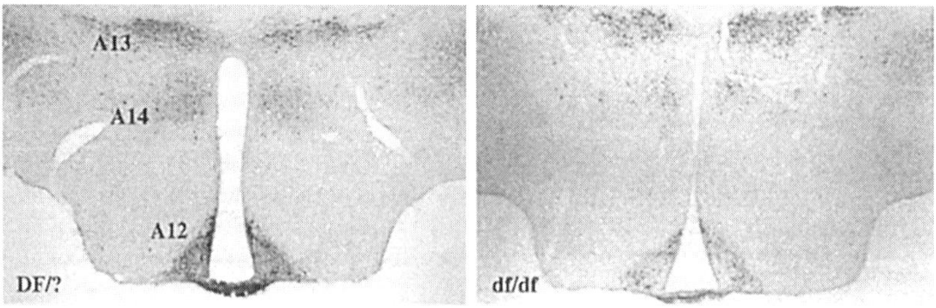

Figure 2-3. TH immunostaining in three dopaminergic neuronal groups in normal (*DF/?*) and Ames dwarf (*df/df*) diencephalon: A13 (medial *zona incerta*), A14 (anterior periventricular), and A12 (arcuate nucleus ME-projecting). Coronal sections, original objective magnification = 4x for both. From Phelps *et al.*, J Neuroendocrinol 1994; 6:145. © Blackwell Science Ltd.

ME terminals and in cell bodies of A12, but not of A13 or A14. Quantification of the reduction in A12 cell numbers has been accomplished for Snell, Ames, and transgenic dwarfs.

Ames dwarf mice

Morgan and Besch (28) reported a decrease in arcuate nucleus TH-immunoreactive (TH+) median cell number per coronal section in female but not male Ames dwarfs (*df/df*). A morphometric analysis of the A12 deficit in Ames dwarfs (29) showed that the deficit in *df/df* A12 TH+ neurons was distributed throughout the area; i.e., numbers of TH+ neurons were reduced significantly at each rostral-to-caudal sampling interval in both male and female *df/df* mice compared with *DF/?* siblings. Total numbers of TH+ neurons in *df/df* mice were reduced to 48% of normal ($p<0.01$) in A12, and also were reduced ($p<0.05$) in A14, while TH+ neuron number in nonhypophysiotropic area A13 (medial *zona incerta*) was unaffected.

Snell dwarf mice

In an early study of Snell (*dw*^{j}) dwarfs, the number of A12 TH+ neurons was reported as only 2% of the number in normal siblings, while nonhypophysiotropic TH+ neurons were comparable for *dw/dw* and *DW/?* mice (30). In a more recent study of A12 development in Snell dwarf mice (31), with improved immunostaining methods, TH+ neurons averaged 23% in adult *dw/dw* compared with *DW/?*.

Transgenic dwarf mice

In transgenic dwarf mice (tg, hGRF, MT-GH, DTA, Bri78) with reduced but not absent PRL cells (22), the reduction in A12 neuron number was less severe - to 60% of that in normal mice (C57Bl6xSJL F_1), and was concentrated in anterior A12. Although A12 DA histofluorescence was virtually absent, TH mRNA in A12, quantified after *in situ* hybridization, was not affected significantly in transgenic dwarfs (33).

A comparison of A12/TIDA and A13 neuron number is illustrated graphically in Figure 2-4, for transgenic, Ames, and Snell dwarfs, and for respective normal mice of the

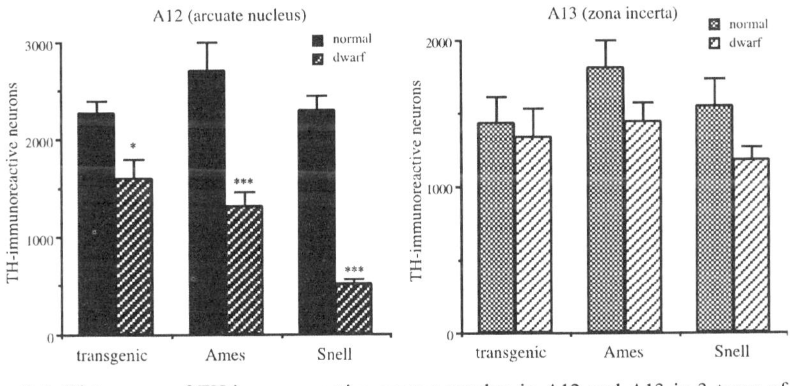

Figure 2-4. Histograms of TH-immunoreactive neuron number in A12 and A13 in 3 types of dwarf mice: transgenic (GH, DT-A; Mt, GHRF; Bri 78), Ames (*df/df*) and Snell (*dw*^{j}*/dw*^{j}), compared with respective normal phenotypes, transgenic control (C57Bl6 x SJL F1), *DF/?* and *DW/?* (n= 5-15). Significant differences between dwarf and normal within each type are indicated as: * = $p < 0.05$; ** = $p < 0.01$; *** = $p < 0.001$. In this and similar figures, columns represent means; vertical lines on each column are SEM. Modified from Phelps and Hurley, Proc Soc Exp Biol Med 1999; 222:39. © Society for Experimental Biology and Medicine.

same strains. Numbers of A13 DA neurons were not different from that in normals in any PRL-deficient dwarf type.

"PROLACTIN" EXCESS: TRANSGENIC ANIMALS

Peripheral expression of human growth hormone

Transgenic mice developed to produce heterologous (human or bovine) GH driven by either MT or PEPCK promoters (23, 39, 40) have been used to study the effects of excessive GH on reproduction as well as growth and metabolism in both sexes (24). The effects of hGH and bGH on reproduction in these mice differ, probably because hGH has lactogenic, as well as somatogenic, properties in rodents (41, 42). An investigation of the possible differential effects of PRL-like hGH *vs.* only-somatogenic bGH on TIDA neurons (34) showed that both catecholamine fluorescence and numbers of neurons detectable by TH ICC were increased significantly in mice bearing either MT-hGH or PEPCK-hGH, but not bGH, transgenes, as shown in Figure 2-5. In the particular lines of mice assessed, hGH levels were higher in PEPCK-hGH than in MT-hGH transgenics; numbers of TIDA neurons were 134% of normal in MT-hGH transgenics, and 147% of normal in the PEPCK-hGH mice. These findings indicate that not only is the stimulatory feedback signal necessary for a normal complement of PRL-regulating TIDA neurons, but that a larger population results if an excess PRL stimulus is present during TIDA development. Because the high circulating levels of hGH ("PRL") are the result of peripheral (liver, gut, kidney) production, these studies, like those of PRL replacement in hypopituitary dwarf mice described below, support a role for circulating, rather than retrogradely transported or hypothalamically produced, PRL in TIDA neuron development. Access of peripheral PRL to the hypothalamus is an important question; pertinent findings include evidence for preferential transport of PRL to CSF (43) and absence of blood-brain barrier in ventromedial MBH (44).

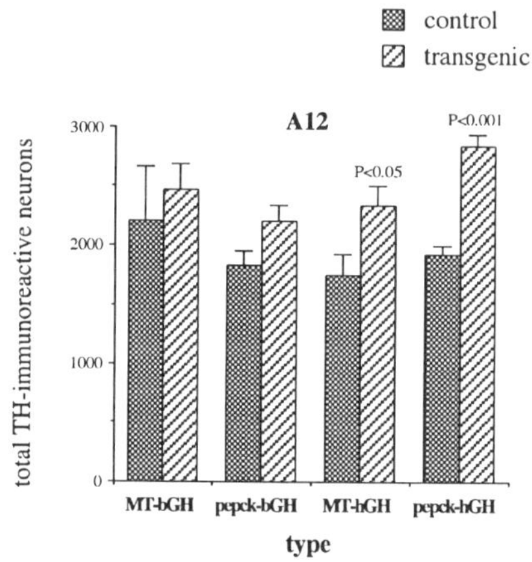

Figure 2-5. TH-immunoreactive neuronal perikarya in A12 (TIDA) of four types of GH-transgenic mice compared with littermate controls. Cell numbers were elevated in hGH transgenics. From Phelps and Bartke, Endocrinology 1997; 138:2849. © The Endocrine Society.

Human growth hormone expressed in hypothalamic neurons

Prompted by the interesting effect of transgenic hGH overexpression on TIDA neurons described above, TIDA neurons were evaluated in rats bearing a transgenic construct consisting of structural hGH "downstream" of a sequence containing the rat GHRH promoter, resulting in hGH expression in GH-releasing hormone (GHRH) neurons (45). The animals were designed to produce GH deficiency by intrahypothalamic feedback suppression of GHRH, resulting in a model that would respond to exogenous GHRH and synthetic GH secretagogues, as opposed to pituitary GH-deficient *dw/dw* rats (46) or mice. These transgenic growth-retarded (Tgr) rats exhibit decreased hypothalamic GHRH expression (45) as opposed to the increased GHRH expression of *dw/dw* rats (47) and *df/df* mice (48).

It was hypothesized that the intrahypothalamic production of hGH in Tgr rats would have a PRL-like stimulatory effect on TIDA neurons, as had circulating hGH in MT- or PEPCK-hGH transgenic mice, especially within neurons that express both GHRH and DA, a subpopulation of the arcuate nucleus that has been demonstrated in rats (49). This background is illustrated by the left cartoon of Figure 2-6. Brains of Tgr, normal (AS parent strain) and *dw/dw* rats prepared by paraformaldehyde/glutaraldehyde (37) perfusion in the laboratory of Dr. I.C.A.F. Robinson at the NIMR, London, were shipped to New Orleans (in a significant adventure with transatlantic couriers and the USDA, wary of British meat) for assessing TIDA CA fluorescence and neuron number by TH ICC. Although increased DA fluorescence in Tgr individual TIDA neurons (as shown in Figure 2-7) indicated stimulation, TH-immunoreactive neuron numbers did not differ between Tgr and AS controls. However, and surprisingly, the TIDA neuronal population was **increased** significantly in dwarf (*dw/dw*) rats. To test whether the 37% increase in TIDA neuron number resulted from increased GHRH expression, a group of *dw/dw* rats was treated with GH to reduce GHRH (50). The treatment did not decrease TIDA neuron numbers (i.e., those with detectable TH content per neuron) significantly, suggesting that a permanent developmental effect had occurred. The results have been interpreted to indicate that, in the *dw/dw* rat, with isolated pituitary GH deficiency, ongoing PRL secretion stimulates differentiation of a normal population of neurons that produce DA only, while the greatly decreased GH negative feedback results in an increased population of GHRH-producing neurons, a portion of which also produce DA, resulting in an increase in total numbers of TIDA neurons. In the Tgr rat hypothalamus, hGH appears to exert a developmental suppression of the GHRH-producing neuronal population, a portion of which are DA-producing. It is likely that the hGH simultaneously exerted a stimulatory effect on TIDA (DA-only) neuronal number, or the resulting TH-immunoreactive population in Tgr rats would have been reduced compared with normal AS rats, rather than comparable. The latter interpretation is supported both by the increased DA fluorescence in TIDA neurons in Tgr rats (Figure 2-7), and by reduced pituitary levels (32) of PRL that would result from increased DA secretion. The interpre-

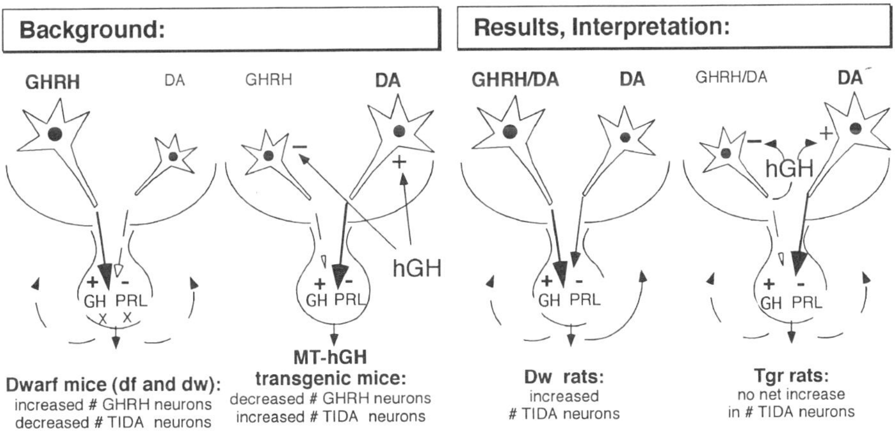

Figure 2-6. The precedent and outcome of studies of TIDA neurons in Tgr and dw dwarf rats. From Phelps and Hurley, Proc Soc Exp Biol Med 1999; 222:39. © Society for Experimental Biology and Medicine.

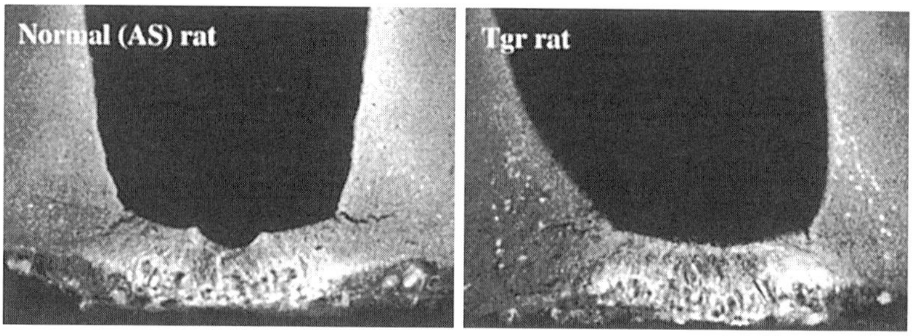

Figure 2-7. Paraformaldehyde-induced DA histofluorescence in normal (parent strain AS) rat (left panel) and Tgr rat (right panel) basal hypothalamus. Sections were coronal, 30 μm thick, and photographed at original objective magnification of 10x. From Phelps and Hurley, Proc Soc Exp Biol Med 1999; 222:39. © Society for Experimental Biology and Medicine.

tation also is illustrated in Figure 2-6. However, GHRH and DA/TH expression have been analyzed only independently in Tgr rats, and confirmation of this interpretation awaits quantification of the neuronal population that expresses both GHRH and DA/TH.

DEVELOPMENT OF TIDA NEURONS IN PRL-DEFICIENT DWARF MICE

AMES DWARF

In order to determine the developmental pattern which leads to reduced hypophysiotropic DA and detectable TIDA neurons, MBH and ventral tegmental mid-brain (SN) CA levels were evaluated morphologically, by histofluorescence, and biochemically, by HPLC assay of dissected areas, in *df/df* and *DF/?* mice at 7, 14, 21, 30 and 90 days of age (26). Phenotype of mice younger than 21 days was verified by GH and PRL ICC of pituitaries, because dwarfism is not manifested grossly until 2-3 weeks of age. Concentration of DA (pg/μg protein) for MBH and SN in *DF/?* and *df/df* mice is illustrated in Figure 2-8. By both morphologic and biochemical methods, A12 DA was

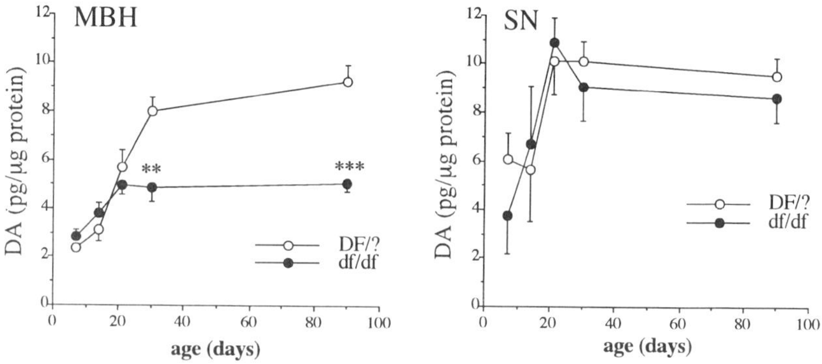

Figure 2-8. DA concentrations in *DF/?* and *df/df* medial basal (MBH) and ventral tegmentum/*substantia nigra* (SN) at 7, 14, 21, 30 and 90 days of age. Each point represents the mean of 8 - 10 animals; vertical lines represent SEM. Asterisks over *df/df* means show significant difference from *DF/?* at the same age; ** = $p < 0.01$, *** = $p < 0.001$. From Phelps *et al.*, Endocrinology 1993; 132:2715. © The Endocrine Society.

comparable for *df/df* and *DF/?* through 14 days of age; after 14 days postnatally, DA levels in MBH in *df/df* mice failed to increase significantly, while levels in *DF/?* mice continued to rise through 30 days of age. Dwarf DA levels in MBH were significantly different from those in *DF/?* at 30 ($p < 0.01$) and 90 ($p < 0.001$) days of age. In contrast, SN DA was comparable for *DF/?* and *df/df* mice at each age examined. Likewise, NE levels were comparable for *DF/?* and *df/df* at each age examined, increasing from birth to plateaus at 21 days in SN and 30 days in MBH (26). The results indicated that A12 DA neurons in dwarfs are comparable to those of normal mice neonatally, and that the adult deficit is a result of failure to develop further.

To complement the study of DA levels, quantification of TH+ neuron number in Ames dwarfs was accomplished by counting cells at 180 μm rostral-to-caudal intervals through areas A12, A13 and A14 of the brains of *df/df* and *DF/?* mice between 7 days and 2 months of age (36). As shown in Figure 2-9, TH ICC intensity in dwarf TIDA A12 neurons appeared to decrease dramatically between postnatal days 14 and 42. Figure 2-10 shows graphically the total neuron number in A12 and A13 as a function of postnatal age in *DF/?* and *df/df* mice. Cell number in A13 did not vary significantly with age or type. The size of the A12 TH+ cell population increased, and was comparable for *df/df* and *DF/?*, through 21 days of age. Thereafter, A12 TH+ neuron number decreased significantly in dwarf mice ($p < 0.01$ for 21 compared with 60 days) to <50% of the number in *DF/?* A12 at 2 months ($p < 0.001$ for means of *DF/?* and *df/df*); the deficit in *df/df* was significant at 30 days of age. Thus, in the case of TIDA TH+ neuron number, the adult condition in dwarfs represents a regression. It is important to note that this reduction in cell number does not necessarily represent cell death and may, instead, show a reduction in perikaryal TH levels below the threshold of ICC detectability.

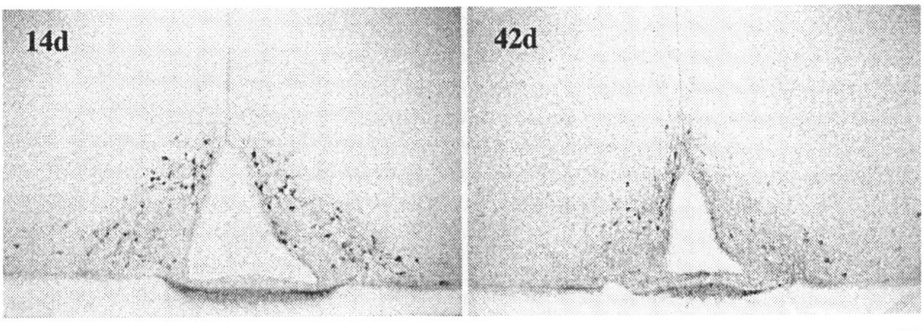

Figure 2-9. TH ICC in A12 of dwarf mice at 14 and 42 ages of age. Original objective magnification = 10x for both. Modified from Phelps, Proc Soc Exp Biol Med 1994; 206:6. © Society for Experimental Biology and Medicine.

SNELL DWARF

A devastating fire at the Jackson Laboratory in 1989 rendered Snell dwarf mice or mutation-bearing breeding pairs unavailable for several years, but subsequent establishment of breeding colonies in individual investigator laboratories allowed renewed studies of *dw/dw,* including hypophysiotropic neuron development. Because an earlier quantification of TIDA neurons in adult Snell dwarf (30) indicated that the population deficit was more severe than that in Ames dwarf (29), it was of interest to compare the developmental patterns of the two dwarf types. A developmental study was conducted for Snell dwarfs and normal siblings, assessing catecholamine fluorescence and TIDA neuron number by TH ICC from 7d through 60d postnatally (31). Qualitatively, TIDA fluorescence and TH immunostaining appeared reduced in *dw/dw* compared with *DW/?* at early ages, which differed from the pattern in *df/df* (26, 36). Neuron numbers were identical for *dw/dw* and *DW/?* at 7d. Subsequently, the TIDA population increased in *DW/?* hypothalamus, reaching the adult complement by 14d. The population size did not increase in dwarfs beyond 7d, was significantly lower than that in *DW/?* at 14d, and, in fact, decreased at 30d and 60d from numbers present at 7 - 21d. The number of TH-immunoreactive TIDA neurons in adult *dw/dw* was 23% the number in *DW/?* mice, as opposed to 48% of *DF/?* in *df/df*. These data are graphed in Figure 2-11, superimposed over the pattern in Ames dwarfs (see Figure 2-10). An explanation for the difference in developmental pattern and severity of deficit between the two mutations is elusive, because the hormone-deficient phenotype is essentially identical for the two mutations. A similarity in the developmental pattern of TIDA neurons in Ames and Snell dwarfs is the precipitous decrease that occurs after 21d in both mutants, albeit from a lower population size in Snell dwarf; it is as if a previous maintenance ceases at that age, when the failure to produce endogenous PRL in dwarfs becomes significant, because PRL is only first detectable at 7-8d in normal mice (26, 35) and circulating levels are low prior to 21d (Romero and Phelps, unpublished results).

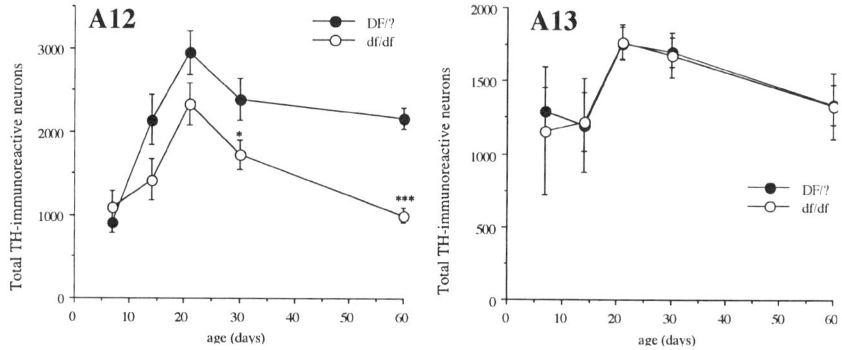

Figure 2-10. Numbers of TH-immunoreactive neurons in hypothalamic dopaminergic areas A12 and A13 of Ames dwarf (*df/df*) and normal (*DF/?*) mice at 7, 14, 21, 30 and 60 days of age. Each point represents the mean of 6 - 10 animals; vertical lines represent SEM. Asterisks indicate *df/df* mean significantly different (* = p < 0.05; *** = p < 0.001) from *DF/?* mean at the same age. Modified from Phelps, Proc Soc Exp Biol Med 1994; 206:6. © Society for Experimental Biology and Medicine.

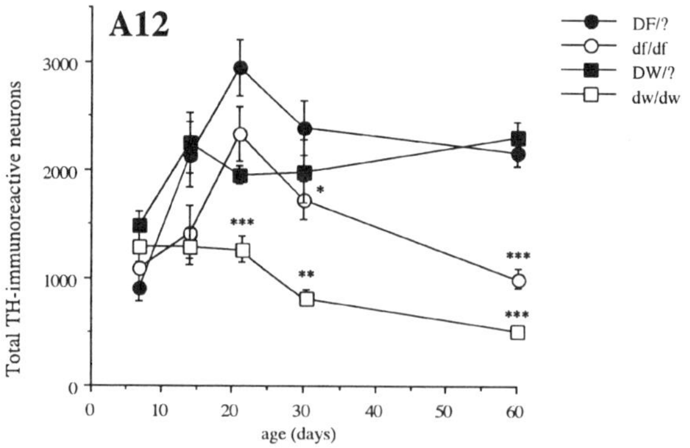

Figure 2-11. The patterns of TIDA neuron development and loss, compared for Snell and Ames dwarf mice and with normal counterparts. Asterisks indicate dwarf means significantly lower than means in normals of the same type (* = p < 0.05; ** = p < 0.01; *** = p < 0.001). From Phelps and Hurley, Proc Soc Exp Biol Med 1999; 222:39. © Society for Experimental Biology and Medicine.

EFFECT OF PROLACTIN REPLACEMENT IN AMES DWARF MICE

TREATMENT IN NEONATAL MICE

The foregoing developmental studies indicated that the onset of the A12 DA and TH deficit occurred after 14 - 21 days of age in dwarf mice. In order to test directly whether the absence of PRL was important in the hypophysiotropic DA decrease, PRL treatment (oPRL, 50 µg/day) of *df/df* mice was initiated at 12 days of age and continued for 30 days (51). The hormone replacement was administered to whole litters because dwarfs are not reliably distinguishable from normals at 12 days, although the dwarf phenotype was recognizable at later ages, even among PRL-treated *df/df* mice. Brains of *df/df* and *DF/?* siblings were assessed by histofluorescence and TH ICC at the end of PRL or vehicle treatment. Perikaryal and ME DA histofluorescence was enhanced markedly in A12 neurons of PRL-treated compared with vehicle-treated dwarfs. A similar enhancement in TH immunoreactivity occurred, as shown in Figure 2-12, which depicts A12 in both *DF/?* and *df/df* treated with vehicle ("sal") or PRL. The coronal level of the normal mouse sections is anterior (with reduced cell number) to that of the sections in dwarfs, but sections are matched for level within type. Figure 2-13 illustrates

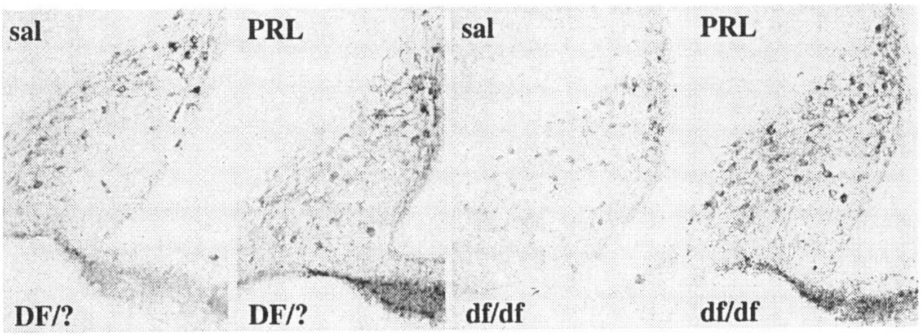

Figure 2-12. TH immunoreactivity in A12/TIDA neurons in phenotypically normal (*DF/?*) and Ames dwarf (*df/df*) mice treated for 30 days, beginning at 12 days of age, with either ovine PRL or saline vehicle (sal). Modified from Romero and Phelps, Endocrinology 1993; 133:1860. © The Endocrine Society.

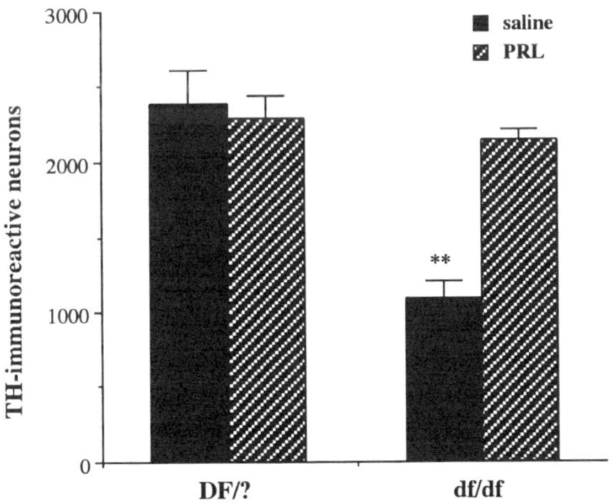

Figure 2-13. TH-immunoreactive cell number in hypothalamic dopaminergic area A12 in normal (*DF/?*) and Ames dwarf (*df/df*) 42 day-old mice treated with either vehicle (sal) or ovine PRL for 30 days beginning at 12 days of age. The *df/df* + sal group differed from all other A12 groups, $p < 0.01$. Modified from Romero and Phelps, Endocrinology 1993; 133:1860. © The Endocrine Society.

total numbers of TH+ cells in A12 in the same four treatment groups. Quantification of TH+ populations showed that A12 neuronal numbers in PRL-treated dwarfs were maintained at a level comparable to that of phenotypically normal siblings. Differential cell counts within A12 showed that the inherent deficit was greater in ventrolateral than in dorsomedial regions, which may suggest differential involvement of these subpopulations in PRL regulation, and in dependence on PRL feedback for normal development. However, normal cell numbers in both regions were maintained by early PRL treatment of dwarfs.

PREPLACEMENT IN ADULT DWARFS

Morgan and coworkers (27) reported that PRL treatment of adult dwarf mice resulted in significantly increased ME DA, but to levels still lower than those in normal mice. A similar result also was reported for studies of DA synthesis in ME after PRL treatment of dwarfs (38). Morgan and Besch (28) reported an increase in TH+ A12 cells for *df/df* females, but not males, after PRL injection or renal pituitary grafts in adult dwarfs. A study in this laboratory (52) of PRL treatment of adult *df/df* mice extended the reports from Morgan's laboratory (27, 28, 38). Dwarf and normal mice 60 days of age were subjected to the same protocol of PRL treatment that was effective in maintaining A12 TH/DA cells when treatment was begun at 12 days (51). Histofluorescence was evaluated for qualitative DA levels, and hypothalamic TH+ cells were quantified in groups of dwarfs treated with PRL (30d), PRL (30d) + saline (15d), PRL (30d) + saline (15d) + PRL (30d), and PRL (30d) + saline (15d) + PRL (30d) + saline (15d). Additional dwarf and normal mice received renal capsule pituitary grafts at 60 days of age, and TIDA neurons were assessed 30 days later. The results were striking qualitatively: PRL treatment repeatedly enhanced intensity of CA histofluorescence and TH immunoreactivity in A12 (as shown for three groups in Figure 2-14), but not in A13, neurons. However, A12 TH+ neuron number was **not** increased after PRL in *df/df* mice, in any treatment group, including pituitary-grafted dwarfs, as shown graphically in Figure 2-15 for the oPRL-treated groups.

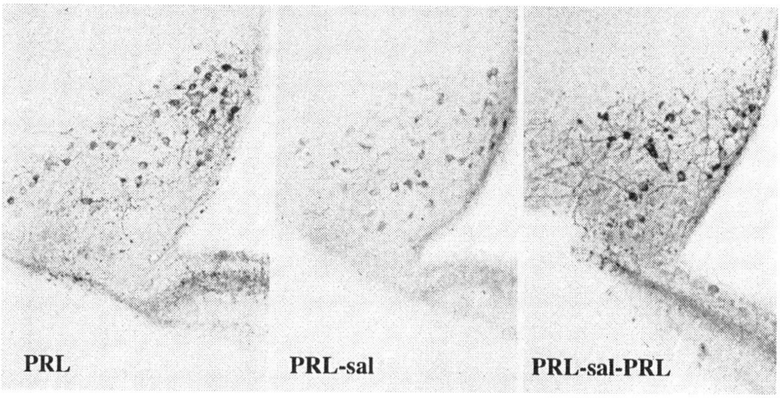

Figure 2-14. TH immunoreactivity in TIDA/A12 area neurons in adult Ames dwarf (*df/df*) mice treated beginning at 60 days of age with ovine (o) PRL for 30 days, oPRL for 30 days followed by saline vehicle for 15 days, or oPRL (30 d), sal (15 d) and oPRL (30 d). All 30 μm coronal sections, 20X original objective magnification. From Phelps, Proc Soc Exp Biol Med 1994; 206:6. © Society for Experimental Biology and Medicine.

Figure 2-15. Histogram of numbers of TH-immunoreactive neurons in hypothalamic dopaminergic area A12 in Ames dwarf mice treated as adults (60 days of age) with PRL and/or vehicle. Treatment groups (5 -7 mice) were: DF/Control = normal mice, df/Control = dwarf mice, both 90 days of age; dwarfs treated with oPRL are indicated by P (prolactin, 30d) and S (saline for 15d), in sequence. A12 cell numbers in all df/df groups were lower than A12 cell numbers in normal mice, $p < 0.01$. Modified from Phelps, Proc Soc Exp Biol Med 1994; 206:6. © Society for Experimental Biology and Medicine.

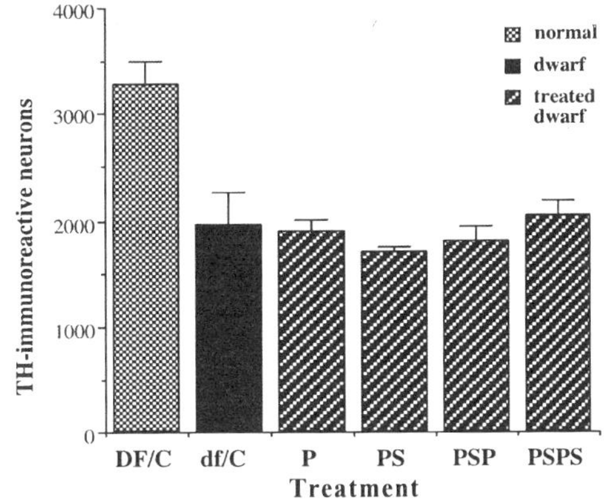

The results of PRL treatment of adult dwarfs suggest that A12 DA neurons are dependent for normal development on PRL feedback at a critical time in postnatal development (between 12 and 60 days). Morgan and Besch (28) suggested that, based on increased TH immunoreactivity in A12 neurons in female *df/df* mice after PRL, A12 neurons in dwarfs were present in nearly normal numbers. Studies in this laboratory suggest, instead, that the threshold or criteria for neuron detectability in untreated dwarfs differed between the two studies. The results from this laboratory indicate that A12 DA neurons in Ames dwarf mice are refractory to "rescue" by PRL treatment at 60 days of age, although TH immunoreactivity and steady-state DA histofluorescence levels are enhanced in extant neurons. The findings indicate that the TH/DA phenotype is lost permanently in 50% of the population, suggesting either cell death or loss of response to PRL.

It had been shown that the decline in numbers of TIDA neurons that occurs in Ames dwarfs between 21 and 60 days of age (36) could be prevented by ovine (o) PRL treatment initiated at 12 (51), but not at 60 (52), days of age. Further studies (Romero and Phelps, unpublished) tested whether PRL replacement initiated at 21d or 30d would prevent the decline in TIDA cell number; it did not. Dwarf mice treated with PRL beginning at 21d or 30d had TIDA populations comparable to those of untreated dwarfs, approximately 50% of the normal population. These findings are summarized in Table 2-2. That PRL treatment beginning at 21d was not effective in maintaining the TIDA neurons suggests that the phenomena responsible for the loss of cells or phenotype (TH and DA production) are already operating at 21d of age.

Table 2-2. Timing and specificity of PRL replacement effects in Ames dwarf (df/df) mice

Treatment	Effects on TIDA neurons
oPRL beg. 12d	↑ perikaryal DA/TH, ↑ cell number [to normal], ↑ DA/TH in ME [to normal] ([51] Romero and Phelps, 1993)
oPRL beg. 21d	no effect on cell number (Romero and Phelps, unpublished)
oPRL beg. 30d	no effect on cell number (Romero and Phelps, unpublished)
oPRL beg. 60d	↑ perikaryal DA/TH, no effect on cell number, no effect on DA/TH in ME ([52] Romero and Phelps, 1995)
mPRL beg. 60d (pituitary graft)	↑ perikaryal DA/TH, no effect on cell number, ↑ DA/TH in ME ([52] Romero and Phelps, 1995)

A distinct qualitative difference in response to PRL treatment in neonates *vs.* adults was observed in DA histofluorescence and TH immunoreactivity in the perivascular zone of external ME. As shown in Figure 2-12, the intensity of ME TH immunostaining in dwarfs treated with PRL beginning at 12 days was at least comparable to that of untreated *DF/?* animals. In dwarfs treated with oPRL as adults (Figure 2-14A), ME TH immunoreactivity remained low. Similar observations were made with regard to DA histofluorescence in the external ME of dwarfs treated with PRL neonatally or as adults. The photomicrographs of TH immunoreactivity in PRL-treated adult dwarfs also showed very low ME intensity in the report of Morgan and Besch (28). The observation of increased intensity of perikaryal CA histofluorescence or TH immunoreactivity after PRL treatment of adult dwarfs, without comparable enhancement in the terminal ME region, suggests an abnormality in A12 axon connectivity in dwarfs. Current tract-tracing studies suggest that A12 ME projections are reduced in dwarfs, within a population that is reduced already in number.

It is important to consider why the initial development of TIDA/A12 neurons in dwarfs is comparable to that of normal mice. Does this observation indicate that early postnatal TIDA neuron development is independent of PRL? One possible answer to this question may be inferred from related studies in rats. Production of PRL by the mouse pituitary is not detectable until 7 - 8 days postnatally (26, 35). In rat pups, significant levels of PRL are present in the circulation from days 1 through 21 of age (53). Although PRL production may be initiated earlier in rats than in mice (54), a majority of PRL in pup serum is likely to be derived from PRL in maternal milk, which is absorbed by pups until at least 14 days of age (55) because maternally-derived PRL has been shown to be important for normal development of TIDA neuron function. Experimental induction of milk PRL deficiency, by treatment of dams with bromocriptine at 2 - 5 days postpartum, has been shown to result in deficient ME DA levels and turnover at 35 days of age (56) and hyperprolactinemia which persisted until 100 days of age (57) in offspring. In rats, the intestinal mucosa undergoes closure *via* formation of intercellular tight junctions at 18-21 days of age, preventing absorption of molecules as large as PRL thereafter (58); at this age, endogenous PRL production is significant and circulating PRL levels rise (53), such that endogenous PRL could supplant the influence of maternal hormone on TIDA neuron development. This substitution of PRL source could not occur in dwarf mice, in which endogenous PRL production is absent, and further influence on TIDA neuron development would cease after removal of maternally-derived PRL.

The collective data in PRL-deficient mice indicate that the DA-producing inhibitory neurons are not only deficient in adults, but exhibit a critical period for maintenance of normal population size and transmitter content by PRL feedback. The results suggest actual cell loss in A12 and aberrant terminal contacts in ME when dwarfs fail to produce the PRL signal.

OR MAYBE NOT: A CONTRAINDICATION

As explained in the Introduction, all the animal models of PRL deficiency or excess are also, or primarily, likewise abnormal in GH expression; in the case of spontaneous dwarf mice, TSH is also deficient. Recent gene disruption technology has produced two transgenic mouse models in which to address the effect of isolated PRL absence. Mice which are "null" (-/-) for the expression of PRL receptor (59) or PRL (60) provide new models for studying specific PRL neurotrophic effects. In the case of PRL-null mice, unpublished results (61) indicate that TIDA neuron differentiation is not compromised by lifelong PRL deficiency. Although CA fluorescence and TH immunoreactivity are reduced markedly in A12, cell number in -/- mice is comparable to that in +/- littermates. Interpretation must invoke, therefore, the currently popular "necessary and sufficient" trial of PRL effect. While replacement studies in panhypopituitary dwarf mice show that PRL (alone, among PRL, GH and TSH deficiencies) is sufficient to maintain TIDA neuron differentiation, PRL evidently is not necessary (or specific) in this effect. Whether other hormones or their mediators can suffice awaits further experimentation, including assessment of TIDA neurons in PRL R-null mice. In PRL-null mice, male fertility is only slightly reduced (62), and there appear to be no deficits in hematopoiesis and maternal behavior (60), although maternal behavior is affected profoundly in PRL R-null mice (63). The findings call into question established tenets of PRL function and indicate experiments to examine redundancy and "cross-talk" among hormones and receptors. Evidently, these questions extend to neurotrophic functions.

In summary, mice with spontaneous or transgenic deficiency in PRL expression exhibit concomitant deficiency in PRL-inhibiting TIDA hypophysiotropic neurons, not only in DA and TH levels, but, importantly, in numbers of properly differentiated neurons. Genetic lifelong excess of "PRL," in the form of transgenic hGH, actually leads to increased numbers of TIDA neurons, above the normal complement. Development of TIDA neurons during PRL deficiency shows initial normal differentiation, followed by regression in cell numbers, which may represent cell death or loss of permanent phenotype or ability to respond to the PRL stimulus. The latter theory is supported by the results of PRL replacement studies in PRL-deficient dwarfs, in which maintenance of TIDA neurons may be accomplished only if PRL therapy is begun prior to 21 days of life in mice, and in which PRL treatment of adult animals stimulates TH and DA in extant neurons, but by which the neuronal population cannot be restored. Indicated further studies include assessment of cell death among TIDA neurons in PRL-deficient dwarfs, and whether PRL serves an anti-apoptotic role in brain, as it has been shown to have in other tissues; the mechanisms that subserve PRL effect in supporting axonal guidance by induction of attractant molecules in the ME; and whether these important functions of PRL can be subserved by other molecules, as results in PRL-null mice would indicate.

ACKNOWLEDGEMENTS

Work in the authors' laboratories was supported by PHS Grant NS25987 to CJP.

REFERENCES

1. Bern H, Nicoll CS. The comparative endocrinology of prolactin. Rec Prog Horm Res 1968; 24:681-720.
2. Ben-Jonathan N. Dopamine: A prolactin-inhibitory hormone. Endocr Rev 1985; 6:564-89.
3. Fuxe K, Hokfelt T. Further evidence for the existence of tubero-infundibular dopamine neurons. Acta Physiol Scand 1966; 66:245-6.
4. Bjorklund A, Moore RY, Nobin A, Stenevi U. The organization of tuberohypophysial and reticuloinfundibular catecholamine neurosystems in the rat brain. Brain Res 1973; 51:171-91.
5. Dahlstrom A, Fuxe K. Evidence of the existence of monoamine containing neurons in the central nervous system. I. Demonstration of monoamines in the cell bodies of brain stem neurons. Acta Physiol Scand 1964; 62, Suppl. 232:1-55.

6. Arbogast LA, Voogt JL. Hyperprolactinemia increases and hypoprolactinemia decreases tyrosine hydroxylase messenger ribonucleic acid levels in the arcuate nuclei, but not the substantia nigra or zona incerta. Endocrinology 1991; 128:997-1005.
7. Phelps CJ. Pituitary hormones as neurotrophic signals: anomalous hypophysiotropic neuron differentiation in hypopituitary dwarf mice. Proc Soc Exp Biol Med 1994; 206:6-23.
8. Phelps CJ, Hurley DL. Pituitary hormones as neurotrophic factors: Update on hypothalamic differentiation in genetic models of altered feedback. Proc Soc Exp Biol Med 1999; 222:39-58.
9. Phelps CJ, Hurley DL, Hypothalamic hypophysiotropic neuron development is abnormal in animal models of pituitary prolactin or growth hormone absence or excess, In: Pandalai SG, ed. Recent Research Developments in Endocrinology: in press.
10. Snell GD. Dwarf, a new mendelian recessive character of the house mouse. Proc Natl Acad Sci USA 1929; 15:733-4.
11. Eicher EM, Beamer WG. New mouse dw allele: Genetic location and effects on lifespan and growth hormone levels. J Hered 1980; 71:187-90.
12. Schaible R, Gowen JW. A new dwarf mouse. Genetics 1961; 46:896.
13. Slabaugh MB, Lieberman ME, Rutledge JJ, Gorski J. Growth hormone and prolactin synthesis in normal and homozygous Snell and Ames dwarf mice. Endocrinology 1981; 109:1040-6.
14. Bartke A. Histology of the anterior hypophysis, thyroid and gonads of two types of dwarf mice. Anat Rec 1964; 149:225-36.
15. Roux M, Bartke A, Dumont F, Dubois MP. Immunohistological study of the anterior pituitary gland - pars distalis and pars intermedia - in dwarf mice. Cell Tissue Res 1982; 223:415-20.
16. Roti E, Christianson D, Harris ARC, Braverman LE, Vagenakis AG. "Short" loop feedback regulation of hypothalamic and brain thyrotropin-releasing hormone content in the rat and dwarf mouse. Endocrinology 1978; 103:1662-7.
17. Li S, Crenshaw EB, III, Rawson EJ, Simmons DM, Swanson LW, Rosenfeld MG. Dwarf locus mutants lacking three pituitary cell types result from mutations in the POU-domain gene *pit-1*. Nature 1990; 347:528-33.
18. Rhodes SJ, Dimattia GE, Rosenfeld MG. Transcriptional mechanisms in anterior pituitary cell differentiation. Curr Opin Genet Dev 1994; 4:709-17.
19. Theill LE, Karin M. Transcriptional control of GH expression and anterior pituitary development. Endocr Rev 1993; 14:670-89.
20. Sornson MW, Wu W, Dasen JS, Flynn SE, Norman DJ, O'Connell SM, Gukovsky I, Carriere C, Ryan AK, Miller AP, Zuo L, Gleiberman AS, Andersen B, Beamer WG, Rosenfeld MG. Pituitary lineage determination by the *Prophet of Pit-1* homeodomain factor defective in Ames dwarfism. Nature 1996; 384:327-33..
21. Mouse Genome Informatics Resource, http://www.informatics.jax.org. The Jackson Laboratory: Bar Harbor, Maine.
22. Behringer RR, Mathews LS, Palmiter RD, Brinster RL. Dwarf mice produced by genetic ablation of growth hormone-expressing cells. Genes Dev 1988; 2:453-61.
23. Steger RW, Bartke A, Parkening TA, Collins T, Buonomo F, Tang K, Wagner TE, Yun JS. Effects of heterologous growth hormones on hypothalamic and pituitary function in transgenic mice. Neuroendocrinology 1991; 53:365-72.
24. Bartke A, Cecim M, Tang K, Steger R, Chandrashekar V, Turyn D. Neuroendocrine and reproductive consequences of overexpression of growth hormone in transgenic mice. Proc Soc Exp Biol Med 1994; 206:345-59.

25. Phelps CJ, Sladek JR, Jr., Morgan WW, Bartke A. Hypothalamic catecholamine histofluorescence in dwarf mice. Cell Tissue Res 1985; 240:19-25.

26. Phelps CJ, Carlson SW, Vaccarella MY, Felten SY. Developmental assessment of hypothalamic tuberoinfundibular dopamine in prolactin-deficient dwarf mice. Endocrinology 1993; 132:2715-22.

27. Morgan WW, Bartke A, Pfiel K. Deficiency of dopamine in the median eminence of Snell dwarf mice. Endocrinology 1981; 109:2069-75.

28. Morgan WW, Besch KC. Effect of prolactin replacement on the number of tyrosine hydroxylase neurons in the arcuate nuclei of Ames dwarf and normal mice. Neuroendocrinology 1990; 52:70-4.

29. Phelps CJ, Carlson SW, Vaccarella MY. Hypothalamic dopaminergic neurons in prolactin-deficient Ames dwarf mice: localization and quantification of deficit by tyrosine hydroxylase immunocytochemistry. J Neuroendocrinol 1994; 6:145-52.

30. Phelps CJ. Isolated deficiency of tyrosine hydroxylase immunoreactivity in tuberoinfundibular neurons in pituitary prolactin-deficient Snell dwarf mice. Brain Res 1987; 416:354-8.

31. Phelps CJ, Romero MA, Malcamp C, Joseph SR. TIDA neuron number in developing Snell dwarf mice. Soc Neurosci Abst 1997; 459.11.

32. Thomas GB, Phelps CJ, Robinson ICAF. Differential regulation in hypothalamic tuberoinfundibular dopamine neurons in two dwarf rat models with contrasting changes in pituitary prolactin. J Neuroendocrinol 1999; 11:229-36.

33. Phelps CJ, Carlson SW, Hurley DL. Hypothalamic dopaminergic neurons in transgenic dwarf mice: Histofluorescence, immunocytochemical, and in situ hybridization studies. Anat Rec 1991; 231:446-56.

34. Phelps CJ, Bartke A. Stimulatory effect of human, but not bovine, growth hormone expression on numbers of tuberoinfundibular dopaminergic neurons in transgenic mice. Endocrinology 1997; 138:2849-55.

35. Slabaugh MB, Lieberman ME, Rutledge JJ, Gorski J. Ontogeny of growth hormone and prolactin gene expression in mice. Endocrinology 1982; 110:1489-70.

36. Phelps CJ, Vaccarella MY, Romero MI, Hurley DL. Postnatal reduction in number of hypothalamic tuberoinfundibular dopaminergic neurons in prolactin-deficient dwarf mice. Neuroendocrinology 1994; 59:189-96.

37. Furness JB, Heath JW, Costa M. Aqueous aldehyde (Faglu) methods for the fluorescence histochemical localization of catecholamines for ultrastructural studies of central nervous tissue. Histochemistry 1978; 57:289-95.

38. Morgan WW, King TS. Monoamine biosynthesis in hypothalamic regions of dwarf mice: Effect of replacement of deficient anterior pituitary hormones. Neuroendocrinology 1986; 42:351-6.

39. Palmiter RD, Norstedt G, Gelinas RE, Hammer RE, Brinster RL. Metallothionein-human GH fusion genes stimulate growth of mice. Science 1983; 222:809-14.

40. McGrane MM, deVente J, Yun J, Bloom J, Park E, Wynshaw A, Wagner T, Rottman FM, Hanson RW. Tissue-specific expression and dietary regulation of a chimeric phosphoenolpyruvate carboxykinase/growth hormone gene in transgenic mice. J Biol Chem 1988; 263:11443-51.

41. Chandrashekar V, Bartke A, Wagner TE. Neuroendocrine function in adult female transgenic mice expressing the human growth hormone gene. Endocrinology 1992; 130:1802-8.

42. Milton S, Cecim M, Li YS, Yun JS, Wagner TE, Bartke A. Transgenic female mice with high human growth hormone levels are fertile and capable of normal lactation without having been pregnant. Endocrinology 1992; 131:536-8.

43. Belchetz PE, Ridley RM, Baker HF. Studies on the accessibility of prolactin and growth hormone to brain: effect of opiate agonists on hormone levels in serial, simultaneous plasma and cerebrospinal fluid samples in the rhesus monkey. Brain Res 1982; 239:310-4.

44. Romero MI, Phelps CJ. Identification of growth hormone-releasing hormone and somatostatin neurons projecting to the median eminence in normal and growth hormone-deficient Ames dwarf mice. Neuroendocrinology 1997; 65:107-16.

45. Flavell DM, Wells T, Wells SE, Carmignac DF, Thomas GB, Robinson ICAF. Dominant dwarfism in transgenic rats by targeting human growth hormone (GH) expression to hypothalamic GH-releasing factor neurons. EMBO J 1996; 15:3871-9.

46. Charlton HM, Clark RG, Robinson ICAF, Porter Goff AEP, Cox BS, Bugnon C, Bloch BA. Growth hormone-deficient dwarfism in the rat: a new mutation. J Endocrinol 1988; 119:51-8.

47. Pellegrini E, Carmignac DF, Bluet-Pajot MT, Mounier F, Bennett P, Epelbaum J, Robinson ICAF. Intrahypothalamic growth hormone feedback: from dwarfism to acromegaly in the rat. Endocrinology 1997; 138:4543-51.

48. Phelps CJ, Dalcik H, Endo H, Talamantes F, Hurley DL. Growth hormone-releasing hormone peptide and mRNA are overexpressed in GH-deficient Ames dwarf mice. Endocrinology 1993; 133:3034-7.

49. Meister B, Hökfelt T, Vale WW, Sawchenko PE, Swanson L, Goldstein M. Coexistence of tyrosine hydroxylase and growth hormone-releasing factor in a subpopulation of tubero-infundibular neurons of the rat. Neuroendocrinology 1986; 42:237-47.

50. Sato M, Frohman LA. Differential effects of central and peripheral administration of growth hormone (GH) and insulin-like growth factor on hypothalamic GH-releasing hormone and somatostatin gene expression in GH-deficient dwarf rats. Endocrinology 1993; 133:793-9.

51. Romero MI, Phelps CJ. Prolactin replacement during development prevents the dopaminergic deficit in hypothalamic arcuate nucleus of prolactin-deficient Ames dwarf mice. Endocrinology 1993; 133:1860-70.

52. Romero MI, Phelps CJ. Prolactin replacement in adult dwarf mice does not reverse the deficit in tuberoinfundibular dopaminergic neuron number. Endocrinology 1995; 136:3238-46.

53. Dohler KD, Wuttke W. Serum LH, FSH, prolactin and progesterone from birth to puberty in female and male rats. Endocrinology 1974; 94:1003-8.

54. Nagata S, Rosenfeld MG, Inoue K. Development of prolactin and growth hormone production in the fetal rat pituitary: An immunochemical study. Dev Growth Diff 1992; 34:473-8.

55. Whitworth NS, Grosvenor CE. Transfer of milk prolactin to the plasma of neonatal rats by intestinal absorption. J Endocrinol 1978; 79:191-9.

56. Shyr SW, Crowley WR, Grosvenor CE. Effect of neonatal prolactin deficiency in prepubertal tuberoinfundibular and tuberohypophyseal dopaminergic neuronal activity. Endocrinology 1986; 119:1227-31.

57. Shah GV, Shyr SW, Grosvenor CE, Crowley WR. Hyperprolactinemia after neonatal prolactin (PRL) deficiency in rats: Evidence for altered anterior pituitary regulation of PRL secretion. Endocrinology 1988; 122:1883-90.

58. Clarke RE, Hardy RN. The use of ^{125}I-polyvinyl pyrrolidine K-60 in the quantitative assessment of the uptake of macromolecular substances by the intestine of young rats. J Physiol 1969; 204:113-25.
59. Ormandy CJ, Camus A, Barra J, Damotte D, Lucas B, Buteau H, Edery M, Brousse N, Babinet C, Binart N, Kelly PA. Null mutation of the prolactin receptor gene produces multiple reproductive defects in the mouse. Genes Dev 1997; 11:167-78.
60. Horseman ND, Zhao W, Montecino-Rodriguez E, Tanaka M, Nakahima K, Engle SJ, Smith F, Markoff E, Dorshkind K. Defective mammopoiesis, but normal hematopoiesis, in mice with a targeted disruption of the prolactin gene. EMBO J 1997; 16:6926-35.
61. Phelps CJ, Horseman ND. Hypophsiotropic dopamine neurons in PRL knockout mice. Proc Endocrine Soc Mtg 1999; P2-492.
62. Steger RW, Chandrashekar V, Zhao W, Bartke A, Horseman ND. Neuroendocrine and reproductive functions in male mice with targeted disruption of the prolactin gene. Endocrinology 1998; 139:3691-5.
63. Lucas BK, Ormandy CJ, Binart N, Bridges RS, Kelly PA. Null mutation of the prolactin receptor gene produces a defect in maternal behavior. Endocrinology 1998; 139:4102-7.

Chapter 3

Mechanisms of Dopamine Action on the Lactotroph

Karen A. Gregerson

University of Maryland, Baltimore, MD

INTRODUCTION

Prolactin (PRL) is a hormone with numerous functions: reproductive, immunological, growth, metabolic, and behavioral. The pattern of PRL release from the anterior pituitary gland changes during development and differs between sexes. In the female rat, PRL release is dynamic, fluctuating rapidly during normal physiological conditions. Episodic release of PRL occurs on the afternoon of proestrus (1,2), in pseudopregnancy (3), during early pregnancy (4) and in response to suckling during lactation (5). PRL secretion appears to be regulated in a complex manner by a variety of hormones and neurotransmitters (for reviews: 6,7). Inhibition of PRL secretion, the major regulatory influence of the hypothalamus, is mediated primarily by dopamine (DA). This DA is released from tuberoinfundibular neurons terminating in the median eminence and delivered to the anterior pituitary gland via the hypophyseal portal vessels (for reviews: 8,9). PRL, in turn, is believed to feed back on the hypothalamic neurons to regulate the release of DA and, ultimately, its own release (10,11). There exists an extensive list of substances that demonstrate a stimulatory action on PRL release (for reviews: 6,7,9). However, demonstration of a physiological role for any of these PRL secretagogues has remained elusive. What is clear, is that normally PRL secretion is under tonic inhibition by hypothalamic DA and the periodic, physiological surges of PRL in females require a withdrawal from this dopaminergic inhibition.

DA acts on the lactotroph at various levels of cell function. In addition to inhibiting exocytotic release of PRL, DA also inhibits transcription of the hormone (12). DA agonists can also effectively stop the growth and reduce the size of prolactinomas (see review: 13). This reduction in tumor size is primarily through loss of cytoplasmic and nuclear volume, with involution of rough endoplasmic reticulum and golgi apparatus, rather than reduction of cell number (14-16). While DA does not appear to induce cell death, the anti-mitotic effect of the DA agonists has been known for some time (17,18). Recent findings in several genetically altered murine models further establish that DA inhibits proliferation of pituitary lactotrophs. These models include the null mutations for the D_2 receptor (19,20) and PRL itself (21). Both models exhibit lactotroph hyperplasia and development, ultimately, of adenomas (22-24). A common phenotype in these animals is the lack of, or dramatic reduction in, dopaminergic input to the pituitary gland

(20,23,25). So, with greatly reduced DA there occurs hyperplasia of the lactotrophs. Correlatively, excessive DA during development leads to lactotroph hypoplasia, as seen in mice with the DA transporter null mutation (26). Thus, in addition to the broad diversity of potential PRL regulators, the actions of this one modulator of lactotroph function are impressively diverse. In this chapter, we will discuss the mechanisms through which DA acts on the lactotroph with attention focused on its action as a regulator of PRL release.

TRANSDUCTION OF DOPAMINERGIC INHIBITION IN LACTOTROPHS

Dopaminergic inhibition of PRL synthesis and release is mediated via D_2-like DA receptors (27,28). Heterogeneity of DA receptors was initially defined on the basis of pharmacological and functional evidence. Through modern molecular cloning techniques, five structurally distinct DA receptors have been identified, D_1-D_5 (see reviews: 29,30). Still, they are grouped into two families, sharing pharmacological and functional, as well as molecular similarities. In general, D_1-like receptors (D_1 and D_5) are defined as being linked to the stimulation of adenylate cyclase. D_2-like receptors (D_2, D_3, and D_4) either inhibit or have no effect on this enzyme. Both the long and short forms of the D_2 receptor (produced by alternative splicing) and the D_4 receptor transcript are present in the anterior pituitary (31,32) but the D_3 receptor is not (31,33).

The D_2-like receptors in the anterior pituitary are both physically and functionally coupled to a pertussis toxin (PTx)-sensitive G protein ($G_{i/o}$) (34-37) and this receptor-G protein coupling plays a necessary role in the dopaminergic inhibition of PRL secretion (37-40). The D_2 receptor-G protein coupling, in turn, can regulate a number of effectors in the lactotrope, including adenylate cyclase, phosphoinositides, calcium and potassium channels.

DOPAMINE AND ADENYLATE CYCLASE/CYCLIC AMP

The inhibitory response of PRL secretion to DA has been shown to be coincident with a PTx-sensitive inhibition of adenylate cyclase activity, a characteristic response of D_2 receptor activation involving G_i-type protein (38). While there is agreement that D_2 receptor stimulation is linked to the inhibition of cAMP in normal lactotropes, the involvement of this phenomenon in the regulation of PRL secretion is controversial at best. Correlative studies suggest a relationship between the two events. However, investigators have also determined that inhibition of cAMP is not required for the inhibition of PRL release by DA. Thorner and colleagues (41) first demonstrated that elevations in cytosolic [Ca^{2+}] using selective ionophores could maintain PRL secretion in the presence of DA despite effective suppression of cAMP. Moreover, DA could inhibit PRL release even when cAMP levels were elevated (41-44). It has also been demonstrated that elevated extracellular K^+ can block the inhibitory effect of DA on PRL release without diminishing the effect of DA on cAMP (45). Further evidence for the dissociation of DA inhibition of cAMP and PRL release was found in rat pituitary cells treated with antisense oligonucleotides recognizing both the long and short forms of the D_2 receptor, but not the D_4 receptor (31). Following this treatment, bromocriptine lost its ability to suppress cAMP accumulation, while inhibition of PRL release was unaffected. Taken together, these data indicate that DA-induced suppression of cyclic AMP is neither sufficient nor necessary for dopaminergic inhibition of PRL release.

DOPAMINE AND PHOSPHOLIPASE/PHOSPHOINOSITOLS

Evidence that DA's inhibitory action on PRL release involved decreases in cytosolic [Ca^{2+}] (41,42) led to the proposal that the D_2 receptor may also be negatively coupled to phospholipase C in lactotrophs (46,47). Inhibition of PLC would reduce phosphoinositol hydrolysis and its product, IP_3, leading to reduced mobilization of intracellular stores of Ca^{2+}. However, acute administration of DA does not alter the basal levels of inositol phosphates in pituitary cells (45,48). DA does reduce inositol phosphate production induced by TRH, but only during the late phase (> 1 min) of this response (49), which is associated with enhanced influx of extracellular Ca^{2+}. The transient generation of IP_3 in response to TRH activation of PLC (5-60 sec) is unaffected by DA (49). In addition, cell-free pituitary membranes exposed to TRH showed no inhibition of inositol phosphate generation by DA (49). These data support the conclusion that D_2-like receptors in lactotrophs are not direct inhibitors of phosphoinositol hydrolysis.

DOPAMINE AND MEMBRANE EXCITABILITY

It is well established that pituitary cells display spontaneous electrical activity, including spontaneous membrane potential fluctuations and calcium-dependent action potentials (50). Furthermore, the ionic conductances underlying this electrical activity have been characterized in lactotrophs (51,52) and the stimulus-secretion coupling processes involved in PRL release appear to involve these electrical events (53,54). Spontaneous PRL release is dependent upon extracellular Ca^{2+} and block of Ca^{2+} influx through voltage-gated Ca^{2+} channels (VGCC) is as potent an inhibitor of this release as DA.

Unchallenged lactotropes fall into two functionally distinct groups (55,56; see Figure 3-1). One population exhibits a stable membrane potential and intracellular [Ca^{2+}]

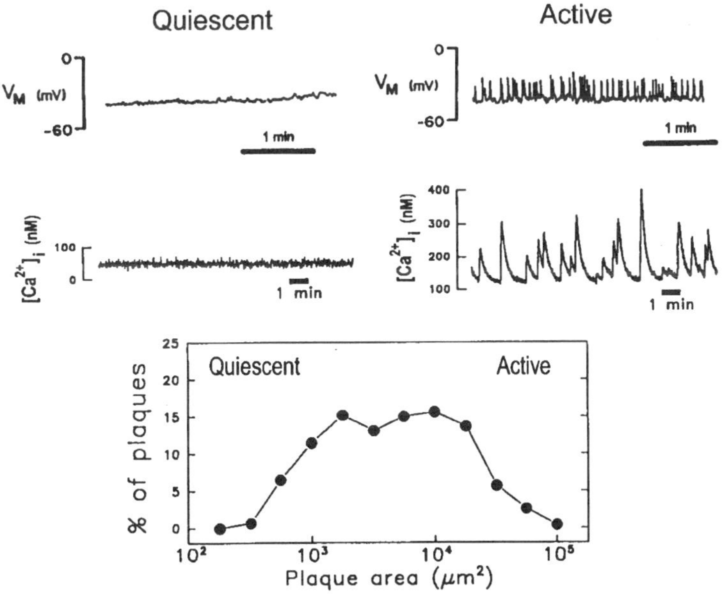

Figure 3-1. Functional heterogeneity of normal lactotrophs. Individual plaque-identified lactotrophs isolated from proestrous female rates were monitored for membrane electrical activity (V_M, *top traces*), cytosolic [Ca^{2+}] (*bottom traces*), and PRL release (*graph*). See text for discussion.

that is not acutely sensitive to extracellular Ca^{2+} ("quiescent" cells). The other exhibits spontaneous membrane depolarizations supporting oscillations in $[Ca^{2+}]_i$ that are dependent upon influx of extracellular Ca^{2+} ("active" cells). This functional difference is also evident in secretory activity of single cells measured by reverse hemolytic plaque assay (RHPA) in which the bimodal frequency distribution of plaque sizes demonstrates two subpopulations of lactotrophs (Figure 3-1). A subpopulation of cells releasing smaller amounts of PRL forms small plaques and a subpopulation releasing greater amounts of PRL forms large plaques (57). Thus, spontaneous electrical activity results in spontaneous influxes of Ca^{2+}, producing elevations of cytosolic $[Ca^{2+}]$ and supporting elevated PRL release (larger plaques).

DOPAMINE AND CALCIUM CHANNELS

DA causes a rapid and dramatic fall in cytosolic $[Ca^{2+}]$ in spontaneously active lactotrophs and direct block of Ca^{2+} channels with verapamil mimics this effect (55,56; Figure 3-2). Neither DA nor verapamil, however, cause further decreases in cytosolic $[Ca^{2+}]$ that is already low and stable (quiescent cells).

One mechanism by which DA could block spontaneous action potential activity and cytosolic $[Ca^{2+}]$ in lactotrophs is by directly inhibiting VGCCs. There is experimental evidence for this in cultured lactotrophs derived from lactating rats (52,58) and in GH_4C_1 cells expressing human D_2 receptors ("GH_4C_1/D_2-DAR"; 59). In these studies, DA treatment of the PRL-secreting cells produced a decrease in calcium current (I_{Ca}) measured under voltage-clamped conditions. A 27% decrease in I_{Ca} was observed after one or more minutes of DA exposure (59) while prolonged treatment (≥ 24 hours) with DA was required for maximal (57-61%) reductions in I_{Ca} (58,59). Voltage-independent inhibition of I_{Ca} is not seen in response to acute applications of DA (60), although this would be expected if the D_2 receptor were directly coupled with VGCCs. Certainly, immediate inhibition of $[Ca^{2+}]_i$ oscillations is observed in response to acute application of DA (55,56; Figure 3-2). The reduction in I_{Ca} observed after longer exposure to DA most likely reflects changes that occur downstream of other actions of DA.

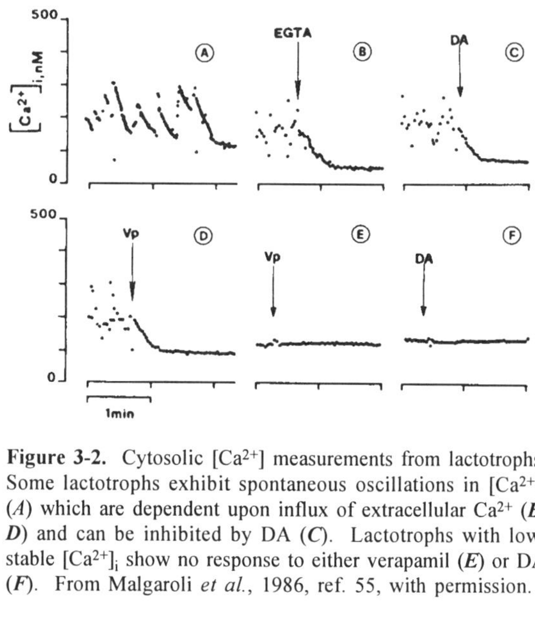

Figure 3-2. Cytosolic $[Ca^{2+}]$ measurements from lactotrophs. Some lactotrophs exhibit spontaneous oscillations in $[Ca^{2+}]_i$ (*A*) which are dependent upon influx of extracellular Ca^{2+} (*B, D*) and can be inhibited by DA (*C*). Lactotrophs with low, stable $[Ca^{2+}]_i$ show no response to either verapamil (*E*) or DA (*F*). From Malgaroli *et al.*, 1986, ref. 55, with permission.

VGCCs have been demonstrated to be a substrate for both cAMP-dependent protein kinase (PKA) and protein kinase C (PKC) (61-63) and dephosphorylation-phosphorylation of the channel protein appears to be one mechanism underlying inactivation-activation properties of VGCCs (64,65). Changes in phosphorylation of the α_1 subunit of L-type Ca^{2+} channels following DA treatment have recently been reported in GH_4C_1/D_2-DAR cells (66). These changes could be mediated through either DA-induced changes in cAMP or $[Ca^{2+}]_i$ or both. However, again these changes cannot account for the acute and rapid inhibition of Ca^{2+}-spiking observed in lactotrophs upon exposure to DA. Thus, DA must be acting through another mechanism to inhibit VGCC gating in lactotrophs.

DOPAMINE AND POTASSIUM CHANNELS

DA could block I_{Ca} indirectly through changes in membrane potential. The VGCCs supporting I_{Ca} require membrane depolarization to open and they close in response to repolarization or hyperpolarization of the membrane. DA-induced membrane hyperpolarizations mediated via a K^+ conductance have been reported using electrophysiological techniques in human PRL-secreting adenoma cells, as well as bovine and rat (lactating) cell populations enriched for PRL-secreting cells (67-70). Other groups (71,72) have reported DA-induced increases in K^+ conductance based on ion flux studies in lactotroph-enriched preparations using ^{86}Rb as a probe of K^+ permeability.

Evidence is gathering that K^+ channels in primary lactotropes are integral to the functional consequences of D_2 receptor activation. The DA-induced increase in K^+ permeability leads to a membrane hyperpolarization that shuts down Ca^{2+}-spiking activity (Figure 3-3; *Note the complete block of Ca^{2+} spiking during each acute hyperpolarization elicited by DA.*). The resulting decline in intracellular $[Ca^{2+}]$ (Figure 3-2) inhibits PRL secretion. This hypothesis predicts that the functional consequences of the hyperpolarization would be greater or more apparent in electrically active cells. Indeed, DA causes a dramatic fall in $[Ca^{2+}]_i$ in cells exhibiting spontaneous Ca^{2+} oscillations (55,56) that are due to Ca^{2+} influx during spontaneous membrane depolarization (73). As demonstrated in the RHPA, DA preferentially suppresses formation of large plaques (57). This causes in a leftward shift of the frequency histogram of plaque area, producing a unimodal distribution of small plaques in contrast to the broader, bimodal distribution observed in the absence of modulators (Figure 3-4). These findings are consistent with the prediction that the DA-induced hyperpolarization exerts a greater or

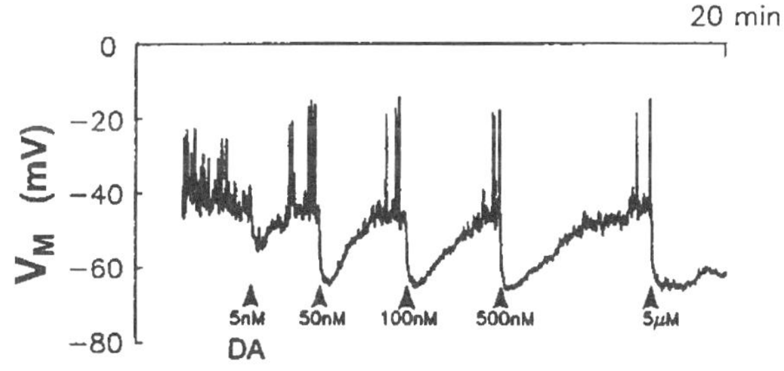

Figure 3-3. Membrane potential responses to varying concentrations of DA. Whole-cell voltage recordings of a plaque-identified lactotroph show rapid, transient hyperpolarizations in response to consecutive 1-sec applications of DA. Maximal voltage changes occurred by 50 and 100 nM DA and recovery of membrane potential became slower with increasing concentrations of DA. Note immediate cessation of spontaneous spiking activity upon hyperpolarization and resumption of spiking with recovery of the resting membrane potential after each DA-induce response. From Einhorn *et al.*, 1993, ref. 84, with permission.

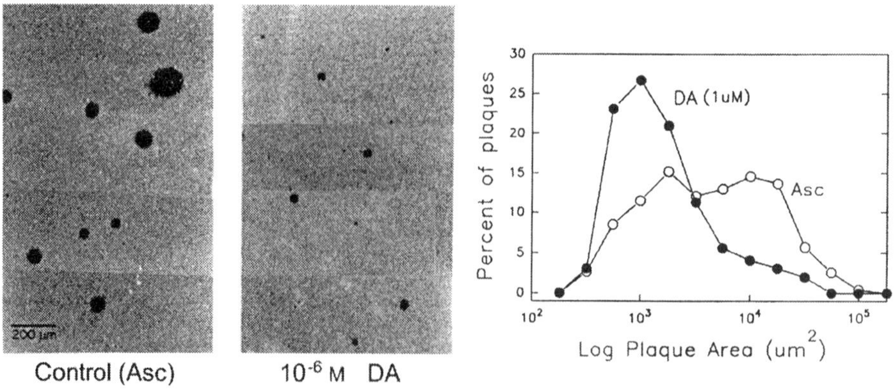

Figure 3-4. Effect of DA (10^{-6} M) on release of PRL from individual lactotrophs as assessed by RHPA. Low magnification photomicrographs of plaques, enhanced by dark-field illumination, illustrate PRL release from single cells in the absence (ascorbic acid control, Asc) and presence of DA. Graph illustrates the effect of DA on the frequency distribution profile of PRL plaque area. See text for discussion.

more apparent inhibitory effect on PRL release from those lactotrophs secreting at a higher rate - the spontaneously active cells.

Cell population release data obtained in perifusion experiments (Figure 3-5) illustrate the role of K^+-mediated hyperpolarization in DA's actions by demonstrating that the inhibition of PRL release by DA correlates with the driving force on K^+. The same correlation is seen with valinomycin, which is a K^+-selective ionophore and therefore directly hyperpolarizes the cell membrane by increasing K^+ permeability. In contrast, the effect of verapamil, which directly blocks calcium channels, is not sensitive to the driving force on K^+. This also supports the idea that D_2 receptor activation in lactotrophs does not directly inhibit Ca^{2+} channels, but that decreases in I_{Ca} are dependent on DA-induced changes in membrane potential.

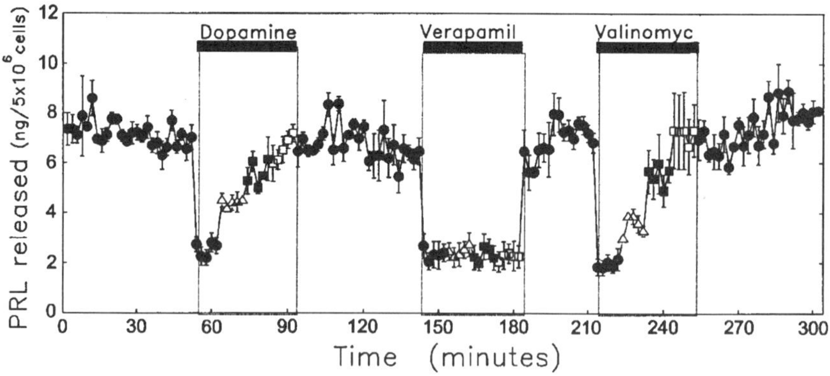

Figure 3-5. PRL release from rat pituitary cells in perifusion. Cells were exposed to DA (100 nM), verapamil (20 µM), or valinomycin (10^{-9} M) while extracellular [K^+] was increased from 5 mM (?), to 15 mM (?), to 30 mM (¦), to 50 mM (?). Perfusate was collected for 2 min intervals and assayed for rat PRL by RIA.

Much effort has been invested into the identification of the specific K^+ channel(s) mediating the DA-induced hyperpolarization. Castelletti *et al.* (72) described two types of DA-activated ^{86}Rb fluxes: one voltage-dependent but calcium-independent, the other calcium-dependent (I_{KCa}). Lledo *et al.* (70) reported that in plaque-identified (by RHPA) lactotrophs DA increased two voltage-dependent outwardly rectifying K^+ currents: I_K (delayed) and I_A (transient). The channels underlying these types of currents all are members of an extended family of voltage-gated K^+ channels ("K*v*" gene nomenclature; 74,75). The most notable feature of these channels is that membrane depolarization and elevated $[Ca^{2+}]_i$ are obligate for activation of these currents. Yet, it is well documented that DA activates a K^+ conductance when lactotrophs are at a polarized resting membrane potential (67-69,76) (Figure 3-3).

In contrast, Gregerson *et al.* (76) have identified and characterized a voltage-*in*dependent K^+ channel activated by DA. In plaque-identified PRL-secreting cells derived from female rats, the application of DA induces a rapidly-activating, robust hyperpolarization in nearly all (>85%) lactotrophs regardless of the cell's functional state (with spontaneous depolarizing action potentials or with a polarized, stable resting membrane potential). Activation of this channel involves a D_2-like subtype of DA receptor and a PTx-sensitive G protein, and occurs over the same range of DA concentrations that inhibit PRL release (5-100 nM) (Figure 3-3) (76). It is independent of intracellular levels of cAMP (77) and Ca^{2+} (76). This voltage-independent K^+ channel also exhibits some inward rectification (78) and sensitivity to block by Ba^{2+}. These biophysical features identify the lactotroph channel as a member of an emerging family of K^+ channels called the Inward Rectifier K^+ (IRK) channels ("K*ir*"gene nomenclature; 79,80). A subfamily (K*ir* 3.0) of the IRK channels are directly activated, through a membrane delimited pathway, by the $\beta\gamma$ subunit of PTx-sensitive G proteins (81-83) as is the case for the K^+ channel present in normal lactotrophs (84).

Three different transcripts for G-protein gated IRK proteins ("GIRKs") are expressed in the female rat anterior pituitary gland: K*ir 3.1*, K*ir 3.2*, and K*ir 3.4* (85). *In situ* hybridization assays on plaque-identified cells demonstrate that all three transcripts are expressed in lactotrophs (86), making the protein products excellent candidate subunits of the DA-activated K^+ channel. Current theory is that G-protein gated IRK channels function as heteromultimers of the GIRK proteins (87,88). Functional analyses of various combinations of GIRK subunits in *Xenopus* oocytes demonstrate that the coexpression of GIRK1 and GIRK4 with the D_2 receptor produces a DA-activated channel whose properties recapitulate many of the properties of the native DA-activated K^+ channel in pituitary lactotropes (85). Indeed, native GIRK1/GIRK4 oligomers exist in the anterior pituitary gland (Figure 3-6).

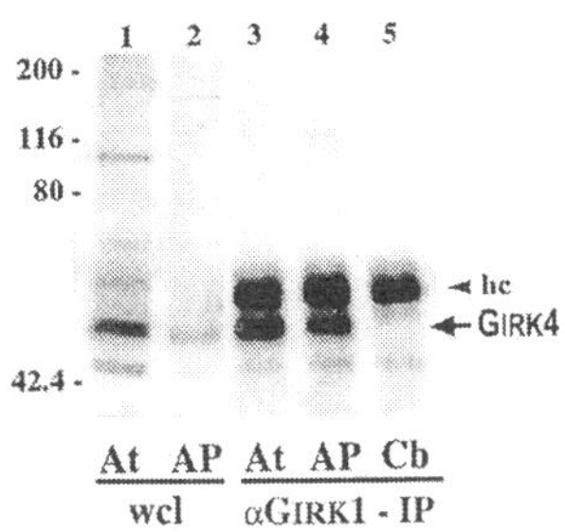

Figure 3-6. Native GIRK1 and GIRK4 proteins are associated in anterior pituitary tissue. Anterior pituitary (AP), cardiac atrium (At), and cerebellum (Cb) were obtained from proestrous female rats. Whole-cell lysates and proteins immunoprecipitated with anti-GIRK1 were subjected to Western blot using anti-GIRK4. GIRK4 (45 kDa) can be recognized in whole cell lysates (*wcl, lanes 1 and 2*) from both atrium and anterior pituitary. GIRK4 co-immunoprecipitated with GIRK1 from atrial and pituitary (*lanes 3 and 4*) but not from cerebellum (*lane 5*). The band migrating at 50 kDa and marked by the arrowhead (*hc*) is the heavy chain of the rabbit anti-GIRK1 used for the ip.

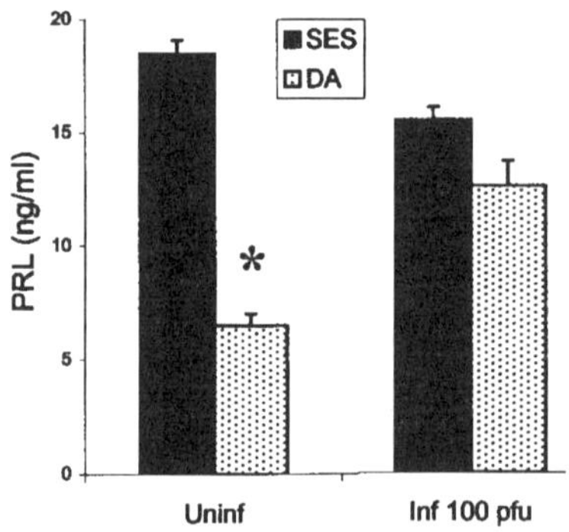

Figure 3-7. Functional "knock-out" of GIRK channels in lactotrophs disrupts DA inhibition of PRL release but not spontaneous PRL release. Pituitary cells derived from proestrous female rats were either infected with 100 pfu of the GIRK1-AAA dominant-negative mutant in an adenoviral vector or left uninfected. PRL release during a 20 min incubation was measured 72 hours after infection in the absence or presence of DA (100 nM).

In the absence of a specific pharmacological blocker of the G-protein gated IRK channel, we designed a molecular approach to selectively disrupt the function of this protein and thereby assess its role in mediating DA's actions. To do this, we exploited the heteromultimeric nature of this channel to design and create a "dominant-negative" mutant to block the function of the channel at the level of the protein. We have made such a variant by mutating the pore region of the protein, GIRK1, encoded by the K*ir* 3.1 gene (*see also* 89). A signature GYG motif in the H5 (pore) region is conserved in all K^+-selective channels and plays a key role in ion selectivity and pore function (90,91). When the GYG sequence is replaced with three alanines (AAA), the channel does not pass current (89). Expression of the GIRK1-AAA mutant in primary cultured pituitary cells reverses dopaminergic inhibition of PRL release without altering basal release (Figure 3-7; 92). Thus, the inward-rectifying, G-protein-gated K^+ channel in lactotropes plays a critical role in the inhibition of PRL release induced by application of DA.

ANOMOLOUS STIMULATION OF PROLACTIN SECRETION BY DOPAMINE

Under certain conditions, changes in DA input to the lactotroph have been demonstrated to produce elevations in PRL secretion. One condition is the withdrawal of DA, or dissociation from its receptor, which leads to a subsequent, transient phase of PRL release. The other is an acute response, in which subnanomolar concentrations of DA directly stimulate PRL release.

DOPAMINE WITHDRAWAL AND PROLACTIN RELEASE

Disruption of dopaminergic tone *in vivo* or withdrawal of DA from lactotrophs *in vitro* elicits transient "surges" of PRL release, as has been demonstrated by numerous groups (93-96). Investigations into the mechanisms by which this phenomenon occurs have demonstrated that the secretory rebound is dependent upon the influx of Ca^{2+} (45,56) and is distinct from the stimulatory actions of low concentrations of DA (97).

Based on the mechanisms known for DA's inhibition of PRL release, it was suggested that the DA-induced changes in membrane potential may play a role in the secretory rebound following its withdrawal (97). Single cell studies demonstrate that recovery from DA-induced membrane hyperpolarization leads to activation of Ca^{2+}-spiking activity in lactotrophs that were electrically quiescent prior to application of DA (97; Figure 3-8). The amplitude of Ca^{2+} action potentials in spontaneously active cells was also increased by ~20% after recovery from DA. Both responses can be

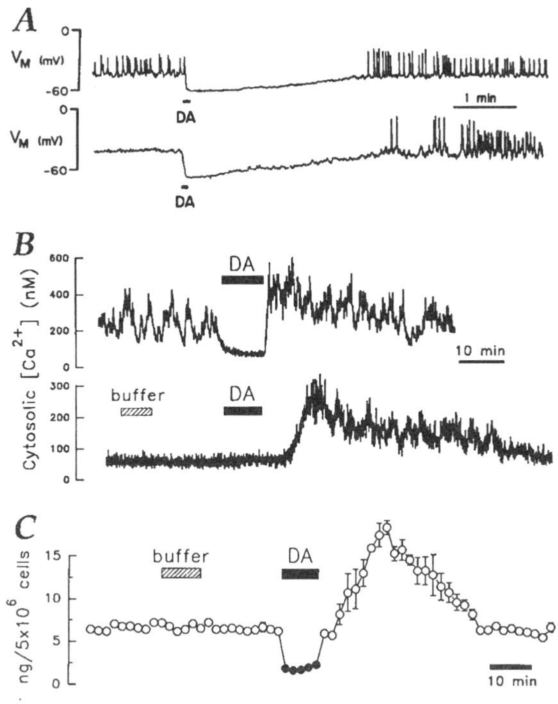

Figure 3-8. Lactotroph responses to DA application and withdrawal. *A.* Membrane potential responses of individual lactotrophs. DA hyperpolarizes both spontaneously active (*upper trace*) and quiescent (*lower trace*) cells, leading to activation of previously quiescent cells after DA withdrawal. *B.* Cytosolic [Ca^{2+}] responses in a spontaneously active (*upper trace*) or quiescent (*lower trace*) lactotroph. *C.* Cell population secretory response illustrating inhibition of PRL release in presence of DA and rebound PRL release following DA withdrawal. See text for discussion.

explained by an increased number of VGCCs available for activation following DA withdrawal, leading to the hypothesis that the DA-induced hyperpolarization "recruits" previously inactivated VGCCs by allowing recovery from inactivation. Upon DA washout and recovery of the resting membrane potential, the enhanced influx of Ca^{2+} through these channels supports the rebound release of PRL (Figure 3-8).

This hypothesis is supported by the demonstration that direct hyperpolarization (bypassing D_2 receptor activation) of the lactotroph membrane mimics both the inhibitory and rebound phases of PRL secretion (97). Hyperpolarization is induced by directly increasing K^+ permeability with the K^+-selective ionophore, valinomycin. Valinomycin inhibits PRL release to the same extent as DA and, upon washout, elicits a secretory rebound with the same amplitude and time course as seen following DA withdrawal. Verapamil also mimics the inhibitory action of DA on PRL release by directly blocking VGCCs *without* altering membrane potential. Thus, washout of verapamil is followed by recovery of basal PRL release but no rebound. Finally, PRL rebound secretion in response to withdrawal of either DA or valinomycin is blocked when these drugs are applied in 50 mM K^+ (97). Elevated $[K^+]_o$ prevents the hyperpolarization by negating the driving force on K^+, but does not alter other DA-induced responses such as changes in cAMP (45).

Although acute administration of DA does not alter basal levels of inositol phosphates in lactotrophs, Martinez de la Escalera and colleagues (48,96) have demonstrated that withdrawal of DA can activate inositol phosphate formation. This increase is accompanied by a reciprocal decrease in polyphosphoinositides, suggesting an involvement of PLC in the response (48). Activation of this pathway appears to be a secondary response and may be related to the fact that PLC can be activated by a rise in intracellular [Ca^{2+}] (98). Indeed, Ca^{2+} ionophores increase inositol phosphate production in pituitary cells, a response that is blocked by the removal of extracellular Ca^{2+} (45,99).

Activation of PLC, albeit indirectly by the influx of extracellular Ca^{2+}, would generate another product, diacylglycerol (DAG), which activates PKC. As mentioned above, VGCCs can be a substrate for this protein kinase and a secondary increase in PKC activity in response to DAG, in synergy with the elevated $[Ca^{2+}]_i$, may mediate the late response of increased phosphorylation of VGCCs in GH_4C_1/D_2-DAR cells observed following DA withdrawal (66). Thus, there is a possibility of a limited positive feedback occurring in which voltage-regulated recruitment of Ca^{2+} channels results in elevated cytosolic [Ca^{2+}] and activation of PKC which then act to prolong the VGCC activity through phosphorylation. However, this phosphorylation has not been demonstrated to be a requirement for the rebound secretion of PRL. In fact, while it has been demonstrated that VGCC activity supports the enhanced PRL release throughout the rebound period, the time course and decay kinetics of the rebound secretion are not different whether Ca^{2+} influx is allowed to occur or not (45). These findings support the hypothesis that voltage is the critical factor in this DA-induced modulation of VGCC activity, again indicating that the DA-activated GIRK channel serves as a primary effector of DA's actions.

LOW CONCENTRATIONS OF DOPAMINE AND PROLACTIN RELEASE

It has been demonstrated by several groups that very low concentrations of DA ($\leq 10^{-10}$ M) can stimulate PRL release acutely (57,100-102). These low concentrations of DA have been shown to induce elevations in cytosolic [Ca^{2+}] (102) that are believed to underlie the stimulatory effect on PRL release. Interestingly, the group who initially described this response (100), demonstrated in a subsequent report that the same concentrations of DA had either no effect or small inhibitory effects on PRL release (93). From all appearances, the two studies used identical sources and preparations of pituitary cells as well as the same conditions (perifusion) for testing PRL release. What is not specifically stated are details on the preparation of the dopamine used in each study. DA oxidizes rapidly at physiological pH (103) and is subject to enzymatic degradation in the presence of cells (104). In concentrations of 10^{-8} M or greater, enough unmetabolized DA should remain, even over periods of hours, to inhibit PRL release through cellular mechanisms discussed above. However, oxidation of lower concentrations of DA would decrease the levels of intact DA below those required for activation of the classic DA receptors. A survey of the literature on stimulatory low concentrations of DA indicates that this phenomenon is studied either under conditions when pituitary cells are exposed to DA for long periods of time (RHPA or static culture) or under conditions when it is unknown how quickly the DA solutions are used after preparation in physiological buffer. Thus, the possibility exists that a metabolite, and not DA *per se*, is acting to stimulate $[Ca^{2+}]_i$ and PRL release. Recently, Fekete and colleagues (105) reported that an enzymatic product, salsalinol, could stimulate PRL release. In contrast, we have found that low concentrations of DA left in physiological buffer for 60 minutes or more prior to exposure to cells, can stimulate rises in $[Ca^{2+}]_i$ while freshly prepared DA either has no effect or reduces cytosolic [Ca^{2+}] (Figure 3-9). Whether the active component is produced by enzymatic or non-enzymatic processes, the possibility that a metabolic product stimulates the same cell functions that are normally inhibited by the parent compound, is an exciting avenue for future research.

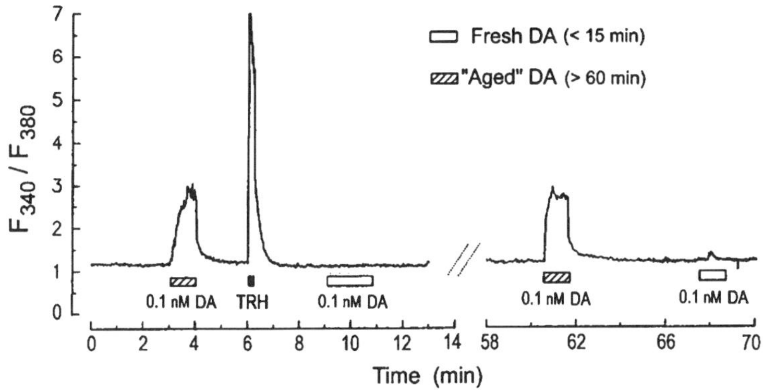

Figure 3-9. Changes in cytosolic [Ca^{2+}] in a quiescent plaque-identified lactotroph in response to 0.1 nM DA. DA was dissolved in PBS (pH 7.4) and applied to the cell within 15 min (*open bars*) or after the solution was allowed to sit at room temperature for at least 60 min (*hatched bars*). The second application of "aged" DA was the same solution as the first application of fresh DA. Cytosolic [Ca^{2+}] response to TRH (100 nM; *solid bar*) is also shown.

SUMMARY

An inwardly rectifying K^+ channel in normal lactotrophs appears to be a critical effector of DA regulation of PRL secretion. The independence of this channel from voltage and Ca^{2+} make its activation by DA "unconditional" rather than conditional upon these known regulators of voltage-gated K^+ channels. Thus, these channels can be activated under basal as well as stimulated conditions, thereby enabling DA to impact on the lactotroph's physiology during either condition. Experimental evidence establishes the functional role of this channel in the regulation of PRL secretion, both in the inhibition of spontaneous PRL release and in the development of PRL rebound following DA removal. Thus, it appears that the DA-activated GIRK channel lies at the top of a hierarchy of effector proteins transducing DA's actions in the lactotroph.

ACKNOWLEDGEMENTS

The author wishes to thank the NIDDK at the National Institutes of Health for continuous support of her own research as well as most research in the field of PRL. The author is also indebted to her former mentors, past and present collaborators and students for continuous challenges, keen insights and always lively discussions.

REFERENCES

1. Butcher RL, Collins WE, Fugo NW 1974 Plasma concentrations of LH, FSH, prolactin, progesterone and estradiol-17b throughout the 4-day estrous cycle of the rat. Endocrinology 94:1704-1708.
2. Smith MS, Freeman ME, Neill JD 1975 The control of progesterone secretion during the estrous cycle and early pseudopregnancy in the rat: prolactin, gonadotropin and steroid levels associated with rescue of the corpus luteum of pseudopregnancy. Endocrinology 92:219-226.
3. Freeman ME, Smith MS, Nazian SJ, Neill JD 1974 Ovarian and hypothalamic control of the daily surges of prolactin secretion during pseudopregnancy in the rat. Endocrinology 94:875-882.

4. Smith MS, Neill JD 1976 Termination at midpregnancy of the two daily surges of plasma prolactin initiated by mating in the rat. Endocrinology 98:696-701.
5. Neill JD 1980 Neuroendocrine regulation of prolactin secretion. IN: Martini L, Ganong ST (eds), Frontiers in Neuroendocrinology, Raven Press, New York, pp 129-155.
6. Leong DA, Frawley LS, Neill JD 1983 Neuroendocrine control of prolactin secretion. Ann Rev Physiol 45:109-127.
7. Lamberts SWJ, MacLeod RM 1990 Regulation of prolactin secretion at the level of the lactotroph. Physiol Rev 70:279-318.
8. Weiner RI, Cronin MJ, Cheung CY, Annunziato L, Faure N, Goldsmith PC 1979 Dopamine: A prolactin inhibitory hormone. IN: Müller EE, Agnoli A (eds), Neuroendocrine Correlates in Neurology and Psychiatry, Elsevier, Amsterdam, pp 41-55.
9. Ben-Jonathan N 1985 Dopamine: a prolactin-inhibiting hormone. Endocr Rev 6:564-589.
10. Hökfelt T, Fuxe K 1972 Effects of prolactin and ergot alkaloids on the tuberoinfundibular dopamine (DA) neurons. Neuroendocrinology 9:100-122.
11. Gudelsky GA, Porter JC 1980 Release of dopamine from tuberoinfundibular neurons into pituitary stalk blood after prolactin or haloperidol administration. Endocrinology 106:526-529.
12. Maurer RA 1980 Dopaminergic inhibition of prolactin synthesis and prolactin messenger RNA accumulation in cultured pituitary cells. J Biol Chem 255:8092-8097.
13. Bevan JS, Webster J, Burke CW, Scanlon MF 1992 Dopamine agonists and pituitary tumor shrinkage. Endocr Rev 13:220-240.
14. Tindall GT, Kovacs K, Horvath E, Thorner MO 1982 Human prolactin-producing adenomas and bromocriptine: a histological, immunocytochemical, ultrastructural, and morphometric study. J Clin Endocrinol Metab 55:1178-1183.
15. Landolt AM, Minder H, Osterwalder V, Landolt TA 1983 Bromocriptine reduces the size of cells in prolactin-secreting pituitary adenomas. Experientia 39:625-626.
16. Saeger W 1992 Effect of drugs on pituitary ultrastructure. Microsc Res Tech 20:162-176.
17. Lloyd HM, Meares JD, Jacobi J 1975 Effects of oestrogen and bromocryptine on *in vivo* secretion and mitosis in prolactin cells. Nature 255:497.
18. Melmed S 1981 Bromocriptine inhibits colony formation by rat pituitary tumor cells in a double-layered agar clonogenic assay. Endocrinology 109:2258-2260.
19. Baik JH, Picetti R, Saiardi A, Thiriet G, Dierich A, Depaulis A, Le Meur M, Borrelli E 1995 Parkinsonian-like locomotor impairment in mice lacking dopamine D2 receptors. Nature 377:424-428.
20. Kelly MA, Rubinstein M, Asa SL, Zhang G, Saez C, Bunzow JR, Allen RG, Hnasko R, Ben-Jonathan N, Grandy DK, Low MJ 1997 Pituitary lactotroph hyperplasia and chronic hyperprolactinemia in dopamine D2 receptor-deficient mice. Neuron 19:103-113.
21. Horseman ND, Zhao W, Engle S, Tanaka M, Markloff E, Dorshkind K 1997 Defective mammopoiesis, but normal hematopoiesis, in mice with a targeted disruption of the prolactin gene. EMBO J 16101-110.
22. Saiardi A, Bozzi Y, Baik J-H, Borrelli E 1997 Antiproliferative role of dopamine: Loss of D_2 receptors causes hormonal dysfunction and pituitary hyperplasia. Neuron 19:115-126.
23. Steger RW, Chandrashekar V, Zhao W, Bartke, Horseman ND 1998 Neuroendocrine and Reproductive functions in male mice with targeted disruption of the prolactin gene. Endocrinology 139:3691-3695.
24. Asa SL, Kelly MA, Grandy DK, Low MJ 1999 Pituitary lactotroph adenomas develop after prolonged lactotroph hyperplasia in dopamine D2 receptor-deficient mice. Endocrinology 140:5348-5355.

25. Phelps DJ, Horseman ND 2000 Prolactin gene disruption does not compromise differentiation of tuberoinfundibular dopaminergic neurons. Neuroendocrinology 72:2-10.
26. Bossé R, Fumagalli F, Jaber M, Giros B, Gainetdinov RR, Wetsel WC, Missale C, Caron MG 1997 Anterior pituitary hypoplasia and dwarfism in mice lacking the dopamine transporter. Neuron 19:127-138.
27. Caron MG, Beaulieu M, Raymond V, Gagne B. Drouin J, Lefkowitz R, Labrie F. Dopaminergic receptors in the anterior pituitary gland. J Biol Chem 253:2244-2253.
28. Enjalbert A, Bockaert J 1983 Pharmacological characterization of the D_2 dopamine receptor negatively coupled with adenylate cyclase in rat anterior pituitary. Molec Pharmacol 23:576-584.
29. Andersen PH, Gingrich JA, Bates MD, Dearry A, Falardeau P, Senogles SE, Caron M 1990 Dopamine receptor subtypes: beyond the D1/D2 classification. Trends Pharmacol Sci 11:231-236.
30. Civelli O, Bunzow JR, Grandy DK, Zhou Q-Y, Van Tol HHM 1991 Molecular biology of the dopamine receptors. Eur J Pharmacol Mol Pharmacol Sect 207:277-286.
31. Valerio A Alberici A, Tinti C, Spano PF, Memo M 1994 Antisense strategy unravels a dopamine receptor distinct from the D2 subtype, uncoupled with adenylyl cyclase, inhibiting prolactin release from rat pituitary cells. J Neurochem 62:1260-1266.
32. Matsumoto M, Hidaka K, Tada S, Tasaki Y, Yamaguchi T 1995 Full-length cDNA cloning and distribution of human dopamine D4 receptor. Mol Brain Res 29:157-162.
33. Sokoloff P, Giros B, Martres M-P, Bouthenet M-L, Schwartz J-C 1990 Molecular cloning and characterization of a novel dopamine receptor (D_3) as a target for neuroleptics. Nature 347:146-151.
34. DeLean A, Kilpatrick B, Caron M 1982 Dopamine receptor of the porcine anterior pituitary gland. Molec Pharmacol 22:290-297.
35. Kilpatrick BF, Caron MG 1983 Agonist binding promotes a guanine nucleotide reversible increase in the apparent size of the bovine anterior pituitary dopamine receptors. J Biol Chem 258:13528-13534.
36. George SR, Watanabe M, Di Paolo T, Falardeau P, Labrie F, Seeman P 1985 The functional state of the dopamine receptor in the anterior pituitary is in the high affinity form. Endocrinology 117:690-697.
37. Senogles SE, Benovic JL, Amlaiky N, Unson C. Milligan G, Vinitsky R, Spiegel AM, Caron MG 1987 The D_2-dopamine receptor of anterior pituitary is functionally associated with a pertussis toxin-sensitive guanine nucleotide binding protein. J Biol Chem 262:4860-4867.
38. Cronin MJ, Myers GA, MacLeod RM, Hewlett EL 1983 Pertussis toxin uncouples dopamine agonist inhibition of prolactin release. Am J Physiol 244:E499-E504.
39. Enjalbert A, Mussett F, Chenard C, Priam M, Kordon C. Heisler S 1988 Dopamine inhibits prolactin secretion stimulated by the calcium channel agonist Bay-K-8644 through a pertussis toxin-sensitive G protein in anterior pituitary cells. Endocrinology 123:406-412.
40. Schofield JG, Khan AI, Wood A 1988 Modification by pertussis toxin of the responses of bovine anterior pituitary cells to acetylcholine and dopamine: effects on hormone secretion and ^{86}Rb efflux. J Endocrinol 116:393-401.
41. Thorner MO, Hackett JT, Murad F, MacLeod RM 1980 Calcium rather than cyclic AMP as the physiological intracellular regulator of prolactin release. Neuroendocrinology 31:390-402.
42. Delbeke D, Scammell JG, Martinez-Campos A, Dannies PS 1986 Dopamine inhibits prolactin release when cyclic adenosine 3',5'-monophosphate levels are elevated. Endocrinology 118:1271-1277.

43. Lafond J, Ducharme JR, Collu R 1986 Inhibition of prolactin release and blockade of adenohypophyseal cell cyclic AMP accumulation are two dissociable effects of dopaminergic and non-dopaminergic drugs. Mol Cell Endocrinol 44:219-225.
44. Ray KP, Gomm JJ, Law GJ, Sigournay C, Wallis M 1986 Dopamine and somatostatin inhibit forskolin-stimulated prolactin and growth hormone secretion but not stimulated cyclic AMP levels in sheep anterior pituitary cell cultures. Mol Cell Endocrinol 45:175-182.
45. Gregerson KA, Chuknyiska R, Golesorkhi N 1994 Stimulation of prolactin release by dopamine withdrawal: role of calcium influx. Am J Physiol 267:E789-E794.
46. Canonico PL, Valdenegro CA, MacLeod RM 1982 Dopamine inhibits 32P incorporation into phosphatidylinositol in the anterior pituitary gland of the rat. Endocrinology 111:347-349.
47. Enjalbert A, Sladeczek F, Guillon G, Bertrand P, Shu C, Epelbaum J, Garcia-Sainz A, Jard S, Lombard C, Kordon C 1986 Angiotensin II and dopamine modulate both cAMP and inositol phosphate productions in anterior pituitary cells. Involvement in prolactin secretion. J Biol Chem 261:4071-4075.
48. Martínez de la Escalera G, Weiner RI 1988 Effect of dopamine withdrawal on activation of adenylate cyclase and phospholipase C in enriched lactotrophs. Endocrinology 123:1682-1687.
49. Vallar L, Vicentini LM, Meldolesi J 1988 Inhibition of inositol phosphate production is a late, Ca^{2+}-dependent effect of D2 dopaminergic receptor activation in rat lactotroph cells. J Biol Chem 263:10127-10134.
50. Ozawa S, Sand O 1986 Electrophysiology of excitable endocrine cells. Physiol Rev 66:887-952.
51. Lingle DJ, Sombati S, Freeman ME 1986 Membrane currents in identified lactotrophs of rat anterior pituitary. J Neurosci 6:2995-3005.
52. Lledo PM, Legendre P, Israel JM, Vincent JD 1990b Dopamine inhibits two characterized voltage-dependent calcium currents in identified rat lactotrophs cells. Endocrinology 127:990-1001.
53. Law GJ, Pachter JA, Dannies PS 1989 Ability of repetitive Ca^{2+} spikes to stimulate prolactin release is frequency dependent. Biochem Biophys Res Commun 158:811-816.
54. Ray KP, Wallis M 1982 Involvement of calcium ions in dopamine inhibition of prolactin secretion from sheep pituitary cells. Mol Cell Endocrinol 28:691-703.
55. Malgaroli A, Vallar L, Elahi FR, Pozzan T, Spada A, Meldolesi J 1986 Dopamine inhibits cytosolic Ca^{2+} increases in rat lactotroph cells. Evidence of a dual mechanism of action. J Biol Chem 262:13920-13927.
56. Ho M-Y, Kao JPY, Gregerson KA 1996 Dopamine withdrawal elicits prolonged calcium rise to support prolactin rebound release. Endocrinology 137:3513-3521.
57. Luque EH, de Toro MM, Smith PF, Neill JD 1986 Subpopulations of lactotrophs detected with the reverse hemolytic plaque assay show differential responsiveness to dopamine. Endocrinology 118:2120-2124.
58. Lledo P-M, Israel JM, Vincent J-D 1991 Chronic stimulation of D_2 dopamine receptors specifically inhibits calcium but not potassium currents in rat lactotrophs. Brain Res 558:231-238.
59. Hernández MM, Ferreiro REG, Garcia DE, Hernández ME, Clapp C, Martínez de la Escalera g 1999 Potentiation of prolactin secretion following lactotrope escape from dopamine action. I. Dopamine withdrawal augments L-type calcium current. Neuroendocrinology 70:20-30.
60. Rendt J, Oxford GS 1994 Absence of coupling between D_2 dopamine receptors and calcium channels in lactotrophs from cycling female rats. Endocrinology 135:501-508.
61. Curtis BM, Catterall WA 1985 Phosphorylation of the calcium antagonist receptor of the voltage-sensitive calcium channel by cAMP-dependent protein kinase. Proc Natl Acad Sci USA 82:2528-2532.

62. Nastainczyk W, Röhrkasten A, Sieber M, Rudolph C, Schächtele C, Marmé D, Hofmann F 1987 Phosphorylation of the purified receptor for calcium channel blockers by cAMP kinase and protein kinase C. Eur J Biochem 169:137-142.
63. Ma J, Gutiérrez LM, Hosey MM, Ríos E 1992 Dihydropyridine-sensitive skeletal muscle Ca^{2+} channels in polarized planar bilayers. 3. Effects of phosphorylation by protein kinase C. Biophys J 63:639-647.
64. Eckert R, Chad JE 1984 Inactivation of Ca channels. Prog Biophys Mol Biol 44:215-267.
65. Armstrong D, Eckert R 1987 Voltage-activated calcium channels that must be phosphorylated to respond to membrane depolarization. Proc Natl Acad Sci USA 84:2519-2522.
66. Hernández ME, Hernández MM, Díaz-Muñoz M, Clapp C, Martínez de la Escalera G 1999 Potentiation of prolactin secretion following lactotrope escape from dopamine action. II. Phosphorylation of the α_1 subunit of L-type voltage-dependent calcium channels. Neuroendocrinology 70:31-42.
67. Israel JM, Jaquet P, Vincent JD 1985 The electrical properties of isolated human prolactin-secreting adenoma cells and their modification by dopamine. Endocrinology 117:1448-1455.
68. Ingram CD, Bicknell RJ, Mason WT 1986 Intracellular recordings from bovine anterior pituitary cells: modulation of spontaneous activity by regulators of prolactin secretion. Endocrinology 119:2508-2518.
69. Israel JM, Kirk C, Vincent JD 1987 Electrophysiological responses to dopamine of rat hypophysial cells in lactotroph-enriched primary cultures. J Physiol 390:1-22.
70. Lledo PM, Legendre P, Zhang J, Israel JM, Vincent JD 1990a Effects of dopamine on voltage-dependent potassium currents in identified rat lactotroph cells. Neuroendocrinology 52:545-555.
71. Memo M, Castelletti L, Missale C, Spano PF 1987 Stimulation of D_2 receptors increases potassium permeability in mammotrophs. Eur J Pharmacol 139:361-362.
72. Castelletti L, Memo M, Missale C, Spano PF, Valerio A 1989 Potassium channels involved in the transduction mechanisms of dopamine D_2 receptors in lactotrophs. J Physiol (Lond) 410:251-265.
73. Schlegel W, Winiger BP, Mollard P, Vacher P, Wuarin F, Zahnd GR, Wolheim CB, Dufy B 1987 Oscillations of cytosolic Ca^{2+} in pituitary cells due to action potentials. Nature (Lond) 329:719-721.
74. Salkoff L, Baker K, Butler A, Covarrubias M, Pak MD, Wei A 1992 An essential "set" of K^+ channels conserved in flies, mice and humans. Trends Neurosci 15:161-166.
75. Chandy KG, Gutman GA 1993 Nomenclature for mammalian potassium channel genes. Trends Pharmacol Sci 14:434.
76. Gregerson KA, Einhorn L, Smith MM, Oxford GS 1989 Modulation of potassium channels by dopamine in rat pituitary lactotrophs: A role in the regulation of prolactin secretion? In: Oxford GS, Armstrong CM (eds), Secretion and Its Control. Rockefeller University Press, New York, pp 123-141.
77. Gregerson KA, Eroshenko D 1995 D2 receptor activation of a K+ conductance in lactotropes is independent of D2 inhibition of adenylate cyclase activity. Abstract, 1995 Endocrine Society.
78. Einhorn LC, Gregerson KA, Oxford GS 1991 D_2 dopamine receptor activation of potassium channels in identified rat lactotrophs: whole-cell and single-channel recording. J Neurosci 11:3727-3737.
79. Aldrich R 1993 Advent of a new family. Nature 362:107-108.
80. Doupnik CA, Davidson N, Lester HA 1995 The inward rectifier potassium channel family. Curr Opin Neurobiol 5:268-277.
81. Krapivinsky G, Krapivinsky L, Wickman K, Clapham DE 1995 G bet gamma binds directly to the G protein-gated K^+ channel, I_{Kach}. J Biol Chem 270:29059-29062.

82. Reuveny E, Slesinger PA, Inglese J, Morales JM, Inlguez-Lluhl, Lefkowitz RJ, Bourne HR, Jan YN, Jan LY 1994 Activation of the cloned muscarinic potassium channel by G protein βγ subunits. Nature 370:143-146.
83. Takao K, Yoshii M, Kanda A, Kokubun S, Nukada T 1994 A region of the muscarinic-gated atrial K^+ channel critical for activation by G protein beta gamma subunits. Neuron 13:747-755.
84. Einhorn LC, Oxford GS 1993 Guanine nucleotide binding proteins mediate D_2 dopamine receptor activation of a potassium channel in rat lactotrophs. J Physiol 462:563-578.
85. Gregerson KA, Flagg TP, O'Neill TJ, Anderson M, Lauring O, Horel JS, Welling PA Identification of G protein-coupled, inward rectifier potassium channel gene products from the rat anterior pituitary gland. Accepted for publication, Endocrinology.
86. Saraswati M, McCarthy MM, Gregerson KA Expression of GIRK1, GIRK2 and GIRK4 transcripts in identified pituitary lactotrophs. Presented at the 29th annual meeting of the Society for Neuroscience, Miami Beach, 1999.
87. Kofuji P, Davidson N, Lester HA 1995 Evidence that neuronal G-protein-gated inwardly rectifying K^+ channels are activated by Gbg subunits and function as heteromultimers. Proc Natl Acad Sci USA 92:6542-6546.
88. Krapivinsky G, Gordon EA, Wickman K, Velimirovic B, Krapivinsky L, Clapham DE 1995 The G-protein-gated atrial K^+ channel I_{Kach} is a heteromultimer of two inwardly rectifying K^+ channel proteins. Nature 374:135-141.
89. Tinker A, Jan YN, Jan LY 1996 Regions responsible for the assembly of inwardly rectifying potassium channels. Cell 87:857-868.
90. Heginbotham L, Lu Z, Abramson R, MacKinnon R 1994 Mutations in the K^+ channel signature sequence. Biophys J 66:1061-1067.
91. Slesinger PA, Patil N, Liao J, Jan YN, Jan LY, Cox DR 1996 Functional effects of the mouse weaver mutation on G protein-gated inwardly rectifying K^+ channels. Neuron 16:321-331.
92. Horel JS, Welling PA, O'Neill TJ, Gregerson KA Functional knock-out of the DA-activated K^+ channel reverses dopaminergic inhibition of acute PRL release. Presented at the annual meeting for the Society for Neuroscience, New Orleans, 2000.
93. Denef C, Baes M, Schramme C 1984 Stimulation of prolactin secretion after short term or pulsatile exposure to dopamine in superfused anterior pituitary cell aggregates. Endocrinology 114:1371-1378.
94. Frawley LS, Neill JD 1984 Brief decreases in dopamine result in surges of prolactin secretion in monkeys. Am J Physiol 247:E778-E780.
95. Haisenleder DJ, Moy JA, Gala RR, Lawson DM 1986 The effect of transient dopamine antagonism on thyrotropin-releasing hormone-induced prolactin release in ovariectomized rats treated with estradiol and/or progesterone. Endocrinology 119:1996-2003.
96. Martinez de la Escalera G, Weiner RI 1992 Dissociation of dopamine from its receptor as a signal in the pleiotropic hypothalamic regulation of prolactin secretion. Endocr Rev 13:241-255.
97. Gregerson KA, Golesorkhi N, Chuknyiska R 1994 Stimulation of prolactin release by dopamine withdrawal: role of membrane hyperpolarization. Am J Physiol 267:E781-E788.
98. Eberhard DA, Holz RW 1988 Intracellular Ca^{2+} activates phospholipase C. Trends Neurosci 11:517-520.
99. Kolesnick RN, Gershengorn MC 1984 Ca^{2+} ionophores affect phosphoinositide metabolism differently than thyrotropin-releasing hormone in GH_3 pituitary cells. J Biol Chem 259:9514-9519.
100. Denef C, Manet D, Dewals R 1980 Dopaminergic stimulation of prolactin release. Nature 285:243-246. -198.

101. Kramer IM, Hopkins CR 1982 Studies on the kinetics of dopamine-regulated prolactin secretion. Mol Cell Endocrinol 28:191
102. Burris TP, Freeman ME 1993 Low concentrations of dopamine increase cytosolic calcium in lactotrophs. Endocrinology 133:63-68.
103. Senoh S, Witkop B 1959 Non-enzymatic conversions of dopamine to norepinephrine and trihydroxyphenethylamines. J Am Chem Soc 81:6222-6231.
104. Senoh S, Creveling CR, Udenfriend S, Witkop B 1959 Chemical, enzymatic and metabolic studies on the mechanisms of oxidation of dopamine. J Am Chem Soc 81:6236-6245.
105. Fekete MIK, Homicsko K, Toth G, Radnai B, Kertesz I, Toth BE, Nagy GM Binding sites of an endogenous tetrahydroisoquinoline, salsalinol, in rat brain regions and in the anterior lobe of the pituitary gland. Presented at the annual meeting of the Endocrine Society, Toronto, 2000.

Chapter 4

Regulation of Lactotrophs and Prolactin Secretion in Non-mammalian Vertebrates

Tom E. Porter and Mohamed E. El Halawani

University of Maryland, College Park, MD and University of Minnesota, Saint Paul, MN

INTRODUCTION

Prolactin (PRL) is a pleiotropic hormone with numerous functions ascribed to it in various species (1). Best known to mammalian researchers is the role of PRL in regulating mammary function and milk production. While PRL serves to regulate maternal functions such as incubation behavior and crop "milk" production in birds, this hormone serves to regulate water and electrolyte balance, the behavioral drive toward water and morphogenesis in other vertebrates. From this diverse set of actions for PRL among vertebrates, it follows that the ontogeny, distribution and regulation of PRL-secreting cells, the lactotrophs, may also vary widely among vertebrate classes. This chapter will summarize the localization of PRL cells within the anterior pituitary, changes in lactotroph abundance during transitions in physiologic state, and the regulation of PRL release in non-mammalian vertebrates. Rather than catalogue research conducted in all vertebrate species, the aim of this chapter is to highlight differences between mammalian and non-mammalian vertebrate classes.

LACTOTROPH DISTRIBUTION WITHIN THE ADENOHYPOPHYSIS

In most adult mammals lactotrophs are distributed as clusters of cells or as single cells throughout the bulk of the *pars distalis*. Lactotroph distribution is more uniform in adult humans and cattle, while in rats they are concentrated in the lateral tips and ventral surface of the anterior pituitary and proximal to the intermediate lobe. In all non-mammalian vertebrate classes, lactotrophs are restricted to one portion of the anterior pituitary. The adenohypophysis in teleosts is divided into three distinct regions, the rostral pro-adenohypophysis, the proximal meso-adenohypophysis and the meta-adenohypophysis, which is equivalent to the intermediate lobe (2,3). The "posterior pituitary" penetrates each region of the adenohypophysis in the form of secretory neurons, a feature unique to this vertebrate class (4). Immunocytochemical localization of PRL within the adenohypophysis of many teleosts has indicated that lactotrophs are located exclusively in the rostral portion of the gland. This localization of lactotrophs to the rostral adenohypophysis has been shown in a wide range of freshwater and saltwater species, including tilapia, salmon, striped bass, sea bream and seabass (3,5-12). However, in the more "primitive" sea lamprey, *Petromyzon marinus* (13), lactotrophs were found in the proximal but not the rostral *pars distalis*. With this exception in mind, localization of the lactotrophs to the rostral pro-adenohypophysis in most teleosts indicates at least a partial separation of this cell type from other anterior pituitary cell types.

In amphibians, the adenohypophysis is divided into a *pars intermedia* and *pars distalis*. Lactotrophs have been localized to the *pars distalis* by both cytological and immunocytochemical techniques (9). In the newt, lactotrophs were immuno-localized to the ventral anterior two-thirds of the gland (14). Similarly, lactotrophs were immunostained in the anterior portion of the adenohypophysis in frogs (15,16). Together, these findings indicate that, as in teleosts, lactotrophs are located in the most anterior portion of the anterior pituitary in amphibians. In reptiles, the *pars distalis* is divided into distinct rostral and caudal lobes, and lactotrophs have been localized exclusively in the rostral lobe in several reptiles, including snakes, lizards, and turtles (17-22). In birds, the anterior pituitary is also clearly divided into two regions, the cephalic and caudal lobes. As in reptiles, the lactotrophs have been localized to the cephalic lobe of the anterior pituitary of numerous avian species, including chickens, ducks, turkeys, quail, sparrows, pigeons, kingfishers, white-backed munia, and black drongos (23-30). Thus, lactotrophs are located in a distinct portion of the anterior pituitary gland in fish, amphibians, reptiles and birds. In each class, the PRL-producing cells are localized in the rostral end of the adenohypophysis, and in each of these non-mammalian vertebrates, the lactotrophs are present in a distinct location apart from the other acidophilic cell type, the growth hormone-producing somatotrophs.

This microanatomical separation of lactotrophs and somatotrophs to distinct anterior pituitary regions would seem to be the norm among vertebrates, and the lack of separation of lactotrophs and somatotrophs in mammals is the exception. At least in birds, the hypothalamo-hypophyseal portal vessels are found in two groups, the anterior and posterior portal vessels (29). This separation of the portal circulation suggests that the localization of lactotrophs and somatotrophs to the cephalic and caudal lobes of the anterior pituitary in birds may result from different concentrations of trophic neurohormones transported separately to the two lobes of the adenohypophysis. The innervation of the adenohypophysis in teleosts would also provide for delivery of divergent signals do distinct regions of the anterior pituitary, which could affect cellular differentiation and localization. Separation of lactotrophs from somatotrophs in most vertebrates calls into question the current model of lactotroph differentiation developed in mammals, wherein PRL cells differentiate from growth hormone cells (31,32). Although evidence from transgenic mice clearly supports this model (33), disparate somatotroph and lactotroph localization in all non-mammalian vertebrates does not support this model for vertebrates in general. Clearly, additional work is required in a non-mammalian model to verify the lactotroph precursor cell type. The separation of lactotrophs and somatotrophs to distinct lobes in birds, with each lobe preferentially supplied by a different set of portal vessels, suggests that lactotroph and somatotroph differentiation may be affected by hypothalamic neurohormones. This possibility is supported by the finding that incubation of chicken anterior pituitary cells with vasoactive intestinal polypeptide (VIP), a potent hypothalamic stimulator of PRL secretion in birds, advanced lactotroph differentiation *in vitro* (34).

LACTOTROPH DIFFERENTIATION DURING DEVELOPMENT

Lactotrophs differentiate around the time of hatching in fish, amphibians and birds. In fish, lactotroph differentiation occurs prior to hatching in the cyprinodont *Cynolebias whitei* (11,12), after hatching at 2 days of age in larval sea bream (6), between days 9 and 15 in larval sea bass (35), and 3 to 4 days after hatching in rainbow trout (36). Although, low water salinity is known to increase PRL secretion, salinity did not affect the timing of lactotroph differentiation in trout (36). Similarly, in *Cynolebias whitei*, although the amount of PRL within lactotrophs was greater in fish reared in freshwater relative to those reared in saltwater (11), and in tilapia, PRL mRNA expression was increased by rearing in freshwater (37), no differences in the initial appearance of PRL mRNA were noted (37). These findings indicate that lactotroph differentiation is not affected by a major environmental signal known to affect PRL secretion in teleosts. In one report in amphibians, lactotrophs differentiated within 2 days after hatching in the frog *Rana dalmatina* (16). This same study failed to detect colocalization of PRL and growth hormone, suggesting that lactotrophs may not differentiate from somatotrophs by way of the dual hormone secreting mammosomatotroph as has been demonstrated in rodents (32,33,38-40). However, two studies have colocalized PRL and growth hormone to the same cells in bullfrog embryos and in pituitary cell cultures from adult bullfrogs, indicating the presence of mammosomatotrophs in amphibians (16,41). The role played by these dual hormone-secreting cells during the lifespan of any amphibian is not clear. Interestingly, lactotrophs also differentiated in pituitary primordium of toads grafted into the tailbud region, indicating that contact with the brain is not necessary for lactotroph differentiation in amphibians (42). However, this finding does not preclude a role for hypothalamic neurohormones in determining lactotroph abundance. No reports exist on the timing of lactotroph differentiation in reptiles. In birds, studies of lactotroph differentiation are restricted primarily to chickens. In this species, lactotroph differentiation occurs between days 16 and 19 of embryonic development, before hatching of the chicks (26,34,43,44). A single report suggests lactotroph differentiation as early as day 6 of chick embryonic development, but by the authors' own admission, detection of PRL at this early age was likely the result of crossreactivity of the PRL antibody with luteinizing hormone or growth hormone (45). Treatment of embryonic chicken pituitary cells *in vitro* with VIP induced the appearance of lactotrophs in cultures derived from as early as day 13 of embryonic development (34). This finding suggests that either VIP induced PRL expression in otherwise undetectable lactotrophs or that VIP induced lactotroph differentiation from their precursor cell type. Regardless of the mechanism or cellular precursor for lactotroph differentiation, lactotrophs appear around the time of hatching in the non-mammalian vertebrates, supporting functional roles for PRL in growth, metabolism or osmoregulation during this period.

PHYSIOLOGIC CHANGES IN LACTOTROPH ABUNDANCE

During the life of vertebrate animals, PRL secretion increases during specific physiologic conditions. At these times, the abundance of lactotrophs within the anterior pituitary also increases. Lactation in rats and humans is associated with elevated levels of PRL in the circulation and an increase in the number of PRL cells in the pituitary (46,47). Similar increases have been noted in other vertebrate classes, during their respective characteristic heights of PRL secretion. In teleosts, the principal function ascribed to PRL is that of osmoregulation (48). PRL stimulates the transport of ions and water across the gill epithelium and within the kidney, and administration of exogenous

PRL improves survivability in hypophysectomized saltwater fish placed in freshwater (1,48). Consistent with these actions, PRL release is regulated by changes in water osmolarity, with PRL secretion increased in many fish exposed to freshwater. Exposure to freshwater is also known to increase the secretory activity of lactotroph cells (3,5), while transfer to saltwater decreases the cellular activity of lactotrophs (49). Moreover, exposure to freshwater appears to increase the number of lactotrophs present in the adenohypophysis (10-12). These findings suggest that the lactotroph population in teleosts is plastic, increasing or decreasing in size depending on the physiologic needs of the animal. This plasticity in lactotroph abundance has been proposed to occur during lactation in mammals (38,39,46,50).

In amphibians, PRL may play a role in regulating metamorphosis and tail reabsorption (1,51). Levels of PRL increase around the time of metamorphosis, although the timing of this increase relative to metamorphosis climax varies among species. The abundance of lactotrophs within the adenohypophysis also increases during this time, with the largest number of PRL cells being detected during prometomorphosis just prior to its climax (15). Thus, lactotroph abundance may also be regulated in amphibians, depending on the physiologic need for PRL by the animal.

Likewise, lactotroph abundance increases in female turkeys and chickens that are incubating their eggs (25,52). Circulating levels of PRL are greatly elevated in incubating birds, and in chickens, the proportion of PRL cells increased from approximately 30% of all pituitary cells in laying hens to more than 50% of all pituitary cells in incubating hens (52). This increase in lactotroph abundance was restricted to the cephalic two-thirds of the pituitary gland. In turkeys, the fraction of sagittal sections stained for PRL was greater in nesting hens than in laying hens (25). This increase in PRL immunoreactive area involved an encroachment of lactotrophs into the caudal lobe of the pituitary, where they appeared to replace the existing somatotrophs in the ventral portion of that lobe. Replacement of somatotrophs with lactotrophs during incubation behavior of hens may involve a mammosomatotroph phenotype, since cells containing both PRL and growth hormone were found in female turkeys and their abundance was greater in incubating hens than in hens laying eggs (24). When nesting behavior of incubating hens was interrupted, the relative abundance of lactotrophs and somatotrophs returned to pre-incubating levels (53). Considered as a whole, the body of evidence across teleosts, amphibians, and birds indicates that, like in mammals, the abundance of the lactotroph population is plastic, fluctuating up and down with changes in the physiologic state of the animal.

REGULATION OF PROLACTIN SECRETION

PRL levels have been shown to increase during fasting in tilapia, supporting a metabolic role for PRL in teleosts (54). PRL is also elevated during incubation of eggs in mouthbrooding fish (55), implicating PRL in maternal behavior in at least some teleosts. Similarly, an increase in PRL during confinement stress has been used to suggest a role for PRL in the response to stress (56). Substantial evidence also exists that PRL secretion is stimulated by a decrease in osmotic pressure *in vivo* and *in vitro*, the latter suggesting the presence of osmoreceptors on teleost lactotrophs (57). Indeed, the most characteristic pattern of PRL secretion in teleosts is the increase in circulating PRL levels following transfer of fish to freshwater, which coincides with the osmoregulatory role of PRL in teleosts (48,58,59). In addition, levels of PRL are generally higher in freshwater fish than in saltwater fish (57). Some evidence exists that calcium content constitutes a major portion of this effect (60). Interestingly, in primitive teleosts, the lactotrophs are arranged as an epithelium that lines cavities within the adenohypophysis, and in some species these cavities communicate with the buccal cavity through an orohypophyseal duct. In one such species, the Japanese anchovy, the lactotrophs lining these cavities are polar. The basal surface exhibits secretory activity, while the luminal surface in contact with the cavity possesses microvilli (61). This arrangement has led to speculation that the luminal surface may serve in a sensory capacity for changes in osmotic pressure. Additional evidence for direct effects of water osmolality on PRL secretion comes from transplantations of the pituitary to an ectopic site. In hypophysectomized tilapia, with their pituitaries grafted onto their optic nerves, placement in freshwater still increased levels of serum PRL and PRL mRNA in the transplanted pituitaries (62). This finding indicates that the lactotrophs responded to changes in water (and likely plasma) osmolality without hypothalamic input. Furthermore, direct effects of changing cell culture medium osmolality on PRL secretion *in vitro* have been reported in tilapia and trout (63,64). Two forms of PRL have been identified in tilapia, each encoded by a separate gene but secreted from the same lactotrophs (8). The effect of water osmolality on PRL synthesis likely begins early during development, as expression of both PRL mRNA's is greater in tilapia raised in freshwater than in those raised in saltwater (37)

In addition to direct osmotic effects on PRL secretion, lactotroph function in teleosts is under hypothalamic control. The neurosecretory fibers that originate in the hypothalamus and terminate in the adenohypophysis are unique to teleosts, and these neurons likely play a major role in the control of PRL secretion in fish (57). Like in mammals, the principal form of hypothalamic control over PRL secretion appears to be inhibitory (4,57). However, evidence does exist for a hypothalamic PRL-releasing factor in fish (65,66), and hypothalamic thyrotropin-releasing hormone (TRH) is a potent stimulator for PRL release in teleosts (57). Inhibitory control of PRL secretion appears to be localized in the nucleus lateral tuberis pars lateralis (67). As in mammals PRL secretion is likely under some inhibitory control by hypothalamic dopamine (DA) (4,60). For example, in trout DA was shown to inhibit PRL secretion *in vivo* and *in vitro* (68). However, in contrast to its role in mammals, DA is not the primary inhibitor of PRL release in at least some teleosts including tilapia (57). Instead, hypothalamic inhibition of PRL secretion seems to be mediated by somatostatin (SRIF). Specifically, SRIF inhibits PRL release in cultured pituitary cells from tilapia and trout (57,60,64,69-71). In tilapia, SRIF neurons emanating from the preoptic nucleus and lateral tuberal nucleus of the hypothalamus terminate adjacent to lactotrophs (57,72). Transfer of tilapia from saltwater to freshwater leads to changes in SRIF immunoreactivity in the hypothalamus and adenohypophysis consistent with diminished SRIF release, a finding compatible with the elevation in PRL secretion found under these same conditions (57). Surprisingly, SRIF failed to decrease PRL mRNA while reducing PRL release from cultures of trout pituitary cells, indicating that the effect of SRIF may be restricted to PRL release and not include PRL synthesis or at least gene expression (73). In addition, the SRIF-related peptide urotensin II has also been shown to inhibit PRL release in tilapia (71).

As in mammals, estrogen increases PRL synthesis in teleosts, and DA can inhibit its release (63,74,75). In addition, estrogen increases the sensitivity of lactotrophs to TRH (57,74), and TRH-induced PRL release is not seen without prior estrogen priming (74). Evidence also exists for a stimulatory role for serotonin (5-HT) on PRL secretion in trout (68). Interestingly, three forms of gonadotropin-releasing hormone (GnRH) have been shown to induce PRL release in tilapia, and this effect was enhanced by co-incubation with estradiol (76). Also, the GnRH-associated peptide (GAP), which is known to inhibit PRL release in mammals, has been shown to stimulate PRL release from cultured pituitary cells of tilapia (70). Insulin like growth factor-I (IGF-I) is also a potent stimulator of PRL secretion in cultures of striped bass pituitary cells (77). Surprisingly, VIP and peptide histidine isoleucine (PHI), which stimulate PRL release in mammals, have been shown to inhibit PRL release in tilapia (78). Thus, regulation of PRL secretion in teleosts includes many unique functions of neuropeptides and other hormones. While the stimulatory and inhibitory effects on PRL by estrogen and DA, respectively, are consistent with their functions in mammals, effects of other secretagogues seem paradoxical if not perplexing. SRIF and IGF-I, which inhibit growth hormone secretion in mammals, have inhibitory and stimulatory effects, respectively, on PRL release in teleosts. Similarly, while 5-HT and TRH induce PRL release in fish as they do in mammals, VIP and PHI inhibit PRL release in teleosts rather than stimulate PRL secretion, as they do in mammals. Thus, the regulation of PRL secretion in teleosts is quite unique and not directly inferable from control of PRL release in mammals. Moreover, with the exception of SRIF, the involvement of these neurohormones in osmotic induced PRL release is not clear, leaving many areas for future research.

PRL serves many functions in amphibians as in other vertebrate classes. PRL secretion and mRNA levels are increased after extended captivity in frogs (79). Levels of circulating PRL increase in newts during the winter (80), the onset of the reproductive season. Seasonal variations in PRL levels in frogs have also been reported, with the greatest plasma PRL concentrations also found during the winter (81). However, this is the non-breeding season for these amphibians. In contrast, PRL levels are greatest during the spring breeding season in toads (82). The best-defined pattern of PRL secretion in amphibians is the increase in PRL levels during metamorphosis. PRL mRNA levels also increase during metamorphosis in frogs, with greatest levels detected after metamorphosis (83). In this same study, it was determined that thyroid hormones increase PRL gene expression and that thyroid hormone synthesis is necessary for the increase in PRL mRNA associated with metamorphosis. Levels of PRL mRNA are also greatest after the climax of metamorphosis in toads (84). However, inhibition of thyroid hormone synthesis in toad tadpoles resulted in dramatic hypertrophy of lactotrophs, suggesting that thyroid hormones inhibit PRL production in toads (85), again indicating contrasting regulatory mechanisms among amphibians. Hypothalamic extracts have been shown to exert primarily a stimulatory effect on PRL release in amphibians (65,66,86). Hypothalamic TRH stimulates PRL release in frogs and toads (86,87). Furthermore, the principal PRL-releasing factor in the hypothalamus of frogs is TRH (88). DA alone did not suppress basal PRL release in frogs, but it did to a small extent in toads (86). However, DA effectively inhibits hypothalamic extract and TRH-induced PRL release in frogs and toads (86). In contrast to results in fish, SRIF failed to suppress basal or TRH-induced PRL secretion in frogs and toads (86). Interestingly, adrenocorticotropin (ACTH) and the N-terminal peptide of the proopiomelanocortin gene product were shown to stimulate PRL release and increase PRL mRNA levels in cultures of frog pituitary cells (89). Much work remains to be done before any complete model can be proposed for the regulation of PRL secretion in any amphibian species.

Even less is known about the regulation of PRL secretion in reptiles. However, like all non-mammalian vertebrates release of PRL is stimulated by hypothalamic extracts in reptiles (65,66,86,90). PRL release in turtles was shown to be induced by TRH (86,91), and TRH-induced PRL secretion was inhibited by DA (86). However, DA did not suppress basal PRL secretion in terrapins. As in amphibians, SRIF did not suppress basal or TRH-induced PRL release (86). Considering that reptiles lie at the phylogenetic branch point for subsequent evolution of birds and mammals, this relative dearth of knowledge concerning PRL regulation in reptiles is problematic for understanding evolutionary influences on mechanisms controlling PRL secretion.

Of all the non-mammalian vertebrates, the regulation of PRL secretion is understood best in birds. In fact, the level of understanding of hypothalamic control of PRL release during nesting in birds is comparable to that during suckling-induced PRL release in mammals. The annual change in daylength is the initial stimulant for increased PRL secretion at the start of the breeding cycle in most avian species. Prior to this increase in daylength, female birds exhibit basal circulating PRL as well as low gonadotropins and gonadal steroids. Within 7-10 days following photostimulation, PRL levels increase 2-3 fold and remain fairly stable until the onset of egg laying (92-94). This elevation in PRL secretion is not dependent on the presence of the predominant gonadal steroids, estrogen and progesterone, which are elevated during this period (95,96). Throughout the period of regular ovulations, PRL secretion gradually increases then plateaus at a concentration 5-10 fold higher than that of photostimulated birds. In contrast to the initial PRL rise at the beginning of the cycle, this elevation is dependent on ovarian steroids. Ovariectomized hens will not exhibit this unless primed with estradiol benzoate, followed by estradiol and progesterone (97). If laying hens are ovariectomized, PRL levels will fall to the photo-induced concentrations. In addition, the magnitude of the PRL plateau during the late phase of egg laying can be attenuated by preventing nesting activity (96). Birds naturally increase the time spent sitting on the eggs or nesting near the termination of egg laying. Therefore nesting stimuli may also be necessary for maintenance of elevated PRL.

Initiation of incubation behavior or continuous nesting activity is associated with a further 6-10 fold increase in PRL, while gonadal steroids and the gonadotropins drop to basal concentrations (92). In many avian species, including the chicken (98) and mallard (99), this third elevation of PRL occurs just prior to or during incubation and is maintained throughout the cycle. It appears to be independent of gonadal steroids, since ovariectomy of incubating turkey hens does not diminish circulating PRL. However, various forms of sensory stimuli associated with nesting are absolutely essential. The attenuation of circulating PRL following the removal of nest and eggs in ring doves (100) or physical removal of incubating turkeys from nests and into cages (101) attests to this requirement. Within 48 hours of nest deprivation of incubating hens, PRL concentrations are comparable to those of laying hens. This phenomenon is attributed to tactile stimuli to the brood patch, a defeathered, hyperemic area of abdominal skin in direct contact with the eggs that is important for transferring heat from the hen to the eggs. Local anesthetization or surgical denervation of the brood patch suppresses PRL in incubating ducks (102). In addition to tactile stimuli, the presence of a mate or young can also influence PRL secretion. Extension of the length of incubation in female ring doves in the absence of the mate will result in depressed PRL despite continued nesting and the transition to the state of photorefractoriness (100). After the hatching of eggs, PRL levels and gene expression in the pituitary decrease precipitously (93,94,99,103).

In summary, fluctuations in PRL secretion during the reproductive cycle of avian species are associated with photostimulation, egg laying and the initiation of incubation and photorefractoriness, with these changes manifested by increases in circulating

PRL and PRL gene expression. Present evidence indicates that the reception of photoperiod length, the presence of ovarian steroids and the sensory stimuli of nesting all serve to modulate secretion at various stages of the cycle.

There is increasing evidence indicating that hypothalamic VIP is the PRL releasing factor in birds (104). VIP immunoreactivity has been localized within the cells and nerve fibers of the mediobasal hypothalamus and external median eminence (105-108). Moreover, VIP has proved to be a potent releaser of PRL *in vivo* (105,109,110) and *in vitro* (110,111) in many avian species. VIP immunoneurtralization decreases PRL levels and expression and prevents the induction of incubation behavior (112-114). Immunocytochemical and *in situ* hybridization studies have shown that increases in hypothalamic VIP immunoreactivity and gene expression within the infundibular nuclear complex (INF) parallel increases in circulating PRL (106,115). Other studies have also shown increases in the number and size of VIP immunoreactive neurons within this region in the domesticated pigeon and ring dove following the initiation of crop milk secretion and feeding of offspring, periods of elevated circulating PRL (109,116).

Considerable evidence indicates that the serotonergic system is a potent stimulator of PRL secretion in birds (92,117). 5-HT seems to act centrally, since it has no effect on PRL secretion when added to pituitary cells *in vitro* (92), and 5-HT receptors are not present in the anterior pituitary (118). However, intraventricular infusion of 5-HT caused plasma PRL to rise in the turkey (119,120). Serotonergic fibers traversing the hypothalamic ventromedial nucleus (121) may stimulate PRL secretion through interneuronal connections to the infundibular nuclear complex, where the majority of VIP immunoreactive neurons are found (106,115). VIP immunoneutralization blocked 5-HT-stimulated PRL secretion (119), a result that is consistent with the view that endogenous VIP mediates the serotonergic stimulation of PRL secretion. The finding that 5-HT-induced PRL secretion is completely abolished by VIP immunoneutralization indicates that no other PRL-releasing factor needs to be invoked to explain the PRL-elevating effect of 5-HT. Nonetheless, whether or not the activation of brain serotonergic mechanisms stimulates VIP release into the hypophysial portal circulation needs to be determined.

Another prominent neurotransmitter that affects PRL secretion in birds is DA. DA has both stimulatory and inhibitory effects on PRL secretion in birds. DA inhibited PRL release from chicken, pigeon, and turkey anterior pituitary cells *in vitro* (122-125) and in turkeys *in vivo* (126). PRL release induced by electrical stimulation of the turkey brain was curtailed by the DA agonist apomorphine and potentiated by the DA antagonist pimozide (126). Intraventricular administration of DA increased PRL levels in nonlaying turkey hens, while it had no effect in incubating ones (127), indicating that changes in responsiveness to DA may contribute to elevated PRL secretion during nesting. DA-binding sites in the anterior pituitary of bantam hens decreased during incubation behavior, when PRL levels were high (105). The dopaminergic D_2 receptor agonist quinpirole inhibited VIP-induced PRL secretion and PRL mRNA levels in turkey anterior pituitary cell cultures (128), suggesting the presence of inhibitory D_2 DA receptors within the avian pituitary. Moreover, the infusion of VIP into the turkey pituitary effected a rapid and substantial increase in plasma PRL that was completely suppressed when DA was infused in conjunction with VIP (129).

In contrast to inhibitory effects in the pituitary, central infusion of low levels of DA into the turkey brain resulted in a significant increase in plasma PRL (129-131). Stimulatory D_1 DA receptors are located within the avian hypothalamus (131,132). When the same amount of DA was infused into the pituitary of the turkey, no rise in plasma PRL was observed. However, PRL release in response to electrical stimulation of the medial preoptic area (ES/POM) was suppressed by the pituitary DA infusion. ES/POM and central infusion of DA appear to increase plasma PRL by releasing VIP, the avian PRL-releasing factor, from the hypothalamus. In turkeys immunized against VIP, neither ES/POM nor central DA infusion would increase plasma PRL (131,133). DA stimulated VIP release from isolated, perifused turkey hypothalami in a dose-dependent manner, and this VIP release was completely suppressed when the D_1 DA receptors were blocked (132).

In addition to the importance of VIP in mediating the effects of such classical neurotransmitters as 5-HT and DA on PRL release, the opioid peptide dynorphin acts centrally to release PRL and/or modulate ongoing secretion (134). Dynorphin increased PRL secretion by activating ? opioid receptors; when these receptors were blocked, exogenous dynorphin was unable to stimulate PRL secretion. Blockade of 5-HT receptors not only prevented 5-HT from stimulating PRL secretion, it also blocked PRL stimulation by infused dynorphin (135). When the D_1 DA receptors were blocked, the PRL-releasing efficacy of not only DA (129), but also 5-HT (136) and dynorphin (135), was suppressed. Thus, 5-HT and ? opioid receptors appear to lie above the synapse containing D_1 DA receptors. In birds, the primary PRL-releasing factor released from the hypothalamus into the hypothalamo-pituitary portal vessels is VIP (104). The ability of dynorphin, 5-HT, and DA to stimulate PRL secretion is contingent upon an intact VIPergic system. When birds were immunized against their own VIP, the infusion of dynorphin, 5-HT (119), or DA (131) could not stimulate PRL secretion. The infusion of VIP into the median eminence of the hypothalamus of immunized birds resulted in a severely curtailed PRL response. Thus, dynorphin, 5-HT, DA, and VIP act to stimulate PRL secretion via a common pathway expressing ? opioid, serotonergic, dopaminergic, and VIPergic receptors at synapses that are arranged successively in that functional order.

TRH stimulated PRL secretion both *in vivo* and *in vitro* in chickens, but only *in vivo* in turkeys (117). The fact that TRH immunoreactivity in neurons within the pigeon hypothalamus increased during lactation-associated hyperprolactinemia (137), suggests a physiological relevance. The possibility exists that TRH may act centrally to modulate the release of other neural agents, which then transmit a hypothalamic message to the lactotrophs. PHI has been shown to be specifically expressed in chicken and turkey hypothalami (138,139). However, its role in avian PRL regulation needs to be clarified, since PHI and PRL levels appear to be uncorrelated. The levels of turkey VIP transcripts containing both the PHI- and VIP-encoding exons did not change in hypothalami of turkey hens in different reproductive states (139). These levels were maintained at about 4-6% of the total turkey VIP transcript level. Even though the minor PHI-VIP transcript exists at a very low steady-state level in the hypothalamus, one cannot rule out the possibility that it might have physiological relevance if it were selectively induced by some factor, or if its stability or translational efficiency differed from those of the VIP transcript lacking the PHI-encoding exon. A preliminary *in vitro* study utilizing synthetic turkey PHI peptide showed that PHI is 100- to 1000-fold less potent in releasing PRL from pituitary cells. Similar findings were reported that utilized mammalian PHI (140).

CONCLUSION

Much has been learned about the location, ontogeny and physiology of lactotrophs in non-mammalian vertebrates. Many unique aspects not common to mammals have been discovered. Still other commonalities among the vertebrate classes have been noted. These findings in the "lower" vertebrates are certainly important to researchers interested in comparative endocrinology, but individuals studying lactotroph regulation and function in mammals also should consider them, as they may provide clues to understanding fundamental aspects of lactotroph differentiation, ontogeny, regulation and function. Although much has been learned in the non-mammalian classes of vertebrates, even more awaits discovery.

REFERENCES

1. Bern HA, Nicoll CS. The comparative endocrinology of prolactin. Recent Prog Horm Res. 1968;24:681-720.
2. Dodd JM, Follett BK, Sharp PJ. Hypothalamic control of pituitary function in submammalian vertebrates. Adv Comp Physiol Biochem. 1971;4:113-223.
3. Sage M, Bern HA. Cytophysiology of the teleost pituitary. International Review of Cytology. 1971;31:339-376.
4. Ball JN. Hypothalamic control of the pars distalis in fishes, amphibians, and reptiles. Gen Comp Endocrinol. 1981;44:135-170.
5. Huang L, Specker JL. Growth hormone- and prolactin-producing cells in the pituitary gland of striped bass (*Morone saxatilis*): immunocytochemical characterization at different life stages. Gen Comp Endocrinol. 1994;94:225-236.
6. Power DM, Canario AVM. Immunocytochemistry of somatotrophs, gonadotrophs, prolactin and adrenocorticotropin cells in larval sea bream (*Sparus auratus*) pituitaries. Cell Tissue Res. 1992;269:341-346.
7. Cambre ML, Verdonick W, Ollevier F, Vandesande F, Batten TF, Kuhn ER. Immunocytochemical identification and localization of the different cell types in the pituitary of the seabass (*Dicentrarchus labrax*). Gen Comp Endocrinol. 1986;61:368-375.
8. Specker JL, Kishida M, Huang L, King DS, Nagahama Y, Ueda H, Anderson TR. Immunocytochemical and immunogold localization of two prolactin isoforms in the same pituitary cells and in the same granules in the Tilapia (*Oreochromis mossambicus*). Gen Comp Endocrinol. 1993;89:28-38.
9. Doerr-Schott J. Immunohistochemical detection, by light and electron microscopy, of pituitary hormones in cold-blooded vertebrates I. Fish and amphibians. Gen Comp Endocrinol. 1976;28:487-512.
10. Follenius E, Doerr-Schott J. Immunocytology of pituitary cells from teleost fishes. International Review of Cytology. 1978;54:193-223.
11. Ruijter AM, Creuwels L. The ultrastructure of prolactin cells in the annual cyprinodont *Cynolebias whitei* during its life cycle. Cell Tissue Res. 1988;253:477-483.
12. Ruijter JM, Kemenade Jv, Bonga SEW. Environmental influences on prolactin cell development in the cyprinodont fish, *Cynolebias whitei*. Cell Tissue Res. 1984;238:595-600.
13. Wright GM. Immunocytochemical study of growth hormone, prolactin, and thyroid-stimulating hormone in the adenohypophysis of the sea lamprey, *Petromyzon marinus L.*, during its upstream migration. Gen Comp Endocrinol. 1984;55:269-274.
14. Campantico E, Guastalla A. Immunocytochemical identification of growth hormone- and prolactin-producing cells in the hypophysis of the newt *triturus cristatus carnifex* laur. Gen Comp Endocrinol. 1992;86:197-202.
15. Miranda LA, Paz DA, Dezi R, Pisano A. Immunocytochemical and morphometric study on the changes of TSH, PRL, GH, and ACTH cells during the development of *Bufo arenarum*. Cell Tissue Res. 1996;283:125-132.
16. Guastalla A, Campantico E, Yamamoto K, Kobayashi T, Kikuyama S. Immunocytochemical and ultrastructural study of *Rana dalmatina* PRL and GH pituitary cells during larval development. Gen Comp Endocrinol. 1993;89:364-377.
17. Doerr-Schott J. Immunohistochemical detection, by light and electron microscopy, of pituitary hormones in cold-blooded vertebrates II. Reptiles. Gen Comp Endocrinol. 1976;28:513-529.

18. Licht P, Nicoll CS. Localization of prolactin in the reptilian pars distalis. Gen Comp Endocrinol. 1969;12:526-535.
19. Mohanty KC, Naik DR. Immunohistochemistry and tinctorial affinity of adenohypophysial cells of the rat snake *ptyas mucosus* (*Colubridae*). Gen Comp Endocrinol. 1997;105:302-313.
20. Naik DR, Sar M, Stumpf WE. Immunohistochemical identification of cells in the pars distalis of the pituitary of the lizard Anolis carolinensis. Histochemistry. 1980;69:19-26.
21. Sotowska-Brochocka J. Seasonal changes of the structure and function of prolactin cells in the pituitary gland of *Vipera Berus* (*L.*). Acta Physiologica. 1977;28:337-352.
22. Pearson AK, Licht P. Morphology and immunocytochemistry of the turtle pituitary gland with special reference to the pars tuberalis. Cell Tissue Res. 1982;222:81-100.
23. Kansaku N, Shimada K, Saito N. Regionalized gene expression of prolactin and growth hormone in the chicken anterior pituitary gland. Gen Comp Endocrinol. 1995;99:60-68.
24. Ramesh K, Solow R, Proudman JA, Kuenzel WJ. Indentification of mammosomatotrophs in the turkey hen pituitary: increased abundance during hyperprolactinemia. Endocrinology. 1998;139:781-786.
25. Ramesh R, Proudman JA, Kuenzel WJ. Changes in pituitary somatotroph and lactotroph distribution in laying and incubating turkey hens. Gen Comp Endocrinol. 1996;104:65-75.
26. Allaerts W, Boonstra-Blom AG, Peeters K, Janse EM, Berghman LR, Jeurissen S. Prenatal development of hematopoietic and hormone-producing cells in the chicken adenohypophysis. Gen Comp Endocrinol. 1999;114:213-224.
27. Mohanty B, Das S, Naik DR. Immunocytochemistry of the Pars tuberalis of the pituitary gland in some Indian wild birds: a comparative study. Gen Comp Endocrinol. 1997;108:109-118.
28. Mikami S-I, Kurosu T, Farner DS. Light-and electron-microscopic studies on the secretory cytology of the adenohypophysis of the Japanese quail, *Coturnix coturnix japanica*. Cell Tissue Res. 1975;159:147-165.
29. Mikami SI, Yamada S. Immunohistochemistry of the hypothalamic neuropeptides and anterior pituitary cells in the Japanese quail. J Exp Zool. 1984;232:405-417.
30. Lopez ME, Hargis BM, Dean CE, Porter TE. Uneven regional distributions of prolactin- and growth hormone-secreting cells and sexually dimorphic proportions of prolactin secretors in the adenohypophysis of adult chickens. Gen Comp Endocrinol. 1995;100:246-254.
31. Rhodes SJ, DiMattia GE, Rosenfeld MG. Transcriptional mechanisms in anterior pituitary cell differentiation. Curr Opin Genet Dev. 1994;4:709-717.
32. Voss JW, Rosenfeld MG. Anterior pituitary development: Short tales from dwarf mice. Cell. 1992;70:527-530.
33. Borrelli E, Heyman RA, Arias C, Sawchenko PE, Evans RM. Transgenic mice with inducible dwarfism. Nature. 1989;339:538-541.
34. Woods KL, Porter TE. Ontogeny of prolactin-secreting cells during chick embryonic development: effects of vasoactive intestinal peptide. Gen Comp Endocrinol. 1998;112:240-246.
35. Cambre M, Mareels G, Corneillie S, Moons L, Ollevier F, Vandesande F. Chronological appearance of the different hypophysial hormones in the pituitary of sea bass larvae (*Dicentrarchus labrax*) during their early development: an immunocytochemical demonstration. Gen Comp Endocrinol. 1990;77:408-415.
36. Shen AC, Leatherland JF. Histogenesis of the pituitary in rainbow trout (*Salmo gairdneri*) in different ambient salinities with particular reference to the rostral pars distalis. Cell Tissue Res. 1978;189:355-363.

37. Ayson FG, Kaneko T, Hasegawa S, Hirano T. Differential expression of two prolactin and growth hormone genes during early development of Tilapia (*Oreochromis mossambicus*) in fresh water and seawater: implications for possible involvement in osmoregulation during early life stages. Gen Comp Endocrinol. 1994;95:143-152.
38. Frawley LS, Boockfor FR. Mammosomatotropes: presence and functions in normal and neoplastic pituitary tissue. Endocr Rev. 1991;12:337-355.
39. Frawley LS. Mammosomatotropes: Current status and possible functions. Trends in Endocrinology and Metabolism. 1989;September/October:31-34.
40. Hoeffler J, Boockfor F, Frawley LS. Ontogeny of prolactin cells in neonatal rats: Initial prolactin secretors also release growth hormone. Endocrinology. 1985;117:187-195.
41. Oguchi A, Aida T, Kobayashi T, Tanaka S, Shioda S, Nakajo S, Kikuyama S. Cosecretion of prolactin and growth hormone by the pituitary of the bullfrog. Ann N Y Acad Sci. 1998;839:453-454.
42. Kikuyama S, Inaco H, Jenks BG, Kawamura K. Development of the ectopically transplanted primordium of epithelial hypophysis (anterior neural ridge) in *Bufo japonicus* embryos. J Exp Zool. 1993;266:216-220.
43. Ishida H, Shimada K, Sato K, Seo H, Murata Y, Matsui N, Zadworny D. Developmental expression of the prolactin gene in the chicken. Gen Comp Endocrinol. 1991;83:463-467.
44. Harvey S, Davidson TF, Chadwick A. Ontogeny of Growth Hormone and prolactin Secretion in the Domestic Fowl (Gallus domesticus). Gen Comp Endocrinol. 1979;39:270-273.
45. Jozsa R, Scanes CG, Vigh S, Mess B. Functional Differentiation of the Embryonic Chicken Pituitary Gland Studied By Immunohistological Approach. Gen Comp Endocrinol. 1979;39:158-163.
46. Porter TE, Hill B, Wiles C, Frawley LS. Is the mammosomatotrope a transitional cell for the functional interconversion of growth hormone- and prolactin-secreting cells? Suggestive evidence from virgin, gestating, and lactating rats. Endocrinology. 1990;127:2789-2794.
47. Goluboff L, Ezrin C. Effect of pregnancy on the somatotroph and the prolactin cell of the human adenohypophysis. Clin Endocrinol (Oxf). 1969;29:1533-1538.
48. Ensor DM, Ball JN. Prolactin and osmoregulation in fishes. Fed Proc. 1972;31:1615-1623.
49. Benjamin M. Cytological changes in prolactin, ACTH, and growth hormone cells of the pituitary gland of Pungitius pungitius L. in response to increased environmental salinities. Gen Comp Endocrinol. 1978;36:48-58.
50. Porter TE, Wiles CD, Frawley LS. Evidence for bidirectional interconversion of mammotropes and somatotropes: rapid reversion of acidophilic cell types to pregestational proportions after weaning. Endocrinology. 1991;129:1215-1220.
51. Huang H, Brown DD. Prolactin is not a juvenile hormone in *Xenopus laevis* metamorphosis. PNAS USA. 2000;97:195-199.
52. Lopez ME, Gregory CC, Porter TE. Cellular basis for elevated prolactin secretion during incubation behavior in bantam chickens: analysis by reverse hemolytic plaque assay. Biol Reprod. 1996;54:826-833.
53. Ramesh R, Kuenzel WJ, Proudman JA. Somatotroph recruitment following disruption of incubation behavior results from proliferation of non-somatotrophic cells in the pituitary gland of turkey hens. Poultry Science. 1999;78 (Suppl. 1): 66.
54. Weber GM, Grau EG. Changes in serum concentrations and pituitary content of the two prolactins and growth hormone during the reproductive cycle in female tilapia, *Oreochromis mossambicus*, compared with changes during fasting. Comp Biochem Physiol C Pharmacol Toxicol Endocrinol. 1999;124:323-335.

55. Tacon P, Baroiller JF, Bail PYL, Prunet P, Jalabert B. Effect of egg deprivation on sex steroids, gonadotropin, prolactin, and growth hormone profiles during the reproductive cycle of the mouthbrooding cichlid fish *Oreochromis niloticus*. Gen Comp Endocrinol. 2000;117:54-65.
56. Auperin B, Baroiller JF, Ricordel MJ, Fostier A, Prunet P. Effect of confinement stress on circulating levels of growth hormone and two prolactins in freshwater-adapted tilapia (*Oreochromis niloticus*). Gen Comp Endocrinol. 1997;108:35-44.
57. Grau EG, Helms LM. The Tilapia prolactin cell-twenty-five years of investigation. Progress in Comparative Endocrinology. 1990:534-540.
58. Fargher RC, McKeown BA. Differential response of plasma prolactin to freshwater transfer of smolts and postmolts of seawater-adapted Coho salmon (*Oncorhynchus Kisutch*). Gen Comp Endocrinol. 1990;78:311-321.
59. Loretz CA, Bern HA. Prolactin and osmoregulation in vertebrates. Neuroendocrinology. 1982;35:292-304.
60. Johnston LR, Wigham T. The effects of ions and hypothalamic factors on the *in vitro* activity of rainbow trout prolactin cells. Gen Comp Endocrinol. 1990;79:253-260.
61. Olsson R, Fijuta H. The follicular prolactin cells of the primitive teleost *Engraulis japonica*. Cell Tissue Res. 1976;172:185-194.
62. Shepherd BS, Sakamoto T, Hyodo S, Nishioka RS, Ball C, Bern HA, Grau EG. Is the primitive regulation of pituitary prolactin ($tPRL_{177}$ and $tPRL_{188}$) secretion and gene expression in the euryhaline tilapia (*Oreochromis mossambicus*) hypothalamic or environmental? J Endocrinol. 1999;161:121-129.
63. Nagahama Y, Nishioka RS, Bern HA, Gunther RL. Control of prolactin secretion in teleost, with special reference to *Gillichthys mirabilis* and *Tilapia mossambica*. Gen Comp Endocrinol. 1975;25:166-188.
64. Grau EG, Ford C-A, Helms LM, Shimoda SK, Cooke IM. Somatostatin and altered medium osmotic pressure elicit rapid changes in prolactin release from the rostral Pars distalis of the Tilapia, *Oreochromis mossambicus*, *in vitro*. Gen Comp Endocrinol. 1987;65:12-18.
65. Singh SP, Singh TP. Regulation of prolactin secretion by hypothalamus in some cold blooded vertebrates. Endokrinologie. 1980;76:51-62.
66. Hall TR, Chadwick A. Hypothalamic control of prolactin and growth hormone secretion in different vertebrate species. Gen Comp Endocrinol. 1979;37:333-342.
67. Peter RE, McKeown BA. Hypothalamic control of prolactin and thyrotropin secretion in teleost, with special reference to recent studies on the goldfish. Gen Comp Endocrinol. 1975;25:153-165.
68. James VA, Wigham T. Evidence for dopaminergic and serotonergic regulation of prolactin cell activity in the trout *Salmo gairdneri*. Gen Comp Endocrinol. 1984;56:231-239.
69. Williams AJ, Wigham T. The regulation of prolactin cells in the rainbow trout (*Oncorhynchus mykiss*) 2. Somatostatin. Gen Comp Endocrinol. 1994;93:398-405.
70. Planas J, Bern HA, Millar RP. Effects of GnRH-associated peptide and its component peptides on prolactin secretion from the Tilapia pituitary *in vitro*. Gen Comp Endocrinol. 1990;77:386-396.
71. Rivas RJ, Nishioka RS, Bern HA. *In vitro* effects of somatostatin and urotensin II on prolactin and growth hormone secretion in Tilapia, *Oreochromis mossambicus*. Gen Comp Endocrinol. 1986;63:245-251.
72. Grau EG, Nishioka RS, Young G, Bern HA. Somatostatin-like immunoreactivity in the pituitary and brain of three teleost fish species: somatostatin as a potential regulator of prolactin cell function. Gen Comp Endocrinol. 1985;59:350-357.
73. Goff PL, Weil C, Valotaire Y, Gonnard JF, Prunet P. Effect of somatostatin on prolactin in rainbow trout (*Oncorhynchus mykiss*) pituitary cells in primary culture. J Mol Endocrinol. 1992;9:137-146.

74. Barry TP, Grau EG. Estradiol-17B and thyrotropin-releasing hormone stimulate prolactin release from the pituitary gland of a teleost fish *in vitro*. Gen Comp Endocrinol. 1986;62:306-314.
75. Williams AJ, Wigham T. The regulation of prolactin cells in the rainbow trout (*Oncorhynchus mykiss*) 1. Possible roles for thyrotropin-releasing hormone (TRH) and oestradiol. Gen Comp Endocrinol. 1994;93:388-397.
76. Weber GM, Powell JF, Park M, Fischer WH, Craig AG, Rivier JE, Nanakorn U, Parhar IS, Ngamvongchon S, Grau EG, Sherwood NM. Evidence that gonadotropin-releasing hormone (GnRH) functions as a prolactin-releasing factor in a teleost fish (*Oreochromis mossambicus*) and primary structures for three native GnRH molecules. J Endocrinol. 1997;155:121-132.
77. Fruchtman S, Jackson L, Borski R. Insulin-like growth factor I disparately regulates prolactin and growth hormone synthesis and secretion: studies using the teleost pituitary model. Endocrinology. 2000;141:2886-2894.
78. Kelley KM, Nishioka RS, Bern HA. Novel effect of vasoactive intestinal polypeptide and peptide histidine isoleucine: inhibition of *in vitro* secretion of prolactin in the Tilapia, *Oreochromis mossambicus*. Gen Comp Endocrinol. 1988;72:97-106.
79. Mosconi G, Nabissi M, Carnevali O, Cardinali M, Polzonettie-Magni AM, Yamamoto K, Takahashi N, Kikuyama S. Prolactin and stress response in frog *Rana esculenta*. Ann N Y Acad Sci. 1998;839:639-641.
80. Mosconi G, Yamamoto K, Kikuyama S, Carnevali O, Mancuso A, Vellano C. Seasonal changes of plasma prolactin concentrations in the reproduction of the Crested Newt (*Triturus carnifex Laur*). Gen Comp Endocrinol. 1994;95:342-349.
81. Mosconi G, Yamamoto K, Carnevali O, Nabissi M, Polzonettie-Magni A, Kikuyama S. Seasonal changes in plasma growth hormone and prolactin concentrations of the frog *Rana esculenta*. Gen Comp Endocrinol. 1994;93:380-387.
82. Ishi S, Yoneyama H, Inoue M, Yamamoto K, Kikuyama S. Changes in plasma and pituitary levels of prolactin in the toad, *Bufo japonicus*, throughout the year with special reference to the breeding migration. Gen Comp Endocrinol. 1989;74:365-372.
83. Buckbinder L, Brown DD. Expression of the *Xenopus laevis* prolactin and thyrotropin genes during metamorphosis. Dev Biol. 1993;90:3820-3824.
84. Takahashi N, Yamamoto K, Kikuyama S. Cloning of a toad prolactin cDNA: expression of prolactin mRNA in larval and adult pituitaries. J Mol Endocrinol. 1993;11:343-349.
85. Miranda LA, Paz DA, Dezi RE, Pisano A. Immunocytochemical and morphometric study of TSH, PRL, GH, and ACTH cells in Bufo arenarum larvae with inhibited thyroid function. Gen Comp Endocrinol. 1995;98:166-176.
86. Hall TR, Chadwick A. Effects of synthetic mammalian thyrotrophin releasing hormone, somatostatin and dopamine on the secretion of prolactin and growth hormone from amphibian and reptilian pituitary glands incubated *in vitro*. Endocrinologie. 1984;102:175-180.
87. Castano J-P, Ramirez J-L, Malagon MM, Gracia-Navarro F. Differential Response of Amphibian PRL and TSH pituitary cells to *in vitro* TRH treatment. Gen Comp Endocrinol. 88;88:178-187.
88. Nakajima K, Uchida D, Sakai M, Takahashi N, Yanagisawa T, Yamamoto K, Kikuyama S. Thyrotropin-releasing hormone (TRH) is the major prolactin-releasing factor in the bullfrog hypothalamus. Gen Comp Endocrinol. 1993;89:11-16.
89. Aida T, Yamamoto K, Kikuyama S. Enhancement by proopiomelanocortin-derived peptides of growth hormone and prolactin secretion by bullfrog pituitary cells. Gen Comp Endocrinol. 1999;115:101-109.
90. Fiorindo RP. Further evidence for a prolactin-stimulating neurohormone in reptiles. Gen Comp Endocrinol. 1980;40:52-58.

91. Preece H, Licht P. Effects of thyrotropin-releasing hormone *in vitro* on thyrotropin and prolactin release from the turtle pituitary. Gen Comp Endocrinol. 1987;67:247-255.
92. El Halawani ME, Fehrer S, Hargis B, Porter T. Incubation behavior in the domestic turkey: physiological correlates. CRC Critical Review in Poultry Biology. 1988;1:285-314.
93. Wong EA, Ferrin NH, Silsby JL, El Halawani ME. Cloning of a turkey prolactin cDNA: expression of prolactin mRNA throughout the reproductive cycle of the domestic turkey (*Meleagris gallopavo*). Gen Comp Endocrinol. 1991;83:18-26.
94. Tong Z, Pitts GR, Foster DN, El Halawani ME. Transcriptional and post-transcriptional regulation of prolactin during the turkey reproductive cycle. J Mol Endocrinol. 1997;18:223-231.
95. El Halawani ME, Silsby JL, Fehrer SC, Behnke EJ. Effects of estrogen and progesterone on serum prolactin and luteinizing hormone levels in ovariectomized turkeys (*Meleagris gallopavo*). Gen Comp Endocrinol. 1983;52:67-78.
96. El Halawani ME, Silsby JL, Behnke EJ, Fehrer SC. Effect of Ambient Temperature on Serum Prolactin and Luteinizing Hormone Levels During the Reproductive Life Cycle of the Female Turkey (*Meleagris gallopavo*). Biol Reprod. 1984;30:809-815.
97. El Halawani ME, Silsby J, Behnke E, Fehrer S. Hormonal induction of incubation behavior in ovariectomized female turkeys (*Meleagris gallopavo*). Biol Reprod. 1986;35:59-67.
98. Lea RW, Dods ASM, Sharp PJ, Chadwick A. The possible role of prolactin in the regulation of nesting behavior and the secretion of luteinizing hormone in broody bantams. J Endocrinol. 1981;91:89-97.
99. Goldsmith AR, Williams DM. Incubation in mallards (*Anas platyrhynchos*): changes in plasma levels of prolactin and luteinizing hormone. J Endocrinol. 1980;86:371-379.
100. Ramsey SM, Goldsmith AR, Silver R. Stimulus requirements for prolactin and LH secretion in incubating ring doves. Gen Comp Endocrinol. 1985;59:246-256.
101. El Halawani ME, Burke WH, Dennison PT. Effects of p-chlorophenylalanine on the rise in serum prolactin associated with nesting in broody turkeys. Biol Reprod. 1980;23:815-819.
102. Hall MR. External stimuli affecting incubation behavior and prolactin secretion in the duck (*Anas platyrhynchos*). Horm Behav. 1987;21:269-287.
103. Sharp PJ, Scanes CG, Williams JB, Harvey S, Chadwick A. Variations in concentrations of prolactin, luteinizing hormone, growth hormone and progesterone in the plasma of broody bantams (*Gallus domesticus*). J Endocrinol. 1979;80:51-57.
104. El Halawani ME, Youngren OM, Pitts GR. Vasoactive intestinal peptide as the avian prolactin releasing factor. In: Etches RJ, Harvey S, eds. Perspectives in Avian Endocrinology. Bristol, UK: Journal of Endocrinology Limited; 1997.
105. Macnamee MC, Sharp PJ, Lea RW, Sterling RJ, Harvey S. Evidence that vasoactive intestinal polypeptide is a physiological prolactin-releasing factor in the bantam hen. Gen Comp Endocrinol. 1986;62:470-478.
106. Mauro LJ, Elde RP, Youngren OM, Phillips RE, El Halawani ME. Alterations in hypothalamic vasoactive intestinal peptide-like immunoreactivity are associated with reproduction and prolactin release in the female turkey. Endocrinology. 1989;125:1795-1804.
107. Mikami MI. Immunocytochemistry of the avian hypothalamus and adenohypophysis. International Review of Cytology. 1986;103:189-248.
108. Silver R, Witkovsky P, Horvath P, Alones V, Barnstable CJ, Lehman MN. Coexpression of opsin- and VIP-like immunoreactivity in CSF-contacting neurons of the avian brain. Cell Tissue Research. 1988;253:189-198.
109. Lea RW, Vowles DM. Vasoactive intestinal polypeptide stimulates prolactin release *in vivo* in the ring dove (*Streptopelia risoria*). Experientia. 1986;42:420-422.
110. Proudman JA, Opel H. Stimulation of prolactin secretion from turkey anterior pituitary cells in culture. Proc Soc Exp Biol Med. 1988;187:448-454.

111. El Halawani ME, Silsby JL, Mauro LJ. Vasoactive intestinal peptide is a hypothalamic prolactin-releasing neuropeptide in the turkey (*Meleagris gallopavo*). Gen Comp Endocrinol. 1990;78:66-73.
112. El Halawani ME, Silsby JL, Rozenboim I, Pitts GR. Increased Egg Production by Active Immunazation against Vasoactive Intestinal Peptide in the Turkey (*Meleagris gallopavo*). Biol Reprod. 1995;52:179-183.
113. Sharp PJ, Sterling RJ, Talbot RT, Huskisson NS. The role of hypothalamic vasoactive intestinal polypeptide in the maintenance of prolactin secretion in incubating bantam hens: Observations using passive immunization radioimmunoassay and immunohistochemistry. J Endocrinol. 1989;122:5-13.
114. El Halawani ME, Pitts GR, Sun S, Silsby JL, Sivanendan V. Active immunization against vasoactive intestinal peptide prevents photo-induced prolactin secretion in turkeys. Gen Comp Endocrinol. 1996;104:76-83.
115. Chaiseha Y, El Halawani ME. Expression of vasoactive intestinal peptide/peptide histidine isoleucine in several hypothalamic areas during the turkey reproductive cycle: relationship to prolactin secretion. Neuroendocrinology. 1999;70:402-412.
116. Cloues R, Ramos C, Silver R. Vasoactive intestinal polypeptide-like immunoreactivity during reproduction in doves: influence of experience and number of offspring. Horm Behav. 1990;24:215-231.
117. Hall TR, Harvey S, Chadwick A. Control of prolactin secretion in birds: A review. Gen Comp Endocrinol. 1986;62:171-184.
118. Macnamee MC, Sharp PJ. The functional activity of hypothalamic 5-hydroxytryptamine neurons in broody bantam hens. J Endocrinol. 1989;120:125-134.
119. El Halawani ME, Youngren OM, Rozenboim I, Pitts GR, Silsby JL, Phillips RE. Serotonergic stimulation of prolactin secretion is inhibited by vasoactive intestinal peptide immunoneutralization in the turkey. Gen Comp Endocrinol. 1995;99:69-74.
120. Pitts GR, Youngren OM, Phillips RE, El Halawani ME. Photoperiod mediates the ability of serotonin to release prolactin in the turkey. Gen Comp Endocrinol. 1996;104:265-272.
121. Youngren OM, El Halawani ME, Phillips RE, Silsby JL. Effects of peroptic and hypothalamic lesions in female turkeys during a photo-induced reproductive cylce. Biol Reprod. 1989;41:610-617.
122. Harvey S, Chadwick A, Border G, Scanes CG, Phillips TG. Neuroendocrine control of prolactin secretion. In: Scanes CG, Ottinger MA, Kenny AD, Balthazart J, Cronshaw J, Jones IC, eds. Aspects of Avian Endocrinology: Practical and Theoretical Implications. Lubbock, Texas: Texas Technological Press; 1982:41-64.
123. Fehrer SC. The role of monoamines and thyrotrpin releasing hormone in prolactin and luteinizing hormone release by the pituitary gland of the young domestice turkey (*Meleagris gallopavo*). In. St. Paul: University of Minnesota; 1984.
124. Hall TR, Chadwick A. Hypothalamic control of prolactin and growth hormone secretion in the pituitary gland of the pigeon and the chicken: *In vitro* studies. Gen Comp Endocrinol. 1983;49:135-143.
125. Hall TR, Chadwick A. Dopaminergic inhibition of prolactin release form pituitary glands of the domestic fowl incubated *in vitro*. J Endocrinol. 1984;103:63-69.
126. El Halawani ME, Youngren OM, Silsby JL, Phillips RE. Involvement of dopamine in prolactin release induced by electrical stimulation of the hypothalamus of the female turkey (*Meleagris gallopavo*). Gen Comp Endocrinol. 1991;84:360-364.
127. Hargis BM, Burke WH. Influence of cerebroventricular injection of dopamine on plasma prolactin and LH levels of post-laying and broody turkey hens. Gen Comp Endocrinol. 1986;61:142-147.

128. Xu M, Proudman JA, Pitts GR, Wong EA, Foster DN, El Halawani ME. Vasoactive intestinal peptide stimulates prolactin mRNA expression in turkey piuitary cells: effects of dopamenergic drugs. Proc Soc Exp Biol Med. 1996;212:52-62.
129. Youngren OM, Chaiseha Y, El Halawani ME. Regulation of prolactin secretion by dopamine and vasoactive intestinal peptide at the level of the pituitary in the turkey. Neuroendocrinology. 1998;68:319-325.
130. Youngren OM, Pitts GR, Phillips RE, El Halawani ME. The stimulatory and inhibitory effects of dopamine on prolactin secretion in the turkey. Gen Comp Endocrinol. 1995;98:111-117.
131. Youngren OM, Pitts GR, Phillips RE, El Halawani ME. Dopaminergic control of prolactin secretion in the turkey. Gen Comp Endocrinol. 1996;104:225-230.
132. Chaiseha Y, Youngren OM, El Halawani ME. Dopamine receptors influence vasoactive intestinal peptide release from turkey hypothalamic explants. Neuroendocrinology. 1997;65:423-429.
133. Youngren OM, Silsby JL, Rozenboim I, Phillips RE, El Halawani ME. Active immunization with vasoactive intestinal peptide prevents the secretion of prolactin induced by electrical stimulation of the turkey hypothalamus. Gen Comp Endocrinol. 1994;95:330-336.
134. Youngren OM, Silsby JL, Phillips RE, El Halawani ME. Dynorphine modulates prolactin secretion in the turkey. Gen Comp Endocrinol. 1993;91:224-231.
135. Youngren OM, Pitts GR, Chaiseha Y, El Halawani ME. An opioid pathway in the hypothalamus of the turkey that stimulates prolactin secretion. Neuroendocrinology. 1999;70:317-323.
136. Youngren OM, Chaiseha Y, El Halawani ME. Serotonergic stimulation of avian prolactin secretion requires an intact dopaminergic system. Gen Comp Endocrinol. 1998;112:63-68.
137. Peczely P, Kiss JZ. Immunoreactivity to vasoactive intestinal polypeptide (VIP) and thyreotropin-releasing hormone (TRH) in hypothalamic neurons of domesticated pigeon (*Columbia livia*). Alterations following lactation and exposure to cold. Cell Tissue Research. 1988;251:485-494.
138. Talbot RT, Dunn IC, Wilson PW, Sang HM, Sharp PJ. Evidence for alternative splicing of the chicken vasoactive intestinal polypeptide gene transcript. J Mol Endocrinol. 1995;15:81-91.
139. You S, Silsby JL, Farris J, Foster DN, El Halawani ME. Tissue-specific alternative splicing of turkey preprovasoactive intestinal peptide messenger ribonucleic acid, its regulation, and correlation with prolactin secretion. Endocrinology. 1995;136:2602-2610.
140. Proudman JA, Opel H. Effects of peptide histidine isoleucine on *in vitro* and *in vivo* prolactin secretion in the turkey. Poultry Science. 1990;69:1209-1214.

Chapter 5

Prolactinomas

Mark E. Molitch

Center for Endocrinology, Metabolism and Nutrition,
Northwestern University Medical School Chicago, IL

Prolactinomas are the most common of the hormone-secreting tumors of the pituitary. There are many causes of hyperprolactinemia which must be excluded and an imaging procedure such as MRI or CT is necessary to demonstrate the size and anatomy of the tumor. Dopamine agonist therapy constitutes the primary treatment for prolactinomas, being effective in shrinking tumors as well as normalizing PRL levels. The most effective and best tolerated dopamine agonist is cabergoline, at present. Transsphenoidal surgery is reserved for tumors not responding to dopamine agonists and radiotherapy is rarely done. Because of its established safety profile for the developing fetus, bromocriptine is the preferred dopamine agonist to use to facilitate ovulation and pregnancy. During pregnancy, close to one-quarter of macroprolactinomas can be expected to enlarge significantly while less than 2% of microprolactinomas do so. Tumor enlargement during pregnancy usually responds readily to reinstitution of the dopamine agonist.

INTRODUCTION

Prolactinomas are the most common of the hormone-secreting pituitary tumors in humans. Rarely life-threatening, they cause symptoms primarily as a result of hyperprolactinemia which will result in alterations in reproductive/sexual function but may also cause symptoms because of mass effects. Fortunately, medical and surgical therapy are excellent and most patients do very well. In addition to prolactinomas, there are a variety of other conditions that may cause hyperprolactinemia and, because of the lack of specificity of imaging techniques, must be considered in evaluating the hyperprolactinemic patient.

ETIOLOGY OF HYPERPROLACTINEMIA

There are a number of causes of hyperprolactinemia (Table 5-1). Most are due to alteration in the neuroendocrine control mechanisms regulating secretion and result in modestly elevated PRL levels (25-150 ng/ml).

A number of medications can cause hyperprolactinemia, generally by interfering with dopamine generation or action. Psychotropic agents causing hyperprolactinemia

Table 5-1. Etiologies of Hyperprolactinemia

Pituitary Disease	Neurogenic	Medications
Prolactinomas	Chest wall lesions	Phenothiazines
Acromegaly	Spinal cord lesions	Haloperidol
"Empty Sella syndrome"	Breast stimulation	Monoamine-oxidase inhibitors
Lymphocytic hypophysitis	**Other**	Tricyclic antidepressants
Cushing's disease	Reserpine	Metoclopramide
Pregnancy	Methyldopa	Amoxepin
Hypothyroidism	Chronic renal failure	Cocaine
	Cirrhosis	Verapamil
Hypothalamic Disease	Pseudocyesis	
Craniopharyngiomas	Adrenal Insufficiency	
Meningiomas	**Idiopathic**	
Dysgerminomas		
Nonsecreting pituitary adenomas		
Other tumors		
Sarcoidosis		
Eosinophilic granuloma		
Neuraxis irradiation		
Vascular		
Pituitary Stalk Section		

Modified from Molitch ME: Pathologic hyperprolactinemia. Endocrinology and Metabolism Clinics of North America 1992;21:877, with permission.

include the neuroleptics (phenothiazines and butyrophenones), tricyclic antidepressants, monoamine oxidase inhibitors, serotonin reuptake inhibitors, opiates and cocaine (1,2). Several antihypertensive medications cause hyperprolactinemia, including alpha-methyldopa, reserpine, and verapamil (3,4).

A variety of medical conditions can also cause hyperprolactinemia. Most patients with end-stage renal disease and one-quarter of those with lesser degrees of renal insufficiency become hyperprolactinemic due to decreased clearance as well as increased production due to disordered hypothalamic regulation of PRL secretion (5,6). When such patients take medications known to alter hypothalamic regulation of PRL, such as methyldopa or metoclopramide, PRL levels may rise to over 2000 ng/ml (6). Basal PRL levels are increased in 5 to 20% of patients with alcoholic and nonalcoholic cirrhosis, possibly due to alterations in hypothalamic dopamine generation (7). Primary hypothyroidism is associated with a modest increase in the level of PRL in 40% of patients, but levels greater than 25 ng/ml are reached in only 10% due to increased TRH production, increased sensitivity of lactotrophs to TRH and possibly increased pituitary VIP generation (8). Patients with adrenal insufficiency may develop PRL elevations due to loss of the normal glucocorticoid suppressive effect on PRL gene transcription and PRL release (9).

Sexual breast stimulation and suckling cause a reflex release of PRL that is mediated, in part, by afferent neural pathways going through the spinal cord. Chest wall and cervical cord lesions have been reported to result in elevated PRL levels and galactorrhea through stimulation of these afferent neural pathways (10). Ectopic production of PRL is exceedingly rare (11).

Hyperprolactinemia caused by lesions of the hypothalamus and of the pituitary stalk is due to disturbance of the neuroendocrine mechanisms that control PRL secretion. From hypothalamic lesion work in animals, it is apparent that this PRL elevation is due to a decrease of the tonic PIF (dopamine) acting at the level of the pituitary lactotrophs, i.e. they are disinhibited. However, many of these patients have normal ACTH and TSH function, implying that there still is significant transmission of hypothalamic releasing factors to the pituitary in most of these cases despite increased PRL levels (12-14). The PRL elevations in such cases are usually < 150 ng/ml (12-14). Thus, when a patient presents with a large macroadenoma and has PRL levels of only about 100 ng/ml, this is likely a nonsecreting tumor rather than a prolactinoma (12,14). This has great importance with respect to therapy, as treatment of such a nonsecreting tumor is surgery, whereas primary treatment of such a prolactinoma would be bromocriptine (see below).

When no specific cause is found, the hyperprolactinemia is of uncertain etiology and has been designated to be idiopathic. It is recognized that in many such cases small prolactinomas may be present that are too small to be detected by current radiologic techniques. In other cases the hyperprolactinemia is due to presumed hypothalamic regulatory dysfunction, but no dysfunction specific to idiopathic hyperprolactinemia has been definitively elucidated. Long-term followup of such patients has found that in about one-third, PRL levels return to normal, in 10 to 15% there is a rise in PRL levels to >50% over baseline, and in the remaining patients prolactin levels remain stable with only about 10% developing microadenomas (15-17).

PATHOGENESIS AND NATURAL HISTORY OF PROLACTINOMAS

Prolactinomas are generally classified clinically by size: microadenomas, < 10 mm in diameter; macroadenomas, >10 mm in diameter; and macroadenomas with extrasellar extension. The direction and degree of extrasellar extension are of obvious clinical importance. In general, serum PRL levels parallel the size of the tumors, although exceptions occur (18). True carcinomas are extremely rare (19).

Some theories have suggested that prolactinomas arise because of underlying hypothalamic dysregulation, either too little DA or too much PRF reaching the pituitary (20). However, several lines of evidence weigh strongly against such underlying hypothalamic dysregulation: (1) Studies have shown that there is little or no hyperplasia in the nontumorous pituitary tissue surrounding prolactinomas as would be expected if hyperplasia was a transition state toward true adenoma formation; (2) Prolactinoma formation does not occur in states of prolonged DA deficits such as occurs with prolonged neuroleptic use or chronic hypothalamic/stalk dysfunction; (3) PRL secretory abnormalities return to normal following selective adenoma resection; (4) The recurrence rate for prolactinomas is relatively low and early after resection, indicating regrowth of tumor remnants rather than new tumor formation; and (5) prolactinomas are monoclonal rather than polyclonal, indicating clonal proliferation of a single mutated cell rather than adenoma formation from hyperplastic tissue stimulated by hypothalamic dysregulation (21-23).

A large number of studies have attempted to determine the precise mutations that could cause prolactinomas. Rearrangement or amplification of oncogenes such as myc, fos, jun, RAS and gsp have not been found with any regularity (23,24). Although loss of heterozygosity (LOH) on the long arm of chromosome 11 has been reported at increased frequency in prolactinomas, especially in invasive ones, implying genetic instability at that locus, the putative tumor suppressor gene located in this region has not yet been identified (24). Mutations in the DA receptor, in the various steps in signal transduction, or in the Pit-1 transcription factor have also not been found with any regularity (25,26). Recently, the heparin-binding secretory transforming gene (*hst*) has been identified in a variety of cancers and has also be found inn a small number of prolactinomas (27,28). Another putative oncogene, pituitary tumor transforming gene (PTTG), has been localized to chromosome 5q33 (29,30). PTTG is overexpressed in rat GH4 pituitary tumor cells and it induces tumor formation in athymic nude mice (29,30). In further studies, Zhang et al (31) found PTTG mRNA to be present in increased amounts in most human pituitary tumors and that the quantity appeared to correlate with tumor invasiveness in hormone-secreting but not nonfunctioning adenomas. The exact roles and potential interrelationships between hst, PTTG and other genes and their products in the pathogenesis of prolactinomas and/or the stimulation of their growth still remain unclear. Further studies may shed additional light on prolactinoma pathogenesis and may provide targets for novel medical therapies.

Data from series of patients with microadenomas being observed for long periods without treatment show that the risk of progression from microadenoma to macroadenoma is only 7% (16, 32-37). In these series, patients with radiologically documented prolactinomas who refused surgery and/or medical treatment for a period of up to 8 years were followed. Because of improvements in techniques, the true rates of progression from sellas with minimal abnormalities to documentation of the presence of a tumor cannot be determined accurately and it is likely that the numbers given above are overestimates rather than underestimates.

Although the high estrogen milieu of pregnancy may stimulate PRL production, elevate PRL levels (38,39) and prolactinoma growth (Figure 5-1), exogenous estrogens in the form of oral contraceptives or postmenopausal hormone replacement therapy have not been demonstrated to be responsible for prolactinoma development (40,41).

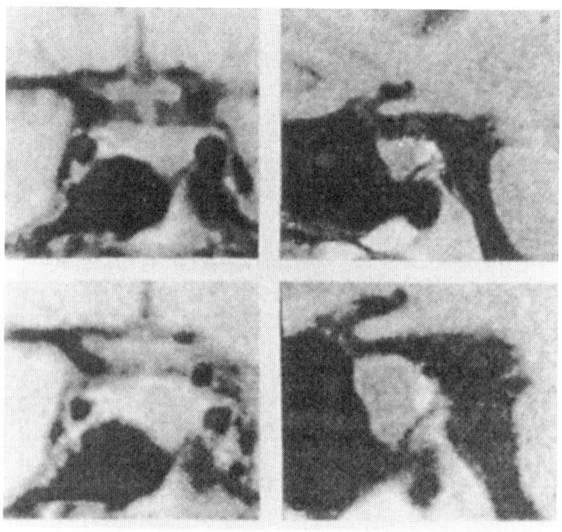

Figure 5-1. Coronal and sagittal MRI scans of an intrasellar prolactin secreting macroadenoma in a woman prior to conception (above) and at 7 months of gestation (below). Note the marked tumor enlargement at the latter point, at which time the patient was complaining of headaches.

SIGNS AND SYMPTOMS OF HYPERPROLACTINEMIA

WOMEN

Milk expressible from one or both breasts more than 1 year after normal delivery and cessation of breast-feeding or its occurrence in the absence of pregnancy generally is taken as a definition of inappropriate lactation or galactorrhea. Data from published series suggest that about one-quarter of women with galactorrhea but normal menses have elevated PRL levels (42). More recent experience, however, suggests that galactorrhea may be found on careful examination in about 5 - 10 % of normally menstruating women and basal PRL levels are normal in more than 90 % of these women.

When amenorrhea or oligo/amenorrhea is associated with galactorrhea, 75% of such women are found to have hyperprolactinemia (42). Although the amenorrhea caused by hyperprolactinemia usually is secondary, it also can be primary if the disorder begins before the usual age of puberty. Of women presenting with primary amenorrhea and low gonadotropin levels, 10 to 30% are generally found to have hyperprolactinemia (43,44). Reduced libido and orgasmic dysfunction, found in most hyperprolactinemic amenorrheic women, can be corrected by normalization of PRL levels. (44).

Infertility also may be a presenting symptom of patients with hyperprolactinemia and is invariable when gonadotropin levels are suppressed with anovulation. In women studied for infertility, about one-third are found to have hyperprolactinemia (45,46). Most of the women had amenorrhea and galactorrhea as well, but in one series of 113 cases of infertility, 5 of the 22 hyperprolactinemic women had neither amenorrhea nor galactorrhea (46). That PRL excess may be important in this type of patient is suggested by the finding that treatment of similar patients with bromocriptine restored fertility (46).

Hyperprolactinemic women have decreased bone mineral density (47-49), but whether this effect is mediated by estrogen deficiency (47) or is a direct effect of the hyperprolactinemia (48) has been controversial. Correction of the hyperprolactinemia results in an increase in bone mass in most but not all patients (47,48). Studies of hyperprolactinemic women who were not amenorrheic and hypoestrogenemic have shown that their bone mineral density is normal (47,49), confirming the initial hypothesis that it is the estrogen deficiency that mediates the bone mineral loss.

The primary mechanism by which hyperprolactinemia interferes with reproductive/ sexual function is through inhibition of pulsatile gonadotropin secretion. Direct measurement of portal vessel GnRH levels in rats showed a marked inhibitory effect of hyperprolactinemia in some studies (50) and in humans the pulsatile release of gonadotropins is inhibited (51). In addition to these effects, hyperprolactinemia in women has been associated with loss of positive estrogen feedback on gonadotropin secretion (52).

MEN

Chronic hyperprolactinemia in males results in impotence and decreased libido in over 90% of cases (53). Galactorrhea in men has been reported in 10 - 20% of cases and is virtually pathognomonic of a prolactinoma (53). In hyperprolactinemic men there is also a decrease in the pulsatile secretion of LH and FSH and testosterone levels are low or are in the lower part of the normal range (53). The testosterone response to stimulation with hCG has been reported to be both decreased (54) and normal (55); in those with decreased responses there is improvement in the response when PRL levels are lowered with bromocriptine (53). If there is sufficient normal pituitary tissue, reduction of elevated PRL levels to normal usually results in a return of normal testosterone levels (53,56,57). Testosterone therapy of hyperprolactinemic men does not always correct the impotence until PRL levels are brought down to normal (53). Sperm counts and motility are decreased with an increase in abnormal forms (55) and histological studies reveal abnormal seminiferous tubule walls and altered Sertoli cell ultrastructure (58). The semen analysis does not always return to normal despite a return to normal of testosterone levels with restoration of normoprolactinemia (56). Between 5 and 25% of males with impotence have been found to be hyperprolactinemic in various series (59,60) but only 1 - 5% of men with infertility have been found to be hyperprolactinemic (61).

DIAGNOSIS OF PROLACTINOMAS

As mentioned above, PRL is secreted episodically and some PRL levels during the day may be above the upper limit of normal established for a given laboratory. Thus the finding of minimally elevated levels in blood requires confirmation in several samples. A careful history and physical examination, screening blood chemistries, thyroid function tests and a pregnancy test will exclude virtually all causes except for hypothalamic-pituitary disease (62,63).

When there is no obvious cause of the hyperprolactinemia from the routine screening, a radiologic evaluation of the hypothalamic-pituitary area is mandatory to exclude a mass lesion. This includes patients with even mild PRL elevations. Magnetic resonance imaging (MRI) with gadolinium enhancement generally provides the best anatomic detail but computed tomography (CT) with intravenous contrast enhancement will suffice to exclude a large mass in the sellar area (64). It should be emphasized here that it is very important to distinguish between a large nonsecreting tumor causing modest PRL elevations (usually < 150 ng/ml, see above) from a PRL secreting macroadenoma (PRL levels usually >>250 ng/ml), as the therapy may be quite different. One potential additional problem in investigating patients with mild hyperprolactinemia is the finding of a false positive CT or MRI scan. Because these techniques are now able to pick incidental nonsecreting tumors, cysts, infarcts, etc., the finding of a "microprolactinoma" on scan in a patient with elevated PRL levels may not always be a true positive finding (65). When there is no obvious cause for the hyperprolactinemia and the MRI or CT scan is normal, the patient is deemed to have idiopathic hyperprolactinemia.

TREATMENT OF PROLACTINOMAS

OBSERVATION

The indications for therapy in patients with prolactinomas may be divided into two categories: effects of tumor size and effects of hyperprolactinemia. In about 93% of patients, microprolactinomas do not enlarge over a 4 - 6 year period of observation. It is very unlikely for a prolactinoma to grow significantly with no increase in serum PRL levels. Therefore, most patients with microadenomas can just be followed with serial PRL levels. If PRL levels rise, then repeat scanning is indicated. Certainly a microadenoma that is documented to be growing demands therapy for the size change alone, as it may be one of the 7% that will grow to be a macroadenoma. Patients with macroadenomas have already indicated a propensity to grow and therefore most should be treated. Local or diffuse invasion or compression of adjacent structures, such as the stalk or optic chiasm, are additional indications for therapy.

Other indications for therapy are relative, being due to the hyperprolactinemia itself. These include decreased libido, sexual dysfunction, menstrual dysfunction, galactorrhea, infertility, hirsutism, and premature osteoporosis (66). In a woman with a microadenoma with normal menses and libido and galactorrhea that is not bothersome, there is no specific reason for therapy. Without therapy PRL levels may return to normal in about one-third of patients (37). The ability to follow a patient closely with PRL levels, CT or MRI scans, and estimations of bone mineral density and rather precise estimates of the efficacy of various modes of therapy (see below) allow a highly individualized way of following patients and choosing the proper timing and mode of therapy.

SURGERY

Transsphenoidal surgery is the surgical procedure used for microadenomas and most macroadenomas. Rarely is craniotomy performed. A summary (67) of surgical results from 34 published series shows that 973/1321 (73.7%) microadenomas and 415/1279 (32.4%) macroadenomas were reported as being curatively resected, i.e. having PRL levels normalized by 1-12 weeks following surgery.

When PRL levels return to normal there is almost a 100% return to normal of gonadal function in both sexes (67). Often normal reproductive function is obtained even with PRL levels slightly above normal. Patients with macroadenomas may be hypopituitary before surgery and may have significant changes in pituitary function postoperatively. In an analysis of 84 patients with macroadenomas (36 were prolactinomas), Nelson et al (68) found that of those with normal preoperative pituitary function, only 78% retained normal function postoperatively; one-third with some pituitary deficits prior to surgery improved these deficits and one-third with such deficits worsened.

Postoperative recurrence of hyperprolactinemia often occurs within the first year following surgery with recurrence rates varying from 0 (32) to 50% (67). From the series compiled above, recurrence rates for microadenomas microadenomas (114/544 = 21.0%) and macroadenomas (50/253 = 19.8%) are similar (67). In a series of 64 women operated upon at the Notre Dame Hospital in Montreal, 58 (90%) of women with microadenomas had normal postoperative PRL levels but after 10-20 years, 25 (43%) had a relapse of hyperprolactinemia (69). Of these 25, 15 were symptomatic from their hyperprolactinemia with amenorrhea and/or galactorrhea; CT scans were performed in 12 of these 15 and 2 were found to have anatomic evidence of tumor recurrence at 5 and 9 years postoperatively while the other 10 had normal scans (69). Therefore, the ultimate, long-term surgical cure rate , using a normal PRL level as the criterion, is about 50-60%for microadenomas and about 25% for macroadenomas. For patients with giant prolactinomas and those with considerable cavernous sinus invasion, the chance for surgical cure is essentially zero.

Complications from transsphenoidal surgery for microadenomas are quite infrequent, the mortality rate being at most 0.27%, the major morbidity rate being about 0.4% (70,71). The mortality rate for transsphenoidal surgery for all types of secreting and nonsecreting macroadenomas is 0.9%, the major morbidity rate being 6-20%, depending upon the experience of the surgeon (70,71). Transient diabetes insipidus (DI) is quite common with transsphenoidal surgery for both micro- and macroadenomas and permanent DI occurs in about 1% of surgeries on macroadenomas (70,71). Surgery involving craniotomy is much more hazardous. Patients with "giant" prolactinomas, i.e. those greater than 4 cm in diameter, are rarely cured by surgery and often end with up with major complications of the procedure, including hemorrhage, tumor swelling with increased intracranial pressure and death (72).

MEDICAL THERAPY

In several large, early studies in the literature (reviewed in ref 42) totaling more than 400 hyperprolactinemic patients treated with bromocriptine (Parlodel®), normoprolactinemia or return of ovulatory menses occurred in 80 - 90% of patients. When both PRL levels and return of menses were studied in the same patients, it was found that substantial reductions in PRL levels to still slightly elevated levels often was enough to restore ovulation and menses despite the fact that normal PRL levels were achieved in only 70 - 80% of treated patients.

There is little intraindividual variability in the absorption and peak blood levels of bromocriptine achieved but there is considerable interindividual variability (73). There is also considerable variability in the PRL lowering effects of a given dose of bromocriptine that does not correlate with serum bromocriptine levels, implying differences in sensitivity to the drug (74). Decreased response to bromocriptine *in vivo* has been shown to correspond to decreased numbers of DA receptors on lactotroph cell membranes (75). Because resistance is associated with a decrease in the relative proportion of the short receptor isoform, it has been postulated that resistance may not only involve defects in receptor expression but also in posttranscriptional splicing (76).

Early *in vitro* studies showed that bromocriptine not only decreases PRL synthesis but also DNA synthesis, cell multiplication and tumor growth (77). Following several early individual case reports, a number of series of patients have been reported showing tumor size reduction in response to bromocriptine. Analysis of data summarized from 24 different series of patients, totaling 302 patients with macroadenomas with macroadenomas (67) shows that 77% had some tumor size decrease with periods of observation ranging from 6 weeks to over 10 years (Table 5-2). Semiquantitative analysis of 112

Table 5-2. Effect of Bromocriptine on Pregnancies*

	Bromocriptine		Normal Population
	n	%	%
Pregnancies	6,239	100.0	100.0
- spontaneous abortion	620	9.9	10 - 15
- terminations	75	1.25	
- ectopic	31	0.5	0.5 - 1.0
- hydatidiform moles	11	0.2	0.05 - 0.7
Deliveries (known duration)	4,139	100.0	100.0
- at term (> 38 weeks)	3,620	87.5	85
- preterm (< 38 weeks)	519	12.5	15
Deliveries (known outcome)	5,120	100.0	100.0
- single births	5,031	9.3	8.7
- multiple births	89	1.7	1.3
Babies (known details)	5,213	100.0	100.0
- normal	5,030	96.5	95.0
- with malformations	3	1.8	3 - 4
- with perinatal disorders	90	1.7	>2

*Data from Krupp P, Monka C, Richter K. Program of the Second World Congress of Gynecology and Obstetrics. Rio de Janeiro, 1988, p. 9.

patients in seven series (67) showed that 45 (40.2%) had a > 50% reduction in tumor size, 32 (28.6%) had a 25 - 50% reduction in tumor size, 14 (12.5%) had a <25% reduction, and 21 (18.7%) had no evidence of any reduction in tumor size (Table 5-2).

The time course of tumor size reduction is variable. Some patients may experience extremely rapid decrease in tumor size, significant changes in visual fields being noted with 24 - 72 hours and significant changes noted on scan with 2 weeks (78). In others, little change may be noted at 6 weeks but scanning again at 6 months may show significant changes (57). A progressive decrease is often noted over several years.

Visual field improvement occurs in 80 - 90% of patients with significant visual field abnormalities with bromocriptine treatment (57). Visual field improvement generally parallels and often precedes the changes seen on scan (57). It is often difficult to determine before treatment whether visual defects are temporary or permanent and only the response to therapy provides a final answer. These studies with medical therapy are reassuring in that a relatively slow chiasmal decompression over several weeks provides excellent restoration of visual fields and that immediate surgical decompression is not necessary. Usually, when there is no significant change in visual fields despite significant evidence of tumor reduction on scan, subsequent surgery also does not improve these fields.

The extent of tumor size reduction does not correlate with basal PRL levels, nadir PRL levels achieved, the percent fall in PRL or whether PRL levels reached normal. Some patients have excellent reduction of PRL levels to normal but only modest changes in tumor size, while others have persistent hyperprolactinemia (although > 75% suppression from basal values) with almost complete disappearance of tumor. A reduction in PRL levels always precedes any detectable change in tumor size and PRL nonresponders are also tumor size nonresponders (57). Once maximum size reduction is achieved, the dose of bromocriptine can often be substantially reduced, gradually (78).

The reduction in tumor size is usually not only accompanied by improved visual fields and reduction of hyperprolactinemia but also by improvement in other pituitary function (57,79). When the prolactinoma is present prepubertally, improved pituitary function allows resumption of normal growth and pubertal development (80).

In some patients, bromocriptine appears to cause perivascular fibrosis to develop in the tumors, resulting in an obliteration of the normal surgical cleavage planes and difficulty in obtaining a surgical cure (81-83). This appears to be more of a problem for macroadenomas than microadenomas (81-83). For macroadenomas, there may be a problem with bromocriptine treatment longer than 6 - 12 weeks, at which point fibrosis may become limiting to complete tumor removal (81,83).

An alternative to late surgical treatment is continued bromocriptine. Prolonged bromocriptine treatment for up to 10 years appears to be well-tolerated (84,85) and the dose can often be reduced considerably (86). Although some tumors expand rapidly when bromocriptine is discontinued in a patient with a macroadenoma that has become reduced in size (87), this is not usually the case with long term therapy. About 10 - 20% of patients can maintain normal PRL levels after stopping treatment and 70 - 80% with marked tumor size reduction may not experience tumor reexpansion with stopping of therapy (88). With patients with very large tumors who have excellent tumor size reduction, stopping therapy must be done very cautiously, if at all. The best approach is probably to reduce the dose gradually, following PRL levels, and only discontinue the drug if there are no increases in PRL levels or tumor size on just 2.5 mg per day.

The most common side effects, occurring in 10 - 15% of patients are nausea and sometimes vomiting; these are usually transient but may recur with each dose increase (67,89). Orthostatic hypotension usually is only a problem when initiating therapy and rarely recurs with dose increases. Limiting nausea and vomiting occurs in 3-5% of patients and digital vasospasm, nasal congestion and depression occur in rare patients when doses less than 7.5 mg/day are used (67). Side effects can be minimized by starting with 1.25 mg daily with a snack at bedtime. The dose is gradually increased to 2.5 mg twice daily with meals over 7 - 10 days and PRL levels checked after 1 - 2 months. Most patients respond within 1 - 2 months if they are going to respond. Dose higher than 7.5 mg per day are usually not necessary except in some patients with very large tumors. Psychotic reactions have been reported in about 1% of patients (90) There are also reports of exacerbation of preexisting schizophrenia and the drug should be given cautiously to such patients (90). It should be noted that phenothiazines given to such patients may blunt the action of bromocriptine on prolactinomas as well (57).

One concerning problem is the tumor that initially shrinks in response to bromocriptine and then enlarges. This is usually due to noncompliance, which is further worsened by the tendency for the patient and physician to resume the full dose instead of gradually restarting. This tends to make side effects worse, further exacerbating the noncompliance. Although extremely rare, tumors that continue to enlarge while being treated with bromocriptine may turn out to be carcinomas.

Bromocriptine has successfully been given intravaginally to reduce nausea. Similar reductions of PRL levels are achieved with oral and intravaginal administration of oral bromocriptine tablets (91,92). However, the drug effect lasts for up to 24 hours with a single dose and gastrointestinal side effects are much less with the intravaginal route, although some patients experience vaginal burning (93). Katz et al (93) reported a woman intolerant of oral bromocriptine with a macroadenoma who responded well with tumor shrinkage to intravaginal bromocriptine.

Another dopamine agonist that went through early trials demonstrating efficacy in the treatment of prolactinomas is pergolide (Permax®), which is approved by the U.S. Food and Drug Administration for the treatment of Parkinson's disease but not for hyperprolactinemia. However, there is considerable experience with its use in prolactinoma patients and several studies have shown comparability to bromocriptine with respect to tolerance and efficacy, including tumor size reduction (94-97). Experience has shown that some patients who do not respond to bromocriptine do so to pergolide and vice-versa. In these series, only 39 patients had sufficient data to quantitate tumor size reduction. Of these, 29 (74%) had > 50% reduction, 4 (10%) had 25-50% reduction, 2 (5%) had < 25% reduction and 4 (10%) had no change in tumor size.

Cabergoline (Dostinex®) is different from the other dopamine agonists in that it has a very long half-life and can be given orally once or twice weekly. The long duration of action stems from its slow elimination from pituitary tissue (98), its high affinity binding to pituitary dopamine receptors (99), and extensive enterohepatic recycling (100). After oral administration, PRL lowering effects are initially detectable at 3 h and gradually increase so that there is a plateau of effect between 48 and 120 hours (100) and with weekly doses there is a sustained reduction of PRL (89).

A number of studies have now shown that cabergoline is at least as effective as and perhaps more effective than bromocriptine in lowering PRL levels but with substantially fewer side effects (89). In a prospective, double-blind comparison study of 459 women

(279 microadenomas, 3 macroadenomas, 167 idiopathic hyperprolactinemia, 10 other), of women treated with cabergoline, 83% achieved normoprolactinemia, 72% achieved ovulatory cycles and 3% discontinued the medication because of adverse effects while of women treated with bromocriptine, 59% achieved normoprolactinemia, 52% achieved ovulatory cycles, and 12% stopped the drug because of adverse effects (101). Rare patients experience limiting nausea and vomiting with cabergoline, and they may be treated with intravaginal cabergoline as well (102).

Several studies have assessed the effect of cabergoline on macroadenoma size (103-113). A total of 320 patients in these series had their tumor size assessed in a semiquantitative way in studies ranging from 3 to 24 months duration of treatment. Of these 320, 91 (28%) experienced a > 50% tumor size reduction, 91 (28%) had a 25-50% reduction, 47 (15%) had a < 25% reduction, and 91 (28%) had no change in tumor size. In many of these series many of the patients had been previously treated with other dopamine agonists, some being in intolerant and others resistant, and that may color the findings. In the study of Colao et al (114), only 8 of the 23 macroadenoma patients had received short courses of bromocriptine previously and were only intolerant and not resistant to bromocriptine; of these 23, 12 (52%) had a < 50% reduction in tumor size, 9 (39%) had a 25-50% reduction in tumor size, and 2 (9%) had a < 25% reduction in tumor size. In a recent series of 27 patients who had all been previously shown to be resistant to bromocriptine or quinagolide (CV205-502), Colao et al (114) showed that cabergoline was able to normalize PRL levels in 15 of 19 patients with macroadenomas and all 8 patients with microadenomas; Tumor shrinkage was documented in 9 of the 19 macroadenomas and 4 of the 8 microadenomas (114).

PREGNANCY IN WOMEN WITH PROLACTINOMAS

EFFECTS OF DOPAMINE AGONISTS ON THE DEVELOPING FETUS

As a general principal, it is advised that fetal exposure to dopamine agonists be limited to as short a period as possible. Mechanical contraception should be used until the first two to three cycles have occurred, so that an intermenstrual interval can be established and a woman will know when she has missed a menstrual period. Thus, the dopamine agonist can be stopped after being given for only about three to four weeks of the gestation. When used in this fashion, bromocriptine has not been found to cause any increase in spontaneous abortions, ectopic pregnancies, trophoblastic disease, multiple pregnancies or congenital malformations (115). Long-term followup studies of 64 children between the ages of 6 months and 9 years whose mothers took bromocriptine in this fashion have shown no ill effects (116). Experience is limited to only just over 100 women, however, with the use of bromocriptine throughout the gestation but no abnormalities were noted in the infants except one with an undescended testicle and one with a talipes deformity (115,117-119). Few data are available on the safety during pregnancy of pergolide. Outcome data available on 265 pregnancies in which cabergoline was administered to facilitate ovulation do not show increased risks of preterm, ectopic, or multiple birth deliveries or malformations (120,121). However, this data is relatively sparse compared to the data in over 6000 pregnancies with bromocriptine, so that at the moment bromocriptine is recommended when fertility is the major reason for treatment. On the other hand, the data for cabergoline is encouraging so that the mother may be reassured if she gets pregnant while taking cabergoline and over the next few years we will likely gain a sufficient amount of data so as to be able to recommend cabergoline for fertility without reservation.

EFFECT OF PREGNANCY ON PROLACTINOMA SIZE

Estrogens have a marked stimulatory effect on PRL synthesis and secretion and the hormonal milieu of pregnancy can stimulate lactotroph cell hyperplasia (38,39). In women with prolactinomas, this can result in tumor growth (Figure 5-1). In a review of 19 series of women with prolactinomas who became pregnant (122), it was found that only 5 of the 376 women (1.3%) with microadenomas had symptoms of tumor enlargement (headaches or visual disturbances or both). In no case was surgical intervention necessary but bromocriptine was reintroduced in one case due to symptoms. These series included 86 patients with macroadenomas who had not had prior surgery or irradiation. Of these 20 (23.3%) had symptomatic tumor enlargement requiring either surgery (in early series) or reinstitution of bromocriptine in many. In these series, 71 women with macroadenomas had been treated with irradiation or surgery before pregnancy and only 2 (2.8%) had symptomatic tumor enlargement. Bromocriptine can be used successfully during pregnancy to reduce symptomatic tumor enlargement in most cases (122). No ill effects on the infant have been observed in such cases. The use of prophylactic bromocriptine throughout the pregnancy likely prevents tumor regrowth during the pregnancy in most cases but no formal studies have been carried out.

RECOMMENDATIONS FOR MANAGEMENT

For the hyperprolactinemic woman with a microadenoma or a macroadenoma that is intrasellar or extends infrasellarly who wishes to become pregnant, bromocriptine is preferred as the primary treatment for such patients because of its efficacy in restoring ovulation and very low (1%) risk of clinically serious tumor enlargement. Such a patient should be carefully followed throughout gestation. PRL levels do not always rise during pregnancy in women with prolactinomas, as they do in normal women. Usually PRL levels rise over the first 6 - 10 weeks after stopping bromocriptine and then do not increase further (123). PRL levels may also not rise with tumor enlargement (124). Therefore, periodic checking of PRL levels is of no benefit. Because of the low incidence of tumor enlargement, routine, periodic visual field testing is not cost effective. Visual field testing and scanning are performed only in patients who become symptomatic. Scans should be repeated routinely after delivery to detect any asymptomatic tumor growth. Clearly in the patient who does not respond to bromocriptine, surgery is the treatment of choice.

In a woman with a larger macroadenoma that may have suprasellar extension, there is a 23% risk of clinically serious tumor enlargement during pregnancy when only bromocriptine is used to allow ovulation and then stopped when pregnancy is diagnosed. There is no clear-cut answer as to the best therapeutic approach and this has to be a highly individualized decision that the patient has to make after a clear, documented discussion of the various therapeutic alternatives. One approach is just to use bromocriptine to allow ovulation, discontinue it when pregnancy is documented and then observe the patient carefully for evidence of tumor growth. Prepregnancy transsphenoidal surgical debulking of the tumor greatly reduces but does not eliminate the risk of serious tumor enlargement. After surgical debulking, bromocriptine is required to restore normal PRL levels and allow ovulation. A third approach, that of giving bromocriptine continuously throughout gestation, has been advocated (119). At this point, however, data regarding the effects of continuous bromocriptine therapy on the developing fetus are still quite meager, and such therapy cannot be recommended without reservation. Should pregnancy at an advanced stage be discovered in a woman taking bromocriptine, the data that exist are reassuring and would not justify therapeutic abortion.

For patients with macroadenomas treated with bromocriptine alone or after surgery, careful follow-up with monthly visual field testing is warranted. Repeat scanning is reserved for patients with symptoms of tumor enlargement or evidence of a developing visual field defect or both, and should also be done routinely after delivery to detect asymptomatic tumor enlargement. Should symptomatic tumor enlargement occur with any of these approaches, reinstitution of bromocriptine is probably less harmful to the mother and child than surgery. Any type of surgery during pregnancy results in a 1.5-fold increase in fetal loss in the first trimester and a 5 fold increase in fetal loss in the 2nd trimester, although there is no risk of congenital malformations from such surgery (125). Thus, bromocriptine reinstitution would appear to be preferable to surgical decompression. However, such medical therapy must be very closely monitored, and transsphenoidal surgery or delivery (if the pregnancy is far enough advanced) should be performed is there is no response to bromocriptine and vision is progressively worsening.

CONCLUSIONS

Prolactinomas are a common cause of reproductive/sexual dysfunction. Once other causes of hyperprolactinemia have been excluded with a careful history and physical examination, routine chemistries and a TSH, imaging with MRI or CT will delineate the size and extent of the tumor. Medical therapy is the initial treatment of choice. When infertility is the primary indication for treatment, bromocriptine use has an extensive safety experience and is preferred. However, for other indications, cabergoline appears to be more efficacious and better tolerated. Transsphenoidal surgery remains an option, especially for patients with microadenomas, when medical therapy is ineffective.

REFERENCES

1. Molitch ME: Prolactin. In S. Melmed, ed.: **The Pituitary**. Blackwell Scientific Publications, Boston, In Press.
2. Davies PH: Drug-related hyperprolactinaemia. Adverse Drug React Toxicol Rev 1997;16:83-94.
3. Steiner J, Cassar J, Mashiter K, Dawes I, Fraser TR, Breckenridge A: Effect of methyl-dopa on prolactin and growth hormone. Brit Med J 1976;1:1186-1188.
4. Kelley SR, Kamal TJ, Molitch ME: Mechanism of verapamil calcium channel blockade-induced hyperprolactinemia. Am J Physiol 270 (Endocrinol Metab 33):E96-E100), 1996.
5. Sievertsen GD, Lim VS, Nakawatase C, & Frohman LA. Metabolic clearance and secretion rates of human prolactin in normal subjects and in patients with chronic renal failure. J Clin Endocrinol Metab 1980;50:846-851.
6. Hou SH, Grossman S, Molitch ME. Hyperprolactinemia in patients with renal insufficiency and chronic renal failure requiring hemodialysis or chronic ambulatory peritoneal dialysis. Am J Kidney Dis 1985 6:245-249.
7. Morgan MY, Jakobovits AW, Gore MB et al Serum prolactin in liver disease and its relationship to gynecomastia. 1978 Gut, 19:170-174.
8. Honbo KS, Herle AJV, Kellett KA. Serum prolactin levels in untreated primary hypothyroidism. Am J Med 1978;64:782-787.
9. Stryker TD, Molitch ME. Reversible hyperthyrotropinemia, hyperthyroxinemia and hyperprolactinemia due to adrenal insufficiency. Am J Med 1985;79:271-276.
10. Boyd AE III, Spare S, Bower B, Reichlin S. Neurogenic galactorrhea-amenorrhea. J Clin Endocrinol Metab 1978 47:1374-1377.

11. Palmer PE, Bogojavlensky S, Bhan AK, Scully RE. Prolactinoma in wall of overian dermoid cyst with hyperprolactinemia. Obstet Gynecol 1990 75:540-543.
12. Molitch ME, Reichlin S. Hypothalamic hyperprolactinemia: neuroendocrine regulation of prolactin secretion in patients with lesions of the hypothalamus and pituitary stalk. Prolactin. Basic and clinical correlates. R.M. MacLeod, M.O. Thorner and U. Scapagnini(eds.) Liviana Press, Padova (1985) pp. 709-719.
13. Arafah MB, Neki KE, Gold RS, Selman WR: Dynamics of prolactin secretion in patients with hypopituitarism and pituitary macroadenomas. J Clin Endocrinol Metab 1995;80:3507-3512.
14. Bevan JS, Burke CW, Esiri MM, Adams CBT. Misinterpretation of prolactin levels leading to management errors in patients with sellar enlargement. Am J Med 1987 82:29-32.
15. Martin TL, Kim M, Malarkey WB. The natural history of idiopathic hyperprolactinemia. J Clin Endocrinol Metab 1985;60:855-858.
16. Schlechte J, Dolan K, Sherman B, et al: The natural history of untreated hyperprolactinemia: a prospective analysis. J Clin Endocrinol Metab 1989;68:412-418.
17. Sluijmer AV, Lappöhn RE. Clinical history and outcome of 59 patients with idiopathic hyperprolactinemia. Fertil Steril 1992;58:72-77.
18. Klijn JGM, Lamberts SWJ, De Jong FH, Docter R, Van Dongen KJ, Birkenhäger JC. The importance of pituitary tumour size in patients with hyperprolactinaemia in relation to hormonal variables and extrasellar extension of tumour. Clin Endocrinol 1980;12:341-355.
19. Pernicone PJ, Scheithauer BW, Sebo TJ et al: Pituitary carcinoma. Cancer 1997;79:804-812.
20. Fine SA, Frohman LA. Loss of central nervous system component of dopaminergic inhibition of prolactin secretion in patients with prolactin-secreting pituitary tumors. J Clin Invest 1978; 61:973-980.
21. Molitch ME: Pathogenesis of pituitary tumors. Endocrinol Metab Clin N Amer 1987;16: 503-527,1987.
22. Klibanski A: Further evidence for a somatic mutation theory in the pathogenesis of human pituitary tumors. J Clin Endocrinol Metab 1990;71:1415.
23. Melmed S: Pathogenesis of pituitary tumors. Endocrinol Metab Clin N Amer 1999;28:1-12.
24. Clayton RN, Boggild M, Bates AS, Bicknell J, Simpson D, Farell W: Tumour suppressor genes in the pathogenesis of human pituitary tumours. Horm Res 1997;47:185-193.
25. Barlier A, Pellegrini-Bouiller I, Caccavelli L et al: Abnormal transduction mechanisms in pituitary adenomas. Horm Res 1997;47:227-234.
26. Pelligrini-Bouiller I, Morange-Ramos A, Barlier A, Gunz G, Enjalbert A, Jaquet P: Pit-1 gene expression in human pituitary adenomas. Horm Res 1997;47:251-258.
27. Gonsky R, Herman V, Melmed S, Fagin J: Transforming DNA sequences present in human prolactin-secreting pituitary tumors. Mol Endocrinol 1991;5:1687-1695.
28. Shimon I, Huttner A, Said J, Spirina OM, Melmed S: Heparin-binding secretory transforming gene (hst) facilitates rat lactrotrope cell tumorigenesis and induces prolactin gene transcription. J Clin Invest 1996;97:187-195.
29. Pei L, Melmed S: Isolation and characterization of a pituitary tumor-specific transforming gene. Mol Endocrinol 1997;11:443-441.
30. Zhang X, Horwitz GA, Prezant TR et al: Structure, expression, and function of human pituitary tumor-transforming gene (PTTG). Mol Endocrinol 1999;13:156-166.
31. Zhang X, Horwitz GA, Heaney AP et al: Pituitary tumor transforming gene (PTTG) expression in pituitary adenomas. J Clin Endocrinol Metab 1999;84:761-767.
32. Von Werder K, Eversmann T, Fahlbusch R, Rjosk H-K. Development of hyperprolactinemia in patients with adenomas with and without prior operative treatment. Excerpta Med Int Congr Ser 1982;584:175-188.

33. March CM, Kletzky OA, Davajan V, *et al.* Longitudinal evaluation of patients with untreated prolactin-secreting pituitary adenomas. Am J Obstet Gynecol 1981;139: 835-844.
34. Weiss MH, Teal J, Gott P, *et al.* Natural history of microprolactinomas: six-year follow-up. Neurosurgery 1983; 12:180-183.
35. Koppelman MCS, Jaffe MJ, Rieth KG, Caruso RC, Loriaux DL. Hyperprolactinemia, amenorrhea, and galactorrhea. Ann Intern Med 1984;100:115-121.
36. Sisam DA, Sheehan JP, Sheeler LR. The natural history of untreated microprolactinomas. Fertil Steril 1987;48:67-71.
37. Jeffcoate WJ, Pound N, Sturrock NDC, Lambourne J: Long-term follow-up of patients with hyperprolactinaemia. Clin Endocrinol 1996;45:299-303.
38. Rigg LA, Lein A, Yen SSC. Pattern of increase in circulating prolactin levels during human gestation. Am J Obstet Gynecol 1977;129:454-456.
39. Scheithauer BW, Sano T, Kovacs KT, Young WF Jr, Ryan N, Randall RV. The pituitary gland in pregnancy. A clinicopathologic and immunohistochemical study of 69 Cases. Mayo Clin Proc 1990;65:461-474.
40. Molitch ME: Clinical features and epidemiology of prolactinomas in women. In *Prolactinomas: Practical Diagnosis and Management,* JM Olefsky and RJ Robbins (eds.), Churchill Livingstone, Inc., New York, 67- 95, 1986.
41. Wingrave SJ, Kay CR, Vessey MP: Oral contraceptives and pituitary adenomas. Br Med J 1980;1:685-686.
42. Molitch ME, Reichlin S. Hyperprolactinemic disorders. Disease-a-Month 1982;28(9):1-58.
43. Mashchak CA, Kletzky OA, Davajan V, Mishell DR. Clinical and laboratory evaluation of patients with primary amenorrhea. Obstet Gynecol 1981;57:715-721.
44. Pepperell RJ, Aust F. Prolactin and reproduction. Fertil Steril 1981 35:267-274.
77. Post KD, Biller BJ, Adelman LS, Molitch ME, Wolpert SM, Reichlin S. Selective transsphenoidal adenomectomy in women with galactorrhea-amenorrhea. JAMA 1979;242:158-162.
45. Skrabanek P, McDonald D, De Valera E, Lanigan O, Powell D. Plasma prolactin in amenorrhoea, infertility, and other disorders: A retrospective study of 608 patients. Irish J Med Sci 1980;149:236-245.
46. Kredentser JV, Hoskins CF, Scott JZ. Hyperprolactinemia-a significant factor in female infertility. Am J Obstet Gynecol 1981;139:264-267.
47. Biller BMK, Baum HBA, Rosenthal DI, Saxe VC, Charpie PM, Klibanski A. Progressive trabecular osteopenia in women with hyperprolactinemic amenorrhea. J Clin Endocrinol Metab 1992;75:692-697.
48. Schlechte J, Walkner L, Kathol M. A longitudinal analysis of premenopausal bone loss in healthy women and women with hyperprolactinemia. J Clin Endocrinol Metab 1992;75:698-703.
49. Ciccarelli E, Savino L, Carlevatto V, Bertagna A, Isaia GC, Camanni F. Vertebral bone density in non-amenorrhoeic hyperprolactinaemic women. Clin Endocrinol 1988;28:1-6.
50. Sarkar DK, Yen SSC. Hyperprolactinemia decreases the luteinizing hormone-releasing hormone concentration in pituitary portal plasma: A possible role ß-endorphin as a mediator. Endocrinology 1985 116:2080-2084.
51. Stevenaert A, Beckers A, Vandalem JL, Hennen G: Early normalization of luteinizing hormone pulsatility after successful transsphenoidal surgery in women with microprolactinomas. J Clin Endocrinol Metab 1986;62:1044-1047.
52. Glass MR, Shaw RW, Butt WR, Edwards RL, London DR. An abnormality of oestrogen feedback in amenorrhoea-galactorrhoea. British Medical J 1975 3:274-275.
53. Perryman RL and Thorner MO. The effects of hyperprolactinemia on sexual and reproductive function in men. J Androl 1981;5:233-242.

54. Heshmati HM, Turpin G, Nahoul K, *et al.* Testicular response to human chorionic gonadotrophin in chronic hyperprolactinaemia. Acta Endocrinol 1985;108:565-569.
55. Luboshitzky R, Rosen E, Trestian S, Spitz M. Hyperprolactinaemia and hypogonadism in men: Response to exogenous gonadotrophins. Clin Endocrinol 1979;11:217-223.
56. Murray FT, Cameron DF Ketchum C. Return of gonadal function in men with prolactin-secreting pituitary tumors. J Clin Endocrinol Metab 1984;59:79-85.
57. Molitch ME, Elton RL, Blackwell RE *et al.* Bromocriptine as primary therapy for prolactin-secreting macroadenomas: results of a prospective multicenter study. J Clin Endocrinol Metab 1985;60:698-705.
58. Cameron DF, Murray FT and Drylie DD. Ultrastructural lesions in testes from hyperprolactinemic men. J Androl 1984;5:283-293.
59. Schwartz MF, Bauman JE and Masters WH. Hyperprolactinemia and sexual disorders in men. Biol Psych 1982;17:861-876.
60. Leonard MP, Nickel CJ, Morales A. Hyperprolactinemia and impotence: Why, when and how to investigate. J Urol 1989;142:992-994.
61. Hargreave TB, Richmond JD, Liakatas J, Elton RA, Brown NS. Searching for the infertile man with hyperprolactinemia. Fertil Steril 1981;36(5):630-632.
62. Biller BMK, Luciano A, Crosignani PG, Molitch M, Olive D, Rebar R, Sanfilippo J, Webster J, Zacur H: Guidelines for the diagnosis and treatment of hyperprolactinemia. J Reprod Med 1999;44(Suppl.):1075-1084.
63. Biller BMK: Diagnostic evaluation of hyperprolactinemia. J Reprod Med 1999;44(Suppl.):1095-1099.
64. Naidich MJ, Russell EJ: Current approaches to imaging of the sellar region and pituitary. Endocrinol Metab Clin N Amer 1999;28:45-79.
65. Molitch ME: Pituitary incidentalomas. Endocrinol Metab Clin N Amer 26:725-740, 1997.
66. Sanfilippo JS: Implications of not treating hyperprolactinemia. J Reprod Med 1999;44(Suppl.):1111-1115.
67. Molitch ME: Prolactinomas. In S. Melmed, ed.: *The Pituitary.* Blackwell Scientific Publications, Boston, In Press.
68. Nelson AT Jr., Tucker HSG Jr., Becker DP. Residual anterior pituitary function following transsphenoidal resection of pituitary macroadenomas. J Neurosurg 1984;61:577-580.
69. Massoud F, Serri O, Hardy J, Somma M, Beauregard H: Transsphenoidal adenomectomy for microprolactinomas: 10 to 20 years of follow-up. Surg Neurol 1996;45:341-346.
70. Ciric I, Ragin A, Baumgartner C et al: Complications of transsphenoidal surgery: results of a national survery, review of the literature, and personal experience. Neurosurgery 1997;40:225-237.
71. Laws ER Jr, Thapar K: Pituitary surgery. Endocrinol Metab Clin N Amer 1999;28:119-131.
72. Molitch ME: Medical treatment of giant pituitary prolactinomas. In: *Controversies in Neurosurgery*, Al-Mefty O (ed.), Theime Medical Publishers, Inc., New York, 2-10, 1996.
73. Schran HR, Bhuta SI, Schwarz HJ, Thorner MO. The pharmacokinetics of bromocriptine in man. In: *Ergot Compounds and Brain Function: Neuroendocrine and Neuropsychiatric Aspects*. Goldstein M, Calne D, Lieberman A, Thorner M. (eds.) Raven Press, NY, 1980, pp. 125-189.
74. Thorner MO, Schran HF, Evans WS, Rogol AD, Morris JL, MacLeod RM. A broad spectrum of prolactin suppression by bromocriptine in hyperprolactinemic women: a study of serum prolactin and bromocriptine levels after acute and chronic administration of bromocriptine. J Clin Endocrinol Metab 1980;50:1026-1033.

75. Pellegrini I, Rasolonjanahary R, Gunz G, *et al.* Resistance to bromocriptine in prolactinomas. J Clin Endocrinol Metab 1989;69:500-509.
76. Caccavelli L, Feron F, Morange I et al: Decreased expression of the two D_2 dopamine receptor isoforms in bromocriptine-resistant prolactinomas. Neuroendocrinology 1994;60:314-322.
77. MacLeod RM, Lehmeyer JE. Suppression of pituitary tumor growth and function by ergot alkaloids. Cancer Res 1973;33:849-855.
78. Thorner MO, Martin WH, Rogol AD, *et al.* Rapid regression of pituitary prolactinomas during bromocriptine treatment. J Clin Endocrinol Metab 1980;51:438-445.
79. Warfield A, Finkel DM, Schatz NJ, Savino PJ, Snyder PJ. Bromocriptine treatment of prolactin-secreting pituitary adenomas may restore pituitary function. Ann Intern Med 1984;101:783-785.
80. Dalzell GW, Atkinson AB, Carson DJ, Sheridan B. Normal growth and pubertal development during bromocriptine treatment for a prolactin-secreting pituitary macroadenoma. Clin Endocrinol 1987;169-172.
81. Landolt AM, Osterwalder V. Perivascular fibrosis in prolactinomas: is it increased by bromocriptine? J Clin Endocrinol Metab 1984;58:1179-1183.
82. Esiri MM, Bevan JS, Burke CW, Adams CBT. Effect of bromocriptine treatment on the fibrous tissue content of prolactin-secreting and nonfunctioning macroadenomas of the pituitary gland. J Clin Endocrinol Metab 1986;63:383-388.
83. Bevan JS, Adams CBT, Burke CW, *et al.* Factors in the outcome of transsphenoidal surgery for prolactinoma and non-functioning pituitary tumour, including pre-operative bromocriptine therapy. Clin Endocrinol 1987;26:541-556.
84. Johnston DG, Prescott RWG, Kendall-Taylor P, *et al.* Hyperprolactinemia: long-term effects of bromocriptine. Am J Med 1983;75:868-874.
85. Ho KY, Smythe GA, Compton PJ, Lazarus L. Long-term bromocriptine therapy may restore the inhibitory control of prolactin release in some patients with pathological hyperprolactinemia. Aust NZ J Med 1985;15:213-219.
86. Liuzzi A, Dallabonzana D, Giuseppe MD, *et al.* Low doses of dopamine agonists in the long-term treatment of macroprolactinomas. N Engl J Med 1985;313:656-659.
87. Thorner MO, Perryman RL, Rogol AD, *et al.* Rapid changes of prolactinoma volume after withdrawal and reinstitution of bromocriptine. J Clin Endocrinol Metab 1981;53:480-483.
88. Johnston DG, Hall K, Kendall-Taylor P, Patrick D, Watson M, Cook DB. Effect of dopamine agonist withdrawal after long-term therapy in prolactinomas. Lancet 1984;2:187-192.
89. Webster J: Dopamine agonist therapy in hyperprolactinemia. J Reprod Med 1999;44(Suppl):1105-1110.
90. Turner TH, Cookson JC, Wass JAH, Drury PL, Price PA, Besser GM. Psychotic reactions during treatment of pituitary tumours with dopamine agonists. Brit Med J 1984;289:1101-1103.
91. Vermesh M, Fossum GT, Kletzky OA. Vaginal bromocriptine: pharmacology and effect on serum prolactin in normal women. Obstet Gynecol 1988;72:693-698.
92. Jasonni VM, Raffelli R, de March A, Frank G, Flamigni C. Vaginal bromocriptine in hyperprolactinemic patients and puerperal women. Acta Obstet Gynecol Scand 1991;70:493-495.
93. Katz E, Schran HF, Adashi EY. Successful treatment of a prolactin-producing pituitary macroadenoma with intravaginal bromocriptine mesylate: a novel approach to intolerance of oral therapy. Obstet Gynecol 1989;73:517-520.
94. Franks S, Lynch SS, Horrocks PM, Butt WR. Treatment of hyperprolactinaemia with pergolide mesylate: acute effects and preliminary evaluation of long-term treatment. Lancet 1981;2:659-661.

95. Blackwell RE, Bradley EL, Kline LB, *et al.* Comparison of dopamine agonists in the treatment of hyperprolactinemic syndromes: a multicenter study. Fertil Steril 1983;39:744-748.
96. Kletzky OA, Borenstein R, Mileikowsky GN. Pergolide and bromocriptine for the treatment of patients with hyperprolactinemia. Am J Obstet Gynecol 1986;154:431-435.
97. Freda PU, Andreadis CI, Khandji AG et al: Long-term treatment of prolactin-secreting macroadenomas with pergolide. J Clin Endocrinol Metab 2000;85:8-13.
98. DiSalle E, Ornati G, Giudici D: A comparison of the *in vivo* and *in vitro* duration of prolactin lowering effect in rats of FCE 21336, pergolide and bromocriptine. J Endocrinol Invest 1984;7 (Suppl1):32.
99. Strolin BM, Doster P, Barone D, Efthymiopoulos C, Peretti G, Roncucci R: *In vivo* interaction of CAB with rat brain dopamine receptors labelled with 3H-N-n-propylinorapomorphine. Eur J Pharmacol 1990;187:399-408.
100. Andreotti AC, Pianezzola E, Persiani S, Pacciarini MS, Strolin Benedetti M, Pontiroli AE: Pharmacokinetics, pharmacodynamics, and tolerability of cabergoline, a prolactin-lowering drug after administration of increasing oral doses (0.5, 1.0, and 1.5 milligrams) in healthy male volunteers. J Clin Endocrinol Metab 1995;80:841.
101. Webster J, Piscitelli G, Polli A, Ferrari CI, Ismail I, Scanlon MF for the Cabergoline Comparative Study Group. N Engl J Med 1994;331:904-909.
102. Motta T, Colombo N, de Vincentiis S, D'Alberton A, Marchini M: Vaginal cabergoline in the treatment of hyperprolactinemic patients intolerant to oral dopaminergics. Fertil Steril 1996;65:440-442.
103. Ciccarelli E, Giusti M, Miola C, Potenzoni F, Sghedoni D, Camanni F, Giordano G: Effectiveness and tolerability of long term treatment with cabergoline, a new long-lasting ergoline derivative, in hyperprolactinemic patients. J Clin Endocrinol 1989;69:725-728.
104. Ferrari C, Mattei A, Melis GB, Paracchi A, Muratori M, Faglia G, Sghedoni D, Crosignani PG: Cabergoline: long-acting oral treatment of hyperprolactinemic disorders. J Clin Endocrinol Metab 1989;68:1201-1206.
105. Ferrari C, Paracchi A, mattei Am, de Vincentils S, D'Alberton A, Crosignani PG: Cabergoline in the long-term therapy of hyperprolactinemic disorders. Acta Endocrinol 1992;126:489-494.
106. Biller BMK, Molitch ME, Vance ML, Cannistraro KB, Davis KR, Simons JA, Schoefelder JR, Klibanski A: Treatment of prolactin-secreting macroadenomas with the once-weekly dopamine agonist cabergoline. J Clin Endocrinol Metab 1996;81:2338-2343.
107. Delgrange E, Maiter D, Donckier J: Effects of the dopamine agonist cabergoline in patients with prolactinoma intolerant or resistant to bromocriptine. Eur J Endocrinol 1996;134:454-456.
108. Motta T, Maggi G, D'Alberton A, Ferrari C, Paracchi A: Twice weekly cabergoline treatment of macroprolactinoma. J Obstet Gynecol 1989;9:334-335.
109. Page SR, Nussey SS: Cabergoline therapy of a large prolactinoma in a bromocriptine-intolerant patient. J Obstet Gynecol 1989;10:156-158.
110. Melis GB, Mais V, Gambacciani M, Sghedoni D, Paoletti AM, Fioretti P: Reduction in the size of prolactin-producing pituitary tumor after cabergoline administration. Fertil Steril 1989;52:412-415.
111. Colao A, DeSarno A, Landi ML et al: Long-term and low-dose treatment with cabergoline induces macroprolactinoma shrinkage. J Clin Endocrinol Metab 1997;82:3574-3579.
112. Ferrari CI, Abs R, Bevan JS et al: Treatment of macroprolactinoma with cabergoline: a study of 85 patients. Clin Endocrinol 1997;46:409-413.
113. Verhelst J, Abs R, Maiter D et al: Cabergoline in the treatment of hyperprolactinemia. J Clin Endocrinol Metab 1999;84:2518-2522.

114. Colao A, DiSarno A, Sarnacchiaro S, et al: Prolactinomas resistant to standard dopamine-agonists respond to chronic cabergoline treatment. J Clin Endocrinol Metab. 1997;82:876-881.
115. Krupp P, Monka C, Richter K. The safety aspects of infertility treatments. Program of the Second World Congress of Gynecology and Obstetrics, Rio de Janeiro, October, 1988.
116. Raymond JP, Goldstein E, Konopka P, Leleu MF, Merceron RE, Loria Y. Follow-up of children born of bromocriptine-treated mothers. Hormone Res 1985;22:239-246.
117. Konopka P, Raymond JP, Merceron RE, Seneze J. Continuous administration of bromocriptine in the prevention of neurologica complications in pregnant women with prolactinomas. Am J Obstet Gynecol 1983;146:935-93
118. Canales ES, García IC, Ruíz JE, Zàrate A. Bromocriptine as prophylactic therapy in prolactinoma during pregnancy. Fertil Steril 1981;36:524-526.
119. Ruiz-Velasco V, Tolis G. Pregnancy in hyperprolactinemic women. Fertil Steril 1984;41:793-805.
120. Robert E, Musatti L, Piscitelli G, Ferrari CI: Pregnancy outcome after treatment with the ergot derivative, cabergoline. Reprod Toxicol 1996;10:333-337.
121. Pharmacia and Upjohn, Inc. Data on file. July, 1999.
122. Molitch ME. Management of prolactinomas during pregnancy. J Reprod Med 1999;44(Suppl):1121-1126.
123. Narita O, Kimura T, Suganuma N, *et al.* Relationship between maternal prolactin levels during pregnancy and lactation in women with pituitary adenoma. Acta Obst Gynaecol Jpn 1985;37:758-762.
124. Divers WA Jr., Yen SSC. Prolactin-producing microadenomas in pregnancy. Obstet Gynecol 1983;61:425-429.
125. Brodsky JB, Cohen EN, Brown BW Jr, Wu ML, Whitcher C. Surgery during pregnancy and fetal outcome. Am J Obstet Gynecol 1980;138:1165-1167.

Chapter 6

The Psychosomatic Interface: Hyperprolactinemia

Luis G. Sobrinho

Department of Endocrinology,
Portuguese Cancer Institute, Lisbon, PORTUGAL

INTRODUCTION

Prolactin is practically ignored in the chapter of psychosomatic disorders of textbooks of psychiatry (1). There are two valid reasons for this omission:

First, psychiatric difficulties associated with inappropriate lactation or hyperprolactinemia (except when induced by drugs) are uncommon. Second, there is no conceptual framework to accommodate such uncommon observations that, when reported, appear either as anecdoctal cases or as unexplained associations. For all practical and theoretical purposes the biology of "psychological stress" has been described exclusively in terms of the hypothalamic-pituitary-adrenal axis and/or the sympathetic nervous system.

Yet, there is unequivocal, albeit sparse, evidence relating psychological variables to inappropriate lactation or to hyperprolactinemia. The paradigmatic example is pseudopregnancy. Other links between psychosomatics (in a broad sense) and prolactin are the following:

1. Galactorrhea or hyperprolactinemia have been reported following psychological stress (2-4);
2. Psychological abnormalities have been consistently described in patients with prolactinoma or idiopathic hyperprolactinemia (5–8). More teasing, however, are observations indicating that environmental conditions may <u>precede</u> the disease. Nunes *et al.* (9) first reported that pathological hyperprolactinemia affects predominantly women brought up under conditions of paternal deprivation. This observation was, subsequently, confirmed by all authors that addressed the question (10-13). Besides, the clinical onset of the disease often follows a major life event (9);

3. Surrogate maternity is sometimes associated with lactation and/or hyperprolactinemia, both in humans and in animals;
4. Population studies associate high prolactin levels with distress in children (14) and with passive coping or unemployment in healthy adults (15-17).

It is my belief that these observations, scattered in the literature, conform to a pattern that makes sense.

In the present chapter I will try to achieve the following goals:

1. To provide some basic information about the actions of prolactin on the brain;
2. To emphasize that prolactin induces behavioral and metabolic effects adaptive to the care of the young but, on the other hand, its secretion can be stimulated when a parental role is required;
3. To bring together observations relating prolactin to psychological difficulties, states, traits and coping strategies;
4. To provide a theoretical framework into which clinical and experimental data can be inserted and that can generate testable hypotheses.

PROLACTIN AND THE BRAIN

Prolactin does act upon the brain and affects behavior. Prolactin binding sites, or the actual receptor, have been identified within the central nervous system in the hypothalamus, choroid plexuses, where the specific binding is twenty fold that in the hypothalamus, and, to a lesser extent, in the substantia nigra and some striatal structures but not in the cerebral cortex, thalamus or the hindbrain (18-20).

Since prolactin does not cross the blood-brain barrier, peripheral prolactin is likely to enter the CSF via the choroid plexuses and/or by retrograde transport through the portal vessels (21). Its concentration in the CSF is about 20% of that in the blood (22).

There is a prolactin synthetic pool in the brain, separate from that of the anterior pituitary. The main, and probably the only, region where prolactin is produced is the hypothalamus, although its concentration there is four orders of magnitude below that found in the pituitary (19,23).

In vitro studies of single neurons reveal that prolactin affects directly the firing rate of only a minority of cells. Application of prolactin to slices of brain or "*in vivo*" (by iontophoresis) affect, predominantly by excitation, the majority of ventromedial hypothalamic neurons (19). It is likely, therefore, that prolactin actions on behavior are predominantly due to its modulatory action on the activity of other transmitters. There is direct evidence that prolactin modulates the activity and receptor density of dopamine (19).

In humans, hyperprolactinemia increases hypothalamic dopaminergic tone as demonstrated by the exaggerated TSH (24) and LH (25) responses to domperidone. Prolactin also appears to increase the central opiate tone. Rats with transplantation-induced hyperprolactinemia present with analgesia, reduced response to electric shock and reduced core temperature, effects that can be reversed by naloxone (26). Opioids markedly reduce hypothalamic dopamine and stimulate prolactin secretion (22,27), suggesting the possibility of a positive feed back system.

PROLACTIN AND BEHAVIOR

PROLACTIN AND MATERNAL BEHAVIOR

The most important behavioral effect of prolactin is stimulation of maternal behavior, most extensively studied in rodents - nursing, licking and retrieving the pups and building nests for them. Prolactin is not necessary for the expression of maternal behavior since such behavior is seen in hypophysectomized, virgin, female rats when exposed to pups. However, the latency period of this effect is considerably shortened by the administration of prolactin (28). In hamsters, one single postpartum administration of bromocriptine induces eating of the pups by the dams (29). The importance and specificity of prolactin as a major regulator of maternal behavior has been emphasized by recent studies in mice carrying a gene with a null mutation for the prolactin receptor. Mutants showed a marked deficiency in pup induced maternal behavior, more intense in homozygotic but also observed in heterozygotic females. Other functions like configurational learning (Morris maze), locomotor activity, exploration, eating, olfaction and sexual behavior were normal in mutants (30).

A major target for the action of prolactin is the medial preoptic area (MPOA) where the effector mechanisms for the expression of maternal behaviour in the rat appears to be centralized. Local bilateral infusions of prolactin, in doses ineffective when infused into the ventricular system, stimulate a rapid onset of maternal behavior. Estradiol and oxytocin applied to the MPOA also facilitate the expression of maternal behavior (31).

OTHER BEHAVIORAL EFFECTS OF PROLACTIN

Other behavioral effects of prolactin such as facilitation of learning capacity (which is reduced either by dopamine or opioid antagonists), induction of novelty-induced grooming behavior, potentiation of amphetamine and apomorphine induced stereotypes, reduction of pain sensitivity and of the incidence of gastric ulcers induced by physical stress (26) may reflect the above-mentioned effects of prolactin on the dopaminergic and opioid systems. Prolactin increases food intake by female rats (32,33), which may play an adaptive role for pregnancy and lactation.

Hyperprolactinemia markedly reduces the levels of sexual behavior in mammals (22). Reduced sexual desire and function, partially reversible by bromocriptine, is common in hyperprolactinemic women (34) and men (35). In normal postmenopausal women prolactin has been found to be negatively associated with sexual desire, sexual arousal and vaginal lubrication during sexual activity (36). The negative action of prolactin on sexual behavior is, at least in part, related to stimulating effect of hyperprolactinemia on the hypothalamic opioid and dopaminergic tones. Both directly suppress the secretion of gonadotrophin releasing hormone (25,37).

ADAPTIVE NON - PUERPERAL LACTATION OR HYPERPROLACTINEMIA

Surrogate maternity

The secretion of prolactin can be activated by environmental stimuli. Upon exposure to pups, the secretion of prolactin by intact female rats is strongly stimulated and so is the expression of maternal behavior (38).

In humans, non puerperal nursing has been reported to occur in most primitive cultures throughout the world, including Europe and North America (39,40). When normal nursing is made impossible by the death of the mother or other circumstances a "wet nurse" has to be secured. Inasmuch as women with milk of their own are at a premium, this function may have to be taken over by a non-lactating woman, often the grandmother. There are convincing reports that a significant amount of milk can thus be produced, at least in some occasions (39,40). The nutritional value of this milk is difficult to assess since children are simultaneously given gruel or other food. The production of milk can be tentatively explained by suckling and, possibly, by some pharmacological effect of the concoctions and rubbings given to the women. However, it has been speculated that the atmosphere surrounding death and childbearing may be an important stimulus to trigger a neuroendocrine response, or to reinforce the suckling reflex, in sensitive women who assume the surrogate maternal role (40). In this context it may be pertinent to quote the case reported by Zeitner et al (4) of a postmenopausal woman in psychotherapy who developed galactorrhea in association with the wish to breast-feed her infant grandchild and re-establish, in fantasy, close symbiotic ties with her own mother.

A more recent study of 240 women who adopted children provides interesting information. These women planned to nurse their adopted children. For this purpose nipple stimulation techniques were used some weeks before the infant's arrival. Half of the women were able to express milk at the end of this preparatory period. Children were given supplementary food, besides breast feeding. However, 63% of the children needed less than full supplementation suggesting that the milk had some nutritional value. Two of the women reported that supplementation was never needed. Amenorrhea followed breast-feeding in 10 of the women (41).

Other observations in primates may be pertinent to this issue. Subordinate female talapoin monkeys, who have no offspring of their own, have higher prolactin levels than those of the dominant class, suggesting a role of this hormone in the preparation for the function of caring for the offspring of the dominant class (42).

Hyperprolactinemia in males during parenting

Increased prolactin secretion as a response to a demand of parental functions has also been observed in males. In a much quoted study, Dixon and George (43) reported that male marmoset monkeys, while carrying their twin offspring have plasma prolactin levels that are five times higher than those from males without infants. Likewise prolactin values are higher in another male primate, the Cotton-Top Tamarin, during the peripartum of their mates (44).

Similar observations have been made in rodents. Male hamsters (*Phodopus*) of a biparental species (*P.campbelli*) behave differently from those of a uniparental species (*P. sungorus*). The first group had a important and significant increase in their prolactin levels at the fifth day of their offspring. No such changes were observed in the males of the uniparental species. Also the P. campbelli males showed significantly more indexes of parental behavior than *P. sungorus* (45).

In conclusion, caring for the young, by providing both food and maternal behavior, is a fundamental biological function. Mothers are specially adapted to this function. However, other members of the group such as non-puerperal females, and even males, can be mustered for the community effort of raising the offspring. This recruitment has been associated either with direct evidence of increased prolactin levels or with lactation.

CLINICAL OBSERVATIONS

PSYCHIATRIC AND NEUROLOGIC SYNDROMES

Basal values and twenty four hour profiles of prolactin have been studied in several psychiatric syndromes. No consistent abnormalities have been found (46,47).

Slight hyperprolactinemia has been reported in the Kleine-Levin syndrome (48) and other hypersomniac states (49). Prolactin rises during epileptic seizures as well as after hypotensive syncope (50). It has been postulated that the prolactin rise depends on activation of the limbic structures. Marked prolactin rises can be observed during the aura of temporal seizures while in the case of extra-limbic partial seizures prolactin only rises after generalization to the limbic region (51). Also, interictal hyperprolactinemia has been reported in intractable temporal epilepsy suggesting that a seizure focus in this region may exert a sustained excitatory influence on prolactin (52).

As prolactin secretion can be stimulated by serotoninergic pathways, dynamic studies using serotonin precursors, agonists or inhibitors of re-uptake have been used. Several authors described an atenuated prolactin response in depressive syndromes. However, doubts have been raised concerning the consistency and interpretation of the results. Detailed information on these studies can be found in a recent review (47).

PSEUDOPREGNANCY

Pseudopregnancy is a psychiatric disorder characterized by the delusional belief of being pregnant, associated with somatic changes characteristic of true pregnancy – rapid weight gain, breast engorgement, production of milk and amenorrhea. Moderate hyperprolactinemia has been described in most cases (53-56). The analysis of the cases described as having normal prolactin levels reveals that the actual values belong to the upper normal range and/or contain high occasional peaks (57-58) or are hyperresponsive to stimulation (59). The situation is reversible. A detailed description of the psychological mechanisms supposed to be involved in pseudopregnancy is beyond the scope of this article and can be found elsewhere (3). Further progress in this area is difficult because pseudopregnancy is rare in industrialized cultures and no studies have been published recently. Extrapolation from studies in animals must be viewed with caution. Besides the usual caveats, the term pseudopregnancy describes different conditions in different species (for review see 3).

ACUTE, PSYCHOGENIC, GALACTORRHEA OR HYPERPROLACTINEMIA

Some reports in the psychoanalytical literature (4,60,61) describe the occurrence of spontaneous lactation in women during periods of emotional difficulties. One of them (60) describes a patient who developed spontaneous galactorrhea after a premature attempt to terminate her analysis and bears the suggestive title of "Lactation as a denial of separation". Acute breast engorgement associated with spontaneous bilateral galactorrhea, occurring after physical or emotional trauma, resolving spontaneously after some weeks and associated with normal or slightly elevated prolactin levels has been described (3,62). These cases are uncommon. In 30 years of endocrine practice I have seen only two such patients. Acute hyperprolactinemia unassociated with symptoms was reported by Reichlin (2) who described the accidental finding of a major prolactin surge (from 40 to 180 ng/ml) in a pregnant patient while arguing with the attending nurse during the course of a metabolic study.

PROLACTIN RESPONSES TO PSYCHOLOGICAL "STRESS"

Prolactin responses to psychological stress in humans have been reported by several authors but are generally modest and inconstant. In most of the reported studies volunteers are submitted to standard conditions – mental arithmetic, public speaking, multiple task tests, interviews, presentation of films and videogames, academic exams, parachute or platform jumping, acrobatic flights, etc. In general, the hormones that most consistently respond to these stimuli are norepinephrine and epinephrine, followed by cortisol and, least consistent of all, prolactin (for review, see 63). While in rodents prolactin responds exuberantly to "psychological stress" (64), threatening situations, in humans, are neither strong nor constant elicitors of a prolactin response.

However, prolactin responds to a specific stimulus – suckling. It rats, stimulation of the cervix triggers an orderly sequence of prolactin surges that, even in the case of non copulatory stimulus, lasts ten days and is responsible for a state of pseudopregnancy (38). Orgasm, both in men (65) and in women (66), is associated with a marked prolactin, but not cortisol, response.

In a systematic study of patients on psychotherapy under a modified state of consciousness we have been observing prolactin or cortisol surges (but not both at the same time) accompanying the evocation of memories with intense emotional content. (Sobrinho et al, unpublished observations). It thus appears that prolactin may respond to specific stimuli, different from those associated with the cortico-adrenal stress response. The characterization of such stimuli is a matter of current research.

PSYCHOSOMATIC ASPECTS OF PROLACTINOMAS AND IDIOPATHIC HYPERPROLACTINEMIA

Mood abnormalities in pathological hyperprolactinemia

Patients with pathological hyperprolactinemia present with an unusual prevalence of emotional difficulties.

Fava et al (5) compared ten women with primary hyperprolactinemia, ten women with amenorrhea and normal prolactin levels and ten normal women matched for age and social class. All the subjects were given the Symptom Questionnaire. The hyperprolactinemic women had significantly higher scores on hostility, anxiety and depression than the other two groups. Of these ten women three met DSM III criteria for major depressive disorder.

Kellner et al (6) compared 14 hyperprolactinemic women, 25 non-psychotic patients attending the psychiatric out-patient clinic for anxiety, depression or both, 29 family practice patients and 26 non-patient employees. The groups were matched for age and social class. The Symptom Rating Test and the Symptom Questionnaire were used. The scores of hyperprolactinemic patients and of psychiatric patients were similar, and significantly higher than family practice patients and employees, on hostility, depression and anxiety. Also, the first two groups had lower scores on the "friendly" and "relaxed" subscales.

Rothchild (8) interviewed and administered the Minnesota Multiphasic Personality Inventory (MMPI) test to 35 patients with prolactinoma, 17 women with galactorrhea but normal prolactin levels and 24 nursing mothers. Depressed mood, loss of interest in usual pleasures, decreased libido and irritability predominated in the adenoma group as compared to the others.

Reavley et al (7) compared 39 patients with prolactinomas and 26 idiopathic hyperprolactinemia to 26 controls with normoprolactinemic non-functioning pituitary tumors or treated acromegalics. The Hospital Anxiety and Depression (HAD) and the 90-item Symptom Checklist (SCL-90) were used. Overall, 54% of the hyperprolactinemic patients were found to have definite or borderline anxiety (HAD scores) as compared to 27% of the controls, a significant difference. A similar trend was observed with the SCL90. Curiously, the hyperprolactinemic patients without tumors scored significantly higher than the prolactinoma group (73% v. 41%; $p<.005$). No differences between groups were seen in the scores for depression.

Amelioration of the sexual dysfunction and in the scores of psychometric tests following successful medical or surgical therapy has been reported by most authors (8,34,67), although some abnormalities persist (8).

Factors predisposing to primary hyperprolactinemia

There are two aspects of the natural history of pathological hyperprolactinemia suggesting that its distribution is not random in the population but, rather, that it develops in predisposed persons when confronted with a psychological difficulty. This is true both for prolactinomas and for idiopathic hyperprolactinemia:

1. Most patients with primary hyperprolactinemia were brought up, as children, in families in which the father was either absent or alcoholic and violent (9-12). To address this issue prospectively, a study of the sisters of patients with prolactinomas (generally exposed to the same environment as the patients) and appropriate controls was undertaken (13). Thirty seven sisters were compared to 72 women matched for age and social class who attended the same Endocrine Clinic because of a benign non-functioning thyroid nodule. Morning prolactin levels were obtained in five different days in all women. The average prolactin values were similar in both sisters and controls (10.6 v. 12.4 ng/ml). However, women with paternal deprivation (from the sisters and control groups pooled together) had significantly higher mean prolactin values than the others (14.7 v. 9.4 ng/ml; $p<0.001$). A closer look into the data revealed that most values in each subgroup were within the normal range (<20 ng/ml). The difference in the mean values was due to the uneven distribution of outliers with abnormally high values (12/50 of the deprived women v. 3/59 of the others; $p<.0.005$). Galactorrhea was also more common in the deprived than in the non-deprived group (14/50 v. 7/59; $p<0.03$). These differences could still be observed when the sisters and the control groups were treated separately (13). The above observations reinforced the association of paternal deprivation in childhood with the development of hyperprolactinemia later in life and, at the same time, deemphasized the epidemiological significance of any possible genetic influence.
2. Often, the onset of the symptoms closely follows an important life event (9). Assies et al (10) did not find a difference between the number of life-events between patients and controls. However, they remarked that, in two of their 14 patients, hyperprolactinemia developed shortly after two very stressful experiences (death of a parent and loss of a son).

Both childhood deprivation and life-events precede the development of the clinical disease and can hardly be consequences of hyperprolactinemia. Therefore, if there are any causal links between early childhood experiences, life events and the development of pathological hyperprolactinemia, the latter must be the secondary event. The above observations apply both to idiopathic hyperprolactinemia and to prolactinomas. Similar observations have been reported for Cushing's disease. In a case-control study of patients with ACTH-producing pituitary tumors Sonino *et al.*(68) found that patients had more life-events with negative impact than controls ($p< 0.001$) during the year preceding the clinical onset of the disease. Again, these observations imply that life events are somehow causally related to functioning pituitary tumors. It is interesting that, in patients with Cushing's syndrome due to adrenal tumors, life events in the year preceding the disease were as common as in controls.

As most, if not all, pituitary adenomas have been shown to be monoclonal it is now widely accepted that pituitary adenomas are primarily due to local, cellular, abnormalities rather than to external overstimulation. However, these two possibilities are not mutually exclusive. Microadenomas, 40% of which are prolactinomas, have been described in about one fourth of the pituitaries studied in autopsy series (69). Therefore, there is no shortage of silent microprolactinomas to promote into clinical relevance. Evolution to nodularity is the rule following stimulation of the endocrine glands. Apparently, all that is required to generate diversity in any newly formed tissue is simple growth (70).

Pituitary growth hormone-secreting adenomas have been induced following prolonged stimulation of the pituitary by growth hormone releasing-hormone (GHRH) in mice transgenic for GHRH (71) and in a patient with a GHRH-producing tumor (72). Mice with knocked out dopamine D2 receptors developed lactotroph hyperplasia and prolactinomas (73). A limited number of male to female transsexuals taking estrogens have been reported to have developed hyperplasia of the lactotrophs or prolactinomas (74-76). On the other hand, neither pregnancy (77) nor prolonged usage of hormonal contraceptives have been demonstrated to induce "*de novo*" appearance of prolactinomas (78).

It can be concluded that the role, if any, of psychological factors in the genesis of prolactinomas is unsettled. The evidence from biographic studies is impressive. As discussed above, it is possible to develop pituitary adenomas by prolonged stimulation. Yet, only exceptionally could pituitary tumors ever be induced in humans. The rare reported cases always required the exposure to intense and prolonged, well defined, stimuli. The stimulus required for the development of prolactinomas in deprived women after losses, if there is any, has not been identified.

The question has some practical relevance. Psychological abnormalities, probably pre-existent to the tumor, persist in many of these patients even when their prolactin levels are brought into the normal range by medical or surgical therapy (8).

Rapid weight gain, prolactin and life-events

It is not rare in clinical practice to come across people, in general women, who complain of having gained a substantial amount of weight over a short period of time without any discernible change in their eating, exercising or smoking habits. When such women, operationally defined as having gained at least 5 Kg in the previous year, were compared to a control population with stable weight matched for age, parity and smoking patterns it was found that patients differed from controls in several ways: 1) Prolactin values were significantly higher in patients at most times of the day (average values of hourly, 24 hour, profiles - 14.60 ng/ml v. 8.84 ng/ml; $p=0.0121$). No differences were found in the 24 hour profiles of cortisol or insulin, nor in basal values of thyroid hormones, TSH, estradiol, gastrin, cholecystokinin, somatostatin, oxytocin or IGF-1. Leptin was higher in patients than in controls but the difference could be explained exclusively by the higher body mass index of the patients; 2) Galactorrhea was more common in patients (11/30 v. 1/29) ; 3) Patients also differed in higher prevalence of sexual dysfunction, higher indexes in several parameters of the MMPI and SCL90 scales (depression, anxiety, phobia, psychoticism, psychastenia, "femininity", schizoidy) and a higher incidence of meaningful life-events in the year preceding the study; 4) Patients and controls did not differ in their estimated and actually observed food intake; 5) Patients had a reduced prolactin response to domperidone, indicating a reduced hypothalamic dopaminergic tone. A closer look into the data revealed that the differences in prolactin concentrations between patients and controls was due to the presence, in the former group, of a minority of outliers with high prolactin values. When this subgroup of outliers was considered separately it was found that, besides higher serum prolactin levels (the defining criterium) it also contained most patients with galactorrhea (Figure 6-1) and the lowest prolactin responses to domperidone (Figure 6-2) (79,80).

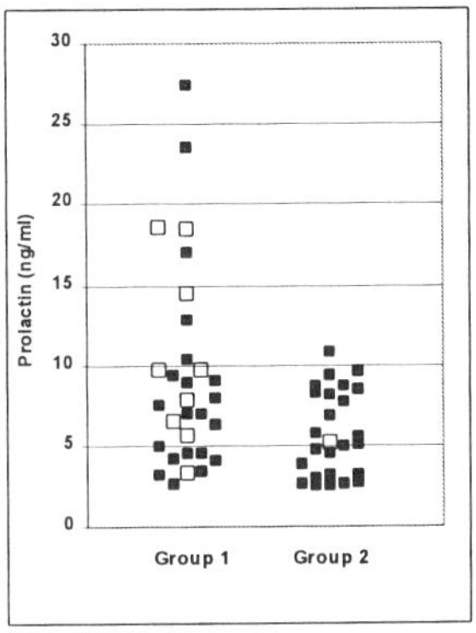

Figure 6-1. Morning prolactin levels in 30 women with rapid weight gain (Group 1) and 26 controls with stable weight (Group 2). Galactorrhea was found in women represented by white squares. (Data pooled from refs. 79 and 80).

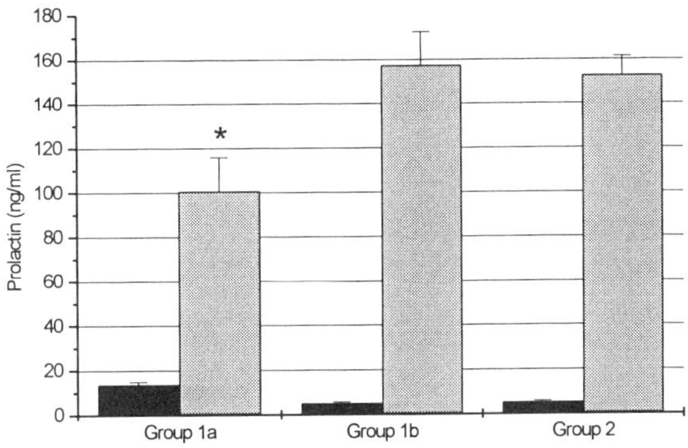

Figure 6-2 . Basal and domperidone-stimulated prolactin values (mean and SEM) in three groups of patients: Group 1a - Women with rapid weight gain (see text) and basal prolactin levels above the range of the controls (n=6); Group 1b - Women with rapid weight gain and basal prolactin levels within the range of the controls (n=11); Group 2 - Control women with stable weight (n=16). * - significantly different from groups 1b and 2, $p < 0.02$. (Data from ref. 80)

Two main conclusions can be drawn from the above observations:

1. Women with rapid weight gain do not constitute an homogeneous group in terms of their prolactin values and responses. About one fourth of them have high basal prolactin levels (higher than any of the values observed in controls), galactorrhea and respond poorly to domperidone. The others are similar to controls in terms of hormonal parameters. It is possible that the first group was studied while still in the active process of gaining weight while the others had already plateaued. Alternatively, the groups may be different from the outstart and the genesis of the weight gain may be distinct.
2. The weight gain follows a major life-event in most cases (both in women with high or normal prolactin levels) suggesting a psychosomatic component in the condition, a suggestion further strengthened by the finding of significantly higher scores in some parameters of the MMPI and SCL90 scales.

EPIDEMIOLOGICAL OBSERVATIONS

PROLACTIN AS A MARKER OF SPECIFIC STATES OR COPING STRATEGIES

A few epidemiological observations suggest that prolactin may be a marker of specific psychological states and of coping strategies.

In a study of 191 infants aged 12 to 23 months who underwent developmental testing (Bayley Scales of Infant Development) Lozoff *et al.* (14) made the intriguing observation that 53% of the children with serum prolactin => 25 ng/ml rated as abnormal in behavior, particularly in the Test Affect factor. Only 20% of those with lower prolactin levels rated abnormally ($p<0.001$). Looking at the same data in another way, infants rated as abnormal in affect had prolactin values significantly higher than the others (23.2 ng/ml v. 14.4; $p<.01$). The infants were observed three times during a three month interval. Prolactin values were remarkably constant for each individual.

Observations in adults under conditions of real life stress suggest that higher prolactin levels are associated with passive coping strategies (15). Four types of observations were made: 1 – Close female relatives of terminal patients who participated in a program of active care were compared to a control group not directly involved in the care of their relatives. The first group had significantly higher cortisol and lower prolactin levels than the controls; 2 – Working men were observed every 3 to 4 months during one year. Blood was drawn and questionnaires regarding their work situation and mood were filled. Men who reported to be sad or depressed at least once during the observation year had more pronounced variations in the morning prolactin values than the others. When the job strain level was low prolactin was low, when the job strain level was high the prolactin levels were high, sometimes even approaching those seen in pregnant women; 3 – Subway drivers who had ran over a person who fell on the railway (an unavoidable experience) had prolactin levels significantly higher 3 weeks after the accident than that of drivers not exposed to the same experience; 4 – Elderly patients who were recovering from a stroke were divided in two groups. The experimental group was given a good deal of responsibility in the conduct of their own therapy whereas the control group was given a standard rehabilitation program. The average prolactin levels were similar for both groups at the start of the experiment. Six and twelve weeks later the prolactin levels had decreased in the experimental group and increased in the control group (15). Also, health personnel of geriatric care caught in a situation of professional instability were divided in two groups – with and without support by a psychologist for 20 weeks. Those with support had a decrease in serum prolactin concentrations while an increase was observed in the others (16). Threat of unemployment was, likewise, associated with slight, but significant, prolactin increases (17).

These observations in adults may have some relationship with the observations in infants described above (14). It is possible that hesitance and wariness in infancy is a manifestation of, or results in, passive coping. However, while Lozoff's work identified prolactin as a marker of a state (or trait), the observations in adults suggest that prolactin values are modifiable, according to the degree of control a subject has in a stressful situation.

THE CONCEPT OF MATERNAL SUBROUTINE

Adaptation to motherhood implies changes in behavior and in energetic metabolism. Maternal behavior, breasts able to produce milk and abundant energetic supplies must be synchronously available. This suggests the existence of a coherent neuroendocrine mechanism such that its activation produces convergent effects, psychological and metabolical, adaptive to the specific purpose of rearing the young.

It seems appropriate to name this mechanism "maternal subroutine" (3). At least in rats, the maternal subroutine is anatomically organized around the MPOA and of its projections (31). Prolactin is a major mediator of the maternal subroutine and, in all studies discussed above, only its concentrations or its effects (galactorrhea) have been considered. However, it is unlikely that prolactin is the sole effector of the maternal subroutine. In some of the examples presented above, such as pseudopregnancy and women with rapid weight gain, there is a marked disproportion between the intensity of the symptoms and the modesty of the increases in prolactin concentrations. This disproportion suggests that prolactin may be only one of the components of a more complex energy sparing system.

Placental and gastrointestinal hormones, oxytocin and the parasympathetic system, are also involved in the biological and psychological adaptation to motherhood (81). Placental estrogens activate prolactin secretion and, along with progesterone, have metabolic and behavioral effects of their own. Oxytocin and cholecystokinin are involved in mother-to-child interaction and stimulate prolactin secretion (81-83). Prolactin controls the production of pancreatic islet cell hormones as discussed elsewhere in this book. (84). A detailed description of the effects of these and other putative components of the maternal subroutine and of their relationship with prolactin is beyond the scope of this chapter.

Whether as part of a broader subroutine or not, prolactin secretion can be activated by external stimuli in the absence of pregnancy. In the examples of surrogate maternity (and paternity) presented above, an adaptive purpose is obvious. Reproductive success is more a need of the community than of the individual. It makes sense that some elements of a group can be recruited - and physiologically prepared - to help caring for the young or even to substitute a missing mother. The mechanisms that lead to the activation of prolactin in such circumstances are unclear. There are no placental hormones to drive the lactotrophs. We are compelled to accept that psychological factors - bereavement, the desire or need to take care of a child, the perception of one's lower rank in a hierarchy, or other - may have the power to activate the secretion of prolactin.

At this stage we can envisage how thin is the frontier between adaptation and disease. Pseudopregnancy is a psychiatric disorder characterized by the delusional belief of being pregnant, associated with activation of the maternal subroutine - an intrinsically normal mechanism. The causes of the extemporaneous activation of the maternal subroutine in pseudopregnancy can only be a matter of speculation at the present time. The psychoanalytical concept of regression may be of some help. Regression has been defined as "the emergence of past, often infantile, trends where such trends are thought to represent the reappearance of modes of functioning which have been abandoned or modified" (85). It is possible that primitive mental constructs such as those that develop during early mother-to-child attachment are closely linked to biological effector mechanisms, as discussed elsewhere (86). Such linkage may underlie the findings of hyperprolactinemia in children with disturbed affect (14), the proneness to develop hyperprolactinemia by women with an history of paternal deprivation (13) and the association of prolactin with passive coping strategies (15).

The maternal subroutine hypothesis proved to have heuristic value. The similarity between the rapid weight gain of pseudopregnancy and rapid weight gain in otherwise healthy women prompted a series of observations revealing that rapid weight gainers differed from controls with stable weight in the frequency of galactorrhea and sexual dysfunction, higher prolactin levels and reduced hypothalamic dopaminergic tone. They also scored differently in the MMPI and SCL90 personality tests (namely in the scales of depression) and in the Paykel life events scale (79,80).

The implications of these findings are relevant to the field of psychosomatics. Rapid weight gain in the absence of other remarkable symptoms, but with some depressive features, may be close to atypical depression with weight gain, an entity recognized in the psychiatric literature (87-89). It is likely that both these entities are part of a spectrum that also includes pseudopregnancy as its most expressive form.

Pseudopregnancy all but disappeared from industrialized countries although it has been reported, in the recent past, to be of epidemiological significance in Africa (54). It appears that the regression to a mother-to-child relationship in the explicit form of a delusion is easier in primitive than in industrialized cultures, and so is the intensity of the activation of the maternal subroutine.

CONCLUSION

Prolactin acts upon the central nervous system and affects behavior in such ways as to make it appropriate for the care of the young. Simultaneously, and serving the same purpose, it promotes weight gain (32,90). On the other hand, prolactin secretion can be activated when the exercise of parental functions is needed, even in the absence of a previous pregnancy. Adaptation to parenthood is also secured by other neuroendocrine mechanisms besides prolactin. Hence the proposed concept of the existence of a maternal subroutine, of which prolactin is one of the effectors and a marker.

The possibility of activating the secretion of prolactin as a response to perceived external needs (however adaptive they may, sometimes, be) provides room for psychosomatic disorders.

The secretion of prolactin (or the entire maternal subroutine) may become conditioned to stimuli different from the original, adaptive, ones. Hyperprolactinemia or galactorrhea have been described in several psychological conditions where some relation to parenthood is sometimes obvious, other times mysterious. In either case, rapid weight gain may occur.

The concept that psychosomatic disorders or symptoms may result from the extemporaneous or exaggerated activation of intrinsically normal neuroendocrine subroutines is a useful one. It underlies the immense extant research on the neuroendocrinology of psychiatric and psychosomatic disorders. The sweeping influence of Cannon´s and Selye´s seminal works on the role of the sympatho-adrenal system and of the hypothalamus-pituitary-adrenal axis in the response to physical and psychological "stress" left little room to study other systems and the consequences of their inappropriate activation. In this chapter, I propose that adaptation to parenthood requires the existence of a specific neuroendocrine subroutine that can be activated out of its original context and provoke disease. The main known marker and effector of this subroutine is prolactin. The maternal subroutine hypothesis, however sketchy its present formulation may be, provides a starting framework for the understanding of clinical observations and for the design of testable hypotheses on the field of prolactin and psychosomatics. .

REFERENCES

1. Rubin RT, King BH Endocrine and metabolic disorders. In Kaplan H.I., Sadock B.J. eds. Comprehenive Textbook of Psychiatry, 6th ed., Baltimore, Williams & Wilkins, 1995: 1541-28.
2. Reichlin S. Prolactin and growth hormone in stress. In Chrousos GP, Loriaux DL, Gold PW eds. Mechanisms of physical and emotional stress. New York: Plenum Press, 1988: 353-76.
3. Sobrinho LG. Neuropsychiatry of prolactin: causes and effects. Baillière's Clinical Endocrinology and Metabolism 1991; 5:119-41.
4. Zeitner RM, Frank MV, Freeman SMD. Pharmacogenic and psychogenic aspects of galactorrhea: a case report. Am J Psychiatry 1980; 137; 111-2.
5. Fava GA, Fava M, Kellner R, Serafini E, Mastrogiacomo I. Depression, hostility and anxiety in hyperprolactinemia. Psychoter Pychosom 1981; 36: 122-8.
6. Kellner R, Buckman MT, Fava GA, Pathak D. Hyperprolactinemia, distress, and hostility. Am J Psychiatry 1984; 141: 759-63.
7. Reavley S, Fisher AD, Owen D, Creed FH, Davis JRE. Psychological distress in patients with hyperprolactinemia. Clin Endocrinol 1997; 47: 343-8.
8. Rothchild E. Psychologic aspects of galactorrhea. J Psychosom Obst Gynecol 1985; 4: 185-96.
9. Nunes MCP, Sobrinho LG, Calhaz-Jorge C, Santos MA, Maurício JC, Sousa MFF: Psychosomatic factors in patients with hyperprolactinemia and/or galactorrhea. Obstetrics & Gynecology 1980; 55: 591.
10. Assies J, Vingerhoets AJJM, Poppelaars K. Psychosocial aspects of hyperprolactinemia. Psychoneuroendocrinology 1992; 17: 673 – 9.
11 Jürgensen O, Bardé B. Psychodynamic findings in women with elevated serum prolactin. In Dennerstein L., De Senarclens M. eds.. The Young Woman, Amsterdam, Excerpta Medica, 1983, 210-16.
12. Rojas LM, Sthory I, Canales ES, Zarate A Factores psicogénicos en el sindroma de amenorrea-galactorrea. Ginecologia y Obstetricia de México 1981; 49: 291-5.
13. Sobrinho LG, Nunes MCP, Calhaz-Jorge C, Afonso AM, Pereira MC, Santos MA Hyperprolactinemia in women with paternal deprivation during childhood. Obstetrics & Gynecology 1984; 64; 465-8.
14. Lozoff B, Felt BT, Nelson EC, Wolf AW, Meltzer HW, Jimenez E. Serum prolactin and behavior in infants. Biological Psychiatry 1995; 37; 4-12.
15. Theorell T. Prolactin – a hormone that mirrors passiveness in crisis situations. Integrative Physiological and Behavioral Science 1992; 27: 32-38.
16. Lökk J Arnetz B. Psychophysiological concomitants of organizational change in health care personnel: effects of controlled intervention study. Psychoter Psychosom 1997; 66: 74-7.
17. Grossi G, Theorell T, Jürisoo M, Setterlind S. Psychophysiological correlates of organizational change of unemployment among police inspectors. Integr Physiol Behav Sci 1999; 34: 30-42.
18. Di Carlo R, Muccioli G, Papotti M, Bussolati G: Characterization of prolactin receptor in human brain and choroid plexus. Brain Res 1992; 570: 341 – 6.
19. Dutt A, Kaplitt MG, Kow L-M, Pfaff DW: Prolactin, central nervous system and behavior: a critical review. Neuroendocrinology 1994; 59: 413–9.
20. Pi XJ; Grattan DR Distribution of prolactin receptor immunoreactivity in the brain of estrogen-treated, ovariectomized rats. J Comp Neurol, 1998; 394: 462-74.

21. Assies J, Schellekens APM, Touber JL. Protein hormones in cerebrospinal fluid: Evidence for retrograde transport of prolactin from the pituitary to the brain in man. Clin Endocrinol 1978; 8: 887-91.
22. Herbert J. Peptides in the limbic system: neurochemical codes for co-ordinated adaptive responses to behavioural and physiological demand. Progress in Neurobiology 1993 41: 723-91.
23. Emanuele NV, Jurgens JK, Halloran MM, Tentler JJ, Lawrence Am, Kelley MR. The rat prolactin gene is expressed in brain tissue; detection of normal and alternatively spliced prolactin messenger. Mol Endocrinol 1992; 2: 35-42.
24. Massara F, Camanni F, Martra M, Dolfin GC, Muller EE, Molinatti GM Reciprocal pattern of the TSH and PRL responses to dopamine receptor blockade in women with physiological or pathological hyperprolactinaemia. Clin Endocrinol 1983; 18: 103-10.
25. Seki K, Kato K. Luteinizing hormone response to domperidone in hyperprolactinaemic women with pituitary microadenomas. Clin Endocrinol 1988; 28: 271-5.
26. Drago F, D'Agata V, Iacona T, Spadaro F, GRassi M, Valerio C, Raffaele R, Astuto C, Lauria N, Vitetta M. Prolactin as a protective factor in stress induced biological changes. J Clin Lab Anal 1989; 3: 340-4.
27. Keverne EB. Central mechanisms underlying the neural and neuroendocrine determinants of maternal behaviour. Psychoneuroendocrinology 1988; 13: 127–41.
28. Bridges RS, Mann PE. Prolactin-brain interactions in the induction of maternal behavior in rats. Psychoneuroendocrinology 1994;19: 611-22.
29. Mc Carthy MM, Curran GH, Siegel HI. Evidence for the involvement of prolactin in the maternal behavior of the hamster. Physiology & Behavior 1994; 55: 181 – 4.
30. Lucas BK, Ormandy CJ, Binart N, Bridges RS, Kelly PA. Null mutation of the prolactin receptor gene produces a defect in maternal behavior. Endocrinology 1998; 139: 4102-7.
31. Numan M: Neural basis of maternal behavior in the rat. Psychoneuroendocrinology 1988; 13: 47 – 62.
32. Byatt JC, Staten NR, Salsgiver WJ, Kostelc JG, Collier RJ. Stimulation of food intake and weight gain in mature female rats by bovine prolactin and bovine growth hormone. Am J Physiol (Endocrinol Metab 27) 1993; 264: E968 – 92.
33. Heil SH. Sex-specific effects of prolactin on food intake in rats. Horm Behav 1999; 35: 47-54.
34. Sobrinho LG, Sá-Melo P, Nunes MCP, Barroco LE, Calhaz-Jorge C, Oliveira LC, Santos MA. Sexual dysfunction in hyperprolactinemic women. Effect of bromocriptine. J Psychosomatic Obstetrics and Gynecology 1987; 6; 43-8.
35. Berezin M, Shimon I, Hadani M. Prolactinoma iin 53 men: Clinical characteristics and modes of treatment (male prolactinoma). Clin Endocrinol 1995; 18: 436-41.
36. Laan E, van Lunsen RH. Hormones and sexuality in postmenopausal women: a psycho-physiological study. J Psychosom Obstet Gynecol 1997; 18: 126-33.
37. Matera C, Freda PU, Ferin M, Wardlaw SL: Effect of chronic opioid antagonism on the hypothalamic-pituitary-ovarian axis in hyperprolactinemic women. J Clin Endocrinol Metab 1995; 80: 540 – 5.
38. Terkel J: Neuroendocrine processes in the establishment of pregnancy and pseudopreg-nancy in rats. Psychoneuroendocrinology 1988; 13: 5 – 28.
39. Wieschhoff HA. Artificial stimulation of lactation in primitive cultures. Bulletin of the History of Medicine 1940; 8: 1403-15.
40. Scarpa A. Etnomedicina. Milan, Franco Lucisano, 1979: 475-86.
41. Auerbach KG and Avery JL. Induced lactation American Journal of Diseases in Children 1981; 135: 340-3.
42. Bowman LA, Dilley SR, Keverne EB. Suppression of estrogen-induced LH surges by social subordination in talapoin monkeys. Nature 1978; 275: 56-8.

43. Dixon AF, George L Prolactin and parental behavior in a male New World primate. Nature 1982; 299: 551-553.
44. Ziegler TE, Wegner FH, Snowdon CT. Hormonal responses to parental and non-parental conditions in male cotton-top tamarins, Saguinus oedipus, a New World primate. Horm Behav 1996; 30: 287-297.
45. Reburn CJ, Wynne-Edwards KE Hormonal changes in males of a naturally biparental and a uniparental mammal. Horm Behav 1999; 35: 163-176.
46. La Fuente JR, Rosenbaum AH. Prolactin in psychiatry. Am J Psychiatry 1981; 138: 1154-60.
47. Nicholas L, Dawkins K, Golden RN. Psychoneuroendocrinology of depression. Prolactin. Psychiatr Clin N Am 1998; 21: 341-58.
48. Chesson A, Levine S, Kong L, Lee S. Neuroendocrine evaluation in Kleine-Levin syndrome: evidence of reduced dopaminergic tone during periods of hypersomnolence. Sleep 1991; 14: 226 – 32.
49. Bamford CR. Menstrual-associated sleep disorder: an unusual hypersomniac variant associated with both menstruation and amenorrhea with a possible link to prolactin and metoclopramide. Sleep 1993; 16: 484-6.
50. Oribe E, Amini R, Nissenbaum E, Boal B: Serum prolactin concentrayios are elevated after syncope. Neurology 1996; 47: 60-2.
51. Meierkord H, Shorvon S, Lightman SL. Plasma concentrations of prolactin, noradrenalin, vasopressin and oxytocin during and after a prolonged epileptic seizure. Acta Neurol Scand 1994; 90: 73-7.
52. Lin YY, Su MS, Yiu CH, Shih YH, Yen DJ, Kwan SY, Wu ZA, Chan SHH. Relationship between mesial temporal seizure focus and elevated serum prolactin in temporal lobe epilepsy. Neurology 1997; 49: 528-32,
53. Devane GW, Vera MI, Buhl WC, Kalra PS. Opioid peptides in pseudocyesis. Obstet Gynecol 1985; 65: 183-8.
54. Osotimehin BO, Ladipo AO, Adejuwon CA, Otolorin EO. Int J Gynaecology and Obstetrics 1981; 19: 399-402.
55. Yen SSC, Rebar RW, Quesenberry W. Pituitary function in pseudocyesis. J Clin Endocrinol Metab 1976; 43: 132-6.
56. Zarate A, Canales ES, Soria J, Jacobs LS, Daughaday WH, Kastin AJ, Schally AV. Gonadotropin and prolactin secretion in human pseudocyesis. Annales d'Endocrinologie (Paris) 1974; 35: 445-50.
57. Bray MA, Muneyyirci-Delale O, Kofinas G, Reyes FI. Circadian, ultradian, and episodic gonadotropin and prolactin secretion in human pseudocyesis. Acta Endocrinol 1991; 124: 501-9.
58. Starkman M, Marshall JC, La Ferla J, Kelch RP. Pseudocyesis: psychologic and neuroendocrine relationships. Psychosomatic Medicine 1985; 47: 46-57.
59. Tulandi T, McInnes RA, Lal S. Altered pituitary hormone secretion in patients with pseudocyesis. Fertil & Steril 1983; 40: 637-41.
60. Aruffo RN. Lactation as a denial of separation. Psychoanal Quart 1971;40: 100-22.
61. Briehl W, Kulka W. Lactation in a virgin. Psychoanal Quart 1935; 4: 484-511.
62. Corenblum B, Whitaker M. Inhibition of stress-induced hyperprolactinemia. Brit Med J 1977;2:1328.
63. Biondi M, Picardi A. Psychological stress and neuroendocrine function in humans: the last two decades of research. Psychoter Psychosom 1999; 68: 114-50.
64. Ratner A, Yelvington DB, Rosenthal M. Prolactin and corticosterone response to repeated footshock stress in male rats. Psychoneuroendocrinology 1989; 14: 393-6.
65. Krüger T, Exton MS, Pawlak C, von zur Mühlen A, Hartman U, Schedlowski M. Neuroendocrine and cardiovascular response to sexual arousal and orgasm in men. Psychoneuroendocrinology 1998; 23: 401-11.

66. Exton MS, Bindert A, Krüger T, Scheller F, Hartman U, Schledowski M. Cardiovascular and endocrine alterations after masturbation-induced orgasm in women. Psychosomatic Medicine 1999; 61: 280-9.
67. Buckman MT, Kellner R. Reduction of distress in hyperprolactiemia with bromocriptine. Am J Psychiatry 1985; 142: 242-4.
68. Sonino N, Fava GA, Boscaro M: A role for life- events in the pathogenesis of Cushing's disease. Clin Endocrinol 1993; 38: 261-4.
69. Burrow GN, Wortzman G, Rewcastle RB, Holgate RC, Kovacs K. Microadenomas of the pituitary and abnormal sellar tomograms in an unselected autopsy series. N. Engl. J Med 1981; 304: 156-8.
70. Studer H, Derwhal M. Mechanisms of nonneoplastic endocrine hyperplasia.a changing concept: a review focused on the thyroid gland. End Rev 1995; 16: 411.
71. Asa S, Kovacs K, Stefaneau L, Horvath E, Billestrup N, Gonzalez-Manchon C, Vale W. Pituitary mammosomatotroph adenomas develop on old mice transgenic for growth-hormone releasing hormone. Proc Soc Exptl Biol Med 1990; 193: 232-5.
72. Zimmerman D, Young WF, Ebersold MJ, Scheithauer BW, Kovacs K, Horvath E, Whitaker MD, Eberhardt NL, Down TR, Frohman LA: Congenital gigantism due to growth hormone-releasing hormone excess and pituitary hyperplasia with adenomatous transformation. J Clin Endocrinol Metab 1993; 76: 216-22,
73. Saiardi A; Bozzi Y; Baik JH; Borrelli E. Antiproliferative role of dopamine: loss of D2 receptors causes hormonal dysfunction and pituitary hyperplasia. Neuron 1997; 19, 115-26.
74. Gooren LJG, Assies J, Asscheman H, De Slegte R, Van Kessel H. Estrogen-induced prolactinoma in a man. J Clin Endocrinol Metab 1988; 66. 444-6.
75. Kovacs K, Stefaneau L, Ezzat S, Smyth HS. Prolactin-producing pituitary adenoma in a male-to-female transsexual patient with protracted estrogen administration. A morphologic study. Arch Path Lab Med 1994; 118: 562-5.
76. Serri O, Noiseux D, Robert F, Hardy J. Lactotroph hyperplasia in an estrogen treated male-to-female transexual patient. J Clin Endocrinol Metab 1996; 81; 3177-9.
77. Scheithauer BW, Sano T, Kovacs KT, Young Jr WF, Ryan N, Randall RV. The pituitary gland in pregnancy: a clinicopathologic and immunocytochemical study of 69 cases. Mayo Clinic Proceedings 1990; 65: 461-74.
78. Pituitary adenoma study group. Pituitary adenomas and oral contraceptives: a multicenter case-control study. Fertil Steril 1983; 39: 753-60.
79. Ferreira MF, Sobrinho LG, Pires JS, Silva MES, Santos MA, Sousa MFF. Endocrine and psychological evaluation of women with recent weight gain. Psychoneuroendocrinology 1995; 20: 53 – 63.
80. Ferreira MF , Sobrinho LG , Santos MA , Sousa MFF, Uvnäs-Moberg K. Weight gain, at least in some women, is an expression of a neuroendocrine state characterized by reduced hypothalamic dopaminergic tone. Psychoneuroendocrinology 1998; 23:1005-13.
81. Uvnäs-Moberg K. Role of efferent and afferent vagal nerve activity during reproduction: integrating function of oxytocin on metabolism and behavior. Psychoneuroendocrinology 1994; 19: 687-695.
82. Liu JW, Ben Jonathan N. Prolactin-releasing activity of neurohypophysial hormones: structure-function relationship. Endocrinology 1994; 134: 114-8 .
83. Dépôt M, Caillé G, Mukherjee J, Katzman MA, Cadieux A, Bradwejn J. Acute and chronic role of 5-HT3 neuronal system on behavioral and neuroendocrine changes induced by intravenous cholecystokinin tetrapeptide administration in humans. Neuropsychopharmacology 1999; 20: 177-87.
84. Sorensen RL. Lactogen control of pancreatic islet cells. This book.

85. Sandler J, Dare C, Holder A. The patient and the analyst. London, Allen and Unwin, 1973: 73.
86. Sobrinho LG, Almeida-Costa JM. Hyperpprolactinemia as a result of immaturity or regression: the concept of maternal subroutine. Psychoter Psychosom 1992; 57: 128-32.
87. Akistal HS Mood disorders: clinical features. In: Kaplan H.I., Sadock B.J. eds. Comprehensive Textbook of Psychiatry, 6th ed., Baltimore, Williams & Wilkins, 1995: 1123-52.
88. Lam RW and Stewart JN. The validity of atypical depression in DSM-IV. Comprehensive Psychiatry 1996; 37: 375-383.
89. American Psychiatric Association: Diagnostic and Statistic Manual of Psychiatric Disorders, 4th ed., American Psychiatric Association, Washington, 1994: 384-6.
90. Cincotta AH, Schiller BC, Meier AH. Bromocriptine inhibits the seasonally occurring obesity, hyperinsulinemia, insulin resistance and impaired glucose tolerance in the Syrian Hamster, Mesocricetus auratus. Metabolism 1991; 40: 639-44.

Chapter 7

Genetics of Prolactinomas

Toni R. Prezant

Cedars-Sinai Medical Center &
UCLA School of Medicine, Los Angeles, CA

BACKGROUND

Prolactinomas are tumors of lactotroph cells in the anterior pituitary that hypersecrete prolactin, resulting in amenorrhea and galactorrhea in women and reduced sexual function in males. Macroadenomas cause additional symptoms, including visual defects and severe headaches, and are present in 90% of males with prolactinomas (1). Prolactinomas are the most common of all pituitary adenomas and occur more frequently in females, with a 3:1 preponderance (2). Mixed tumors expressing both prolactin and growth hormone also occur, and are derived from the earlier precursor mammosomatotrophs, which diverge into lactotroph and somatotroph lineages in normal development.

Like other types of pituitary tumors, prolactinomas are usually due to monoclonal expansion of progenitor cells that have sustained tumorigenic initiating event(s), with further growth influenced by extrinsic hypothalamic hormones (3). Evidence in support of this "intrinsic defect" hypothesis came from X-inactivation studies. In females, one of the two X chromosomes is inactivated and methylated, and can be distinguished by restriction fragment length polymorphism analysis, using methylation-sensitive restriction enzymes. Of seven female patients with hyperprolactinemia, the X chromosomes had a nonrandom methylation pattern in prolactinomas (n=4), corresponding to the expansion of a single cell (3). However, hyperprolactinemia associated with mixed or plurihormonal adenomas (n=3) was not monoclonal, raising the possibility that these polyclonal pituitary tumors were instead initiated by extrinsic changes in hypothalamic factors. Another possibility is that the elevated PRL levels were due to causes other than prolactinoma, such as pituitary stalk compression, which blocks dopamine inhibition (1).

Prolactinomas are generally sporadic, occurring in patients without a family history or without additional neuroendocrine tumors. They are a common feature, though, in the familial tumor syndrome Multiple Endocrine Neoplasia Type I (MEN I), and some MEN I "variant" families have been reported that have frequent prolactinomas, but altered frequencies of the other classic tumor types. Despite its importance in familial prolactinomas, however, the *men1* gene does not commonly have mutations or altered expression in sporadic prolactinomas.

Other genes whose products regulate prolactin synthesis or secretion, and lactotroph development or growth, have been screened for mutations or abnormal expression in prolactinomas. Most of these candidates have been excluded from a role in tumorigenesis, except for the receptors for dopamine D2 and estrogen, and basic fibroblast growth factor.

Animal models have implicated additional genes in tumorigenesis: *DRD2*, *hst*, *TGFα*, $p27^{Kip1}$, possibly *NGF*, and the recently discovered oncogene *PTTG* (Pituitary Tumor Transforming Gene). Many studies ruled out involvement of well-known oncogenes (*ras, c-myc, c-myb, c-fos*) and tumor suppressor genes (*p53*, *Rb*, *nm23*) in the majority of pituitary tumors, although alterations are sometimes associated with invasiveness. Finally, two cell cycle regulatory proteins (p16, p27) have significantly decreased expression in the majority of pituitary adenomas, involving DNA methylation and ubiquitin-mediated proteolysis pathways in tumorigenesis.

Table 7-1 outlines the functions of these and other genes and the evidence for or against their involvement in prolactinoma formation. The experimental evidence is elaborated further in the following sections, which highlight candidate genes and their roles in the multistep process of prolactinoma initiation, promotion and progression.

GENES INHIBITING LACTOTROPH GROWTH OR PRL SECRETION

DRD2

Prolactin secretion and lactotroph growth are inhibited by dopamine and its agonists, acting through the dopamine D2 receptor (see Chapter 4). Blocking the dopamine inhibition by pituitary stalk compression, neovascularization (5), or gene disruption (6) results in lactotroph hyperplasia, hyperprolactinemia and prolactinoma formation. Female transgenic *DRD2* knockout mice develop ~10-fold larger pituitaries with multifocal prolactinomas after 17-18 months (6). The prolonged lactotroph hyperplasia that preceded adenoma development suggested that removing dopamine inhibition allowed an increased pool of lactotrophs to acquire initiating tumorigenic changes in these female mice. However, with a similar age of onset, male *DRD2* knockout mice developed multifocal microprolactinomas without an increase in pituitary size. This could suggest that loss of DRD2 signalling activates two different gender-specific tumorigenic pathways. Alternatively, hyperplasia might not be a prerequisite for tumorigenesis.

Although the *DRD2* gene is a strong candidate for involvement in human pituitary tumorigenesis, no mutations were detected in coding sequences of the *DRD2* gene in 46 human prolactinomas and 19 mixed GH/PRL adenomas (4). However, dopamine receptor expression is reduced in prolactinomas that are nonresponsive to bromocriptine (7, 8).

GENES PROMOTING LACTOTROPH GROWTH OR PRL SECRETION

ESTROGEN RECEPTOR

Estrogen stimulates the proliferation of normal and transformed lactotrophs and gonadotrophs. High circulating levels of estradiol can account for the female preponderance of prolactinomas and their increased size during pregnancy when estrogen levels rise (2). In addition, pharmacologic doses of estrogen cause lactotroph hyperplasia and adenomas in rats (9). Estrogen stimulates the prolactin promoter and activates the expression of other genes implicated in tumorigenesis (*PTTG*, *TGFα*, described below).

Table 7-1. Candidate Genes in Prolactinoma Tumorigenesis

Gene	Function	Evidence for involvement	Evidence against involvement
men1	menin tumor suppressor protein, interacts with JunD represses transcription	PRLomas occur frequently in MEN I (Multiple Endocrine Neoplasia Type I)	mutations are infrequent & mRNA levels normal in most sporadic PRLomas
DRD2	dopamine receptor subtype 2, inhibits PRL secretion and lactotroph growth	*DRD2* -/- mice develop PRLomas dopamine agonists shrink PRLomas, decreased expression in nonresponders	no *DRD2* mutations in human PRLomas
NGF	nerve growth factor	reduces tumorigenicity of 1° PRLoma cultures (after 90 days *in vitro*)	ectopically expressed *NGF* induces lactotroph hyperplasia in transgenic mice
bFGF	basic fibroblast growth factor, FGF-2, angiogenic protein	highly expressed in PRLomas, induced early after estrogen treatment	
hst	heparin binding secretory transforming gene, fibroblast growth factor, FGF-4	in "transforming DNA" from 2 human PRLomas, transfected GH4 cells are more tumorigenic, expressed in invasive PRLomas	other DNA sequences also present, transforming changes unknown, normal in Southern blot of human PRLomas
ERα, ERβ	estrogen receptor	estrogen induces PRLomas in rats, human PRLomas express ER and constitutively active ER splice variants	

Table 7-1 continued. Candidate Genes in Prolactinoma Tumorigenesis

Gene	Function	Evidence for involvement	Evidence against involvement
TGFα	transforming growth factor-α, mitogenic protein	expressed in lactotrophs, induced early by estrogen, PRLomas in female transgenic mice	
H-, K-, N-*ras*	ras protein family, transduces signals from tyrosine kinase receptors	activating mutations found in 2 highly invasive PRLomas	no mutations in most benign or invasive PRLomas
PTTG1	pituitary tumor transforming gene 1, securin	overexpression is tumorigenic, induc bFGF, induced early by estrogen, increased in pituitary tumors and cancer, regulates chromosome segregation?	
PTTG2	pituitary tumor transforming gene 2	increased proportion of *PTTG2* mRNA in some pituitary tumors	
p53	checkpoint protein, senses DNA damage and blocks cell cycle progression	immunodetectable in all pituitary carcinomas and some invasive adenomas	no mutations, conflicting reports of immunodetectable p53 in less invasive adenomas
Rb	retinoblastoma protein, tumor suppressor gene, controls cell cycle progression	*Rb* knockout mice develop tumors of intermediate lobe, 13q14 LOH in invasive or malignant pituitary tumors	no mutations, hypophosphorylated Rb (active form) is immunodetected

Table 7-1 continued. Candidate Genes in Prolactinoma Tumorigenesis

Gene	Function	Evidence for involvement	Evidence against involvement
p27 Ki pl	cyclin dependent kinase inhibitor protein, p27, blocks phosphorylation of Rb protein to keep it in active form	*p27* knockout mice develop tumors of intermediate lobe, decreased or absent protein in most human pituitary tumors	
pl 6 INK4a	cyclin dependent kinase inhibitor protein, pl 6, blocks phosphorylation of Rb protein to keep it in active form	protein undetectable in all pituitary tumors, mRNA usually decreased	
nm23	purine-binding factor, tumor suppressor in rodent models decreased expression in metastatic cell lines	decreased expression of H2 isoform protein and mRNA in invasive pituitary tumors	

Estrogen acts through the estrogen receptor (ER), a ligand-activated transcriptional factor in the steroid receptor superfamily. Prolactinomas are the pituitary tumor subtype that most strongly express estrogen receptors, shown by immunohistochemistry, binding assays and ribonuclease protection assays (10, 11).

ER is encoded by two genes, *ERα* and *ERβ*. *ERα* is expressed in 70-100% of prolactinomas (12), while *ERβ* is detectable in 60% (13). *In vitro* experiments showed that binding of estradiol leads to ER dimerization and activation of estrogen-responsive genes, with a stronger estrogen response due to ERα (14-fold induction) than ERβ (4-fold induction) (13).

Alternative splicing of *ERα* mRNAs can produce estrogen receptor isoforms with altered responsiveness to estrogens and antiestrogens, suggesting an attractive mechanism for tumorigenesis (12, 13). In particular, deletion of exon 2 results in a stimulatory isoform and deletion of exon 5 (ERα-delta5) produces a constitutively active isoform that is hormone independent. In contrast, deletions of exon 3 or exon 7 produce dominant negative isoforms of ERα, which lack DNA binding or transactivation functions, respectively. Transcripts containing the exon 2 and exon 5 deletions were detected in nearly all prolactinomas (12, 13), while exon 7 deletions were much less common (6 of 14), although the relative proportions of mutant and full-length ERα cDNAs were not determined. ERβ, while not strongly activating by itself and not expressed in all prolactinomas, can nonetheless interact with ERα-delta5 to increase expression of a luciferase reporter gene controlled by an estrogen response element (13).

PROLACTIN RELEASING PEPTIDE

Prolactin secretion is stimulated by the recently discovered hypothalamic hormone, prolactin releasing peptide (PrRP), the ligand which binds to a pituitary G-protein coupled receptor, PrRP-R (14). As expected, normal human pituitary expresses *PrRP-R* mRNA, but, surprisingly, *PrRP* is also transcribed in normal pituitary and in a subset of each type of pituitary adenoma (15). No role has yet been shown for this peptide in prolactinoma formation, although *PrRP*-positive prolactinomas (5 of 11) were dopamine responsive.

TRH-R

In addition to stimulating TRH release from thyrotrophs, hypothalamic TRH stimulates PRL release from lactotrophs (16). The stimulation is mediated by the pituitary G-protein coupled TRH receptor. The *TRH-R* gene was screened for activating mutations that could stimulate lactotrophs in the absence of ligand (17). No mutations were detected by single-strand conformational polymorphism in 15 prolactinomas.

GENES INVOLVED IN LACTOTROPH DEVELOPMENT

TRANSCRIPTION FACTORS

The pituitary-specific transcription factor Pit-1 is a transactivator of the *GH*, *PRL* and *TSHβ* genes, and is required for proliferation of the cells that express these hormones (reviewed in 8). To determine whether increased expression or activating mutations of this transcription factor could be involved in tumorigenesis, *Pit-1* mRNA was quantitated in 15 pituitary adenomas (including 5 PRL and 2 PRL/GH tumors), and found to be increased 2.5- to 5-fold compared to normal pituitary (18). However, this expression level was attributed to the greater predominance of Pit-1 dependent cell types in these tumors. The *Pit-1* sequence was also normal in 2 prolactinomas, ruling out *Pit-1* activating mutations as a common mechanism of tumorigenesis (18).

Alternative splicing of *Pit-1* mRNA can produce different isoforms with altered DNA-binding and transactivation properties, and was tested as a possible mechanism for cell-specific expression of pituitary hormones. RT-PCR analyses did not reveal major quantitative differences of *Pit-1* isoforms in 15 PRL, 7 PRL/GH and 8 GH tumors (8). However, lower *Pit-1* and *DRD2* mRNAs were correlated (significantly) in prolactinomas with decreased response to dopaminergic agonist.

The Prophet of Pit-1 gene (*PROP-1*) is a transcription factor expressed in early pituitary development (reviewed in 19). In humans, *PROP-1* mutations result in combined pituitary hormone deficiency (deficits of GH, PRL, TSH and gonadotropins). Nonquantitative RT-PCR experiments showed that *PROP-1* is expressed in both prolactinomas and in normal pituitary, so altered expression is probably not associated with tumorigenesis.

GROWTH FACTORS

HST

The heparin-binding secretory transforming gene (*hst*), also known as fibroblast growth factor 4 (FGF-4), was identified in "transforming DNA" from two of five human prolactinomas (20). Genomic DNAs were transfected into NIH 3T3 cells, and transformed foci were selected. After a second round of transfection and selection, the human DNA was screened by Alu PCR and hybridization with known oncogenes. The presence of *hst* among the human sequences was confirmed by PCR and partial sequence analyses. It was not determined what, if any, oncogenic changes had occurred to the *hst* gene, or if other transforming sequences were present. The *hst* gene appeared normal by Southern blot screening in 7 prolactinomas (21).

A more direct demonstration of *hst* involvement in tumorigenesis came from stable transfections of GH4 cells with human *hst* cDNA (22). The transfected cells formed more aggressive tumors in Wistar-Furth rats. In addition to its role in lactotroph proliferation, recombinant FGF-4 treatment of GH4 or GC cells, or primary rat pituitary culture, resulted in dose- and time-dependent increases in prolactin transcription and secretion (22).

Finally, strong FGF-4 immunoreactivity was detected in 36% of human prolactinomas (5 of 14) and correlated with invasiveness, assessed by the Ki-67 index, a proliferation related nuclear antigen (23). FGF-4 protein was not expressed in normal anterior pituitary or in most other pituitary tumor subtypes. Taken together, these studies implicate *hst* in prolactinoma formation, although the mechanism of gene activation is not known.

BFGF

Basic fibroblast growth factor, or bFGF, is a potent mitogen for neuroectoderm cells *in vitro* and stimulates angiogenesis *in vivo*, and has highest expression in the pituitary and brain (reviewed in 24). bFGF is secreted, although it lacks a classical amino-terminal signal sequence. Normally bFGF is undetectable in human serum, but in 7 of 12 MEN I patients with prolactinomas, immunoreactive bFGF was detectable before surgery or dopamine agonist treatment and decreased after treatment (surgery (n=4) or bromocriptine (n=4)) (24).

Elevated plasma bFGF might be specific to MEN I, since it was not detectable in three patients with sporadic macroprolactinomas. However, in seven sporadic prolactinomas, intratumor bFGF expression was positively correlated with expression of the *PTTG* oncogene (25). Overexpression of the human *PTTG* proto-oncogene in NIH 3T3 cells causes increased synthesis and secretion of bFGF (26), and transcription of both genes was increased in a time- and dose-dependent manner in rats treated with estrogen, prior to development of prolactinomas (25).

NGF

Nerve growth factor, or NGF, might play a role in prolactinoma tumorigenesis. Through binding to cell surface tyrosine kinase receptors, NGF is involved in growth, differentiation, and survival of neurons. NGF and its receptors are also expressed in the anterior pituitary, specifically in the mammosomatotroph lineage (27). Experimental manipulation of primary cultures of human prolactinomas showed that *NGF* and its receptor were not expressed in dopamine nonresponders, but both genes could be induced by NGF treatment. The dopamine responders, which express *NGF*, had lower tumorigenic potential than nonresponders (according to soft agar colony assay, ^{3}H-thymidine uptake and tumor formation in nude mice). These phenotypes could be reversed in cultured prolactinoma cells by inducing *NGF* and its receptor (with NGF treatment) or inhibiting their expression (with antibody to NGF or antisense oligonucleotides). A caveat in interpreting these experiments is that the cultures were maintained for 90 days, and the authors did not state that prolactin secretion was measured throughout this time period to confirm lactotroph cell phenotype. Otherwise, the experiments suggest that *NGF* participates in proliferation and progression of prolactinomas through an autocrine loop.

An apparently opposite effect of NGF was observed in transgenic mice bearing *NGF* cDNA under the transcriptional control of the rat prolactin promoter (28). By 12 months the pituitaries of female transgenic mice were 5-fold larger than in age-matched controls, and by 15-18 months the pituitaries were 10-100 times larger than normal. Histologically, the pituitaries had lactotroph hyperplasia without adenomas, and prolactin and NGF antibodies generally did not co-stain the same cells. It is not yet clear why *NGF* overexpression *in vivo* leads to lactotroph hyperplasia, while *in vitro* induction of *NGF* expression suppresses tumorigenesis.

TGFα

Transforming growth factor-alpha, TGFα, is a mitogenic protein that is found in conditioned medium from retrovirally transformed cells, and is expressed in fetal development and in several adult tissues, including lactotrophs (reviewed in 29). In ovariectomized adult rats treated with estrogen, *TGFα* mRNA levels increase before lactotroph hyperplasia begins, suggesting it might be involved in prolactinoma formation. Female transgenic mice containing the *TGFα* cDNA under the transcriptional control of the rat PRL promoter developed lactotroph hyperplasia by 6 months, and prolactinomas by 12 months; no pituitary disease appeared in transgenic males (29). This growth factor did not induce tumors of other pituitary cell subtypes, indicating a specific role in prolactinoma tumorigenesis.

ONCOGENES

RAS

Ras proteins participate in signal transduction across the plasma membrane, interacting with tyrosine kinase receptors and transmitting signals to the nucleus through a protein kinase cascade. An activating mutation was identified in the H-*ras* gene in two highly invasive prolactinomas (30, 31). Pei *et al.* (31) found that the H-*ras* mutation arose *de novo* in the metastatic tumor. To determine whether *ras* activating mutations played a general role in prolactinoma tumorigenesis, Cai *et al.* (32) screened the N-, H-, and K-*ras* genes for activating mutations in codons 12, 13 and 61, by allele-specific oligonucleotide hybridization. In this larger study, no activating mutations were detected in 59 prolactinomas and 13 invasive prolactinomas, thereby ruling out *ras* activating mutations in these three codons as a common occurrence in initiating prolactinoma tumorigenesis, but *ras* activation may play a role in tumor progression.

C-MYC, C-MYB, C-FOS

Altered expression of the nuclear oncogenes *c-myb* and *c-fos* does not appear to be a common mechanism for pituitary tumorigenesis, although *c-myc* mRNA was overexpressed in 9 of 30 pituitary tumors, including 1 of 5 prolactinomas (33). However 2 invasive prolactinomas had normal *c-myc* mRNA levels, so there was not a correlation between *c-myc* expression and proliferative state.

PTTG

Pituitary tumor transforming gene, or *PTTG*, was isolated by differential display because of its increased expression in the rat GH4 pituitary tumor cell line, relative to normal rat pituitary (34). Overexpression of the cDNAs for rat or human *PTTG* results in transformation of NIH 3T3 cells and tumors in athymic mice (26, 34). While *PTTG* is mainly expressed in normal adult testis, *PTTG* mRNA is increased dramatically in cancer cell lines and primary neoplasia (26, 35, 36). *PTTG* mRNA was also increased in 10/10 prolactinomas, in some cases more than 10-fold, compared to 5 normal pituitary controls, as assessed by semi-quantitative RT-PCR (37). *PTTG* transcript levels correlated with invasiveness for these and other secretory pituitary tumors.

PTTG overexpression in NIH 3T3 cells led to increased transcription and secretion of bFGF (26). The carboxyl half of human PTTG protein has two Pro-X-X-Pro motifs, that potentially bind to SH3 proteins, involved in signal transduction. Disruption of these motifs abolishes the bFGF induction, as well as the transforming activity of PTTG protein.

In the rat model of estrogen induced prolactinomas, both *PTTG* mRNA and bFGF protein are induced early (24 hrs.), preceding lactotroph hyperplasia and progression to tumors (25). This study also demonstrated a direct correlation between *PTTG* and bFGF mRNA levels in human pituitary tumors. Taken together, *PTTG* is an estrogen-inducible proto-oncogene, that induces bFGF, an angiogenic factor, early in lactotroph tumorigenesis.

PTTG protein co-immunoprecipitates with separin, a vertebrate protein involved in chromosome segregation, so Zou *et al.* (38) gave the alias "securin" to *PTTG*. PTTG protein levels vary during the cell cycle, disappearing before cells enter anaphase. These investigators hypothesize that premature chromosome segregation occurs if *PTTG* is dysregulated, leading to aneuploidy and cancer.

PTTG2

While determining the chromosomal location of human *PTTG* by radiation hybrid mapping (5q33) (26), we consistently amplified an intronless form of *PTTG, PTTG2,* that mapped to a different chromosome, 4p12 (39). The genes are highly homologous, with

90% identity in the encoded proteins. Expression of the two genes could be distinguished by gene-specific oligonucleotide hybridization in a PCR-ELISA. We found that intronless *PTTG2* is expressed in normal and tumourous pituitary, ranging from <1% to 8.7% of total *PTTG* cDNA, and is not subtype specific. Although *PTTG2* is transcribed in pituitary tumors and in some normal tissues with low levels of *PTTG* mRNA, it is not appreciably expressed in five cancer cell lines or in normal testis. Thus, this proto-oncogene family appears to differentially participate in tumor-specific pathogenesis.

TUMOR SUPPRESSOR GENES

Tumor suppressor genes (tsg's) restrain cell division, and their loss by homozygous inactivation or lack of expression results in progression through the cell cycle. The major tsg's implicated or tested for roles in pituitary tumorigenesis are summarized in Figure 7-1.

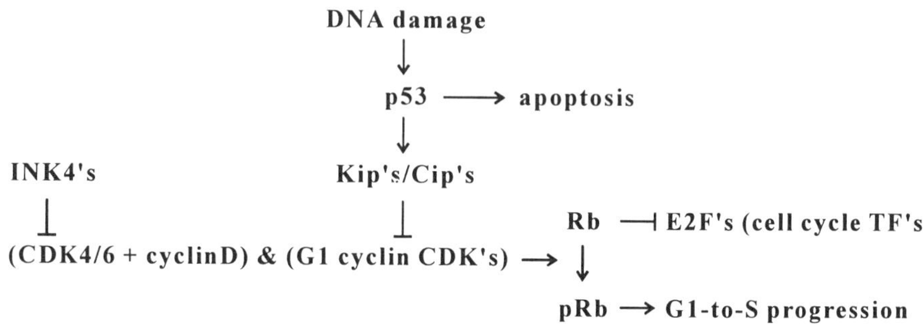

Figure 7-1. The regulation of the G1-to-S transition by tumor suppressor gene products. Cyclin dependent kinases (CDK's) phosphorylate and inactivate Rb, allowing release of inhibition of E2F transcription factors (TF's). Rb phosphorylation is inhibited indirectly by blocking CDK binding to cyclin (INK4's) or by blocking cyclin kinase activity (Kip's). One of the Kip proteins is activated by p53, in response to DNA damage. Adapted from Hanahan & Weinberg (40).

P53

p53 encodes a checkpoint protein that prevents cell cycle progression when DNA is damaged, and is the most commonly mutated gene found in human tumors (41, 42). Missense mutations usually occur in exons 5 through 8 (41), and may be associated with the nuclear accumulation of mutant forms of p53 that have a longer half-life than wild-type p53, although the correlation is not absolute (43, 44). Most mutational studies included few prolactinomas, but in general, *p53* mutations are not common in pituitary adenomas. Herman *et al.* (21) found no mutations in 22 GH and 22 nonfunctioning tumors, Levy *et al.* (43) found no mutations in 29 adenomas, including 1 prolactinoma, and Pei *et al.* (31) identified a *p53* mutation in one PRL carcinoma.

p53 expression studies are less concordant. Some investigators reported p53 immunopositive non-invasive adenomas (17%, (43); 23%, (45)), while Thapar *et al.* (44) found a clear correlation between p53 nuclear immunopositivity and tumor aggressiveness. In their study of 70 pituitary adenomas and 7 carcinomas (including 4 PRL tumors), p53 was expressed in 100% of carcinomas, in 15% of invasive adenomas and none of 37 noninvasive adenomas.

RB

The retinoblastoma *Rb* tumor suppressor gene participates in the control of the cell cycle, and is differentially phosphorylated (inactivated) by cell cycle dependent kinases and cyclins in response to growth signals (40, 46), summarized in Figure 7-1. *Rb* was the classic tumor suppressor gene that led to the Knudson two-hit hypothesis: germline mutations predispose to bilateral multifocal tumors because one of two recessive mutations is already present, whereas sporadic tumors require two somatic mutations and are more rare (47). The second mutation is often a heterozygous deletion that unmasks the mutant allele on the corresponding chromosome, visualized as a loss of heterozygosity, or LOH, for polymorphic DNA markers flanking the gene.

Transgenic mice with *Rb* disruptions develop pituitary tumors, but the tumors originate from the intermediate lobe, not the anterior pituitary, in ACTH producing cells (48, 49). Because pituitary tumors were found in *Rb* knockout mice, it was a prime candidate for involvement in human pituitary tumors. However, no intragenic LOH was found in more than 50 benign adenomas (50, 51), or in 2 invasive prolactinomas (52). A higher frequency of LOH was suggested by Pei *et al.* (53), who examined aggressive pituitary tumors for *Rb* deletions and loss of expression. One *Rb* allele was presumed lost in 13/13 highly invasive or malignant pituitary tumors, including two metastatic PRL carcinomas, since the tumor DNA was homozygous for an intragenic polymorphic marker that is heterozygous in 94% of controls (53). In one of these PRL carcinomas, lymphocyte DNA was available to directly demonstrate LOH of the *Rb* locus, with retention of heterozygosity for flanking markers on chromosomal region 13q14, while two benign prolactinomas retained both *Rb* alleles.

Despite the potential LOH at the *Rb* locus, Rb protein was immunodetectable for 10/10 tumors (53). Similar results were obtained by Woloschak *et al.* (54), who ruled out *Rb* mutations in the "pocket domain" and furthermore detected hypophosphorylated active Rb protein in 24 pituitary tumors. Thus, another gene on 13q14 is implicated in the progression, but not the initiation, of aggressive human pituitary tumors.

P27

The eukaryotic cell cycle is tightly controlled by cyclin-dependent kinases that promote progression through G1 to S phase, and their inhibitors, which block cell cycle progression and act as tumor suppressors (Figure 7-1). The CDK inhibitors fall into two classes. Ink4 proteins, with ankyrin repeats ($p16^{Ink4a}$, $p15^{Ink4b}$, $p18^{Ink4c}$, $p19^{Ink4d}$), block the interaction of CDK4 and CDK6 kinases with cyclin D. Cip/Kip proteins ($p21^{Cip1}$, $p27^{Kip1}$, $p57^{Kip2}$) bind to cyclin-CDK complexes and block their activation or kinase activity (55).

p27 knockout mice develop multiorgan hyperplasia and frequent tumors of POMC-positive cells of the intermediate lobe of the pituitary (56-58). Thus, like *Rb*, the *p27* gene is a candidate for a role in pituitary tumorigenesis. Bamberger *et al.* (59) screened 54 pituitary tumors for p27 expression by immunohistochemistry. p27 protein was underexpressed or absent in the majority of these tumors, and in 8 of the 9 prolactinomas. By double-staining for p27 and pituitary hormones, Lidhar *et al.* (60) observed a decrease in the percentage of cells expressing p27 in all types of pituitary tumors compared to the same cell types in normal pituitary (including 7 PRL adenomas). p27 protein was especially reduced in carcinomas (including 2 PRL). Although prolactinomas were not included in the 28 tumors investigated, Ikeda *et al.* (61) ruled out mutations in the *p27* and *p21* genes by SSCP analysis. Others (62, 63) showed that *p27* expression is regulated by post-transcriptional and post-translational mechanisms, including ubiquitin dependent protein degradation.

In summary, *p27* behaves as a tumor suppressor gene *in vivo*, where disruption of the gene leads to increased cell divisions in the organs that express p27. In contrast to *Rb*, which also leads to pituitary tumors in knockout mice, *p27* appears to be involved in human pituitary tumorigenesis, because its expression is reduced in the majority of pituitary tumors of all subtypes and is absent in pituitary carcinomas. *p27* is not generally inactivated by mutations or reduced transcription, but rather by increased proteolysis.

P16

$p16^{Ink4a}$ is a tumor suppressor gene that is inactivated in many human tumor cell lines and primary tumors (reviewed in 64). Its product normally prevents the phosphorylation/inactivation of Rb by inhibiting CDK4. Mutation or decreased expression of *p16* allows Rb to become phosphorylated, and hence cell cycle transcription factors become active. Woloschak *et al.* (64) found that p16 was undetectable by Western blot analysis in 25/25 pituitary tumors, including 3 prolactinomas. Mutations were not detected by SSCP in any of the 25 tumors, although the investigators state that 3 of the 25 tumors had homozygous *p16* gene deletions, so perhaps contaminating normal DNA was present in these three tumors. The loss of p16 protein correlated with decreased or undetectable mRNA in 93% of tumors (n=14). The 5' CpG island of the *p16* gene is extensively methylated in most human pituitary tumors, suggesting a mechanism for its transcriptional inactivation (65).

NM23

The purine-binding factor, *nm23*, has decreased expression in highly metastatic melanoma cell lines and behaves as a tumor suppressor when expressed in rodent tumor models (reviewed in 66). *nm23* gene has two isoforms, H1 and H2, which were screened for expression in 22 pituitary tumors (including 2 prolactinomas) by ribonuclease protection assay and by immunostaining (66). Decreased expression of the H2 isoform was observed in the invasive tumors, relative to the benign adenomas.

FAMILIAL PROLACTINOMAS

MEN1

MEN I is a familial syndrome in which tumors form in more than one endocrine organ, most often in the parathyroid gland (95-100%) and endocrine pancreas (73-80%), and less frequently in the anterior pituitary (16-50%) (67, 68). For unknown reasons, the majority of MEN I pituitary tumors are prolactinomas (69). Although MEN I is inherited as an autosomal dominant disorder or can arise *de novo*, it has reduced penetrance because an additional somatic mutation in the *men1* tumor suppressor gene is required for phenotypic expression. In this way, MEN I is similar to other familial cancer syndromes that conform to the Knudson two-hit model of tumorigenesis (47). Affected family members inherit a germline mutation in one chromosome that becomes unmasked by a "second hit," usually a deletion of the normal gene in the second chromosome, resulting in LOH for polymorphic DNA markers spanning the *men1* gene.

Before the cloning of the *men1* gene, an interesting case of discordant development of prolactinoma in one of two monozygotic twins provided elegant proof that a second hit was required to express the *men1* mutation (70). These brothers shared the mutant *men1* chromosome transmitted through three generations of their family, but only one of them developed a prolactinoma. Therefore, since they were genetically identical, the tumor was due to a somatic (non-inherited) mutation, most likely in the normal *men1* gene.

Linkage analysis localized *men1* to chromosomal region 11q13 (71), while finer mapping was achieved by LOH analysis to clone the *men1* gene (72), which encodes a 610 amino acid nuclear protein, menin (73). Menin interacts with the transcription factor JunD, to repress transactivation of other genes (74). The majority of mutations in *men1* are truncating (72), while many of the missense mutations have been shown to disrupt the interaction of menin with JunD (74).

Tumor suppressor genes identified in familial cancer syndromes are usually implicated in their corresponding sporadic tumors. For example, the genes associated with retinoblastoma (Rb) and familial adenomatous polyposis coli (APC) are also mutated in sporadic retinoblastoma (75) and colon carcinoma (76, 77), respectively. The *men1* gene fits this general rule for parathyroid (78) and pancreatic tumors (79), but its role in sporadic prolactinomas is less clear.

The frequency of LOH for 11q13 in sporadic pituitary tumors ranges from 0 in small studies to ~30%. In a screen of 88 pituitary tumors, the overall LOH for 11q13 was 18%, with a 12% occurrence in prolactinomas (2 of 17) (80). When pituitary tumors were stratified according to tumor grade, LOH at 11q13 occurred more often (31%) in invasive tumors (n=47, including 7 prolactinomas) than in noninvasive tumors (2.5%, n=42, including 3 prolactinomas) (81). Thus, 11q13 chromosome disruptions appear to be involved in tumor progression, rather than initiation, and they are not specific to pituitary tumor subtype. Although suggestive, these chromosomal analyses do not identify pathogenic changes in *men1*, nor do they rule out other potential tumor suppressor genes in the same region.

Sporadic pituitary tumors were screened for mutations in the *men1* tumor suppressor gene. Zhuang *et al.* (82) detected LOH in 1 of 8 prolactinomas. One nonfunctional tumor that immunostained for PRL and GH had LOH and an additional mutation of the remaining *men1* allele. Our study of 45 sporadic pituitary tumors included 4 PRLomas and 1 PRL/GH tumor (83). These five tumors retained both alleles of *men1*, with no mutations detected by dideoxy fingerprinting. The study of Farrell *et al.* (84) examined pituitary tumors previously identified as having LOH for the neighboring microsatellite marker PYGM. Two prolactinomas with LOH had no detectable mutation in the coding sequence of the remaining allele.

While *men1* mutations are rare in sporadic prolactinomas, it was possible that altered *men1* expression could be tumorigenic. Our study showed that the majority of sporadic pituitary tumors have approximately equal expression of both *men1* alleles, which rules out genomic imprinting as a mechanism to silence this chromosomal region (83). Others have performed comparative RT-PCR experiments, normalizing the cDNAs for *men1* to housekeeping genes, *PGK-1* (85) or *GAPDH* (86), finding that *men1* mRNA levels were normal in 11/12 and 4/4 prolactinomas, respectively. In cases with LOH, *men1* levels were decreased in one prolactinoma and one GH/PRL tumor (85).

In summary, *men1* mutations are rare in sporadic prolactinomas, even with LOH. However, expression may be affected by a decreased copy number of the *men1* gene.

OTHER FAMILIAL PROLACTINOMA GENES?

Four pedigrees with isolated prolactinoma in two family members were described (87). The patients and their families had normal serum calcium, PTH and gastrin, suggesting that MEN I was not the cause of the pituitary tumors. However, molecular genetic analysis was not performed, so it is not known whether these families have a variant of MEN I or a new disease.

In MEN I(Burin), families develop prolactinomas, parathyroid tumors and carcinoids, but rarely pancreatic tumors (88). The disease was linked to 11q13 in four large related kindreds from the Burin Peninsula of Newfoundland (88). Until the *men1* gene was identified, the investigators could not distinguish between two mechanisms for the unique ratios of endocrine tumors: mutations in two closely linked genes on 11q13, or a family-specific allele of *men1*. Indeed, the family has a germline nonsense mutation in *men1* (89). Thus, family specific alleles of *men1* can produce different phenotypes, although it is not clear why a truncating *men1* mutation should result in these proportions of MEN I related tumors.

Genetic analysis of the Tasman I MEN I genealogy provided evidence for a likely modifying gene that segregates independently from the *men1* locus (67). The Tasman 1 pedigree contains over 160 individuals with MEN I, all related to an original English settler in Tasmania. While prolactinomas occurred in 18 of 90 individuals affected with MEN I, the patients were not distributed evenly among the family branches. Most of the prolactinomas occurred in two branches of the family, affecting 50% and 63% of the MEN I carriers (who developed other MEN I tumors or transmitted the mutant gene). Two other branches of the pedigree had a much lower penetrance of the *men1* mutation, with 0% and 8% of the obligate MEN I carriers having prolactinomas. Thus, an additional gene(s) is postulated to affect the manifestation, or penetrance, of the *men1* mutation in this pedigree.

CONCLUSIONS

While true “initiating events” have not been identified, some of the early changes in prolactinoma tumorigenesis are becoming better understood. The recently identified menin protein interacts with JunD to regulate transcription. Presumably, altered expression of the target genes that occurs in the absence of functional menin is tumorigenic, and some of these genes may be identified in prolactinoma formation. Decreases in expression of the growth restraining genes *p16* and *p27* are a universal finding in pituitary tumors. Loss of their proteins dysregulates the cell cycle, by allowing the Rb protein to become inactivated. Many prolactinomas become insensitive to dopamine inhibition and have increased expression of growth factors (bFGF, FGF-4, TGF-α). In addition to raising levels of bFGF and TGF-α, estrogen induces *PTTG*, which is a potent oncogene when overexpressed. *PTTG* further induces bFGF, promoting angiogenesis even before lactotroph hyperplasia is evident. Finally, dysregulation of *PTTG* may result in aberrant chromosome segregation, leading to additional changes in the multistep progression of prolactinomas.

REFERENCES

1. Blevins LS, Shore D, Weinstein J and Isaacs S. Clinical presentation of pituitary tumors. In: Krisht AF and Tindall GT, eds. Pituitary Disorders, Comprehensive Management. Baltimore: Lippincott Williams & Wilkins, 1999: 145-63.

2. Holmgren U, Bergstrand G, Hagenfeldt K and Werner S. Women with prolactinoma - effect of pregnancy and lactation on serum prolactin and on tumour growth. Acta Endocrinol (Copenh) 1986; 111: 452-9.
3. Herman V, Fagin J, Gonsky R, Kovacs K and Melmed S. Clonal origin of pituitary adenomas. J Clin Endocrinol Metab 1990; 71: 1427-33.
4. Friedman E, Adams EF, Hoog A, Gejman PV, Carson E, Larsson C, De Marco L, Werner S, Fahlbusch R and Nordenskjold M. Normal structural dopamine type 2 receptor gene in prolactin-secreting and other pituitary tumors. J Clin Endocrinol Metab 1994; 78: 568-74.
5. Schechter J, Goldsmith P, Wilson C and Weiner R. Morphological evidence for the presence of arteries in human prolactinomas. J Clin Endocrinol Metab 1988; 67: 713-9.
6. Asa SL, Kelly MA, Grandy DK and Low MJ. Pituitary lactotroph adenomas develop after prolonged lactotroph hyperplasia in dopamine D2 receptor-deficient mice. Endocrinol 1999; 140: 5348-55.
7. Pellegrini I, Rasolonjanahary R, Gunz G, Bertrand P, Delivet S, Jedynak CP, Kordon C, Peillon F, Jaquet P and Enjalbert A. Resistance to bromocriptine in human prolactinomas. J Clin Endocrinol Metab 1989; 69: 500-9.
8. Pellegrini-Bouiller I, Morange-Ramos I, Barlier A, Gunz G, Figarella-Branger D, Cortet-Rudelli C, Grisoli F, Jaquet P and Enjalbert A. Pit-1 gene expression in human lactotroph and somatotroph pituitary adenomas is correlated to D2 receptor gene expression. J Clin Endocrinol Metab 1996; 81: 3390-6.
9. Lloyd RV. Estrogen-induced hyperplasia and neoplasia in the rat anterior pituitary gland. Am J Pathol 1983; 113: 198-206.
10. Friend KE, Chiou YK, Lopes MBS, Laws Jr ER, Hughes KM and Shupnik MA. Estrogen receptor expression in human pituitary: correlation with immunohistochemistry in normal tissue, and immunohistochemistry and morphology in macroadenomas. J Clin Endocrinol Metab 1994; 78: 1497-1504.
11. Jaffrain-Rea ML, Petrangeli E, Ortolani F, Fraioli B, Lise A, Esposito V, Spagnoli LG, Tamburrano G, Frati L and Gulino A. Cellular receptors for sex steroids in human pituitary adenomas. J Endocrinol 1996; 151: 175-84.
12. Shupnik MA, Pitt LK, Soh AY, Anderson A, Lopes MB and Laws Jr ER. Selective expression of estrogen receptor a and b isoforms in human pituitary tumors. J Clin Endocrinol Metab 1998; 83: 3965-72.
13. Chaidarun SS, Swearingen B and Alexander JM. Differential expression of estrogen receptor-b (ERb) in human pituitary tumors: functional interactions with ERa and a tumor-specific splice variant. J Clin Endocrinol Metab 1998; 83: 3308-15.
14. Hinuma S, Habata Y, Fujii R, Kawamata Y, Hosoya M, Fukusumi S, Kitada C, Masuo Y, Asano T, Matsumoto H, Sekiguchi M, Kurokawa T, Nishimura O, Onda H and Fujino M. A prolactin-releasing peptide in the brain. Nature 1998; 393: 272-6.
15. Zhang X, Danila DC, Katai M, Swearingen B and Klibanski A. Expression of prolactin-releasing peptide and its receptor messenger ribonucleic acid in normal human pituitary and pituitary adenomas. J Clin Endocrinol Metab 1999; 84: 4652-5.
16. Kaplan SL, Grumbach MM, Friesen HG and Gostom BH. Thyrotropin-releasing factor (TRF) effect on secretion of human pituitary prolactin and thyrotropin in children and in idiopathic hypophysiotropic hormone deficiencies. J Clin Endocrinol 1972; 35: 825-30.
17. Faccenda E, Melmed S, Bevan JS and Eidne KA. Structure of the thyrotrophin-releasing hormone receptor in human pituitary adenomas. Clin Endocrinol 1996; 44: 341-7.
18. Pellegrini I, Barlier A, Gunz G, Figarella-Branger D, Enjalbert A, Grisoli F and Jaquet P. Pit-1 gene expression in the human pituitary and pituitary adenomas. J Clin Endocrinol Metab 1994; 79: 189-96.
19. Nakamura S, Ohtsuru A, Takamura N, Kitange G, Tokunaga Y, Yasunaga A, Shibata S and Yamashita S. Prop-1 gene expression in human pituitary tumors. J Clin Endocrinol Metab 1999; 84: 2581-4.

20. Gonsky R, Herman V, S Melmed and J Fagin. Transforming DNA sequences present in human prolactin-secreting pituitary tumors. Mol Endocrinol 1991; 5: 1687-95.
21. Herman V, Drazin NZ, Gonsky R and Melmed S. Molecular screening of pituitary adenomas for gene mutations and rearrangements. J Clin Endocrinol Metab 1993; 77: 50-5.
22. Shimon I, Huttner A, Said J, Spirina OM and Melmed S. Heparin-binding secretory transforming gene (hst) facilitates rat lactotrope cell tumorigenesis and induces prolactin gene transcription. J Clin Invest 1996; 97: 187-95.
23. Shimon I, Hinton DR, Weiss MH and Melmed S. Prolactinomas express human heparin-binding secretory transforming gene (hst) protein product: marker of tumour invasiveness. Clin Endocrinol 1998; 48: 23-9.
24. Zimering MB, Katsumata N, Sato Y, Brandi ML, Aurbach GD, Marx SJ and Friesen HG. Increased basic fibroblast growth factor in plasma from multiple endocrine neoplasia type 1: relation to pituitary tumor. J Clin Endocrinol Metab 1993; 76: 1182-7.
25. Heaney AP, Horwitz GA, Wang Z, Singson R and Melmed S. Early involvement of estrogen-induced pituitary tumor transforming gene and fibroblast growth factor expression in prolactinoma pathogenesis. Nat Med 1999; 5: 1317-21.
26. Zhang X, Horwitz GA, Prezant TR, Valentini A, Nakashima M, Bronstein MD and Melmed S. Structure, expression, and function of human pituitary tumor-transforming gene (PTTG). Mol Endocrinol 1999; 13: 156-66.
27. Missale C, Losa M, Sigala S, Balsari A, Giovanelli M and Spano PF. Nerve growth factor controls proliferation and progression of human prolactinoma cell lines through an autocrine mechanism. Mol Endocrinol 1996; 10: 272-85.
28. Borrelli E, Sawchenko PE and Evans RM. Pituitary hyperplasia induced by ectopic expression of nerve growth factor. Proc Natl Acad Sci USA 1992; 89: 2764-8.
29. McAndrew J, Paterson AJ, Asa SL, McCarthy KJ and Kudlow JE. Targeting of transforming growth factor-a expression to pituitary lactotrophs in transgenic mice results in selective lactotroph proliferation and adenomas. Endocrinol 1995; 136: 4479-88.
30. Karga HJ, Alexander JM, Hedley-Whyte ET, Klibanski A and Jameson JL. Ras mutations in human pituitary tumors. J Clin Endocrinol Metab 1992; 74: 914-9.
31. Pei L, Melmed S, Scheithauer B, Kovacs K and Prager D. H-ras mutations in human pituitary carcinoma metastases. J Clin Endocrinol Metab 1994; 78: 842-6.
32. Cai WY, Alexander JM, Hedley-Whyte ET, Scheithauer BW, Jameson JL, Zervas NT and Klibanski A. Ras mutations in human prolactinomas and pituitary carcinomas. J Clin Endocrinol Metab 1994; 78: 89-93.
33. Woloschak M, Roberts JL and Post K. c-myc, c-fos, and c-myb gene expression in human pitutary adenomas. J Clin Endocrinol Metab 1994; 79: 253-7.
34. Pei L and Melmed S. Isolation and characterisation of a pituitary tumor-transforming gene (PTTG). Mol Endocrinol 1997; 13: 156-66.
35. Dominguez A, Ramos-Morales F, Romero F, Rios RM, Dreyfus F, Tortolero M and Pintor-Toro JA. *hpttg*, a human homologue of rat *pttg*, is overexpressed in hematopoietic neoplasms. Evidence for a transcriptional activation function of hPTTG. Oncogene 1998; 17: 2187-93.
36. Heaney AP, Singson R, McCabe CJ, Nelson V, Nakashima M and Melmed S. Expression of pituitary-tumour transforming gene in colorectal tumours. Lancet 2000; 355: 716-9.
37. Zhang X, Horwitz GA, Heaney AP, Nakashima M, Prezant TR, Bronstein MD and Melmed S. Pituitary tumor transforming gene (PTTG) expression in pituitary adenomas. J Clin Endocrinol Metab 1999; 84: 761-7.
38. Zou H, McGarry TJ, Bernal T and Kirschner MW. Identification of a vertebrate sister-chromatid separation inhibitor involved in transformation and tumorigenesis. Science 1999; 285:418-22.

39. Prezant TR, Kadioglu P and Melmed S. An intronless homolog of human proto-oncogene hPTTG is expressed in pituitary tumors: evidence for hPTTG family. J Clin Endocrinol Metab 1999; 84: 1149-52.
40. Hanahan D & Weinberg RA. The hallmarks of cancer. Cell 2000; 100: 57-70.
41. Hollstein M, Sidransky D, Vogelstein B and Harris CC. p53 mutations in human cancers. Science 1991; 253: 49-53.
42. Levine AJ, Momand J and Finlay CA. The p53 tumour suppressor gene. Nature 1991; 351:453-6.
43. Levy A, Hall L, Yeudall WA and Lightman SL. p53 gene mutations in pituitary adenomas: rare events. Clin Endocrinol 1994; 41: 809-14.
44. Thapar K, Scheithauer BW, Kovacs K, Pernicone PJ and Laws ER Jr. p53 expression in pituitary adenomas and carcinomas: correlation with invasiveness and tumor growth fractions. Neurosurg 1996; 38: 763-71.
45. Green VL, White MC, Hipkin LJ, Jeffreys RV, Foy PM and Atkin SL. Apoptosis and p53 suppressor gene protein expression in human anterior pituitary adenomas. Eur J Endocrinol 1997; 136: 382-7.
46. Weinberg RA. The retinoblastoma protein and cell cycle control. Cell 1995; 81: 323-30.
47. Knudson AG Jr. Mutation and cancer: statistical study of retinoblastoma. Proc Natl Acad Sci USA 1971; 68: 820-3.
48. Jacks R, Fazeli A, Schmitt EM, Bronson RT, Goodell MA and Weinberg RA. Effects of an Rb mutation in the mouse. Nature 1992; 359: 295-300.
49. Hu N, Gutsmann A, Herbert DC, Bradley A, Lee WH and Lee EY. Heterozygous Rb-1 delta 20/+ mice are predisposed to tumors of the pituitary gland with a nearly complete penetrance. Oncogene 1994; 9: 1021-7.
50. Woloschak M, Roberts JL and Post K. Loss of heterozygosity at the retinoblastoma locus in human pituitary tumors. Cancer 1994; 693-6.
51. Zhu J, Leon SP, Beggs AH, Busque L, Gilliland DG and Black PM. Human pituitary adenomas show no loss of heterozygosity at the retinoblastoma gene locus. J Clin Endocrinol Metab 1994; 78: 922-7.
52. Cryns VL, Alexander JM, Klibanski A and Arnold A. The retinoblastoma gene in human pituitary tumors. J Clin Endocrinol Metab 1993; 77: 644-6.
53. Pei L, Melmed S, Scheithauer B, Kovacs K, Benedict WF and Prager D. Frequent loss of heterozygosity at the retinoblastoma susceptibility gene (RB) locus in aggressive pituitary tumors: evidence for a chromosome 13 tumor suppressor gene other than RB. Cancer Res 1995; 55: 1613-6.
54. Woloschak M, Yu A, Xiao J and Post KD. Abundance and state of phosphorylation of the retinoblastoma gene product in human pituitary tumors. Int J Cancer 1996; 67: 16-9.
55. Sherr CJ and Roberts JM. Inhibitors of mammalian G1 cyclin-dependent kinases. Genes Dev 1995; 9: 1149-63.
56. Fero ML, Rivkin M, Tasch M, Porter P, Carow CE, Firpo E, Polyak K, Tsai LH, Broudy V, Perlmutter RM, Kaushansky K and Roberts JM. A syndrome of multiorgan hyperplasia with features of gigantism, tumorigenesis, and female sterility in $p27^{Kip1}$-deficient mice. Cell 1996; 85: 733-44.
57. Kiyokawa H, Kineman RD, Manova-Todorova KO, Soares VC, Hoffman ES, Ono M, Khanam D, Hayday AC, Frohman LA and Koff A. Enhanced growth of mice lacking the cyclin-dependent kinase inhibitor function of $p27^{Kip1}$. Cell 1996; 85: 721-32.
58. Nakayama K, Ishida N, Shirane M, Inomata A, Inoue T, Shishido N, Horii I, Loh DY and Nakayama KI. Mice lacking $p27^{Kip1}$ display increased body size, multiple organ hyperplasia, retinal dysplasia, and pituitary tumors. Cell 1996; 85: 707-20.
59. Bamberger CM, Fehn M, Bamberger AM, Ludecke DK, Beil FU, Saeger W and Schulte HM. Reduced expression levels of the cell-cycle inhibitor $p27^{Kip1}$ in human pituitary adenomas. Eur J Endocrinol 1999; 140: 250-5.

60. Lidhar K, Korbonits M, Jordan S, Khalimova Z, Kaltsas G, Lu X, Clayton RN, Jenkins PJ, Monson JP, Besser GM, Lowe DG and Grossman AB. Low expression of the cell cycle inhibitor $p27^{Kip1}$ in normal corticotroph cells, corticotroph tumors, and malignant pituitary tumors. J Clin Endocrinol Metab 1999; 84: 3823-30.
61. Ikeda H, Yoshimoto T and Shida N. Molecular analysis of p21 and p27 genes in human pituitary adenomas. Br J Cancer 1997; 76: 1119-23.
62. Pagano M, Tam SW, Theodoras AM, Beer-Romero P, Del Sal G, Chau V, Yew PR, Draetta GF and Rolfe M. Role of the ubiquitin-proteasome pathway in regulating abundance of the cyclin-dependent kinase inhibitor p27. Science 1995; 269: 682-5.
63. Hengst L and Reed SL. Translational control of $p27^{Kip1}$ accumulation during the cell cycle. Science 1996; 272: 1861-4.
64. Woloschak M, Yu A, Xiao J and Post KD. Frequent loss of the $p16^{INK4a}$ gene product in human pituitary tumours. Cancer Res 1996; 56: 2493-6.
65. Woloschak M, Yu A and Post KD. Frequent inactivation of the p16 gene in human pituitary tumors by gene methylation. Mol Carcinog 1997; 19: 221-4.
66. Takino H, Herman V, Weiss M and Melmed S. Purine-binding factor (nm23) gene expression in pituitary tumors: marker of adenoma invasiveness. J Clin Endocrinol Metab 1995; 80:1733-8.
67. Burgess JR, Shepherd JJ, Parameswaran V, Hoffman L and Greenaway TM. Prolactinomas in a large kindred with multiple endocrine neoplasia type 1: clinical features and inheritance pattern. J Clin Endocrinol Metab 1996; 81: 1841-5.
68. Calender A, Giraud S, Cougard P, Chanson P, Lenoir G, Murat A, Hamon P and Proye C. Multiple endocrine neoplasia type 1 in France: clinical and genetic studies. J Intern Med 1995; 238: 263-8.
69. Larsson C, Calender A, Grimmond S, Giraud S, Hayward NK, Teh B and Farnebo F. Molecular tools for presymptomatic testing in multiple endocrine neoplasia type 1. J Int Med 1995; 238: 239-44.
70. Flanagan DEH, Armitage M, Clein GP and Thakker RV. Prolactinoma presenting in identical twins with multiple endocrine neoplasia type 1. Clin Endocrinol 1996; 45: 117-20.
71. Larsson C, Skogseid B, Oberg K, Nakamura Y and Nordenskjold M. Multiple endocrine neoplasia type 1 gene maps to chromosome 11 and is lost in insulinoma. Nature 1988; 332: 85-7.
72. Chandrasekharappa SC, Guru SC, Manickam P, Olufemi SE, Collins FS, Emmert-Buck MR, Debelenko LV, Zhuang Z, Lubensky IA, Liotta LA, Crabtree JS, Wang Y, Roe BA, Weisemann J, Boguski MS, Agarwal SK, Kester MB, Kim YS, Heppner C, Dong Q, Spiegel AM, Burns AL and Marx SJ. Positional cloning of the gene for multiple endocrine neoplasia-type 1. Science 1997; 276: 404-7.
73. Guru SC, Goldsmith PK, Burns AL, Marx SJ, Spiegel AM, Collins FS and Chandrasekharappa SC. Menin, the product of the *men1* gene, is a nuclear protein. Proc Natl Acad Sci USA 1998; 95: 1630-4.
74. Agarwal SK, Guru SC, Heppner C, Erdos MR, Collins RM, Park SY, Saggar S, Chandrasekharappa SC, Collins FS, Spiegel AM, Marx SJ and Burns AL. Menin interacts with the AP1 transcription factor JunD and represses JunD-activated transcription. Cell 1999; 96: 143-52.
75. Cavenee WK, Dryja TP, Phillips RA, Benedict WF, Godbout R, Gallie BL, Murphree AL, Strong LC and White RL. Expression of recessive alleles by chromosomal mechanisms in retinoblastoma. Nature 1983; 305: 779-84.
76. Solomon E, Voss R, Hall V, Bodmer WF, Jass JR, Jeffreys AJ, Lucibello FC, Patel I and Rider SH. Chromosome 5 allele loss in human colorectal carcinomas. Nature 1987; 328: 616-9.
77. Miyoshi Y, Nagase H, Ando H, Horii A, Ichii S, Nakatsuru S, Aoki T, Miki Y, Mori T and Nakamura Y. Somatic mutations of the APC gene in colorectal tumors: mutation cluster region in the APC gene. Hum Mol Genet 1992; 1: 229-33.

78. Heppner C, Kester MB, Agarwal SK, Debelenko LV, Emmert-Buck MR, Guru SC, Manickam P, Olufemi SE, Skarulis MC, Doppman JL, Alexander RH, Kim YS, Saggar SK, Lubensky IA, Zhuang Z, Liotta LA, Chandrasekharappa SC, Collins FS, Spiegel AM, Burns AL and Marx SJ. Somatic mutation of the *men1* gene in parathyroid tumours. Nat Genet 1997; 16: 375-8.
79. Zhuang Z, Vortmeyer AO, Pack S, Huang S, Pham TA, Wang C, Park WS, Agarwal SK, Debelenko LV, Kester M, Guru SC, Manickam P, Olufemi SE, Yu F, Heppner C, Crabtree JS, Skarulis MC, Venzon DJ, Emmert-Buck MR, Spiegel AM, Chandrasekharappa SC, Collins FS, Burns AL, Marx SJ, Lubensky IA. Somatic mutations of the *men1* tumor suppressor gene in sporadic gastrinomas and insulinomas. Cancer Res 1997; 57: 4682-6.
80. Boggild MD, Jenkinson S, Pistorello M, Boscaro M, Scanarini M, McTernan P, Perrett CW, Thakker RV and Clayton RN. Molecular genetic studies of sporadic pituitary tumors. J Clin Endocrinol Metab 1994; 78: 387-92.
81. Bates AS, Farrell WE, Bicknell EJ, Talbot AJ, Broome JC, Perrett CW, Thakker RV and Clayton RN. Allelic deletion in pituitary adenomas reflects aggressive biological activity and has potential value as a prognostic marker. J Clin Endocrinol Metab 1997; 82: 818-24.
82. Zhuang Z, Ezzat SZ, Vortmeyer AO, Weil R, Oldfield EH, Park WS, Pack S, Huang S, Agarwal SK, Guru SC, Manickam P, Debelenko LV, Kester MB, Olufemi SE, Heppner C, Crabtree JS, Burns AL, Spiegel AM, Marx SJ, Chandrasekharappa SC, Collins FS, Emmert-Buck MR, Liotta LA, Asa SL and Lubensky IA. Mutation of the *men1* tumor suppressor gene in pituitary tumors. Cancer Res 1997; 57: 5446-51.
83. Prezant TR, Levine J and Melmed S. Molecular characterization of the *men1* tumor suppressor gene in sporadic pituitary tumors. J Clin Endocrinol Metab 1998; 83: 1388-91.
84. Farrell WE, Simpson DJ, Bicknell J, Magnay JL, Kyrodimou E, Thakker RV and Clayton RN. Sequence analysis and transcript expression of the *men1* gene in sporadic pituitary tumours. British J Cancer 1999; 80: 44-50.
85. Asa SL, Somers K and Ezzat S. The men-1 gene is rarely down-regulated in pituitary adenomas. J Clin Endocrinol Metab 1998; 83: 3210-12.
86. Satta MA, Korbonits M, Jacobs RA, Bolden-Dwinfour DA, Kaltsas GA, Vangeli V, Adams E, Fahlbusch R and Grossman AB. Expression of menin gene mRNA in pituitary tumours. Eur J Endocrinol 1999; 140: 358-61.
87. Berezin M and Karasik A. Familial prolactinoma. Clin Endocrinol 1995; 42: 483-6.
88. Petty EM, Green JS, Marx SJ, Taggart RT, Farid N and Bale AE. Mapping the gene for hereditary hyperparathyroidism and prolactinoma (MEN I Burin) to chromosome 11q: evidence for a founder effect in patients from Newfoundland. Am J Hum Genet 1994; 54: 1060-6.
89. Olufemi SE, Green JS, Manickam P, Guru SC, Agarwal SK, Kester MB, Dong Q, Burns AL, Spiegel AM, Marx SJ, Collins FS and Chandrasekharappa SC. Common ancestral mutation in the *men1* gene is likely responsible for the prolactinoma variant of MEN I (MEN I Burin) in four kindreds from Newfoundland. Hum Mutat 1998; 11: 264-9.

Chapter 8

Rodent Prolactin Family and Pregnancy

Michael J. Soares and Daniel I.H. Linzer

University of Kansas Medical Center, Kansas City, KS and Northwestern University, Evanston, IL

INTRODUCTION

Rodents have exploited the ancestral PRL gene to evolve a large number of closely related genes encoding PRL-like proteins. These PRL-related genes are all expressed at the implantation site, either in the trophoblasts of the placenta or in the decidual cells of the uterus. Some of these hormones have bioactivities indistinguishable from PRL, whereas most others have distinct molecular and cellular targets, and as a consequence distinct activities. This chapter will review what is currently known about this family of hormones.

DISCOVERY OF THE RODENT PROLACTIN FAMILY

The rodent placenta has long been recognized as an endocrine organ that produces hormones related to PRL. Both luteotropic and mammotropic activities, indistinguishable from the effects of PRL, were reported for rat placental extracts well before the responsible factors were characterized (1, 2). The assignment of these activities in extracts to specific protein hormones was made possible by the development of a radioreceptor assay for ligand binding to the PRL receptor (PRL-R; 3). Surprisingly, this assay revealed two distinct peaks of activity from the rat and mouse placenta during gestation, one peak at mid-pregnancy and a second peak in late-pregnancy (4, 5). That these two activities were not identical was demonstrated by the development of specific radioimmunoassays (6, 7), and later by the isolation of cDNA clones for the two hormone mRNAs (8-11). Based on the source of these proteins and their ability to stimulate milk production, they were named placental lactogen I (PL-I) and II (PL-II). As expected from bioassays and receptor binding assays, PL-I and PL-II are closely related to PRL, in the mouse sharing 32 and 37 percent amino acid sequence identity with PRL, respectively.

Between the time of discovery of PL-I and PL-II as proteins secreted by the placenta, and the cloning of the PL-I and PL-II cDNAs, two other PRL-related proteins were found. A search in mouse fibroblasts for factors that might be involved in cell cycle regulation led to the isolation of several serum-inducible mRNAs (12). The analysis of one of these cDNA clones represented one of the first examples of successful database comparison and placement of a novel gene within a known gene family (13). William

Pearson, then a postdoctoral fellow at The Johns Hopkins University School of Medicine, had just written an initial version of the sequence similarity analysis program known as FASTA, and was seeking a sequenced cDNA clone encoding an unknown protein. A partial cDNA clone, isolated and sequenced in an adjacent laboratory, was available to test FASTA. What was also needed was a sequence library, but no nucleotide databases had yet been compiled in 1983. Instead, Pearson had access to a minimal protein sequence database that was maintained on a computer at Stanford University. That database included sequences obtained by direct protein sequencing, and was therefore limited to proteins such as insulin, cytochrome C, and PRL and GH. The inclusion of PRL and GH in the database resulted from the pioneering efforts of C.H. Li at the University of California. Despite the low probability of success of analyzing a partial cDNA clone with a new algorithm by comparison to an extremely minimal database, a significant sequence similarity was found between the test sequence and PRL. As a result of the sequence similarity to *prol*actin, and because of the *prolifer*ation-dependent pattern of expression in mouse fibroblasts, the encoded protein was named *proliferin* and abbreviated PLF (14). Several years later, it became recognized that PLF and a "mitogen-regulated protein" or MRP found previously in cultured mouse fibroblasts were the same (15).

Taken together, the earlier characterization of PL-I and PL-II proteins as close relatives of PRL and the identification of a novel cDNA clone encoding a protein also related to PRL suggested that PLF might actually correspond to PL-I or PL-II. Consistent with this hypothesis, a PLF-hybridizing mRNA was detected specifically in the placenta (16) and with a time course similar to that of PL-I synthesis. To confirm that the placental mRNA hybridizing to the PLF cDNA was in fact the same as that isolated from mouse fibroblasts, the placental mRNAs were also isolated as clones. This analysis revealed that PLF is expressed in the placenta, but also led to two unexpected findings. First, the PLF cDNA clones were found to be nearly but not completely identical; subsequent analysis has shown that several PLF genes are present in the mouse genome and are transcribed into apparently functional mRNAs encoding slightly different proteins (17, 18). Second, a weakly hybridizing cDNA clone was also isolated and was found to encode yet another protein similar to PRL but distinct from PLF; this protein was therefore named PLF-related protein or PRP (19). PRP was also found to be expressed specifically in the placenta, with a time course similar to that of PL-II.

The hypothesis that the PLF and PRP cDNA clones correspond to the mouse protein hormones PL-I and PL-II was attractive, but no PRL-like bioactivity was detected for PLF or PRP (20, 21). Indeed, subsequent cloning of the mouse PL-I and PL-II mRNAs (9, 10) demonstrated that these hormones are distinct from PLF and PRP. Thus, it seemed that these two cDNA clones might define a new class of PRL-related hormones that had escaped prior detection in assays for proteins with receptor binding properties and actions similar to those of PRL or GH. What was completely unexpected was the vast numbers of these "orphan" PRL-related hormones that would be discovered in the mouse and rat in the following years. These hormones, which now number approximately two dozen, were discovered during the characterization of the available members of the PRL family at protein, cDNA, and genomic levels, by differential display and screening of placental RNAs, and, most recently, by computer searches of expressed-sequence tag (EST) databases. Upon completion of the mouse genome sequencing project, the actual number of distinct genes in the family and their relative chromosomal positions will be defined, but variations in post-translational modification of the hormone products (and potentially in the primary translation products as a result of alternative mRNA splicing) are expected to add even more complexity to the spectrum of PRL-related placental hormone structures and functions.

NOMENCLATURE

The PRL family consists of a group of structurally-related proteins. Nomenclature for members of the PRL family reflects biological activities (PLs), structural relationships with PRL (PRL like proteins, PRL related proteins), or association with proliferation (PLF). Shared functional characteristics with pituitary PRL are not a prerequisite for inclusion in the PRL family. Over a dozen different laboratories have participated in the discovery and naming of members of the rodent PRL family. Accordingly, it is not surprising that terminology for some PRL family members is confusing. In Table 8-1, we have attempted to clarify ambiguities associated with the names for members of the rodent PRL family that currently exist in the literature. Once we have a better appreciation for the size of the PRL family and the interrelationships of its members, it will be advantageous and warranted to devise a unified and more appropriate nomenclature.

STRUCTURAL CHARACTERISTICS

Hormones in the PRL/GH family share several features. All are synthesized as preproteins with co-translational cleavage of an amino-terminal, hydrophobic secretion signal sequence of approximately 30 residues. In each case, the remainder of the primary translation product is then secreted as an intact polypeptide chain of approximately 200 amino acids in length. Many (but not all) of these hormones include consensus sites for N-linked glycosylation. Although variations in glycoforms has not been shown to give rise to significant differences in the bioactivities of these placental hormones, ovine, porcine, and human PRL glycoforms have been reported to differ in activity (50-54). No post-translational modifications of these proteins other than glycosylation have been reported, although it remains possible that other modifications occur such as cleavage or phosphorylation to generate hormones with distinct bioactivities, as has been found for PRL (55, 56). The primary amino acid sequence also includes several highly conserved cysteine residues, but variations have been found in the pattern of cysteines and in the formation of disulfide linkages.

The crystal structure of GH reveals it to be a four helix bundle protein (57). The primary amino acid sequences of the placental hormones are compatible with four alpha-helical regions, and the recent crystal structure for ovine PL complexed to the PRL receptor is consistent with this prediction (58). In the two independent regions of the hormone known to be responsible for contact of PRL and GH with their receptors in a 1 hormone : 2 receptor complex (58-60), critical residues are found to be common among PRL, PL-I, and PL-II, and absent in PLF and PRP (61). It is not known, though, if variations of amino acid sequence in these regions confer distinct receptor specificities for PLF and PRP, or if receptor interaction occurs at unrelated sites in these hormones. While the former prediction would be most consistent with a common structural basis for binding of cytokines to cytokine receptors, the ability of PLF to interact (at least in part) through its carbohydrate moiety with the insulin-like growth factor II/mannose 6-phosphate receptor (21) suggests that the topology of the PRL/PRL-R and GH/GH-R interactions may not translate to the complexes of all of the other related hormones and their corresponding receptors.

The alpha-helical and putative receptor binding domains do not equate to individual gene exons, and in fact the receptor binding domains for GH actually correspond

Table 8-1. Rat and Mouse Prolactin Family

Member	Other names	Species	GenBank Accession	Refs
Classical (4)				
PRL*		Rat, mouse	V01249, X02892	22, 23
PL-I	Midpregnancy lactogen, PL-Im	Rat, mouse	M55269, D21103, M35662	10, 11, 24, 25
PL-Iv		Rat	M55269, U32679	26-28
PL-II	PL	Rat, mouse	M13749, M14647	8, 9
Nonclassical (17)				
PLF subfamily				
PLF	MRP	Mouse	K02245, X16009	14, 29, 30
PRP	PLF-RP	Rat, mouse	AF139809, X02594, AF226609	19, 31, 32
PLP-A		Rat, mouse	M13750, AF011383, AF015562	33-35
PLP-B		Rat, mouse	M31155, AF011384, AF015563	34-36
PLP-C subfamily				
PLP-C		Rat	M76537	37
PLP-Cv		Rat	U93351	38
PLP-D		Rat	AB000107	39
PLP-H	PLP-E	Rat	AB009889	40
D/tPRP	DPRP		L06441, AF015729, AF011385	34, 41, 42
PLP-C∀		Mouse	AF090140	43
PLP-E/F subfamily				
PLP-E	PLP-G	Mouse	AF011381, AF020525	44, 45
PLP-F		Rat, mouse	AF139808, AF011382, AF02054	31, 44, 45
PLP-I		Rat	AB019791	46
PLP-J	PLP-I, Decidualin	Rat, mouse	AB019945, AB019119, AF150741, AF234638, AB019118, AF150740, AF234637	46-49
PLP-K		Rat, mouse	AB022882, AF234635	46, 49
PLP-L		Rat, mouse	AB022883, AF226607, AF226611	45, 32
PLP-M		Rat, mouse	AF226608, AF226610, AF234636	32, 49

*Abbreviations: PRL, prolactin; PL, placental lactogen; PLP, prolactin-like protein; d/tPRP or dPRP, decidual/trophoblast prolactin-related protein; PLF, proliferin; MRP, mitogen regulated protein; PRP or PLF-RP, proliferin related protein.

to three dimensional structures that form from discontinuous stretches of the polypeptide chain (60). Nevertheless, the number and size of the exons, and the locations of interruptions in the genes by introns, are highly conserved with one major exception. While most of the genes encoding hormones in the family consist of five exons, genes for PRP and the PLP-C subfamily (see Table 8-1) include an extra, small third exon that encodes a segment of approximately 12 amino acids (38, 43, 62, 63). All of the genes for the placental PRL family of hormones in mice have been mapped to the same region of chromosome 13 that contains the PRL gene (see the Mouse Genome Database at *www.informatics.jax.org* for the most current information). The proximity of these genes to the PRL gene, and the closer relationship between the sequences of the placental hormone genes and the PRL gene than the GH gene, argue that an ancestral PRL gene gave rise to multiple gene copies and that these gene copies subsequently diverged to encode distinct protein products.

EXPRESSION

During pregnancy the expression of the PRL family is precisely orchestrated in spatial and temporal patterns (see Tables 8-2 and 8-3). There are three main lineages of cells involved in the biosynthesis of members of the PRL family: *i*) lactotropes of the anterior pituitary, *ii*) decidual cells of the uterus, and *iii*) trophoblast cells of the placenta. The cell biology of the lactotrope is presented elsewhere in this volume. Below we present an overview of the organization of the uteroplacental compartment, describe cellular and temporal expression patterns, and discuss possible mechanisms controlling expression of the PRL family. We will focus on expression characteristics relevant to the biology of each hormone as a regulator or potential regulator of the gestational state. Factors controlling the expression of members of the PRL family are tightly coupled to

Table 8-2. Tissue Distribution

Tissue Distribution	PRL Family Member	Refs
Maternal compartment		
Anterior pituitary	PRL	64
Decidual cell	PLP-B, d/tPRP, PLP-J, PRL	34, 35, 41, 42, 48, 49, 65-68
Extraembryonic compartment		
Trophoblast giant cell (TGC)	PL-I, PL-II, PLF, PLP-E	44, 69-75
Spongiotrophoblast cells	PLP-B, PLP-F (mouse)	66, 72, 34, 35
TGC & spongiotrophoblast cells	PL-Iv, PLP-A, PLP-C, PLP-Cv, d/tPRP, PLP-D, PLP-H, PLP-C∀, PRP (mouse), PLP-F (rat), PLP-K (mouse), PLP-M	26, 31, 38-40, 43, 49, 67, 71, 72, 75-77
Labyrinthine trophoblast	PL-II, PRP (rat), PLP-K	26, 31, 49
Endovascular trophoblast	PRP, PLP-L, PLP-M	32
Fetal compartment		
Anterior pituitary	PRL	78-81

Table 8-3. Temporal Patterns of Expression

Member	Temporal profile				Refs
	Pre-implant	Post-implant	Mid gestation	Late gestation	
Classical members					
PRL	X	X		X	64, 78, 80
PL-I		X	X		10, 73, 82, 83
PL-Iv			X	X	26, 27, 84
PL-II			X	X	8, 9, 71, 73
Nonclassical members					
PLF subfamily					
PLF		X	X		16, 70
PRP			X	X	19, 76
PLP-E/F subfamily					
PLP-E		X	X		44, 45
PLP-F			X	X	31, 44, 45
Decidual family					
PLP-B		X	X		36, 65, 66
PLP-J		X	X		48, 49
D/TPRP		X	X	X	67, 85
PLP-A			X	X	33, 71
PLP-K			X	X	49
PLP-M			X	X	49
PLP-C subfamily					
PLP-C				X	37, 77
PLP-Cv				X	38
PLP-D				X	39
PLP-H				X	40
PLP-Cα				X	43

*X denotes expression of the PRL family member as monitored by immunocytochemistry, in situ hybridization, Northern blotting, Western blotting, and/ or immunoassays.

the differentiation of the cells responsible for their production. Thus, insight into the control of PRL family gene expression potentially provides clues into the regulation of lactotrope, decidual cell, and/or trophoblast cell differentiation. Conversely, information regarding the differentiation of these cell types similarly enhances our understanding of the control of PRL family gene expression.

ORGANIZATION OF THE UTEROPLACENTAL COMPARTMENT

In order to proceed with a discussion of expression patterns for members of the PRL family it is necessary to introduce terminology pertinent to the morphology of the mouse and rat uteroplacental unit. Our account of these organizational features is based on earlier reports (86-91). Features of the organization of the uteroplacental compartment are depicted in Figure 8-1.

In the rodent, trophoblast and decidual lineages characterize pregnacy. The trophoblast lineage first arises after the morula stage of embryogenesis. Cells on the outer layer of the morula, *trophectoderm*, become distinguishable from those in the inner core, *inner cell mass* (ICM). Trophectoderm cells are the progenitors of trophoblast cells of the placenta, while the ICM will go on to differentiate into embryonic and other extraembryonic tissues. At the stage of blastocyst implantation, *mural* trophectoderm cells, located farthest from the

ICM, initially adhere to *antimesometrial* (farthest from the incoming uterine blood supply) uterine epithelial cells and then invade into the uterine stroma initiating a transformation of the stromal cells termed *decidualization.*

Decidualization is dependent upon the hormonal milieu (estrogen and especially progesterone) and can be induced by the blastocyst resulting in the formation of *decidual tissue*. Trophoblast lineages go on to contribute to the formation of two placental structures in the rodent: *i*) the *choriovitelline* placenta and *ii*) the *chorioallantoic* placenta. The choriovitelline placenta develops first, and is comprised of mural trophoblast cells adherent to a prominent basement membrane, *Reichert's membrane* (secreted by *parietal endoderm* cells) and is associated with decidual cells, termed *decidua capsularis*. The choriovitelline placenta eventually degenerates and disappears shortly after midpregnancy. The chorioallantoic placenta develops before the demise of the choriovitelline placenta and is located in the *mesometrial* region of the uterus (closest to the incoming blood supply). The chorioallantoic placenta is comprised of *polar* trophoblast cells which initially organize into a structure referred to as the *ectoplacental cone* and then as gestation progresses contribute to the formation of two structurally and functionally distinct regions: *i*) *junctional zone* and *ii*) *labyrinth zone*. *Basal zone* is a term sometimes used synonymously with junctional zone. The junctional zone is located adjacent to the *decidua basalis* and contains trophoblast cells and maternal vascular channels but is devoid of fetal vessels, while the labyrinth zone is located adjacent to the developing embryo and contains trophoblast cells, maternal vascular channels, and fetal vessels. The size and complexity of these placental zones change as gestation proceeds. A number of differentiated trophoblast cells can be identified in the junctional zone (*trophoblast giant cells*, *spongiotrophoblast cells*, *glycogen cells*) and in the labyrinth zone (*labyrinthine trophoblast stem cells*, *labyrinthine trophoblast giant cells*, *syncytiotrophoblast cells*). An additional population of cells derived from the trophoblast lineage and associated with the chorioallantoic placenta is referred to as *endovascular trophoblast*. This lineage of trophoblast cells is invasive and supplants the endothelium of uterine blood vessels.

PATTERNS OF EXPRESSION

Individual PRL family members can be expressed by a single cell type or by multiple cell types. Expression of the same PRL family member in more than one cell type has important implications. Cellular source directly impacts posttranslational processing and thus potentially the distribution and activity of the ligand. Sites of ligand biosynthesis also affect the ligand's access to potential cellular targets.

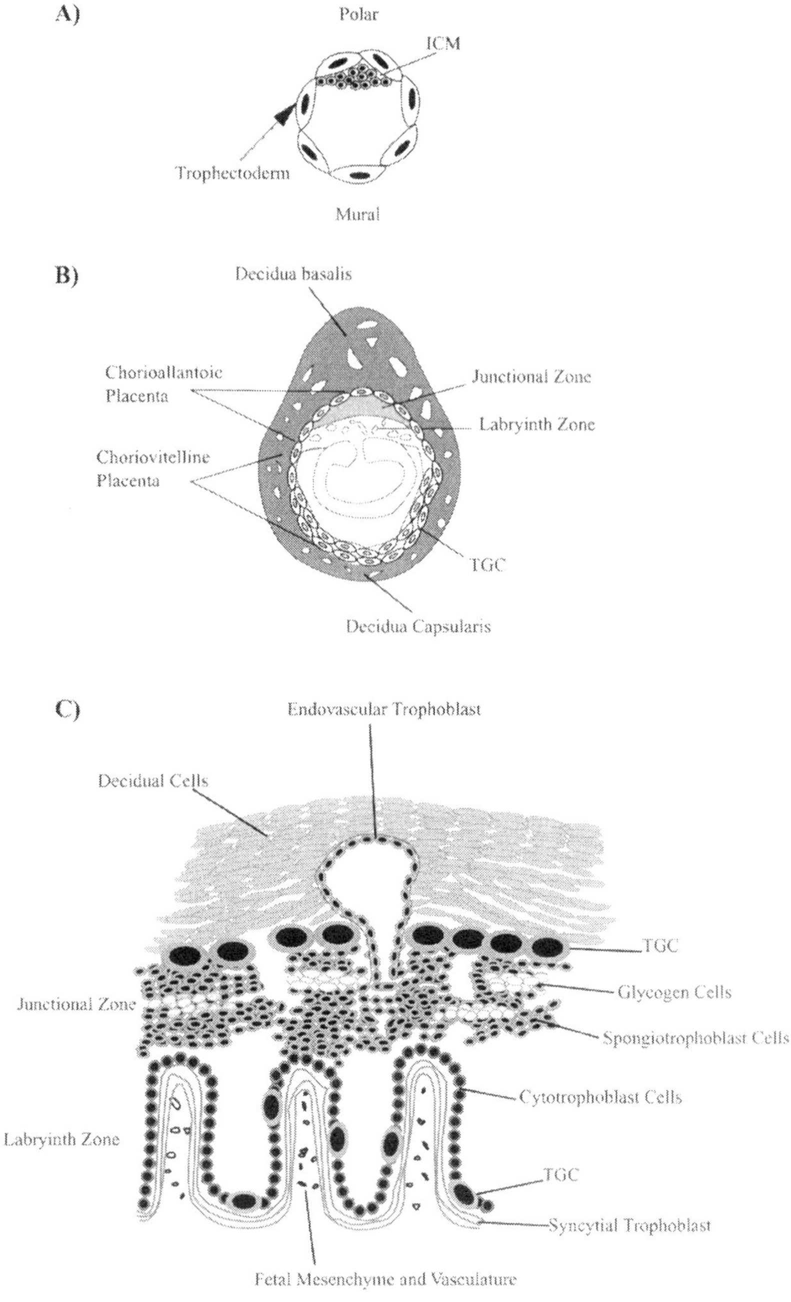

Figure 8-1. Schematic representations of the organization of the blastocyst and the uteroplacental compartment at two developmental phases during pregnancy in the rat or mouse. A) Blastocyst; B) Midgestation conceptus; C) Portion of late gestation chorioallantoic placenta. Abbreviations: ICM, inner cell mass; TGC, trophoblast giant cell.

DECIDUAL CELL

Decidual cells arise from uterine stromal cells under the control of progesterone and contribute to the production of four members of the PRL family: PRL-like protein-B (PLP-B; 34, 35, 65, 66), decidual/trophoblast PRL-related protein (d/tPRP; 34, 41, 42, 67, 85), PLP-J (47-49), and PRL (68). The site of expression for each of these members is primarily confined to the antimesometrial deciduum and is independent of the presence of embryonic or extraembryonic structures. The time course of expression follows the life span of decidua. While decidual cells appear to be the sole source of PLP-J; PLP-B and d/tPRP are also produced by trophoblast cells of the chorioallantoic placenta and of course, the primary source of PRL is the anterior pituitary.

A few insights are available regarding the regulation of a couple of decidual PRL family genes. Decidual cell expression of PRL is stimulated by progesterone and cyclic AMP and inhibited by PRL (68). Since progesterone and cyclic AMP promote decidual cell differentiation, they will also likely impact the biosynthesis of PLP-B, d/tPRP, and PLP-J (92). These decidual cell regulators also stimulate PRL gene expression in human decidual cells (92). The gene for rat d/tPRP has been cloned and its 5' flanking DNA has been partially characterized (62, 93). Putative activator protein-1 (AP-1) and Ets elements within the 93 bp proximal promoter participate in the transcriptional control of the d/tPRP gene in decidual cells (93). Decidual proteins binding to these regulatory sites have not been identified, nor have the upstream pathways controlling d/tPRP gene expression. We would anticipate that at least some transcriptional controls will be shared among each PRL family gene expressed in decidual cells.

TROPHOBLAST GIANT CELL

Trophoblast giant cells are the most versatile of the cellular sources, possessing the capacity to express most members of the PRL family and are the exclusive sources of PL-I, PL-II, PLF, and PLP-E (69, 70, 72, 73, 44). PL-I, PL-II, and PLF have been co-localized to trophoblast giant cells (94). The behavior of trophoblast giant cells differs depending upon their intraplacental location and developmental state. Expression patterns involve trophoblast giant cells residing in the choriovitelline placenta and in both the junctional and labyrinth regions of the chorioallantoic placenta. PL-I and PLF first appear immediately postimplantation (73, 75, 95). PL-I terminates at midgestation (11, 73, 75, 95), whereas PLF continues through the second half of gestation at reduced levels of production (16, 70, 75). PL-II expression begins at midgestation and continues until the termination of pregnancy (71-73). Although, PL-I, PLF, and PL-II share cell-specific requirements for expression, they appear to be optimally activated in response to distinct factors in the trophoblast giant cell environment.

Some limited information is available on the regulation of trophoblast giant cell restricted members of the PRL family. *In vivo* manipulations such as hypophysectomy, gonadectomy, and fetectomy influence circulating PL-I and/or PL-II levels (96-103). Removal of the anterior pituitary or ovaries results in an increase in circulating PLs, whereas removal of the fetus decreases serum PL concentrations. Anterior pituitary factors appear to affect PL clearance (104) and ovarian and fetal factors affect the development and organization of the placenta (105). A variety of hormones, growth factors, cytokines, and other activators of signal transduction cascades have been shown to modulate the production of PL-I, PL-II, and PLF *in vitro* (63, 106).

Transcriptional control of PRL family gene expression by trophoblast giant cells has been investigated using transgenic and gene targeted mutant mouse models and through the use of a rat trophoblast cell line, Rcho-1, which faithfully mimics many of the features of trophoblast giant cells developing in situ (107-110). Investigations have focused on the transcriptional regulation of PL-I, PLF, PL-II, PLP-A, PLP-Cv, and d/tPRP. MAP kinase signaling pathways and transcription factors interacting with AP-1 and GATA elements participate in the control of the PL-I gene (108, 111-113). Consistent with these observations, GATA-2 or GATA-3 null mutant mice exhibit a profound diminution in their expression of PL-I and PLF (112). Putative AP-1 elements and Ets binding sites have also been shown to participate in the regulation of PL-II and d/tPRP gene transcription (93, 114). Additional mapping of DNA regulatory regions controlling trophoblast-specific expression of PLP-A and PLP-Cv genes in Rcho-1 trophoblast cells has been reported (38, 115, 116). Thus far, only 5' flanking DNA associated with the PL-II gene has been reported to effectively target the *in vivo* expression of reporter genes to the placenta (117-119).

Significant progress has been made in understanding the control of trophoblast giant cell differentiation and some linkages have been made between pathways controlling trophoblast giant cell differentiation and the expression of members of the PRL gene family (120, 121). The basic helix-loop-helix transcription factor, Hand1, plays an essential role in the development of trophoblast giant cells (122). Other proteins with the capacity to modulate Hand1 activities (Id1, Id2, E-factors, Mash2, mSna, and I-mfa) have also been implicated in the control of trophoblast giant cell formation (119, 120, 123, 124).

SPONGIOTROPHOBLAST CELL

Clearly, the most common expression pattern for members of the PRL family involves dual expression by both spongiotrophoblast and trophoblast giant cells. PLP-B and PLP-F represent exceptions. Within the chorioallantoic placenta PLP-B expression is restricted to spongiotrophoblast cells (66, 72, 125). Similarly in the mouse chorioallantoic placenta PLP-F expression is confined to spongiotrophoblast cells (31, 44). Interestingly, in the rat, PLP-F follows the more typical pattern and is expressed in both spongiotrophoblast and trophoblast giant cell populations (31). Spongiotrophoblast cells first arise at midgestation in association with the formation of the chorioallantoic placenta and at this juncture it appears that their main function is the production of PRL family proteins (89, 126). Hormone/cytokine expression by spongiotrophoblast cells is restricted to the second half of gestation.

Only modest information is available on the regulation of spongiotrophoblast cell production of PRL family members. Retinoic acid attenuates PLP-C and PL-Iv production by primary cultures of spongiotrophoblast cells (127). There is also some evidence for the separation of DNA regulatory elements controlling spongiotrophoblast versus trophoblast giant cell gene activation within the PLP-Cv promoter (116). The development of spongiotrophoblast cells is dependent upon the helix-loop-helix transcription factor, Mash2 (128), and the epidermal growth factor receptor (129, 130). Consequently, these factors are involved in regulating the expression of PRL family genes, if not directly then indirectly by controlling the formation of the spongiotrophoblast cell lineage.

LABYRINTHINE TROPHOBLAST CELL

The labyrinthine region of the chorioallantoic placenta is chiefly involved in regulating bidirectional nutrient transport between maternal and fetal compartments (131). However, during the last week of pregnancy, trophoblast cells within the labyrinth zone begin to produce a subset of PRL family members [PL-II, PLF-RP (rat only), PLP-K; 26,

31, 49, 71]. The cell principally responsible for these activities is situated at the interface with the maternal blood space and appears to be related to the trophoblast giant cell. These labyrinthine trophoblast giant cells are phenotypically distinct from giant cells derived from the Rcho-1 trophoblast cell line which appear to more closely represent junctional zone trophoblast giant cells (31). Labyrinthine trophoblast hormone/cytokine production may facilitate hormone/cytokine entry into the fetal compartment. Similarly, the regulation of labyrinthine PRL family expression is likely directly linked to signals emanating from the fetus.

ENDOVASCULAR TROPHOBLAST CELL

Trophoblasts throughout the placenta come into direct contact with maternal blood, a property shared with vascular endothelial cells. A subset of rodent trophoblasts, though, form an invasive column that penetrates deep into the deciduum, and surrounds a large maternal blood space to form the "placental vessel" (90). These trophoblasts express several hormones in the PRL family (32), and thus these hormones serve as markers of this invasive cell population. The production of these hormones in the invasive trophoblasts presumably provides an efficient means for their delivery into the maternal circulation, and perhaps also indicates that concentrations of extracellular factors regulating the synthesis of these hormones are high in this region.

TRANSPORT AND TISSUE DISTRIBUTION

Presently we have a few examples for different modes of hormone delivery. A number of members of the PRL family are present in maternal circulation and a few have been detected in the fetus (106, 132). Cytokine access and availability to target cells can be influenced by posttranslational modifications and interactions with various transport and binding proteins. Glycosylation appears to be the most common posttranslational modification for members of the PRL family. The nature of protein glycosylation appears to be cell type and protein specific (133). Addition of carbohydrate to PL-I by trophoblast giant cells is apparently responsible, at least in part, for its lengthened half-life in circulation relative to the short half-life of the unglycosylated PL-II (4). Specific associations of members of the PRL family with transport or binding proteins have also been observed. Mouse PL-II, in part, circulates bound to "α_2-macroglobulin (134). Transport proteins such as "α_2-macroglobulin can specifically direct ligands to their target cells or effectively protect the maternal environment from the actions of the ligand. In the rat, PLP-A achieves very high concentrations in blood and is known to circulate as a high molecular weight complex bound to another protein(s) (135, 136). The nature of the circulating PLP-A binding protein has not been reported. Some members of the PRL family appear to be restricted in their distribution because of specific interactions with components of the extracellular matrix. D/tPRP binds to heparin containing molecules and is present in the decidual extracellular matrix (137). Such a location is ideal for gaining access to cells that traverse the decidual compartment (trophoblast, immune, endothelial, decidual) and in limiting exposure to extrauterine sites. A few members of the PRL family have been reported to gain access to the fetal circulation, whereas others are selectively excluded from the fetus (132, 138-140). Mechanism(s) responsible for a ligand gaining access to or being excluded from the fetal compartment have not yet been determined but may involve posttranslational modifications or interactions with carrier molecules.

The modes of hormone delivery discussed above are based on only a few members of the PRL family and thus provide us with only a hint of what to expect. We have much to learn about posttranslational processing, tissue distribution, and transport characteristics for the majority of PRL family members.

FUNCTIONS AND MECHANISMS OF ACTION

Initial investigations into the function of PRL family members were focused on their abilities to mimic the actions of PRL. However, in 1994 it became apparent that members of the PRL family could act on apparently unique cellular targets yielding unique biological responses (141). These observations gave rise to a classification of the PRL family based on function. Members of the PRL family utilizing the PRL receptor signaling pathway were categorized as *classical* and those that did not as *nonclassical*. In the next few paragraphs we outline our current understanding of the function and mechanisms of action for several members of the PRL family.

PRL MIMICRY

Much of PRL family biology grew out of classic observations implicating the rodent uterus and placenta as sources of factors influencing mammary gland development and corpus luteum function. Rodent placental hormones possessing both mammotrophic and luteotrophic actions have been isolated and characterized. PL-I and PL-II stimulate various parameters of mammary epithelial cell growth and differentiation (20, 25, 142, 143) and promote the biosynthesis of progesterone by the corpus luteum (144, 145). PL-I and PL-II are part of a regulatory network that also includes PRL and PL-Iv. The expression of these maternal (PRL), trophoblastic (PL-I, PL-Iv, and PL-II), and fetal (PRL) hormones are orchestrated in a precise temporal pattern that ensures the presence of activators of the PRL receptor signaling system throughout gestation (7, 83, 84, 146-148). PL-Iv is a weak PRL receptor agonist when compared to PRL, PL-I, or PL-II (28, 84). In addition to the mammary gland and corpus luteum, there are several other targets for these ligands in both maternal and fetal compartments. Evidence has arisen for their involvement in the control of maternal behavior (149, 150), the regulation of PRL (151) and insulin secretion (152, 153), and the modulation of immune cells (10, 25, 28, 146, 147). The significance of multiple ligands activating the PRL receptor signaling pathway likely provides for complementarity and presumably some form of specialization (154).

Considerable information has been generated regarding PRL signal transduction (see elsewhere in this volume). However, knowledge about intracellular events subsequent to PRL receptor activation by members of the placental PRL family is minimal. PL-I, PL-II, and PL-Iv are all capable of binding to various PRL receptor preparations (20, 25, 28, 55, 84,139, 147,156-159). The only apparent disparity in binding characteristics is a lower affinity of PL-Iv for the PRL receptor (28, 84). This lower affinity correlates with a reduced biopotency of PL-Iv in assays for PRL-like bioactivities (28, 84). It is assumed that PL-I, PL-II, and PL-Iv activate signaling pathways downstream of the PRL receptor similar to those activated by PRL. Beyond the abilities of PL-I and PL-Iv to stimulate Jak2-STAT5 pathway in Nb2 lymphoma cells (28, 40,160), we lack information on intracellular signaling pathways triggered by classical members of the placental PRL family. It would appear likely that at least some aspects of PRL signaling are mimicked by PL-I, PL-II, and PL-Iv; however, whether these three placental ligands reproduce the full spectrum of PRL action within their target cells is unknown.

ANGIOGENESIS

The ability of hormones in the PRL family to act on endothelial cells and regulate angiogenesis was first shown for a proteolytic fragment of PRL (55). A tantalizing result was obtained for PLF by Nelson Horseman, while testing this hormone for PRL-like bioactivity in the classical pigeon crop sac assay; he found no induction of crop sac growth, but he did see increased vascularization at the site of injection of PLF (141). Subsequent analysis of PLF and PRP indicated that they also have activities on endothelial cells, with PLF stimulating, and PRP inhibiting, endothelial cell migration in cell culture and *in vivo* (141). Since PLF is secreted earlier in gestation than PRP, it seems likely that PLF contributes to the growth of the placental vasculature and PRP later limits that growth (161). Consistent with these roles, the mid-gestation placenta secretes net angiogenic activity, most of which is removed with an antibody against PLF, whereas the later gestation placenta secretes anti-angiogenic activity, most of which is blocked by an antiserum against PRP (141). Furthermore, disruption of the gene encoding the GATA-2 transcription factor, which is essential for PLF expression *in vivo*, results in a significant decrease in mid-gestation placental secretion of angiogenic activity and in a decrease in neovascularization at the implantation site (112).

PLF is able to bind to endothelial cells at the implantation site and in the fetus, as well as to some non-endothelial cell types in the fetus, including developing vertebrae (140, 141, 162). Lee and Nathans (21) had reported that this hormone interacts with the insulin-like growth factor II/mannose 6-phosphate receptor (IGF-II/M6PR), and that this interaction depends on a mannose 6-phosphate modification to PLF's N-linked carbohydrate. In contrast, Nilsen-Hamilton and colleagues (163), found that PLF (which they refer to as mitogen regulated protein) binds a distinct receptor in the uterus that does not involve mannose 6-phosphate. To determine which, if either, of these receptors is relevant for PLF binding to endothelial cells, the ability of mannose 6-phosphate to block hormone binding and action was examined; this phosphorylated sugar was found to compete for PLF binding sites on endothelial cells, to inhibit PLF-induced endothelial cell migration in culture, and to prevent PLF-stimulated neovascularization *in vivo* in a corneal implant assay (162). Thus, the IGF-II/M6PR appears to be a relevant component of the binding and signaling complex in PLF-regulated angiogenesis.

Despite the ability to bind IGF-II, a potent activator of cell proliferation, the IGF-II/M6PR seems to be incapable of transmitting signals that lead to DNA synthesis and cell cycle progression; this growth factor function of IGF-II is instead mediated by binding of the factor to the IGF-I receptor tyrosine kinase (164). The effects of PLF on endothelial cells, though, implicates a distinct signaling pathway that leads to cytoskeletal reorganization and cell migration. The IGF-II/M6PR may itself signal in this pathway, or upon PLF binding may interact with a signaling receptor protein as part of a larger complex. An event that occurs downstream of PLF binding to the IGF-II/M6PR receptor includes a weak activation of MAP Kinase through a pertussis toxin-sensitive step, suggesting that PLF signaling is mediated by an inhibitory G protein that communicates with the MAP Kinase pathway; activation of this pathway is required for PLF-induced endothelial cell migration (165).

The discovery that PLF is both angiogenic and expressed in a growth-regulated manner in mouse fibroblast cell lines (12, 166) suggests that reexpression of this placental angiogenic hormone might also occur under certain conditions outside of the placenta. Fibroblasts are intimately involved in wound healing, a process that includes the growth of blood vessels, but no PLF gene reactivation was detected in fibroblasts at wound sites (J. Groskopf and D. Linzer, unpublished results). Another possibility is that PLF expression may recur in tumor cells as they progress from small, non-angiogenic masses to aggressive and metastatic cancers; some mouse tumor models do show PLF expression (167, 168), but it is not yet known what contribution PLF makes to the growth of these or other tumors.

In contrast to the presumed stimulatory effect of PLF on tumor growth by increasing tumor vascularization, the anti-angiogenic hormone PRP would be predicted to restrict tumor growth by reducing tumor vascularization. Indeed, tumor cells engineered to secrete PRP are significantly reduced in their growth *in vivo*, but are unaffected in their growth in culture where a blood supply is irrelevant (N. Bengtson and D. Linzer, manuscript submitted). The identification of PRP in rodents other than the mouse, including the hamster (169) and rat (31, 32), suggest that the effects of PRP may be conserved. The mechanism by which PRP blocks angiogenesis is not known; neither the receptor nor any downstream signaling components have yet been implicated.

HEMATOPOIESIS

As members of the PRL family, these placental hormones are also in the cytokine superfamily, and as such may be expected to act in hematopoiesis. In mammalian reproduction, maternal blood volume increases (170), and as a result proliferation and differentiation in the hematopoietic system must increase to maintain the concentrations of the many types of blood cells. Differentiation of one of these cell types, megakaryocytes (MK cells), involves numerous rounds of DNA synthesis without cell division until the cell fragments into blood platelets, which in turn mediate wound and vascular repair. Platelet levels increase in pregnant rodents (171), which suggests that increased levels of cytokines that act on MK cells are present.

One cytokine that targets MK cells during pregnancy is PLP-E (172). PLP-E binds to a specific receptor on MK cells and stimulates primary MK cell differentiation, both in terms of increased cell size and increased DNA content. In combination with interleukin 3 (IL-3), a cytokine that stimulates the proliferation of MK cell precursors (colony forming unit - megakaryocyte, or CFU-MK), PLP-E enhances colony formation in semi-solid culture, suggesting that PLP-E acts to stimulate CFU-MK proliferation, to improve CFU-MK survival, or to promote the commitment of multipotential precursors to the MK lineage. PLP-E's effects are similar to those of the IL-6 family of cytokines, although addition of IL-3, PLP-E, and IL-6 together has an even greater effect on CFU-MK colony formation than IL-3 with either of the other factors alone. Another similarity is that both PLP-E and IL-6 activate a signal transduction pathway that requires the signaling co-receptor gp130.

The action of PLP-E is not restricted to MK cells, as this hormone is able to stimulate the formation of granulocyte/macrophage cell colonies (CFU-GM) in primary bone marrow cell cultures (173). The ability of PLP-E to act on both CFU-MK and CFU-GM suggests that this hormone may target a common, myeloid precursor cell. The actions of PLP-E may overlap with PLP-F, which is expressed later in gestation. PLP-F and PLP-E appear to compete for the same receptor on MK cells (J. Lin, H. Lum, and D. Linzer, unpublished results), but it is not yet known if PLP-E and PLP-F function as agonists or antagonists.

IMMUNE CELL MODULATION

PRL is a known modulator of immune cell function (see elsewhere in this volume). In this section we present evidence for the interaction of other PRL family members with components of the immune system during pregnancy.

Hemochorial placentation, as occurs in both primates and rodents, results in the establishment of a close connection between maternal and fetal tissues (90). The close connection facilitates the exchange of nutrients and wastes at the expense of an increased risk of attack by the maternal immune system. This is of special relevance in that the uterine stroma is an abundant reservoir of immune cells (174). Trophoblast and decidual cells and their secretory products likely provide the signaling system that coordinates activities within the uterine immunologic milieu. The nature of the controls on the maternal immune system must be transitory and local (intrauterine); otherwise they may compromise the mother's ability to respond to infections. Through the use of an alkaline phosphatase (AP)-ligand tagging strategy, eosinophils and natural killer (NK) cells have been identified as targets for two members of the PRL family, d/tPRP (175) and PLP-A (176), respectively.

D/TPRP-EOSINOPHIL INTERACTIONS

D/tPRP was originally isolated during a search for a decidual luteotropin (41). Although, d/tPRP exhibits significant structural similarities to PRL, it does not interact with PRL receptors and thus its putative luteotropic actions are suspect (137, 175). D/tPRP is predominantly expressed by decidual cells during the first half of pregnancy and subsequently by spongiotrophoblast and trophoblast giant cells of the chorioallantoic placenta during the second half of pregnancy (34, 41, 42, 67, 85). Once secreted from its cellular sources d/tPRP readily associates with heparin containing molecules and accumulates in the extracellular matrix (137, 175). Thus, d/tPRP is in an excellent position to interact with various populations of cells traversing the uteroplacental compartment, including immune cells. Additionally, these heparin-binding attributes also restrict d/tPRP's actions to the uterine compartment. In a xenotransplantation model, CHO cells expressing d/tPRP more readily form tumors in athymic mice than do control CHO cells (137). The *in vivo* growth differences in the two CHO cell populations cannot be accounted for by *in vitro* growth rates and are independent of effects of d/tPRP on the host vasculature. These observations suggested that d/tPRP participates in the regulation of host immune cell responses. More recent insights regarding d/tPRP have come from the use of an AP-d/tPRP fusion protein. Using this approach, d/tPRP has been shown to interact with eosinophils (175).

Eosinophils contribute actively to inflammatory responses through their secretion of a variety of chemical mediators (177). These cells are actively recruited to sites of allergic responses (178), tumors (179), and parasite infestation (180). Eosinophil dynamics are precisely controlled within the uterus by the hormonal milieu, and exhibit dramatic fluctuations during the reproductive cycle and pregnancy (181-183). For example, estrogen stimulates the infiltration of eosinophils into the uterus, whereas uterine exposure to progesterone during pregnancy results in eosinophil exit and death (184, 185). Since d/tPRP is a product of the progesterone-dependent decidua and it specifically interacts with eosinophils, this cytokine is a candidate mediator of the anti-inflammatory actions of progesterone during gestation. The nature and mechanism of d/tPRP effects on eosinophil function have not yet been described.

PLP-A AND NK CELLS

PLP-A was one of three cDNAs, structurally related to PRL, originally identified during a search for the PL-II cDNA (33). Expression of PLP-A protein and mRNA are restricted to spongiotrophoblast cells and trophoblast giant cells of the junctional zone of the rat and mouse chorioallantoic placenta (34, 35, 71, 72). Although, PLP-A has significant structural homologies to PRL it is not capable of binding to PRL receptors or activating the PRL receptor signaling pathway (136).

The generation of AP-PLP-A fusion proteins led to the identification of specific PLP-A binding to NK cells within the mesometrial compartment of the uterus from pregnant rats and mice (176). These observations were supported by the co-distribution of PLP-A targets with cells expressing the rat NK cell surface marker, gp42, the absence of PLP-A binding in conceptuses from NK cell deficient tge26 mice, and the specific interaction of PLP-A with rat NK cells. Furthermore PLP-A effectively suppresses rat NK cell cytolytic activities. Uterine NK cells have been proposed to participate in immunological adjustments to pregnancy and in the establishment of the chorioallantoic placenta (174, 186). During pregnancy, NK cells undergo considerable morphological and functional changes creating, in effect, an NK cell with a unique phenotype. Midgestation uterine NK cells are distinct in their bio-effector secretory profile (cytokines and nitric oxide) and their relative absence of cytolytic activities (174, 186).

PLP-A represents an intriguing paracrine signal participating in a new paradigm of embryonic-maternal communication between trophoblast cells and uterine NK cells. Suppression of NK cell killing activity facilitates survival of genetically disparate extraembryonic and embryonic tissues. NK cells also release nitric oxide which directly influences uterine vasculature (187, 188), facilitating the delivery of nutrients to the developing placenta, while their elaboration of cytokines (189) potentially promotes the growth and maturation of the chorioallantoic placenta. Whether PLP-A influences these latter aspects of uterine NK cell function remain to be determined. Effects of PLP-A on extrauterine NK cells are likely obviated by the association of PLP-A with serum binding proteins (135, 136). NK cell surface targets and intracellular signaling pathways activated by PLP-A have not been reported.

SPECULATIONS ABOUT OTHER PRL FAMILY MEMBERS

We have only scratched the surface in understanding the biology of the PRL family. Currently we have an appreciation of the actions or cellular targets for ten of the approximately two dozen currently identified members (Table 8-4). Although the present state of our knowledge is fragmentary, it is sufficient to recognize that we have a very compelling biological story. Members of the PRL family are being used to enlist maternal tissue responses that directly benefit prenatal and postnatal development.

Over the next few years, we will gain new insights into the biology of each member of the PRL family. As our analyses expand, we are likely to discover additional maternal targets and extraembryonic and/or fetal tissue responses dependent on members of the PRL family. The AP-tagging strategy will provide an objective means of identifying cellular targets for orphan members of the PRL family. Gene targeting represents a complementary approach to directly evaluate the significance of each family member in the physiology of pregnancy. The extensive specialization among PRL family members regarding structure, expression patterns, and/or cellular targets portends well for the identification of phenotypes in null mutant mice. Receptor signaling pathways respon-

Table 8-4. TARGET TISSUES FOR MEMBERS OF THE PRL FAMILY

Hormone	Cellular targets	Refs
PRL, PL-I, PL-II	Mammary gland, ovary, liver, pancreas, brain, T-cells	25, 139, 144, 149-153, 190, 191
PL-Iv	Ovary, liver, T-cells	28, 84
PLF	Endothelial cells, uterine cells, muscle cells	141, 163, 192, 193
PRP	Endothelial cells	141
PLP-A	Natural killer cells	176
PLP-E, PLP-F	Hemapatopoietic precursors, megakaryocytes	172, 194
D/tPRP	Eosinophils, trophoblast	175

sible for transducing each ligand's action will also be resolved. At present, it is not clear whether specific receptor signaling systems have co-evolved with each of the PRL family members, whether PRL family members utilize signaling pathways for other known ligands, or whether PRL family members act through receptor-independent mechanisms (e.g. transport proteins, binding proteins, etc.). In the future, we may observe examples of each of these possibilities.

EVOLUTIONARY CONSIDERATIONS

Even a cursory analysis of the PRL family opens interesting evolutionary issues. Members of the PRL family have been identified in several species (195) and significant molecular information exists in a few rodent species (rat, mouse, hamster; 154), a few ruminants (cow, sheep, goat; 196, 197), and a couple of primate species (human, rhesus monkey; 198). Unlike PRL, whose structure, expression patterns, and functions have been relatively well conserved, it has been difficult to discern species conservation for other members of the PRL family. Orthologous relationships are apparent within closely related species (e.g. mouse versus rat; cow versus sheep) but are not evident when examining more distantly related species (e.g. rodent versus ruminant versus primate). Some species are noteable for their use of different ancestral templates, while others are noteable for their apparent lack of more than a single member of the PRL family. Key among the evolutionary issues related to the PRL family is the apparent species diversity.

SPECIES DIFFERENCES ARE APPARENT BUT ARE THEY REAL?

Ancestral gene templates

The PRL family arose by duplication from ancestral genes (199). Two different templates have been used for the generation of PRL related pregnancy-specific regulators. Primates evolved a family of proteins expressed in the placenta more closely related to GH, while rodents and ruminants used an ancestral PRL related gene as an evolutionary prototype. The utilization of GH as a template in primates has a functional corollary. Primate GHs are unique among GHs in their ability to activate the PRL receptor signaling pathway (200). Whether primates also possess a family of proteins more closely related to PRL is not known. The evolution of protein families from PRL ancestral prototypes for the purpose of regulating pregnancy must be related to the well-recognized functional versatility of PRL (201). Over one hundred different biological actions have been attributed to PRL. The structural motifs within the PRL backbone responsible for this versatility are not well understood.

Absence of placental lactogenic activities in some species

Some species such as the pig, horse, rabbit, cat, and dog do not appear to possess a classic PL (195). This may represent divergent evolution or alternatively it may relate to limited experimentation. Some of the apparent non-expressing species may actually possess a classic PL with a restricted expression pattern that has not been adequately tested. Alternatively, nonclassical members of the PRL family may be expressed by uteroplacental tissues in these species. In fact, the driving force(s) for derivation of the PRL family may relate to more elemental aspects of viviparity, such as vascular connectivity and maternal-fetal tolerance rather than the regulation of ovarian or mammary gland function.

Deficits in our research

Our present information about the PRL family is biased by the small group of species studied and the obstacles associated with studying pregnancy-related events in some species. The mouse and the rat are particularly tractable experimental models, especially in relation to the duration of their gestation. Research in the mouse and rat has further benefited from the mouse and rat genome projects and the availability of ESTs from an assortment of uterine and extraembryonic cDNA libraries representing a variety of different phases of pregnancy. Thus the absence of establishing true cross-species homologies may be real or the result of our limited current state of knowledge. Over the past few years, we have accumulated many new tools for evaluating PRL families. Consequently, there is considerable merit in revisiting a comparative approach. Completion of existing genome projects and expansion of genome characterization to other species will also help resolve these uncertainties.

What are the implications of the apparent species diversity?

It is essential that we place the PRL family in proper context. As is evident from the preceding discussion, the PRL family is a collection of hormones/cytokines that regulate pregnancy. Problems associated with young developing within the female reproductive tract are similar for all species. Adequate supplies of nutrients must be delivered to the embryo/fetus without compromising the mother. However, the solutions utilized by individual species vary widely. Among viviparous species there are striking differences in the organization of the maternal-fetal interface, the length of gestation, and the progression of embryonic/fetal development. Hence, it is not surprising that factors controlling the gestational state differ in a species-dependent manner. The PRL family represents a newly evolved collection of regulators directly linked to viviparity. Functional homologies among species will exist and may include the ligands, their cellular targets, and/or components of their signaling pathways.

DEVELOPMENT OF NEW THERAPEUTIC STRATEGIES

An exciting possibility is that rodent PRL-related hormones will reveal new mechanisms in the regulation of physiology, an expectation suggested by the unusual physiological changes that occur during pregnancy and the importance of the uterus and placenta in the endocrine control of pregnancy. If new molecular targets in signal transduction or new ways to combine known signaling components are revealed by analyzing the mechanisms of action of the uteroplacental hormones, then presumably these hormones may become valuable tools in the design of new clinical therapies. For *ex vivo* therapies, for example in the expansion of human bone marrow in cell culture for autologous transplantation, novel hematopoietic effects of these mouse hormones might be exploited without concern for the delivery or immunological problems that would occur if these proteins were to be administered *in vivo*. For clinical interventions requiring *in vivo* treatment, for example in controlling autoimmune effects caused by NK cells or in restricting tumor growth by preventing tumors from attaining a blood supply, hormones such as PLP-A and PRP may help to identify receptors, and these receptors could then be used to screen small molecule libraries for ligands which would be more compatible with use in humans.

ACKNOWLEDGEMENTS

We would like to thank past and present members of our laboratories and our collaborators. Research from the authors' laboratories has been supported by grants from the National Institute of Child Health and Human Development.

REFERENCES

1. Astwood EB, Greep RO. A corpus luteum-stimulating substance in the rat placenta. Proc Soc Exp Biol Med 1938; 38:713-716.
2. Cerruti RA, Lyons WR. Mammogenic activities of the midgestational placenta. Endocrinology 1960; 884-887.
3. Shiu RPC, Kelly PA, Friesen HG. Radioreceptor assay for prolactin and other lactogenic hormones. Science 1973; 180:968-970.
4. Kelly PA, Shiu RPC, Robertson MC, Friesen HG.Characterization of rat chorionic mammaotropin. Endocrinology 1975; 96:1187-1195.
5. Kelly PA, Tsushima T, Shiu RPC, Friesen HG. Lactogenic and growth hormone-like activities in pregnancy determined by radioreceptor assays. Endocrinology 1976; 99:765-774.
6. Robertson MC, Friesen HG. Two forms of rat placental lactogen revealed by radioimmunoassay. Endocrinology 1981; 108:2388-2390.
7. Soares MJ, Colosi P, Talamantes F. The development and characterization of a homologous radioimmunoassay for mouse placental lactogen. Endocrinology 1982; 110:668-670.
8. Duckworth ML, Kirk KL, Friesen HG. Isolation and identification of a cDNA clone of rat placental lactogen II. J Biol Chem 1986; 261:10871-10878.
9. Jackson LL, Colosi P, Talamantes F, Linzer DIH. Molecular cloning of mouse placental lactogen cDNA. Proc Natl Acad Sci USA 1986; 83:8496-8500.
10. Colosi P, Talamantes F, Linzer DIH. Molecular cloning and expression of mouse placental lactogen I complementary deoxyribonucleic acid. Mol Endocrinol 1987; 1:767-776.
11. Robertson MC, Croze F, Schroedter IC, Friesen HG. Molecular cloning and expression of rat placental lactogen-I complementary deoxyribonucleic acid. Endocrinology 1990; 127:702-710.
12. Linzer DIH, Nathans D. Growth-related changes in specific mRNAs of cultured mouse cells. Proc Natl Acad Sci USA 1983; 80:4271-4275.
13. Pearson WR. Using the FASTA program to search protein and DNA sequence databases. Meth Mol Biol 1994; 24:307-331.
14. Linzer DIH, Nathans D. Nucleotide sequence of a growth-related member of the prolactin-growth hormone family. Proc Natl Acad Sci USA 1984; 81:4255-4259.
15. Nilsen-Hamilton M, Hamilton RT, Alvarez-Azaustre E. Relationship between mitogen-regulated protein (MRP) and proliferin (PLF), a member of the prolactin/growth hormone family. Gene 1987; 51:163-170.

16. Linzer DIH, Lee S-J, Ogren L, Talamantes F, Nathans, D. Identification of proliferin mRNA and protein in mouse placenta. Proc Natl Acad Sci USA 1985; 82:4356-4359.
17. Wilder EL, Linzer DIH. Expression of multiple proliferin genes in mouse cells. Mol Cell Biol 1986; 6:3283-3286.
18. Fang Y, Lepont P, Fassett JT, Ford SP, Mubaidin A, Hamilton RT, Nilsen-Hamilton M. Signaling between the placenta and the uterus involving the mitogen-regulated protein/ proliferins. Endocrinology 1999; 140:5239-5249.
19. Linzer DIH, Nathans D. A new member of the prolactin-growth hormone family expressed in mouse placenta. EMBO J 1985; 4:1419-1423.
20. Colosi P, Ogren L, Southard JN, Thordarson G, Linzer DIH, Talamantes F. Biological, immunological, and binding properties of recombinant mouse placental lactogen-I. Endocrinology 1998; 123:2662-2667.
21. Lee S-J, Nathans D. Proliferin secreted by cultured cells binds to mannose 6-phosphate receptors. J Biol Chem 1988; 263:3521-3527.
22. Cooke NE, Coit D, Weiner RI, Baxter JD, Martial JA. Structure of cloned DNA complementary to rat prolactin messenger RNA. Jbiol Chem 1980; 255:6502-6510.
23. Linzer DIH, Talamantes F. Nucleotide sequence of mouse prolactin and growth hormone mRNAs, and expression of these mRNAs during pregnancy, J Biol Chem 1985; 260:9574-9579.
24. Hirosawa M, Miura R, Min K-S, Hattori N, Shiota K, Ogawa T. A cDNA encoding a new member of the rat placental lactogen family, PL-I mosaic (PL-Im). Endocr J 1994; 41:387-397.
25. Dai G, Imagawa W, Liu B, Levan G, Szpirer C, Kwok SCM, Soares MJ. Rcho-1 trophoblast cell placental lactogens: complementary DNAs, heterologous expression, and biological activities. Endocrinology 1996; 137:5020-5027.
26. Deb S, Faria TN, Roby KF, Larsen D, Kwok SCM, Talamantes F, Soares MJ. Identification and characterization of a new member of the prolactin family, placental lactogen-I variant. J Biol Chem 1991; 266:1605-1610.
27. Robertson MC, Schroedter IC, Friesen HG. Molecular cloning and expression of rat placental lactogen-Iv, a variant of rPL-I present in late pregnant rat placenta. Endocrinology 1991; 129:2746-2756.
28. Cohick CB, Dai G, Xu L, Deb S, Kamei T, Levan G, Szpirer C, Szpirer J, Kwok SCM, Soares MJ. Placental lactogen-I variant utilizes the prolactin receptor signaling pathway. Mol Cell Endocrinol 1996; 116:49-58.
29. Parfett CL, Hamilton RT, Howell BW, Edwards DR, Nilsen-Hamilton M, Denhardt DT. Characterization of a cDNA clone encoding murine mitogen-regulated protein: regulation of mRNA levels in mortal and immortal cell lines. Mol Cell Biol 1985; 5:3289-3292.
30. Connor AM, Waterhouse P, Khokha R, Denhardt DT. Characterization of a mouse mitogen-regulated protein/proliferin gene and its promoter: a member of the growth hormone/prolactin gene superfamily. Biochim Biophys Acta 1989; 1009:75-82.
31. Sahgal N, Knipp GT, Liu B, Chapman BM, Dai G, Soares MJ. Identification of two new nonclassical members of the rat prolactin family. J Mol Endocrinol 2000; 24:95-108.
32. Toft DJ, Linzer DIH. Identification of three prolactin-related hormones as markers of invasive trophoblasts in the rat. Biol Reprod 2000; in press.
33. Duckworth ML, Peden LM, Friesen HG. Isolation of a novel prolactin-like cDNA clone from developing rat placenta. J Biol Chem 1986; 261:10879-10884.
34. Lin J, Poole J, Linzer DIH. Three new members of the mouse prolactin/growth hormone family are homologous to proteins expressed in the rat. Endocrinology 1997; 138:5541-5549.
35. Müller H, Ishimura R, Orwig KE, Liu B, Soares MJ. Homologues for prolactin-like protein-A and B are present in the mouse. Biol Reprod 1998; 58:45-51.

36. Duckworth ML, Peden LM, Friesen HG. A third prolactin-like protein expressed by the developing rat placenta: complementary deoxyribonucleic acid sequence and partial structure of the gene. Mol Endocrinol 1988; 2:912-920.
37. Deb S, Roby KF, Faria TN, Szpirer C, Levan G, Kwok SCM, Soares MJ. Molecular cloning and characterization of prolactin-like protein-C complementary deoxyribonucleic acid. J Biol Chem 1991; 266:23027-23032.
38. Dai G, Liu B, Levan G, Szpirer C, Kwok SCM, Soares MJ. Prolactin-like protein-C variant: complementary DNA, unique six exon gene structure, and trophoblast cell-specific expression. Endocrinology 1996; 137:5009-5019.
39. Iwatsuki K, Shinozaki M, Hattori N, Hirasawa K, Itagaki S-I, Shiota K, Ogawa T. Molecular cloning and characterization of a new member of the rat placental prolactin (PRL) family, PRL-like protein D (PLP-D). Endocrinology 1996; 137:3849-3855.
40. Iwatsuki K, Oda M, Sun W, Tanaka S, Ogawa T, Shiota K. Molecular cloning and characterization of a new member of the rat placental prolactin (PRL) family, PRL-like protein H. Endocrinology 1998; 139:4976-4983.
41. Roby KF, Deb S, Gibori G, Szpirer C, Levan G, Kwok SCM, Soares MJ. Decidual prolactin related protein: identification, molecular cloning, and characterization. J Biol Chem 1993; 268:3136-3142.
42. Orwig KE, Ishimura R, Müller H, Liu B, Soares MJ. Identification and characterization of a mouse homolog for decidual/trophoblast prolactin-related protein. Endocrinology 1997; 139:5511-5517.
43. Dai G, Chapman BM, Liu B, Wang D, Orwig KE, White RA, Preuett B, Soares MJ. A new mouse member of the prolactin-like protein-C subfamily: structure and expression. Endocrinology 1998; 139:5157-5163.
44. Lin J, Poole J, Linzer DIH. Two novel members of the prolactin/growth hormone family are expressed in the mouse placenta. Endocrinology 1997; 138:5535-5540.
45. Müller H, Orwig KE, Soares MJ. Identification of two new members of the mouse prolactin gene family. Biochim Biophys Acta 1998; 251-258.
46. Ishibashi K, Imai M. Identification of four new members of the rat prolactin/growth hormone gene family. Biochem Biophys Res Commun 1999; 262:575-578.
47. Hiraoka Y, Ogawa M, Sakai Y, Takeuchi Y, Komatsu N, Shiozawa M, Tanabe K, Aiso S. PLP-I: a novel prolactin-like gene in rodents. Biochim Biophy Acta 1999; 1447:291-297.
48. Toft DJ, Linzer DIH. Prolactin (PRL)-like protein J, a novel member of the PRL/growth hormone family, is exclusively expressed in maternal decidua. Endocrinology 1999; 140:5095-5101.
49. Dai G, Wang D, Liu B, Kasik JW, Müller H, White RA, Hummel GS, Soares MJ. Three novel paralogs of the rodent prolactin gene family. J Endocrinol 2000; in press.
50. Frigeri LG, Lewis LJ, Teguh K, Lewis UJ. Glycosylation abolishes an *in vitro* insulin-like action of prolactin. Biochem Biophys Res Comm 1986; 134:764-769.
51. Markoff E, Sigel MB, Lacour N, Seavey BK, Friesen HG, Lewis UJ. Glycosylation selectively alters the biological activity of prolactin. Endocrinology 1988; 123:1303-1306.
52. Lewis UJ, Singh RN, Lewis LJ. Two forms of glycosylated human prolactin have different pigeon crop sac-stimulating activities. Endocrinology 1989; 124:1558-1563.
53. Young KH, Buhi WC, Horseman N, Davis J, Kraeling R, Linzer D, Bazer FW. Biological activities of glycosylated and nonglycosylated porcine prolactin. Mol Cell Endocrinol 1990; 71:155-162.
54. Hoffmann T, Penel C, Ronin C. Glycosylation of human prolactin regulates hormone bioactivity and metabolic clearance. J Endocrinol Invest 1993; 16:807-816.
55. Ferrara N, Clapp C, Weiner R.The 16K fragment of prolactin specifically inhibits basal or fibroblast growth factor stimulated growth of capillary endothelial cells. Endocrinology 1991; 129:896-900.

56. Wang YF, Walker AM. Dephosphorylation of standard prolactin produces a more biologically active molecule: evidence for antagonism between nonphosphorylated and phosphorylated prolactin in the stimulation of Nb2 cell proliferation. Endocrinology 1993; 133:2156-2160.
57. Abdel-Meguid SS, Shieh HS, Smith WW, Dayringer HE, Violand BN, Bentle LA. Three-dimensional structure of a genetically engineered variant of porcine growth hormone. Proc Natl Acad Sci USA 1987; 84:6434-6437.
58. Christinger HW, Elkins PA, Sandowski Y, Sakal E, Gertler A, Kossiakoff AA, de Vos AM. Crystallization of ovine placental lactogen in a 1:2 complex with the extracellular domain of the rat prolactin receptor. Acta Crystallog Section D-Biol Crystallog 1998; 54:1408-1411.
59. Cunningham BC, Ultsch M, de Vos AM, Mulkerrin MG, Clauser KR, Wells JA. Dimerization of the extracellular domain of the human growth hormone receptor by a single hormone molecule. Science 1991; 254:821-825.
60. de Vos AM, Ultsch M, Kossiakoff AA. Human growth hormone and extracellular domain of its receptor: crystal structure of the complex. Science 1992; 255:306-312.
61. Davis JA, Linzer DIH. Mutational analysis of a lactogenic hormone reveals a region essential for receptor binding and mitogenic activity. Mol Endocrinol 1989; 3:1987-1995.
62. Orwig KE, Dai G, Rasmussen CA, Soares MJ. Decidual/trophoblast prolactin related protein: characterization of gene structure and cell-specific expression. Endocrinology 1997; 138:2491-2500.
63. Soares MJ, Müller H, Orwig KE, Peters TJ, Dai, G. The uteroplacental prolactin family and pregnancy. Biol Reprod 1998; 58:273-284.
64. Gunnet JW, Freeman ME. The mating-induced release of prolactin: a unique neuroendocrine response. Endocr Rev 1983; 4:44-61.
65. Croze F, Kennedy TG, Schroedter IC, Friesen HG. Expression of rat prolactin-like protein B in deciduoma of pseudopregnant rat and in decidua during early pregnancy. Endocrinology 1990; 127:2665-2672.
66. Cohick CB, Xu L, Soares MJ. Placental prolactin-like protein-B: heterologous expression and characterization of placental and decidual species. J Endocrinol 1997; 152:291-302.
67. Rasmussen CA, Orwig KE, Vellucci S, Soares MJ. Dual expression of prolactin-related protein in decidua and trophoblast tissues during pregnancy. Biol Reprod 1997; 55:647-654.
68. Prigent-Tessier A, Tessier C, Hirosawa-Takamori M, Boyer C, Ferguson-Gottschall S, Gibori G. Rat decidual prolactin. Identification, molecular cloning, and characterization. J Biol Chem 1999; 274:37982-37989.
69. Hall J, Talamantes F. Immunocytochemical localization of mouse placental lactogen in the mouse placenta. J Histochem Cytochem 1984; 32:379-382.
70. Lee S-J, Talamantes F, Wilder E, Linzer DIH, Nathans D. Trophoblastic giant cells of the mouse placenta as the site of proliferin synthesis. Endocrinology 1988; 1761-1768.
71. Campbell WJ, Deb S, Kwok SCM, Joslin J, Soares MJ. Differential expression of placental lactogen-II and prolactin-like protein-A in the rat chorioallantoic placenta. Endocrinology 1989; 125:1565-1574.
72. Duckworth ML, Schroedter IC, Friesen HG. Cellular localization of rat placental lactogen-II and rat prolactin-like proteins A and B by in situ hybridization. Placenta 1990; 11:143-155.
73. Faria TN, Deb S, Kwok SCM, Talamantes F, Soares MJ. Ontogeny of placental lactogen-I and placental lactogen-II expression in the developing rat placenta. Dev Biol 1990; 141:279-291.
74. Faria TN, Ogren L, Talamantes F, Linzer DIH, Soares MJ. Localization of placental lactogen-I in trophoblast giant cells of the mouse placenta. Biol Reprod 1991; 44:327-331.

75. Carney EW, Prideaux V, Lye SJ, Rossant J. Progressive expression of trophoblast specific genes during formation of mouse trophoblast giant cells *in vitro*. Mol Reprod Dev 1993; 34:357-368.
76. Colosi P, Swiergiel JJ, Wilder EL, Oviedo A, Linzer DIH. Characterization of proliferin-related protein. Mol Endocrinol 1988; 2:579-586.
77. Deb S, Roby KF, Faria TN, Larsen D, Soares MJ. Identification and immunochemical characterization of a major placental secretory protein related to the prolactin-growth hormone family, prolactin-like protein-C. Endocrinology 1991; 128:3066-3072.
78. Chatelain A, Dupuoy JP, Dubois MP. Ontogenesis of cells producing polypeptide hormones (ACTH, MSH, LPH, GH, Prolactin) in the fetal hypophysis of the rat: influence of the hypothalamus. Cell Tissue Res 1979; 196:409-427.
79. Watanabe YD, Daikoku S. An immunohistochemical study on the cytogenesis of adenohypophysial cells in fetal rats. Dev Biol 1979; 68:559-567.
80. Tong Y, Zhao H, Labrie F, Pelletier G. Ontogeny of prolactin mRNA in the rat pituitary gland as evaluated by in situ hybridization. Mol Cell Endocrinol 1989; 67:11-16.
81. Simmons DM, Voss JW, Ingraham HA, Holloway JM, Broide RS, Rosenfeld MG, Swanson LW. Pituitary cell phenotypes involve cell-specific Pit-1 mRNA translation and synergistic interactions with other classes of transcription factors. Genes Dev 1990; 4:695-711.
82. Colosi P, Ogren L, Thordarson G, Talamantes F. Purification and partial characterization of two prolactin-like glycoprotein hormone complexes from the midpregnant mouse conceptus. Endocrinology 1987; 120:2500-2511.
83. Ogren L, Southard JN, Colosi P, Linzer DIH, Talamantes F. Mouse placental lactogen-I: RIA and gestational profile in maternal serum. Endocrinology 1989; 125:2253-2257.
84. Robertson MC, Cosby H, Shiu RPC. Rat placental lactogen-I variant (rPL-Iv), product of an unique gene, is biologically different from rPL-I. Endocrinology 1996; 137:5242-5249.
85. Gu Y, Soares MJ, Srivastava RK, Gibori G. Expression of decidual prolactin-related protein in the rat decidua. Endocrinology 1994; 135:1422-1427.
86. Jollie WP. Radioautographic observations on variations in desoxyribonucleic acid synthesis in rat placenta with gestational age. Am J Anat 1964; 114:161-171.
87. Enders AC. A comparative study of the fine structure of the trophoblast in several hemochorial placentas. Am J Anat 1965; 116:29-67.
88. DeFeo VJ. Decidualization. In: Wynn RM (ed) Cellular Biology of the Uterus. New York: Appleton-Century-Crofts, 1967:191-290.
89. Davies J, Glasser SR. Histological and fine structural observations on the placenta of the rat. Acta Anat 1968; 69:542-608.
90. Enders AC, Welsh AO. Structural interactions of trophoblast and uterus during hemochorial placenta formation. J Exper Zool 1993; 266:578-587.
91. Rossant J. Development of the extraembryonic lineages. Sem Dev 1995; 6:237-247.
92. Telgmann R, Gellersen B. Marker genes of decidualization: activation of the decidual prolactin gene. Human Reprod Update 1998; 4:472-479.
93. Orwig KE, Soares MJ. Transcriptional activation of the decidual/trophoblast prolactin-related protein gene. Endocrinology 1999; 140:4032-4039.
94. Yamaguchi M, Ogren L, Endo H, Thordarson G, Bigsby RM, Talamantes F. Co-localization of placental lactogen-I, placental lactogen-II, and proliferin in the mouse placenta at midpregnancy. Biol Reprod 1994; 51:1188-1192.
95. Nieder GL, Jennes L. Production of mouse placental lactogen-I by trophoblast giant cells in utero and *in vitro*. Endocrinology 1990; 126:2809-2814.
96. Robertson MC, Owens RE, Klindt J, Friesen HG. Ovariectomy leads to a rapid increase in rat placental lactogen secretion. Endocrinology 1984; 114:1805-1811.
97. Robertson MC, Owens RE, McCoshen JA, Friesen HG. Ovarian factors inhibit and fetal factors stimulate the secretion of rat placental lactogen. Endocrinology 1984; 114:22-30.

98. Soares MJ, Bartke A, Colosi P, Talamantes F. Identification of a placental lactogen in pregnant Snell and Ames dwarf mice. Proc Soc Exp Biol Med 1984; 175:106-108.
99. Soares MJ, Talamantes F. Placental lactogen secretion in the mouse: *in vitro* responses and ovarian and hormonal influences. J Exp Zool 1985; 234:97-104.
100. Day JR, Ogren L, Talamantes F. The effect of hypophysectomy on serum placental lactogen and progesterone in the mouse. Endocrinology 1996; 119:898-903.
101. Kishi K, Ogren L, Southard JN, Talamantes F. Pituitary factors regulating mouse placental lactogen-II (mPL-II) secretion during the last half of pregnancy in mice. Endocrinology 1988; 122:2309-2317.
102. Lopez MF, Carrion FA, Talamantes F. Pituitary-placental interaction: hypophysectomy modulates secretion of mouse placental lactogen-I. Endocrinology 1992; 129:2325-23.
103. Lopez MF, Ogren L, Linzer DIH, Talamantes F. Pituitary placental interaction during pregnancy: regulation of prolactin-like proteins. Endocr J 1993; 1:513-518.
104. Pinon I, Kishi K, Talamantes F. The kinetics of disappearance of endogenous mouse placental lactogen-II in intact and hypophysectomized pregnant mice. Mol Cell Endocrinol 1988; 55:45-51.
105. Roby KF, Soares MJ. Trophoblast cell differentiation and organization: role of fetal and ovarian signals. Placenta 1991; 14:529-545.
106. Soares MJ, Dai G, Cohick CB, Müller H, Orwig KE. The rodent placental prolactin family and pregnancy. In Bazer FW, ed. The Endocrinology of Pregnancy. Totowa, NJ: Humana Press, 1998:145-176.
107. Faria TN, Soares MJ. Trophoblast cell differentiation: establishment, characterization, and modulation of a rat trophoblast cell line expressing members of the placental prolactin family. Endocrinology 1991; 129:2895-2906.
108. Shida MM, Ng YK, Soares MJ, Linzer DIH. Trophoblast-specific transcription from the mouse placental lactogen-I gene promoter. Mol Endocrinol 1993; 7:181-188.
109. Hamlin GP, Lu X-J, Roby KF, Soares MJ. Recapitulation of the pathway for trophoblast giant cell differentiation *in vitro*: stage-specific expression of members of the prolactin gene family. Endocrinology 1994; 134:2390-2396.
110. Peters TJ, Chapman BM, Soares MJ. Trophoblast differentiation. An *in vitro* model for trophoblast giant cell development. In Tuan RS, Lo CW, eds. Methods in Molecular Biology Vol 137: Developmental Biology Protocols, Vol. III. Totowa, NJ: Humana Press, 2000:301-311.
111. Ng YK, George KM, Engel JD, Linzer DIH. GATA factor activity is required for the trophoblast-specific transcriptional regulation of the mouse placental lactogen I gene. Development 1994; 120:3257-3266.
112. Ma GT, Roth ME, Groskopf JC, Tsai F-Y, Orkin SH, Grosveld F, Engel JD, Linzer DIH. GATA-2 and GATA-3 regulate trophoblast-specific gene expression *in vivo*. Development 1997; 124:907-914.
113. Peters TJ, Chapman BM, Wolfe MW, Soares MJ. Placental lactogen-I gene activation in differentiating trophoblast cells: extrinsic and intrinsic regulation involving mitogen-activated protein kinase signaling pathways. J Endocrinol 2000; 165:143-156.
114. Sun Y, Duckworth ML. Identification of a placental-specific enhancer in the rat placental lactogen II gene that contains binding sites for members of the Ets and AP-1 (Activator Protein 1) families of transcription factors. Mol Endocrinol 1999; 13:385-399.
115. Vuille J, Cattini PA, Bock ME, Verstuyf A, Schroedter IC, Duckworth ML, Friesen HG. Rat prolactin-like protein A partial gene and promoter structure: promoter activity in placental and pituitary cells. Mol Cell Endocrinol 1993; 96:91-98.
116. Dai G, Wolfe MW, Soares MJ. Distinct regulatory regions from the prolactin-like protein C variant promoter direct trophoblast giant cell versus spongiotrophoblast cell-specific expression. Endocrinology 1999; 140:4691-4698.

117. Shida MM, Ross SR, Linzer DIH. Placental-specific expression in transgenic mice from the mouse placental lactogen-II gene promoter. Proc Natl Acad Sci USA 1992; 89:3864-3868.
118. Lin J, Linzer DIH. Identification of trophoblast-specific regulatory elements in the mouse placental lactogen II gene. Mol Endocrinol 1998; 12:418-427.
119. Shah P, Sun Y, Szpirer C, Duckworth ML. Rat placental lactogen II gene: characterization of gene structure and placental-specific expression. Endocrinology 1998; 139:967-973.
120. Cross JC, Flannery ML, Blanar MA, Steingrimsson E, Jenkins NA, Copeland NG, Rutter WJ, Werb Z. Hxt encodes a basic helix-loop-helix transcription factor that regulates trophoblast cell development. Development 1995; 121:2513-2523.
121. Scott IC, Anson-Cartwright L, Riley P, Reda D, Cross JC. The HAND1 basic helix-loop-helix transcription factor regulates trophoblast differentiation via multiple mechanisms. Mol Cell Biol 2000; 20:530-541.
122. Riley P, Anson-Cartwright L, Cross JC. The Hand1 helix-loop-helix transcription factor is essential for placentation and cardiac morphogenesis. Nat Genet 1998; 18:271-275.
123. Kraut N, Snider C-M, Chen A, Tapscott SJ, Groudine M. Requirement of I-mfa gene for placental development and skeletal patterning. EMBO J 1998; 17:6276-6288.
124. Nakayama H, Scott IC, Cross JC. The transition to endoreduplication in trophoblast giant cells is regulated by the Sna zinc-finger transcription factor. Dev Biol 1998; 199:150-163.
125. Ogilvie S, Duckworth ML, Larkin LH, Buhi WC, Shiverick KT. *De novo* synthesis and secretion of prolactin-like protein-B by rat placental explants. Endocrinology 1990; 126:2561-2566.
126. Soares MJ, Chapman BM, Rasmussen CA, Dai G, Kamei T, Orwig KE. Differentiation of trophoblast endocrine cells. Placenta 1996; 17, 277-289.
127. Lu X-J, Deb S, Soares MJ. Spontaneous differentiation of trophoblast cells along the spongiotrophoblast pathway: expression of the placental prolactin gene family and modulation by retinoic acid. Dev Biol 1994; 163:86-97.
128. Guillemot F, Nagy A, Auerbach A, Rossant J, Joyner A. Essential role of Mash-2 in extraembryonic development. Nature 1994; 371:333-336.
129. Sibilia M, Wagner EF. Strain-dependent epithelial defects in mice lacking the EGF receptor. Science 1995; 269:234-238.
130. Threadgill DW, Dlugosz AA, Hansen LA, Tennenbaum T, Lichti U, Yee D, LaMantia C, Mourton T, Herrup K, Harris RC, Barnard JA, Yuspa SH, Coffey RJ, Magnuson T. Targeted disruption of mouse EGF receptor: effect of genetic background on mutant phenotype. Science 1995; 269:230-234.
131. Knipp GT, Audus KL, Soares MJ. Nutrient transport across the placenta. Adv Drug Delivery Rev 1999; 38:41-58.
132. Ogren L, Talamantes F. Prolactins of pregnancy and their cellular source. Int Rev Cytol 1988; 112:1-65.
133. Manzella SM, Dharmesh SM, Cohick CB, Soares MJ, Baenziger JU. Developmental regulation of a pregnancy-specific oligosaccharide structure, NeuAca2,6GalNAcß1,4GlcNAc on select members of the rat placental prolactin family. J Biol Chem 1997; 271:4775-4782.
134. Southard JN, Talamantes F. High molecular weight forms of placental lactogen: evidence for lactogen-macroglobulin complexes in rodents and humans. Endocrinology 1989; 125:791-800.
135. Deb S, Soares MJ. Characterization of placaental prolactin-like protein-A in intracellular and extracellular compartments. Mol Cell Endocrinol 1990; 74:163-172.
136. Deb S, Hamlin GP, Roby KF, Kwok SCM, Soares MJ. Heterologous expression and characterization of prolactin-like protein-A: identification of serum binding proteins. J Biol Chem 1993; 268:3298-3305.

137. Rasmussen CA, Hashizume K, Orwig KE, Xu L, Soares MJ. Decidual prolactin-related protein: heterologous expression and characterization. Endocrinology 1996; 137:5558-5566.
138. Deb S, Youngblood T, Rawitch A, Soares MJ. Placental prolactin-like protein-A: identification and characterization of two major glycoprotein species with antipeptide antibodies. J Biol Chem 1989; 264:14348-14353.
139. Freemark M, Kirk K, Pihoker C, Robertson MC, Shiu RPC, Driscoll P. Pregnancy lactogens in the rat conceptus and fetus: circulating levels, distribution of binding, and expression of receptor messenger ribonucleic acid. Endocrinology 1993; 133:1830-1842.
140. Jackson D, Linzer DIH. Proliferin transport and binding in the mouse fetus. Endocrinology 1997; 138:149-155.
141. Jackson D, Volpert O, Bouck N, Linzer DIH. Stimulation and inhibition of angiogenesis by placental proliferin and proliferin-related protein. Science 1994; 266:1581-1584.
142. Thordarson G, Villalobos R, Colosi P, Southard J, Ogren L, Talamantes F. Lactogenic response of cultured mouse mammary epithelial cells to mouse placental lactogen. J Endocrinol 1986; 109:263-274.
143. Fielder PJ, Thordarson G, English A, Rosenfeld RG, Talamantes F. Expression of a lactogen-dependent insulin-like growth factor-binding protein in cultured mouse mammary epithelial cells. Endocrinology 1992; 131:261-267.
144. Galosy SS, Talamantes F. Luteotropic actions of placental lactogens at midpregnancy in the mouse. Endocrinology 1995; 136:3993-4003.
145. Thordarson G, Galosy S, Gudmundsson GO, Newcomer B, Sridaran R, Talamantes F. Interaction of mouse placental lactogens and androgens in regulating progesterone release in cultured mouse luteal cells. Endocrinology 1997; 138:3236-3241.
146. Robertson MC, Gillespie B, Friesen HG. Characterization of the two forms of rat placental lactogen (rPL): rPL-I and rPL-II. Endocrinology 1982; 111:1862-1866.
147. Robertson MC, Cosby H, Fresnoza A, Cattini PA, Shiu RPC, Friesen HG. Expression, purification, and characterization of recombinant rat placental lactogen-I: a comparison with the native hormone. Endocrinology 1994; 134:393-400.
148. Soares MJ, Colosi P, Ogren L, Talamantes F. Identification and partial characterization of a lactogen from the midpregnant mouse conceptus. Endocrinology 1983; 112:1313-1317.
149. Bridges RS, Robertson MC, Shiu RPC, Friesen HG, Stuer AM, Mann PE. Endocrine communication between conceptus and mother: placental lactogen stimulation of maternal behavior. Neuroendocrinology 1996; 64:57-64.
150. Bridges RS, Roberston MC, Shiu RPC, Sturgis JD, Henriquesz BM, Mann PE. Central lactogenic regulation of maternal behavior in rats: steroid dependence, hormone specificity, and behavioral potenicies of rat prolactin and rat placental lactogen I. Endocrinology 1997; 138:756-763.
151. Voogt JL, Soares MJ, Robertson MC, Arbogast LA. Rat placental lactogen-I abolishes nocturnal prolactin surges in the pregnant rat. Endocrine 1996; 4:233-238.
152. Brelje TC, Scharp DW, Lacy PE, Ogren L, Talamantes F, Robertson M, Friesen HG, Sorenson RL. Effect of homologous placental lactogens, prolactins, and growth hormones on islet B-cell division and insulin secretion in rat, mouse, and human islets: implication for placental lactogen regulation of islets function during pregnancy. Endocrinology 1993; 132:879-887.
153. Kawai M, Kishi K. *In vitro* studies of the stimulation of insulin secretion and B-cell proliferation by rat placental lactogen-II during pregnancy in rats. J Reprod Fertil 1997; 109:145-152.
154. Soares MJ, Faria TN, Roby KF, Deb S. Pregnancy and the prolactin family of hormones: coordination of anterior pituitary, uterine, and placental expression. Endocr Rev 1991; 12: 402-423.

155. Robertson MC, Friesen HG. The purification and characterization of rat placental lactogen.Endocrinology 1975; 97:621-629.
156. Colosi P, Marr G, Lopez J, Haro L, Ogren L, Talamantes F. Isolation, purification, and characterization of mouse placental lactogen. Proc Natl Acad Sci USA 1982; 79:771-775.
157. Harigaya T, Smith WC, Talamantes F. Hepatic placental lactogen receptors during pregnancy in the mouse. Endocrinology 1988; 122:1366-1372.
158. MacLeod KR, Smith WC, Ogren L, Talamantes F. Recombinant mouse placental lactogen-I binds to lactogen receptors in mouse liver and ovary: partial characterization of the ovarian receptor. Endocrinology 1989; 125:2258-2266.
159. Sakal E, Robertson MC, Chapnik-Cohen N, Tchelet A, Gertler A. Interaction of recombinant rat placental lactogen-I with extracellular domains of prolactin receptors from three species. Receptors Signal Transduction 1996; 6:35-42.
160. Hattori N, Nukada T, Oda M, Tanaka S, Ogawa T, Shiota K. Evaluation of the role of N-linked oligosaccharides in rat placental lactogen action by site-directed mutagenesis. Endocrine J 1998; 45:659-674.
161. Linzer DIH, Fisher SJ. The placenta and the prolactin family of hormones - regulation of the physiology of pregnancy. Mol Endocrinol 1999; 13:837-840.
162. Volpert O, Jackson D, Bouck N, Linzer DIH. The insulin-like growth factor II/mannose 6-phosphate receptor is required for proliferin-induced angiogenesis. Endocrinology 1996; 137:3871-3876.
163. Nelson JT, Rosenzweig N, Nilsen-Hamilton M. Characterization of the mitogen-regulated protein (proliferin) receptor. Endocrinology 1995; 136:283-288.
164. O'Dell SD, Day IN. Insulin-like growth factor II (IGF-II). Int J Biochem Cell Biol 1998; 30:767-771.
165. Groskopf JC, Syu L-J, Saltiel AR, Linzer DIH. Proliferin induces endothelial cell chemotaxis through a G protein-coupled, MAP kinase-dependent pathway. Endocrinology 1997; 138:2835-2840.
166. Nilsen-Hamilton M, Shapiro JM, Massoglia SL, Hamilton RT. Selective stimulation by mitogens of incorporation of 35S-methionine into a family of proteins released into the medium by 3T3 cells. Cell 1980; 20:19-28.
167. Linzer DIH, Nathans D. Changes in specific mRNAs following serum stimulation of cultured mouse cells: increase in a prolactin-related mRNA. In: Levine AJ, Vande Woude GF, Topp WC, Watson JD, eds. Cancer Cells 1/ The Transformed Phenotype. Cold Spring Harbor, NY: Cold Spring Harbor Laboratory, 1984:111-115.
168. Gil-Torregrosa B, Urdiales JL, Lozano J, Mates JM, Sanchez-Jimenez F.Expression of different mitogen-regulated protein/proliferin mRNAs in Ehrlich carcinoma cells. FEBS Lett 1994; 349:343-348.
169. Barnes SW, Renegar RH. Identification, characterization, and expression of a new prolactin-like molecule in the hamster placenta. Biol Reprod 1996; 55:370-378.
170. Davison JM, Lindheimer MD. Volume homeostasis and osmoregulation in human pregnancy. Baillieres Clin Endocrinol Metab 1989; 3:451-472.
171. Jackson CW, Steward SA, Ashmun RA, McDonald TP. Megakaryocytopoiesis and platelet production are stimulated during late pregnancy and early postpartum in the rat. Blood 1992; 79:1672-1678.
172. Lin J, Linzer DIH. Induction of megakaryocyte differentiation by a novel pregnancy-specific hormone. J Biol Chem 1999; 274:21485-21489.
173. Lin J, Toft DJ, Bengtson NW, Linzer DIH. Placental prolactins and the physiology of pregnancy. Rec Prog Horm Res 2000; 55: in press.
174. Croy BA, Whitelaw PF, Engelhardt H. The influences of immune cells on the success of pregnancy. In: Bazer FW, ed, The Endocrinology of Pregnancy, Totowa, NJ: Humana Press Inc, 1998:229-289.

175. Wang D, Ishimura R, Walia DS, Müller H, Dai G, Hunt JS, Lee NA, Lee JJ, Soares MJ. Eosinophils are cellular targets of the novel uteroplacental heparin-binding cytokine, decidual/trophoblast prolactin-related protein. J Endocrinol 2000; submitted.
176. Müller H, Liu B, Croy BA, Head JR, Hunt JS, Dai G, Soares MJ. Uterine natural killer cells are targets for a trophoblast cell-specific cytokine, prolactin-like protein-A. Endocrinology 1999; 140:2711-2720.
177. Rothenberg ME. Eosinophilia. New England J Med 1998; 338:1592-1600.
178. Kay AB, Barata L, Meng Q, Durham SR, Ying S. Eosinophils and eosinophil-associated cytokines in allergic inflammation. Int Arch Allergy Immunol 1997; 113:196-199.
179. Tepper RI. An eosinophil-dependent mechanism for the antitumor effect of interleukin-4. Science 1992; 257:548-551.
180. Pearlman E. Immunopathology of onchocerciasis: a role for eosinophils in onchocercal dermatitis and keratitis. Chem Immunol 1997; 66:26-40.
181. Bassett EG. Infiltration of eosinophils into the modified connective tissue of oestrous and pregnant animals. Nature 1962; 194:1259-1261.
182. McMaster MT, Newton RC, Dey SK, Andrews GK. Activation and distribution of inflammatory cells in the mouse uterus during the preimplantation period. J Immunol 1992; 148:1699-1705.
183. Jeziorska M, Salamonsen LA, Woolley DE. Mast cell and eosinophil distribution and activation in human endometrium throughout the menstrual cycle. Biol Reprod 1995; 53:312-320.
184. Bjersing L, Borglin NE. Effect of hormones on incidence of uterine eosinophilia in rats. Acta Path Microbiol Scand 1964; 60:353-364.
185. Howe RS, Lee YH, Fischokoff SA, Teuscher C, Lyttle CR. Glucocorticoid and progestin regulation of eosinophil chemotactic factor and complement C3 in the estrogen-treated rat uterus. Endocrinology 1990; 126:3193-3199.
186. Liu C-C, Parr EL,Young JD-E. Granulated lymphoid cells of the pregnant uterus: morphological and functional features. Int Rev Cytol 1994; 153:105-136.
187. Guimond M-J, Luross JA, Wang B, Terhorst C, Danial S, Croy BA. Absence of natural killer cells during pregnancy is associated with reproductive compromise in tge26 mice. Biol Reprod 1997; 56:169-179.
188. Hunt JS, Miller L, Vassmer D, Croy BA. Expression of the inducible nitric oxide synthase gene in mouse uterine leukocytes and potential relationships with uterine function during pregnancy. Biol Reprod 1997; 57:827-836.
189. Croy BA, Guilbert LJ, Browne MA, Gough NM, Stinchcomb DT, Reed N, Wegmann TG. Characterization of cytokine production by the metrial gland and granulated metrial gland cells. J Reprod Immunol 1991; 19:149-166.
190. Pihoker C, Robertson MC, Freemark M. Rat placental lactogen-I binds to the choroid plexus and hypothalamus of the pregnant rat. J Endocrinol 1993; 139:235-242.
191. Müller H, Dai G, Soares MJ. Placental lactogen-I (PL-I) target tissues identified with an alkaline phosphatase-PL-I fusion protein. J Histochem Cytochem 1998; 46:737-744.
192. Wilder EL, Linzer DIH. Participation of multiple factors, including proliferin, in the inhibition of myogenic differentiation. Mol Cell Biol 1989; 9:430-441.
193. Muscat GE, Gobius K, Emery J. Proliferin, a prolactin/growth hormone-like peptide represses myogenic-specific transcription by the suppression of an essential serum response factor-like DNA-binding activity. Mol Endocrinol 1991; 5:802-814.
194. Bittorf T, Jaster R, Soares MJ, Seiler J, Brock J, Friese K, Müller H. Induction of erythroid proliferation and differentiation by a trophoblast-specific cytokine involves activation of the Jak/STAT pathway. J Mol Endocrinol 2000; Submitted.
195. Talamantes F, Ogren L, Markoff E, Woodard S, Madrid J. Phylogenetic distribution, regulation of secretion, and prolactin-like effects of placental lactogens. Fed Proc 1980; 39:2582-2587.

196. Schuler LA, Kessler MA. Bovine placental prolactin-related hormones. Trends Endocrinol Metab 1992; 3:334-338.
197. Anthony RV, Liang R, Kayl EP, Pratt SL. The growth hormone/prolactin gene family in ruminant placentae. J Reprod Fertil 1995; Suppl 49:83-95.
198. Cooke NE, Liebhaber SA. Molecular biology of the growth hormone-prolactin gene system. Vitamins Horm 1995; 50:385-459.
199. Wallis M. The expanding growth hormone/prolactin family. J Mol Endocrinol 1992; 9:185-188.
200. Goffin V, Shiverick KT, Kelly PA, Martial JA. Sequence-function relationships within the expanding family of prolactin, growth hormone, placental lactogen, and related proteins in mammals. Endocr Rev 1996; 17:385-410.
201. Nicoll CS. Ontogeny and evolution of prolactin's functions. Fed Proc 1980; 39:2563-2566.

Chapter 9

Human Uteroplacental Lactogens: Physiology and Molecular Biology

Stuart Handwerger and Anoop Brar

Division of Endocrinology, Children's Hospital Medical Center and Department of Pediatrics, University of Cincinnati College of Medicine, Cincinnati, OH

INTRODUCTION

The human uteroplacental unit synthesizes and secretes three lactogenic hormones - placental lactogen (hPL), growth hormone-variant (hGH-V) and decidual prolactin (PRL). The three hormones, which evolved from a common ancestral molecule, have striking homologies in their chemical and biological properties. The hormones bind to both lactogenic (prolactin) and somatotrophic (growth hormone) receptors and exert many physiologic actions that are critical for the regulation of maternal and fetal metabolism and the growth of the fetus. HPL and hGH-V are synthesized and secreted by the multinucleated syncytiotrophoblast cells of the placenta, while decidual PRL is synthesized and secreted by decidual cells, the predominant cells that line the uterus during pregnancy.

In this chapter, we will review some aspects of the physiology and molecular biology of the hGH/hPL/hPRL gene family, emphasizing the regulation of the gene family during uteroplacental differentiation. Extensive reviews of the hormones have recently been presented elsewhere (1-5).

PHYSIOLOGY OF HPL AND HGH-V

PLASMA CONCENTRATIONS

HPL is first detected in maternal plasma at about six weeks of pregnancy (6). The concentration then increases at a linear rate until about the thirtieth week of gestation when peak concentrations of 5000 to 7000 ng/ml are attained. During gestation, plasma hPL concentrations in the mother greatly exceed those of pituitary growth hormone (hGH-N), hGH-V and pituitary PRL (7). The plasma concentration of hPL in the mother throughout pregnancy correlates with placental mass and is greater in multiple than in singleton gestations. The secretion rate near term is about 1.0 gm/day, a rate considerably greater than that of any other protein hormone (8). Aberrations of hPL secretion are present in several pathologic conditions of pregnancy associated with abnormal fetal growth, including pre-eclampsia, hypertensive vascular disease and diabetes mellitus (9-11).

HGH-V is first detected in maternal plasma at 10 to 12 weeks of pregnancy and reaches a maximum of 15 to 25 ng/ml in the first trimester (12). During the latter half of pregnancy, hGH-V becomes the predominant form of growth hormone in the maternal circulation when plasma pituitary hGH-N concentrations decrease to undetectable levels, probably due to suppression of hormone expression at the level of the pituitary. Although little is known about maternal plasma hGH-V concentrations in pathologic conditions of pregnancy, preliminary studies suggest that maternal HGH-V concentrations are also decreased in pregnancies complicated by intrauterine growth retardation (13).

The plasma concentrations of hPL and hGH-V are considerably different in the fetus than in the mother (14). HPL is detected in fetal plasma in the first trimester, reaching a concentration as high as 300 ng/ml. The plasma concentration in the fetus at term is in the range of 20 to 30 ng/ml. In contrast to hPL, hGH-V is not detected in the fetus (15). HGH-N is present in fetal plasma at relatively high concentrations, but the hormone has little or no biologic actions in the fetus due to a lack or paucity of membrane receptors (see below).

BINDING TO MEMBRANE RECEPTORS

HPL binds to the PRL receptor with an affinity similar to that of hGH-N and even greater than that of hPRL (2). HPL, however, binds to the growth hormone receptor with low affinity, suggesting that hPL is a less potent somatotrophic hormone than growth hormone (2). Like hGN-N, hGH-V binds to both somatotrophic and lactogenic plasma membrane receptors and has biological activities similar to PRL and GH. However, the affinity of hGH-V for the somatotrophic receptor is approximately 7-fold greater than that of hGH-N, while the lactogenic activity of hGH-V is less than that of hPRL and hPL (16). These finding suggests that hGH-V is a more potent growth-promoting factor and a less potent lactogenic factor than hGH-N.

The human fetus has abundant PRL receptors, but only a paucity of GH receptors. This paucity of fetal GH receptors accounts for the limited role of the hormone in fetal linear growth and metabolism. The biologic actions of hPL in the fetus appear to be mediated by the PRL receptor and a distinct fetal PL receptor (17). However, attempts to isolate a fetal PL receptor have so far been unsuccessful.

ACTIONS IN MOTHER

HPL plays a major role in the metabolic adaptation to pregnancy. Since plasma hPL concentrations increase markedly during gestation while hGH-N concentrations are suppressed, Grumbach and co-workers (8) many years ago presented a cogent argument for a major role of hPL as a maternal "GH" of the second half of pregnancy. Like hGH-N, hPL increases food intake and stimulates glucose uptake, glucose oxidation and the incorporation of glucose into glycogen, glycerol and fatty acids. The hormone also reduces insulin sensitivity and induces carbohydrate metabolism. In addition, hPL stimulates insulin gene transcription and glucose-dependent insulin secretion. The hormone also affects lipid metabolism by increasing the rate of lipolysis and stimulating plasma concentrations of non-esterified fatty acids, ketones and glycerol. Like GH, hPL promotes nitrogen retention. The net effect of these effects on metabolism appear during fasting and may facilitate the mobilization and utilization of maternal free fatty acids for energy and thereby spare maternal glucose for the fetus. These metabolic changes promote the transport of glucose and amino acids to the fetus.

Several lines of evidence strongly suggest that hPL as well as hGH-V are responsible, at least in part, for the marked increase in maternal plasma IGF-I concentrations that normally occurs during the third trimester of pregnancy. Firstly, ovine PL (oPL) stimulates IGF-I production in hypophysectomized immature rats (18). Secondly, the increase in IGF-I concentrations during pregnancy parallels the increase in hPL concentrations (19). Thirdly, pregnant sheep (20) and rats (21) hypophysectomized during pregnancy maintain normal plasma IGF-I concentrations until after delivery of the placenta. Shortly after delivery, IGF-I concentrations in the hypophysectomized animals decrease rapidly to concentrations typically observed in GH deficient animals. The increase in maternal IGF-1 concentrations most likely accounts, at least in part, for the growth of many maternal tissues during pregnancy, including the uterus and breasts.

Since hGH-V binds to the same receptors as hGH-N, it is very likely that hGH-V has similar biologic actions as hGH-N and hPL during pregnancy.

ACTIONS IN FETUS

HPL also affects fetal growth by acting directly on fetal tissues (for review, see 2,3). For example, oPL has been shown to stimulate ornithine decarboxylase activity (ODC) in fetal liver (22), amino acid transport in fetal muscle (23), and DNA synthesis and IGF production in mouse embryo fibroblasts (24). HPL has also been shown to stimulate the uptake of neutral amino acids and stimulate IGF production by human fetal fibroblasts and myoblasts (25). The observation that PL stimulates ornithine decarboxylase (ODC) activity and IGF production in fetal tissues has important implications for understanding the mechanisms by which PL affects fetal growth. ODC is the rate-limiting step in the synthesis of the polyamines, which are critical in the regulation of protein, RNA and DNA synthesis. Furthermore, many lines of evidence implicate IGF in the regulation of fetal growth. HPL has also been shown to stimulate insulin production from fetal pancreatic islets.

PL stimulates glycogen synthesis in isolated hepatocytes of fetal lambs and rats (26). The glycogenic effect of PL is potentiated by insulin, suggesting that certain metabolic effects of insulin in the fetus may result from synergism of the actions of insulin and PL (26). PL also has been shown to inhibit the glycogenolytic actions of glucagon on fetal hepatocytes by inhibiting glucagon-induced activation of phosphorylase A. This observation suggests that PL promotes glycogen storage in the fetal liver by antagonizing the actions of glucagon as well as by stimulating the synthesis of glycogen. The accumulation of hepatic glycogen in late pregnancy is of critical importance to the fetus since hepatic glycogen is the major source of glucose for the brain and red cells of the neonate. The abrupt decrease in PL concentrations after delivery may contribute to the mobilization of hepatic glycogen immediately after birth.

Although GH stimulates ODC activity, amino acid transport, DNA synthesis and IGF production in postnatal tissues, GH has no effects on these parameters in fetal tissues. This finding is compatible with clinical observations indicating that fetal growth is not impaired by GH deficiency due to anencephaly or congenital absence of the pituitary gland (27,28). Failure of GH to affect fetal growth is also supported by the finding that transgenic mice bearing metallothionein-GH fusion genes are normal size at birth and do not demonstrate excessive size until about three weeks of age (29). As

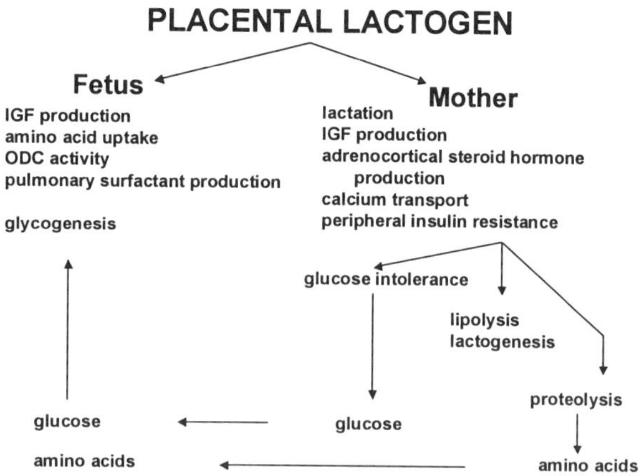

Figure 9-1. Schematic representation of the physiological actions of hPL in the mother and fetus during pregnancy. HPL has direct actions on maternal and fetal tissues. Adapted from (106)

indicated above, the failure of GH to affect fetal growth and metabolism appears to be due, at least in part, to an absence or relative deficiency of GH receptors on fetal tissues.

HPL acting via PRL receptors may play a role in modulating fetal adrenocortical steroid hormone production, including the surge in fetal cortisol synthesis near term (30), and in stimulating production of pulmonary surfactant (31). The hormone, like PRL, may also be important in preparing the breast for lactation. A possible role for hPL in calcium transport is suggested by the observation that hPL stimulates the production of parathyroid hormone related peptide (PTH-RP) (32). A recent investigation demonstrating that rat PL-1 stimulates the growth of whole rat embryos *in vitro* suggests a possible role of hPL in fetal embryonic development (33). A summary of the biologic actions of hPL in the mother and fetus during pregnancy is shown in Figure 9-1.

PATHOLOGIC PREGNANCIES

Several patients have been described with deletions of the hPL-A, hPL-B and hGH-V genes (34-36). Many of these infants were normal size at birth and had no apparent clinical abnormalities. However, there has been a recent report of a child with deletions of the hPL-A, hPL-B and hGH-V genes and severe intrauterine growth retardation (37). In addition, a growth-retarded infant with the classical features of the Russell-Silver syndrome has also been described with a heterozygous deletion of the hPL-A gene (38). Although it is unclear why infants with deletions in the hPL-hGH gene cluster have different phenotypes, it is possible that the loss of hPL-A, and hPL-B and hGH-V in some pregnancies may be compensated, at least in part, by PRL and hPL-L gene products.

REGULATION OF SYNTHESIS, SECRETION AND GENE EXPRESSION

Although the biologic actions of hPL are strikingly homologous to those of hGH-N, the factors that regulate the secretion of the two hormones appear to be distinct. For example, glucocorticoids, estrogens, oxytocin, prostaglandins, epinephrine, thyrotropin-releasing hormone, gonadotropin-releasing hormone or L-DOPA have been shown to have no consistent effects on the synthesis or secretion of hPL *in vitro* (for review see (14). In addition, changes in plasma free fatty acid concentrations have no effect on hPL secretion, and arginine and dexamethasone (which affect pituitary hGH secretion) are also without effect. Furthermore, most investigators have found that changes in plasma glucose concentrations cause no consistent changes in plasma hPL concentrations. HPL concentrations have been shown to increase 30-40% in women fasted 84-90 hours during weeks 16-22 of gestation (prior to therapeutic abortion), but the specific factors contributing to the increase in PL concentrations is unknown. However, acute and chronic changes in extracellular glucose concentrations have no consistent effects on hPL secretion *in vitro*. Angiotensin II (39)and IGF-I have been reported to stimulate hPL release. Unlike hGH-N, hPL is not stored in large secretory granules and is not released in a pulsatile manner (40).

Investigations from our laboratory strongly suggest a novel physiologic role for high density lipoproteins (HDL) in the regulation of hPL gene expression during pregnancy (41). HDL stimulates the synthesis and release of hPL from cultured trophoblast cells (42) and stimulates a significant increase in plasma PL concentrations when infused into pregnant ewes (43). In addition, plasma HDL concentrations are significantly decreased in several pathologic conditions of pregnancy associated with low plasma hPL concentrations and intrauterine growth retardation (44), suggesting that the lower than normal HDL concentrations may contribute to the pathophysiology of these conditions.

Although HDL plays a critical role in lipid metabolism, the effect of HDL on hPL expression is not due to the lipid constituents of the HDL particle but to apolipoprotein (apo) A-I (the major apoprotein constituent of HDL) and to a much lesser extent, apo-A-II and apo C. Investigations of the different forms of apo A-I in plasma suggest that the stimulatory activity of HDL is primarily to pre-beta HDL, a minor component of HDL in that is smaller in size and contains a higher ratio of A-1/lipid than α-HDL, the major circulating form of HDL (45). Pre-beta HDL concentrations in plasma increase from 3 to 4% of total HDL in the first trimester to nearly 20% at term. Investigations of the mechanisms of action of apo A-I on hPL synthesis and release demonstrated that the stimulation of hPL synthesis is due to an increase in gene transcription that is mediated by multiple signal transduction pathways. There appears to be a complex interaction of the adenylate cyclase (46), phospholipase C-mediated phosphoinositide hydrolysis (47) and MAP kinase pathways (48) in apo A-I-mediated hPL expression.

IL-6 and IL-1β also stimulate the synthesis and release of hPL from trophoblast cells (49). LPS, which is known to stimulate the production of cytokines by the placenta, has a similar effect. The action of IL-6 on hPL synthesis is due to the stimulation of hPL gene transcription by a NF-IL6-dependent mechanism (50). Transfection of an expression plasmid for NF-IL6 into choriocarcinoma cells along with the plasma containing the hPL promoter coupled to a CAT reporter gene resulted in a significant increase in CAT activity. Deletion analysis studies of the hPL promoter showed that the transactivation of the promoter in response to IL-6 is due to DNA elements between -1.1 and -2.3 kb, a region of the promoter that contains three consensus NF-IL-6 sites.

Other investigations from our laboratory strongly suggest roles for 1,25-dihydroxyvitamin D_3 (51), retinoic acid (52)and thyroid hormone (52) in the regulation of hPL gene expression. The studies with 1,25-dihydroxyvitamin D_3 and IL-6 are of interest since syncytiotrophoblast cells synthesize and secrete both factors, indicating that the action of these factors on hPL gene expression are mediated, at least in part, by an autocrine/paracrine mechanism. In studies with retinoic acid, both all-*trans* retinoic acid and 9-*cis* retinoic acid were shown to stimulate hPL gene expression, acting through RARα and RXRα nuclear hormone receptors. Computer analysis of the hPL-A promoter reveals three regions of the hPL promoter between -500 to -1100 bp that contain several half-site motifs that closely resemble the response elements for thyroid hormone, all-trans retinoic acid and vitamin D_3 but with unusual spacing between the half-site repeats (53). Site A consists of three putative half-site motifs arranged as direct repeats with one nucleotide spacing (DR 1), which corresponds to a high affinity binding site for homodimeric RXR. Site B has four possible half-site motifs, one of which is a VDRE. A putative AP-1 site is "imbedded" within the half-sites of region B. Site C contains a DR 7 and also a putative SP-1 site adjacent to one of the half-site motifs. Transfection studies in BeWo choriocarcinoma cells indicate that site B is response to RARα but not TRβ; site B is responsive to TRβ, RARα, RXR and VDR; and site C is not response to any of the nuclear hormones. Although the role of the AP-1 element in site B is unknown, the observation that activation of protein kinase C by phorbol esters stimulates the synthesis and release of hPL suggests that the AP-1 site may be important in regulation of hPL expression. A summary of the factors implicated in the regulation of hPL secretion is shown in Figure 9-2.

Although little is known about the regulation of the synthesis and release of hGH-V, it appears that the regulation of hGH-V is different than that of hPL and hGH-N. For example, hGH-N is secreted episodically while hGH-V is secreted tonically. In addition, growth hormone-releasing hormone does not stimulate hGH-V release *in vitro* from human trophoblast cells (54). The release of hGH-V is stimulated by cyclic AMP (55)

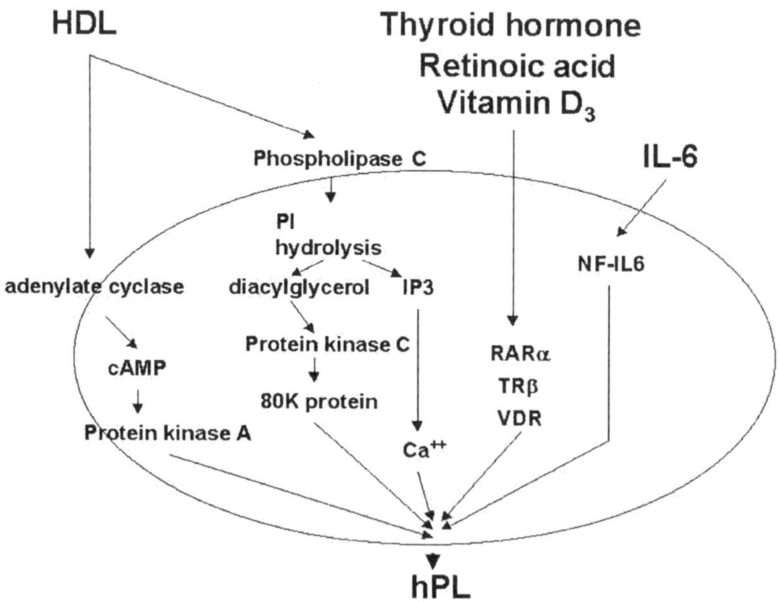

Figure 9-2. Schematic representation of the regulation of hPL expression. HDL also activates the MAP kinase signal transduction pathway. Adapted from (1)

and thyroid hormone (56). Unlike hPL, hGH-V secretion is inhibited *in vitro* and *in vivo* by glucose (5,57), suggesting that hGH-V may protect the fetus against nutrient deficiency.

MOLECULAR BIOLOGY OF THE HGH/HPL GENE CLUSTER

STRUCTURE

The hPL and hGH genes form a superfamily consisting of five closely related genes that are localized contiguously on chromosome 17 (Figure 9-3). A single PRL gene is located on chromosome 6. The three hPL genes (hPL-A, hPL-B and hPL-L) and the two hGH genes (hGH-N and hGH-V) are organized in the same transcriptional orientation in a gene cluster which spans 66 kB. Each of the genes is composed of five exons and four introns, and the genes are ordered from 5' to 3' as hGH-N, hPL-L, hPL-A, hGH-V and hPL-B. The overall activity of the gene cluster has been shown to be regulated by a locus control region located about 23 kB upstream of the cluster (58).

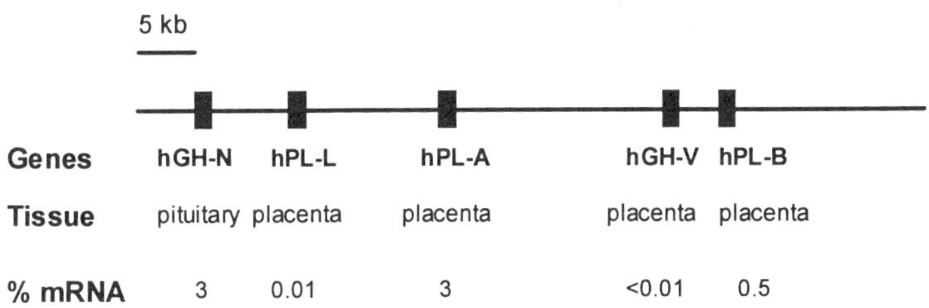

Figure 9-3. Schematic representation of the growth hormone/placental lactogen gene cluster on human chromosome 17. All of the genes are organized in the same transcriptional orientation, and each gene consists of five exons and four introns. Adapted from (7)

The hPL-A and hPL-B genes have 92.5% sequence homology throughout their exons. The signal peptides of the two proteins differ in only a single amino acid, and the mature secreted proteins have identical amino acid sequences. The hPL-A gene is expressed 3 to 6 times more than the hPL-B gene, probably due to differences in stability of the two mRNAs. The mRNAs for the two hormones comprise about 3.5% of the total mRNA of the placenta. The proximal 5′ untranslated regions of the two genes share 94% homology, strongly suggesting that the transcriptional regulation of the two genes are identical or nearly identical. A transcriptional enhancer has also been identified in a 1.0 kB region located about 3 kB downstream of the hPL-B gene. Since the enhancer has been shown to increase the activity of several promoters, including hPL, hGH and SV40, it has been suggested that elements of the enhancer interact with ubiquitous transcription factors of the transcription initiation complex rather than with promoter-specific factors. Multiple sites of protein binding within the enhancer DNA region have been detected by DNase I footprinting. A 22-bp region overlapping the TEF-1 binding site has been detected within 138 bp of enhancer DNA. In addition, a repressor site and a derepressor site have been identified within the enhancer region. Recent studies from our laboratory also indicate that the enhancer is activated by NF-IL6.

The hPL-L gene product has not been identified in maternal product, although the gene produces several alternatively spliced mRNA transcripts with leader sequences. The predominant hGH-V gene product, which is expressed only in the placenta, is a 22-kDa protein that differs by only 13 amino acids from pituitary-specific 22-kD hGH-N (12,59). A second hGH-V protein (26 kDa hGH-V2) is predicted to be encoded by an alternately spliced hGH-V mRNA that retains intron 4 (60). The hGH-V2 gene product, however, is not secreted by cells transfected with the hGH-V gene and appears to remain cell-associated.

REGULATION OF HPL AND HGH-V EXPRESSION DURING UTEROPLACENTAL DIFFERENTIATION

At present, very little is known about the factors that regulate the induction of hPL gene expression during trophoblast differentiation and the expression of hPL from fully differentiated syncytiotrophoblast cells. Specific factors such as IGF-1, CSF-1, GM-CSF-1, LIF, EGF, TGF-β, hCG and cyclic AMP are known to stimulate trophoblast differentiation *in vitro*. However, the mechanisms by which these factors regulate trophoblast differentiation are poorly understood. Earlier investigations by Walker and co-workers clearly demonstrated the importance of a SP-1 site in the basal expression of hPL. Recent studies from our laboratory strongly suggest that the transcription factor activator protein-2 (AP-2) plays a critical role in the induction of trophoblast differentiation by binding to two AP-2 elements on the distal promoter (61). Mutation of either of the two AP-1 binding sites results in a 50% loss of hPL promoter activity, while mutation of both sites almost completely inhibits promoter activity. Studies from our laboratory strongly suggest roles for 1,25-dihydroxyvitamin D3, retinoic acid, thyroid hormone and IL-6 in the induction of hPL gene expression during differentiation. Using semi-quantitative RTPCR, we have demonstrated that RARα, RARβ and RXRα mRNA levels increase markedly during trophoblast differentiation, while the levels of RARγ mRNA remain relatively constant (62). The increases in retinoid receptor mRNA expression in each case preceded the increase in the release of hPL, suggesting a possible link between the upregulation of retinoid receptor gene expression and hPL gene expression.

PHYSIOLOGY OF DECIDUAL PROLACTIN

CONCENTRATIONS DURING PREGNANCY

Decidual PRL is synthesized and secreted by decidual cells during the luteal phase of the menstrual cycle when endometrial stromal cells, under the influence of progesterone, differentiate into decidual cells by a process known as decidualization (63). If pregnancy ensues, the decidual cells become the predominant cell type in the endometrium and comprise the maternal component of the placenta. Uterine implantation, however, is not necessary for the increase in decidual PRL content during pregnancy since a similar increase in PRL content occurs in the decidua of women with tubal pregnancies (64). PRL is also detected in endometrial tissues of women receiving progesterone or combined progesterone/estrogen therapy and (65) in endometrial tissue exposed *in vitro* to progesterone (63,66). Numerous studies strongly suggest that the decidua is the source of the large amounts of PRL in amniotic fluid (67-69). In *in vitro* studies, PRL synthesized by decidual tissue is transported across the fetal membranes to the luminal side of the amnion (i.e., towards the amniotic cavity). Although PRL is transported from decidual tissue to amniotic fluid, there is no evidence for the transport of decidual PRL to either the maternal or fetal circulations. There is a significant correlation between the decidual content of PRL during pregnancy and amniotic fluid PRL concentrations and between the amounts of PRL released by decidual explants and amniotic fluid PRL concentrations (70). The amounts of PRL in decidual tissue and amniotic fluid reach maximal concentrations during the first half of gestation, while maternal and fetal plasma PRL concentrations reach maximal concentrations near term. Maximal amniotic fluid PRL concentrations of approximately 4000 ng/ml are attained at 16-20 weeks of gestation when maternal and fetal PRL concentrations are approximately 50 and 10 ng/ml, respectively (71). Abnormalities of amniotic fluid PRL concentrations and decidual PRL production have been noted in several pathologic conditions of pregnancy. These include diabetes mellitus, intrauterine growth retardation, moderately severe toxemia, moderately severe erythroblastosis, chronic hypertension and idiopathic polyhydramnios (70,72).

BIOLOGIC ACTIONS

Numerous investigations suggest physiologic roles for decidual PRL during pregnancy. Since PRL affects water and ion transport in lower vertebrates and binds specifically to amniotic membranes with high affinity, amniotic fluid PRL may play a role in the regulation of amniotic fluid volume and ion content. *In vitro* and *in vivo* studies suggest that amniotic fluid PRL may protect the fetus from sudden changes in the extracellular fluid content of water and electrolytes that might occur as a consequence of altered amniotic fluid tonicities (73). Amniotic fluid PRL may also be one of the factors involved in the regulation of surfactant synthesis in the fetus (74). PRL has also been shown to inhibit uterine contractility, suppress the immune response, and be involved in the maintenance of T-cell immunocompetence (75). Since PRL is present in decidual tissue throughout gestation, decidual PRL may therefore act locally on the uterus to inhibit myometrial contractility and prevent immunologic rejection of the blastocyst and fetus. The observation that human decidual tissue contains specific membrane receptors for PRL further supports an autocrine/paracrine role for decidual PRL.

REGULATION OF DECIDUAL PROLACTIN RELEASE AND SYNTHESIS

Although decidual PRL has chemical and biological properties that are identical to those of pituitary PRL, the factors involved in the release of decidual PRL are different than those involved in the release of pituitary PRL (Figure 9-4). Comprehensive reviews of the factors involved in the regulation of the synthesis and release of decidual PRL have been recently published. TRH, dopamine and bromocriptine, each of which affects the synthesis and release of pituitary PRL, do not affect either the synthesis or release of decidual PRL - even at concentrations considerably greater than the half-maximal

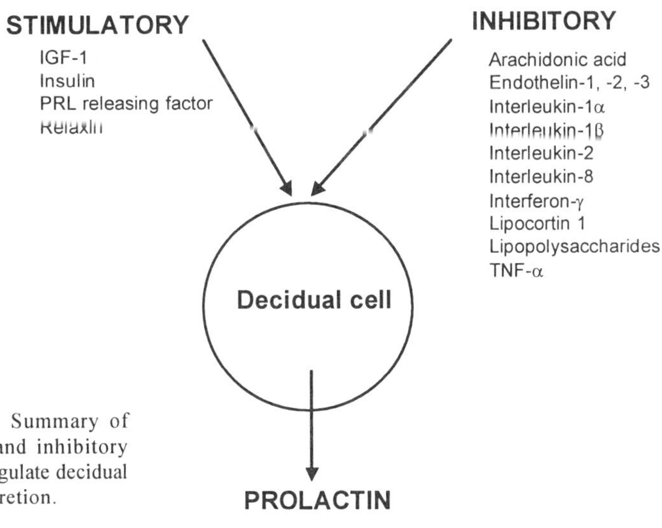

Figure 9-4. Summary of stimulatory and inhibitory factors that regulate decidual prolactin secretion.

concentrations that affect pituitary PRL release (76). The observation that bromocriptine does not affect the release of decidual PRL is consistent with clinical studies that indicate that bromocriptine therapy during pregnancy has no effect on amniotic fluid PRL concentrations even though maternal and fetal plasma PRL concentrations are markedly inhibited. Although PRL may regulate ion and water transport across the amnion, it does not appear that changes in the ionic environment affect the synthesis and/or release of decidual PRL. Exposure of decidual explants incubated to hyperosmolar medium of greater than 350 mosm/L results in inhibition of PRL release, but such high osmolalities are considerably greater than those observed *in vivo* even under pathologic conditions. Changes in tissue culture medium within the normal physiologic range have no effect on the release of PRL from human decidual explants (77).

The release of decidual PRL is stimulated by decidual PRL-releasing factor (dPRL-RF), a 23.5 K M_r protein that has been purified to homogeneity from conditioned medium of human placental explants and from extracts of human placenta (78). The releasing factor, however, has no effect on the release of pituitary PRL or other pituitary hormones. Decidual PRL release is acutely stimulated by physiologic concentrations of free α molecules, forms of α glycoprotein subunits that no longer combine with gonadotropin β subunits (79). IGF-I, which is synthesized and released by the placenta, stimulates decidual PRL release (80). Insulin also stimulates the synthesis and release of decidual PRL release (81). However, the stimulation by insulin does not result from the binding of insulin to the IGF-I receptor. Relaxin, which is synthesized by both decidual and placental tissues and has homologies in chemical structure to both IGF-I and insulin, also stimulates PRL release (82). Since relaxin has not been observed to bind to insulin or IGF-I receptors in other cells, the stimulation of PRL release by the three hormones appears to be due to binding to three distinct receptors. The release of decidual PRL is inhibited by lipocortin I, one of six members of the lipocortin/annexin group of calcium- and phospholipid-binding proteins that are present in the placenta (83). Since recent studies indicate that human decidual cells also synthesize and release lipocortin I, the inhibition of decidual PRL release by lipocortin I may result from an autocrine as well as a paracrine effect.

Bone marrow-derived cells, which release abundant amounts of cytokines, are major cellular components of human decidua, with macrophages comprising about 30% of the cells in term tissue (84). Studies of decidual release from decidual cells in culture have shown that physiologic amounts of TNF-α, IL-1α, IL-1β, IL-2, IL-8, TNFα, interferon-γ and TGFβ cause dose-dependent inhibition of basal PRL release, while IL-6 has no effect (85-87). Initial inhibition by each of the active cytokines occurs after 24 h of exposure, and maximal inhibition occurs after 3 or 4 days. In each instance, the cytokine-induced decrease in decidual PRL release was accompanied by a decrease in PRL synthesis and PRL mRNA levels. Each of the cytokines also inhibited stimulation of the synthesis and release of PRL in response to insulin. Bacterial lipopolysaccharides (LPS), which bind to specific cell surface receptors on macrophages and stimulate cytokine release, also inhibited the synthesis and release of decidual PRL (88). Taken together, these findings strongly suggest that cytokines released by decidual macrophages and other bone marrow-derived cells may have a paracrine role in the regulation of decidual PRL expression.

Endothelin-1 (ET-1), which stimulates the release of pituitary PRL inhibits the basal synthesis and release of PRL from isolated decidual cells and markedly inhibits the stimulation of the synthesis and release of decidual PRL in response to insulin and

IGF-1 (89). The ET isotypes ET-2 and ET-3 and by the ET-1 precursor big ET have similar effects on basal and insulin and IGF-I-mediated decidual PRL release. Agonists for the ETβ receptor mimic the effects of the endothelins while agonist for the ETα receptor are inactive. Since decidual cells synthesize endothelins, and decidual tissue contains specific receptors for ET, the action of ET on PRL release may result from an autocrine and/or paracrine effect mediated through the ETβ receptor.

Both the protein kinase A and protein kinase C signal transduction pathways are involved in the regulation of decidual PRL release (90,91). In addition, arachidonic acid, which is released from membrane phospholipids by the activation of phospholipase A2, may act as a second messenger in the release of decidual PRL (92). The action of arachidonic acid does not appear to be secondary to its metabolism to prostaglandins, thromboxanes or lipooxygenase products since cyclooxygenase and lipoxygenase inhibitors do not prevent the inhibition of PRL release by arachidonic acid.

MOLECULAR BIOLOGY OF DECIDUAL PROLACTIN

STRUCTURE

The nucleotide sequence of the cDNA for decidual PRL is identical to the cDNA for human pituitary PRL except for the addition of a non-coding exon in the decidual PRL gene. Consequently the protein coding sequence of decidual PRL cDNA is identical to the pituitary PRL sequence (93) and the amino acid sequence of the mature secreted form of decidual PRL is identical to that of pituitary PRL. The 3′-untranslated region (UTR) of the decidual PRL cDNA, however, exhibits a 7-14-nucleotide elongation and the 5′-UTR has an extension of 41 base pairs upstream of the normal transcription start site for the human pituitary PRL gene. This extension is preceded by an additional sequence that is not homologous to the DNA found 5′ to the pituitary PRL cap site. The decidua-specific segment of the 5′- UTR is located as a new 5′-noncoding exon (exon 1a) that is 5.3 kb pairs upstream of the human pituitary PRL gene (93). Consequently, the mRNA for decidual PRL is approximately 150 nt longer than the pituitary transcript (94). These studies indicate that decidual PRL is regulated by a promoter distinct from the pituitary PRL promoter. A schematic representation of the decidual PRL gene is shown in Figure 9-5. More recent studies indicate that exon 1a is also present in other non-pituitary cells that express PRL (for example, T-lymphocytes, brain, myometrium, dermal fibroblasts and lacrimal gland), strongly suggesting that the expression of PRL in these cells is regulated by the "decidual type" promoter.

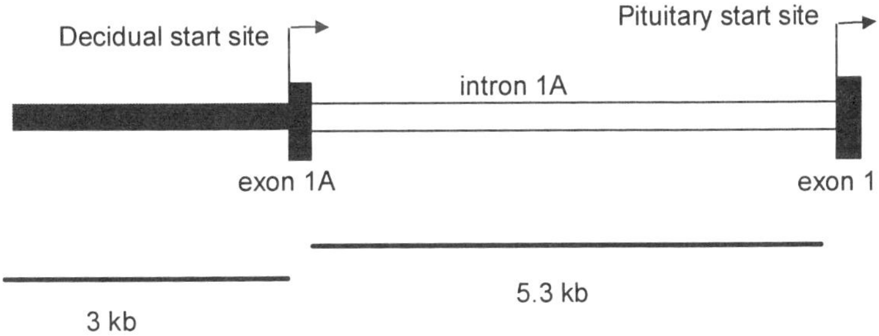

Figure 9-5. Schematic representation of the human prolactin gene. In the decidua, transcription of the gene is initiated at a site (decidual start site) located 5.3kb upstream (within intron 1A) of the initiation start site in the anterior pituitary (pituitary start site). Approximately 3 kb of the 5' region flanking the decidual start site has been characterized.

REGULATION OF DECIDUAL PROLACTIN GENE EXPRESSION

As indicated above, decidual PRL is first expressed by decidual cells during the luteal phase of the menstrual cycle when endometrial stromal cells, under the influence of progesterone, differentiate into decidual cells. The decidualized endometrial stromal cells not only synthesize and secrete PRL but also many other differentiation-specific proteins such as IGFBP-1 and TIMP-3. Transfection studies indicate that the promoter is transcriptionally active in decidualized endometrial stromal cells but not in undecidualized endometrial stromal cells, strongly suggesting that endometrial stromal cells must undergo decidualization before the gene is expressed (95). At present very little is known about the transcription factors and DNA elements on the promoter that regulate decidual PRL gene expression during decidualization. Decidualization can be induced *in vitro* by incubating primary cultures of endometrial stromal cells with medroxyprogesterone acetate (MPA) in combination with estradiol or relaxin, or with high levels of cyclic AMP in the absence of exogenous hormone. PRL gene expression can also be induced in primary cultures of human decidual fibroblasts (96) and in a permanent cell line, St-2, of human endometrial fibroblasts that was immortalized by infection with simian virus 40 large T antigen (97). Although the promoter contains consensus response elements for Pit-1, a transcription factor critical in pituitary PRL gene expression, Pit-1 does not have a role in transactivation of the decidual PRL promoter since it is not present in decidual cells (95).

The mechanisms by which progesterone induces decidualization and PRL gene expression are poorly understood. Several studies from our laboratory over the past few years strongly suggest that the effects of progesterone on decidualization and PRL gene expression are mediated by the cAMP and phosphoinositide hydrolysis/protein kinase C (PKC) signal transduction pathways. In initial experiments relating to the cAMP pathway, we observed that intracellular cAMP levels in human endometrial stromal cells are markedly increased by progesterone treatment and remain elevated as long as the cells are exposed to the steroid. High doses of dibutyryl-cAMP (500 μM), a cAMP analog, induced decidualization and PRL release in the absence of steroids (98); and low doses of dibutyryl-cAMP (50 μM) potentiated the effects of progesterone and estradiol. In addition, the protein kinase A inhibitor, 8-bromoadenosine-3′,5′-cyclic monophosphorothioate, significantly suppressed progesterone-dependent PRL expression in a dose-dependent manner. The maximal inhibition of PRL release was about 50%, strongly suggesting that the action of progesterone is mediated, in part, by other second messenger pathways. Gellersen and coworkers have recently reported studies that corroborate these findings (99). They have shown that decidualization of endometrial stromal cells results in a permanently elevated level of intracellular cAMP that is accompanied by a post transcriptional down regulation of PKA regulatory subunit RIα and a paradoxical and as yet unexplained increase in the isoforms of the cAMP-inducible early repressor (ICER). Taken together, our findings and those of Gellersen strongly suggest that progesterone-induced decidualization and PRL gene expression is mediated, in part, by activation of the cAMP signal transduction pathway.

Gellersen and coworkers recently demonstrated by deletion analysis of the decidual PRL promoter that the first 332 bp of the promoter is sufficient to induce promoter activity in endometrial cells decidualized *in vitro* by dibutyryl cAMP (100). However, this region of the decidual PRL promoter does not contain a classical cAMP response element (CRE). Several investigations have shown that the cAMP signal cascade is activated during *in vitro* decidualization (100,101). Telgmann et al have demonstrated that activation of protein kinase A is required for induction of decidual PRL gene transcription (100). Within 12 hours of treatment with 8-Br-cAMP, a weak induction of gene expression occurs that is mediated by an imperfect cAMP response element at position –12 relative to the transcription start site. A strong induction occurs after 12 hours that is dependent on a region of the decidual PRL promoter at –332/-270 (100). More recently, CCAAT/enhancer binding proteins in endometrial stromal cell extracts have been shown to be important in transducing the cAMP signal to the decidual PRL promoter (102). PGE_2 markedly enhances the effects of progesterone and estradiol on decidualization and PRL gene expression; and the enhancement is due, at least in part, to the enhancement of intracellular cAMP levels. During decidualization, there is a significant increase in the PGE_2 receptor subtype EP_2, which is coupled to the cAMP signal transduction pathway. In contrast to PGE_2, $PGF_2\alpha$ treatment has no effect on PRL expression in steroid-treated cells. *In vitro* decidualization of endometrial stromal cells in response to progesterone is also accompanied by a marked increase in the long form of the PRL receptor. PGE_2, which by itself has no effect on the long form of the receptor, potentiated the upregulation of the PRL receptor. These studies of the PRL receptor support the hypothesis that PRL may have an autocrine effect on decidualization and that the marked enhancement of decidualization by PGE_2 may be due, at least in part, to upregulation of the PRL receptor.

In contrast to PGE_2, retinoic acid and IL-1β inhibit decidual PRL expression by cAMP-dependent mechanisms (103,104). The inhibition of decidualization by retinoic acid and IL-1β is accompanied by a reduction in intracellular cAMP levels and the addition of dibutyryl cAMP to retinoic acid-exposed cells partially reverses the suppression of decidualization and PRL release. Laminin, a major constituent of endometrial cell basement membranes that increases during decidualization, causes a paradoxical decrease in PRL and IGFBP-1 gene expression during *in vitro* decidualization of human endometrial stromal cells (105). The amounts of PRL and IGFBP-1 released from cells cultured on a laminin matrix are significantly less than that of cells cultured on type I collagen, type IV collagen or fibronectin. The mechanism by which laminin inhibits PRL and IGFBP-1 is unclear but is not due to differences in initial cell attachment efficiency or final DNA content.

More recent studies from our laboratory have also demonstrated a role for the protein kinase C (PKC) signal transduction pathway in progesterone-induced decidualization. We observed that exposure of progesterone and estradiol-treated endometrial stromal cells with a PKC inhibitor attenuated decidualization (as assessed by PRL expression and morphology) by about 50%. In addition, desensitization of PKC by prolonged exposure of the endometrial cells to phorbol 12-myristate 13-acetate (PMA) also attenuated decidualization by 50%.

SUMMARY

In summary, the human utero-placental unit synthesizes and secretes three lactogenic hormones - hPL, hGH-V and decidual PRL. This is in sharp contrast to the situation in the mouse where approximately 20 PRL-like proteins have been identified. While relatively little is known about the biologic role of the PRL-like proteins in the mouse and rat, there is considerable evidence indicating that the utero-placental lactogen hormones are important in the regulation of maternal and fetal growth and metabolism and in the preparation of the breast for lactation.

ACKNOWLEDGMENTS

Supported by NIH grants HD-07447 and HD-15201. We thank the many students and postdoctoral fellows who have contributed to the studies from our laboratory.

REFERENCES

1. Handwerger S FM. The roles of placental growth hormone and placental lactogen in the regulation of human fetal growth and development. Journal of Pediatric Endocrinology and Metabolism. 2000;13:343-356.
2. Freemark M. The roles of growth hormone, prolactin and placental lactogen in human fetal development. In: Handwerger S, ed. Molecular and Cellular Pediatric Endocrinology. Totowa, NJ: Human Press; 1999:57-84.
3. Richards RGH, S. Physiology and molecular biology of placental lactogen in human pregnancy. In: Handwerger S, ed. Molecular and Cellular Pediatric Endocrinology. Totowa, NJ: Human Press; 1999:253-270.
4. Ogren L, Talamantes F. Prolactins of pregnancy and their cellular source. Int Rev Cytol. 1988;112:1-65.
5. Alsat E, Guibourdenche J, Couturier A, Evain-Brion D. Physiological role of human placental growth hormone. Mol Cell Endocrinol. 1998;140:121-7.
6. Kaplan SL, Grumbach MM, Shepard TH. The ontogenesis of human fetal hormones. I. Growth hormone and insulin. Journal of Clinical Investigation. 1972;51:3080-93.
7. Handwerger S. Clinical counterpoint: the physiology of placental lactogen in human pregnancy. Endocr Rev. 1991;12:329-36.
8. Grumbach MM, Kaplan SL, Sciarra JJ, Burr IM. Chorionic growth hormone-prolactin (CGP): secretion, disposition, biologic activity in man, and postulated function as the “growth hormone” of the 2d half of pregnancy. . 1968;148:501-531.
9. Ursell W, Brudenell M, Chard T. Placental lactogen levels in diabetic pregnancy. Br Med J. 1973;2:80-2.
10. Josimovich JB. Placental protein hormones in pregnancy. Clin Obstet Gynecol. 1973;16:46-65.
11. Lindberg BS, Nilsson BA. Human placental lactogen (HPL) levels in abnormal pregnancies. J Obstet Gynaecol Br Commonw. 1973;80:1046-53.
12. Frankenne F, Closset J, Gomez F, Scippo ML, Smal J, Hennen G. The physiology of growth hormones (GHs) in pregnant women and partial characterization of the placental GH variant.J. Clin .Endocrinol .Metab. 1988;66:1171-1180.
13. Mirlesse V, Frankenne F, Alsat E, Poncelet M, Hennen G, Evain-Brion D. Placental growth hormone levels in normal pregnancy and in pregnancies with intrauterine growth retardation. Pediatr Res. 1993;34:439-42.
14. Handwerger S. Clinical counterpoint: the physiology of placental lactogen in human pregnancy. Endocrine Reviews. 1991;12:329-336.
15. Hill DJ, Freemark M, Strain AJ, Handwerger S, Milner RD. Placental lactogen and growth hormone receptors in human fetal tissues: relationship to fetal plasma human placental lactogen concentrations and fetal growth. Journal of Clinical Endocrinology & Metabolism. 1988;66:1283-90.
16. MacLeod JN, Worsley I, Ray J, Friesen HG, Liebhaber SA, Cooke NE. Human growth hormone-variant is a biologically active somatogen and lactogen. Endocrinology. 1991;128:1298-1302.
17. Freemark M, Comer M, Korner G, Handwerger S. A unique placental lactogen receptor: implications for fetal growth. Endocrinology. 1987;120:1865-72.
18. Hurley TW, D’Ercole AJ, Handwerger S, Underwood LE, Furlanetto RW, Fellows RE. Ovine placental lactogen induces somatomedin: a a possible role in fetal growth. Endocrinology. 1977;101:1635-1638.

19. Furlanetto RW, Underwood LE, Van Wyk JJ, Handwerger S. Serum immunoreactive somatomedin-C is elevated late in pregnancy. J Clin Endocrinol Metab. 1978;47:695-8.
20. Gluckman PD, Uthne K, Styne DM, Kaplan SL, Rudolph AM, Grumbach MM. Hormone ontogeny in the ovine fetus. IV. Serum somatomedin activity in the fetal and neonatal lamb and pregnant ewe: correlation with maternal and fetal growth hormone, prolactin, and chorionic somatomammotropin. Pediatr Res. 1979;13:194-6.
21. Daughaday WH, Kapadia M. Maintenance of serum somatomedin activity in hypophysectomized pregnant rats. Endocrinology. 1978;102:1317-20.
22. Butler SR, Hurley TW, Schanberg SM, Handwerger S. Ovine placental lactogen stimulation of ornithine decarboxylase activity in brain and liver of neonatal rats. Life Sciences. 1978;22:2073-2078.
23. Freemark M, Handwerger S. Ovine placental lactogen stimulates amino acid transport in rat diaphragm. Endocrinology. 1982;110:2201-3.
24. Adams SO, Nissley SP, Handwerger S, Rechler MM. Developmental patterns of insulin-like growth factor-I and -II synthesis and regulation in rat fibroblasts. Nature. 1983;302:150-153.
25. Hill DJ, Crace CJ, Milner RD. Incorporation of [3H]thymidine by isolated fetal myoblasts and fibroblasts in response to human placental lactogen (HPL): possible mediation of HPL action by release of immunoreactive SM-C. J Cell Physiol. 1985;125:337-44.
26. Freemark M, Handwerger S. Synergistic effects of oPL and insulin on glycogen metabolism in fetal rat hepatocytes. American Journal of Physiology. 1984;247:E714-8.
27. Gluckman PD. The role of pituitary hormones, growth factors and insulin in the regulation of fetal growth. Oxford Reviews of Reproductive Biology. 1986;8:1-60.
28. Gluckman PD, Gunn AJ, Wray A, Cutfield WS, Chatelain PG, Guilbaud O, Ambler GR, Wilton P, Albertsson-Wikland K. Congenital idiopathic growth hormone deficiency associated with prenatal and early postnatal growth failure. The International Board of the Kabi Pharmacia International Growth Study. Journal of Pediatrics. 1992;121:920-923.
29. Palmiter RD, Norstedt G, Gelinas RE, Hammer RE, Brinster RL. Metallothionein-human GH fusion genes stimulate growth of mice. Science. 1983;222:809-14.
30. Pepe GJ, Albrecht ED. Regulation of baboon fetal adrenal androgen production by adrenocorticotropic hormone, prolactin, and growth hormone. Biology of Reproduction. 1985;33:545-50.
31. Mendelson CR, Snyder JM. Role of prolactin, cortisol, and insulin in the regulation of surfactant synthesis by the human fetal lung. Pediatr.Pulmonol. 1985;1:S91-S98.
32. Dvir R, Golander A, Jaccard N, Yedwab G, Otremski I, Spirer Z, Weisman Y. Amniotic fluid and plasma levels of parathyroid hormone-related protein and hormonal modulation of its secretion by amniotic fluid cells. Eur J Endocrinol. 1995;133:277-82.
33. Seyoum G, Robertson MC, Persaud TV, Paterson JA, Shiu RP. Influence of rat placental lactogen-I on the development of whole rat embryos in culture. J Endocrinol. 1999;160:231-7.
34. Nielsen PV, Pedersen H, Kampmann EM. Absence of human placental lactogen in an otherwise uneventful pregnancy. Am J Obstet Gynecol. 1979;135:322-6.
35. Wurzel JM, Parks JS, Herd JE, Nielsen PV. A gene deletion is responsible for absence of human chorionic somatomammotropin. DNA 1982;1:251-7.
36. Wohlk P, Nexo E, Jorgensen EH, Chemnitz J, Nielsen PV, Parks JS. Low or absent serum placental lactogen hormone in 2 normal pregnancies. Ugeskr Laeger. 1984;146:727-9.
37. Rygaard K, Revol A, Esquivel-Escobedo D, Beck BL, Barrera-Saldana HA. Absence of human placental lactogen and placental growth hormone (HGH- V) during pregnancy: PCR analysis of the deletion. Hum Genet. 1998;102:87-92.

38. Eggermann T, Eggermann K, Mergenthaler S, Kuner R, Kaiser P, Ranke MB, Wollmann HA. Paternally inherited deletion of CSH1 in a patient with Silver-Russell syndrome. J Med Genet. 1998;35:784-6.
39. Petit A, Guillon G, Tence M, Jard S, Gallo-Payet N, Bellabarba D, Lehoux JG, Belisle S. Angiotensin II stimulates both inositol phosphate production and human placental lactogen release from human trophoblastic cells. J.Clin.Endocrinol.Metab. 1989;69:280-286.
40. Handwerger S, Wilson SP, Tyrey L, Conn PM. Biochemical evidence that human placental lactogen and chorionic gonadotropin are not stored in cytoplasmic granules. Molecular & Cellular Endocrinology . 1987;37:28-32.
41. Handwerger S, Richards RG, Myers SE. Novel regulation of the synthesis and release of human placental lactogen by high density lipoproteins. Trophoblast research; 1994:339-354.
42. Handwerger S, Myers S, Richards R, Richardson B, Turzai L, Moeykins C, Meyer T, Anantharamahiah GM. Apolipoprotein A-I stimulates placental lactogen expression by human trophoblast cells. Endocrinology. 1995;136:5555-5560.
43. Grandis A, Jorgensen V, Kodack L, Quarfordt S, Handwerger S. High-density lipoproteins (HDL) stimulate placental lactogen secretion in pregnant ewes: further evidence for a role of HDL in placental lactogen secretion during pregnancy. Journal of Endocrinology. 1989;120:423-427.
44. Rosing U, Samsioe G, Olund A, Johansson B, Kallner A. Serum levels of apolipoprotein A-I, A-II and HDL-cholesterol in second half of normal pregnancy and in pregnancy complicated by pre-eclampsia. Hormone & Metabolic Research. 1989;21:376-382.
45. Handwerger S, Datta G, Richardson B, Schmidt CM, Siddiqi T, Turzai L, Anantharamaiah GM. Pre-beta-HDL stimulates placental lactogen release from human trophoblast cells. Am J Physiol. 1999;276:E384-9.
46. Wu YQ, Jorgensen EV, Handwerger S. High density lipoproteins stimulate placental lactogen release and adenosine 3',5'-monophosphate (cAMP) production in human trophoblast cells: evidence for cAMP as a second messenger in human placental lactogen release. Endocrinology. 1988;123:1879-1884.
47. Wu YQ, Handwerger S. High density lipoproteins stimulate molecular weight 80K protein phosphorylation in human trophoblast cells: evidence for a protein kinase-C-dependent pathway in human placental lactogen release. Endocrinology. 1992;131:2935-2940.
48. Kanda Y, Richards RG, Handwerger S. Apolipoprotein A-I stimulates human placental lactogen release by activation of MAP kinase. Mol Cell Endocrinol. 1998;143:125-31.
49. Stephanou A, Handwerger S. Interleukin-6 stimulates placental lactogen expression by human trophoblast cells. Endocrinology. 1994;135:719-23.
50. Stephanou A, Handwerger S. The nuclear factor NF-IL6 activates human placental lactogen gene expression. Biochem Biophys Res Commun. 1995;206:215-22.
51. Stephanou A, Ross R, Handwerger S. Regulation of human placental lactogen expression by 1,25- dihydroxyvitamin D3. Endocrinology. 1994;135:2651-2656.
52. Stephanou A, Handwerger S. Retinoic acid and thyroid hormone regulate placental lactogen expression in human trophoblast cells. Endocrinology. 1995;136:933-8.
53. Stephanou A, Handwerger S. Identification of a composite steroid hormone response element on the human placental lactogen promoter. Mol Cell Endocrinol. 1995;112:123-9.
54. De Zegher F, Vanderschueren-Lodeweyckx M, Spitz B, Faijerson Y, Blomberg F, Beckers A, Hennen G, Frankenne F. Perinatal growth hormone (GH) physiology: effect of GH-releasing factor on maternal and fetal secretion of pituitary and placental GH. Journal of Clinical Endocrinology & Metabolism. 1990;71:520-522.
55. Golos T, Durning M, Fisher J, Johnson K, Nadeau K, Blackwell R. Developmentally regulated expression of the rhesus placental growth hormone-variant gene: cloning and cAMP-stimulated transcription in primary syncytiotrophoblast cultures. Endocrine. 1994;2:537-545.

56. Nickel BE, Bock ME, Nachtigal MW, Cattini PA. Differential expression of human placental growth hormone variant and chorionic somatomammotropin genes in choriocarcinoma cells treated with methotrexate. Mol Cell Endocrinol. 1993;91:159-66.
57. Patel N, Alsat E, Igout A, Baron F, Hennen G, Porquet D, Evain-Brion D. Glucose inhibits human placental GH secretion, *in vitro*. J Clin Endocrinol Metab. 1995;80:1743-6.
58. Jones BK, Monks BR, Liebhaber SA, Cooke NE. The human growth hormone gene is regulated by a multicomponent locus control region. Mol Cell Biol. 1995;15:7010-21.
59. Igout A, Scippo ML, Frankenne F, Hennen G. Cloning and nucleotide sequence of placental hGH-V cDNA. Arch Int Physiol Biochim. 1988;96:63-7.
60. Cooke NE, Ray J, Emery JG, Liebhaber SA. Two distinct species of human growth hormone-variant mRNA in the human placenta predict the expression of novel growth hormone proteins. Journal of Biological Chemistry. 1988;263:9001-9006.
61. Richardson BD, Langland RA, Bachurski CJ, Richards RG, Kessler CA, Cheng Y, Handwerger S. Activator protein-2 regulates human placental lactogen gene expression. Mol Cell Endocrinol. 2000;160:183-192.
62. Stephanou A, Sarlis NJ, Richards R, Handwerger S. Expression of retinoic acid receptor subtypes and cellular retinoic acid binding protein-II mRNAs during differentiation of human trophoblast cells. Biochem Biophys Res Commun. 1994;202:772-80.
63. Daly DC, Maslar IA, Riddick DH. Term decidua response to estradiol and progesterone. Am.J.Obstet.Gynecol. 1983;145:679-683.
64. Maslar IA, Kaplan BM, Luciano AA, Riddick DH. Prolactin production by the endometrium of early human pregnancy. J.Clin.Endocrinol.Metab. 1980;51:78-83.
65. Meuris S, Soumenkoff G, Malengreau A, Robyn C. Immunoenzymatic localization of prolactin-like immunoreactivity in decidual cells of the endometrium from pregnant and nonpregnant women. J.Histochem.Cytochem. 1980;28:1347-1350.
66. Maslar IA, Ansbacher R. Effects of progesterone on decidual prolactin production by organ cultures of human endometrium. Endocrinology. 1986;118:2102-2108.
67. Golander A, Hurley T, Barrett J, Hizi A, Handwerger S. Prolactin synthesis by human chorion-decidual tissue: a possible source of prolactin in the amniotic fluid. Science. 1978;202:311-3.
68. Riddick DH, Kusmik WF. Decidua: a possible source of amniotic fluid prolactin. American Journal of Obstetrics & Gynecology. 1977;127:187-190.
69. Riddick DH, Luciano AA, Kusmik WF, Maslar IA. Evidence for a nonpituitary source of amniotic fluid prolactin. Fertility & Sterility. 1979;31:35-39.
70. Josimovich J. The role of pituitary prolactin in fetal and amniotic fluid water and salt balance. In: Grosignani PG RC, ed. Prolactin and Human Reproduction. London: Academic Press; 1977:27-41.
71. Handwerger S, Richards RG, Markoff E. The physiology of decidual prolactin and other decidual protein hormones. Trends Endocrinol.Metab. 1992;3:91-95.
72. Golander A, Kopel R, Lazebnik N, Frenkel Y, Spirer Z. Decreased prolactin secretion by decidual tissue of pre-eclampsia *in vitro*. Acta Endocrinol (Copenh). 1985;108:111-3.
73. Pullano JG, Cohen-Addad N, Apuzzio JJ, Ganesh VL, Josimovich JB. Water and salt conservation in the human fetus and newborn. I. Evidence for a role of fetal prolactin. J.Clin.Endocrinol.Metab. 1989;69:1180-1186.
74. Mendelson CR, Johnston JM, MacDonald PC, Snyder JM. Multihormonal regulation of surfactant synthesis by (human) fetal lung *in vitro*. J.Clin.Endocrinol.Metab. 1981;53:307-317.
75. Handwerger S, Freemark M. Role of placental lactogen and prolactin in human pregnancy. Adv.Exp.Med.Biol. 1987;219:399-420.
76. Golander A, Barrett J, Hurley T, Barry S, Handwerger S. Failure of bromocriptine, dopamine, and thyrotropin-releasing hormone to affect prolactin secretion by human decidual tissue *in vitro*. Journal of Clinical Endocrinology & Metabolism. 1979;49:787-789.

77. Markoff E, Barry S, Handwerger S. Influence of osmolality and ionic environment on the secretion of prolactin by human decidua *in vitro*. Journal of Endocrinology. 1982;92:103-110.
78. Handwerger S, Barry S, Markoff E, Barrett J, Conn PM. Stimulation of the synthesis and release of decidual prolactin by a placental polypeptide. Endocrinology. 1983;112:1370-1374.
79. Blithe DL, Richards RG, Skarulis MC. Free alpha molecules from pregnancy stimulate secretion of prolactin from human decidual cells: a novel function for free alpha in pregnancy. Endocrinology. 1991;129:2257-9.
80. Thrailkill KM, Golander A, Underwood LE, Handwerger S. Insulin-like growth factor I stimulates the synthesis and release of prolactin from human decidual cells. Endocrinology. 1988;123:2930-2934.
81. Thrailkill KM, Golander A, Underwood LE, Richards RG, Handwerger S. Insulin stimulates the synthesis and release of prolactin from human decidual cells. Endocrinology. 1989;124:3010-3014.
82. Huang JR, Tseng L, Bischof P, J„nne OA. Regulation of prolactin production by progestin, estrogen, and relaxin in human endometrial stromal cells. Endocrinology. 1987;121:2011-2017.
83. Pihoker C, Feeney RJ, Su JL, Handwerger S. Lipocortin-I inhibits the synthesis and release of prolactin from human decidual cells: evidence for autocrine/paracrine regulation by lipocortin-I. Endocrinology. 1991;128:1123-1128.
84. Bulmer JN, Pace D, Ritson A. Immunoregulatory cells in human decidua: morphology, immunohistochemistry and function. Reproduction, Nutrition, Developpement. 1988;28:1599-1613.
85. Jikihara H, Handwerger S. Tumor necrosis factor-alpha inhibits the synthesis and release of human decidual prolactin. Endocrinology. 1994;134:353-357.
86. Kanda Y, Jikihara H, Markoff E, Handwerger S. Interleukin-2 inhibits the synthesis and release of prolactin from human decidual cells [In Process Citation]. J Clin Endocrinol Metab. 1999;84:677-81.
87. Handwerger S, Richards R, Markoff E. Autocrine/paracrine regulation of prolactin release from human decidual cells. Ann N Y Acad Sci. 1991;622:111-9.
88. Chao HS, Poisner AM, Poisner R, Handwerger S. Lipopolysaccharides inhibit prolactin and renin release from human decidual cells. Biol Reprod. 1994;50:210-4.
89. Chao HS, Myers SE, Handwerger S. Endothelin inhibits basal and stimulated release of prolactin by human decidual cells. Endocrinology. 1993;133:505-10.
90. Handwerger S, Harman I, Costello A, Markoff E. Cyclic AMP inhibits the synthesis and release of prolactin from human decidual cells. Mol.Cell Endocrinol. 1987;50:99-106.
91. Harman I, Costello A, Ganong B, Bell RM, Handwerger S. Activation of protein kinase C inhibits synthesis and release of decidual prolactin. Am J Physiol. 1986;251:E172-7.
92. Handwerger S, Barry S, Barrett J, Markoff E, Zeitler P, Cwikel B, Siegel M. Inhibition of the synthesis and secretion of decidual prolactin by arachidonic acid. Endocrinology. 198: 109:2016-2021.
93. DiMattia GE, Gellersen B, Duckworth ML, Friesen HG. Human prolactin gene expression. The use of an alternative noncoding exon in decidua and the IM-9-P3 lymphoblast cell line. J.Biol.Chem. 1990;265: 16412-16421.
94. Gellersen B, DiMattia GE, Friesen HG, Bohnet HG. Prolactin (PRL) mRNA from human decidua differs from pituitary PRL mRNA but resembles the IM-9-P3 lymphoblast PRL transcript. Mol Cell Endocrinol. 1989;64: 127-30.
95. Gellersen B, Kempf R, Telgmann R, DiMattia GE. Nonpituitary human prolactin gene transcription is independent of Pit-1 and differentially controlled in lymphocytes and in endometrial stroma. Molecular Endocrinology. 1994: 8:356-73.

96. Richards RG, Brar AK, Frank GR, Hartman SM, Jikihara H. Fibroblast cells from term human decidua closely resemble endometrial stromal cells: induction of prolactin and insulin- like growth factor binding protein-1 expression. Biology of Reproduction. 1995; 52:609-615.
97. Brosens JJ, Takeda S, Acevedo CH, Lewis MP, Kirby PL, Symes EK, Krausz T, Purohit A, Gellersen B, White JO. Human endometrial fibroblasts immortalized by simian virus 40 large T antigen differentiate in response to a decidualization stimulus. Endocrinology. 1996; 137:2225-31.
98. Tang B, Guller S, Gurpide E. Cyclic adenosine 3',5'-monophosphate induces prolactin expression in stromal cells isolated from human proliferative endometrium. Endocrinology. 1993;133:2197-2203.
99. Gellersen B, Kempf R, Telgmann R. Human endometrial stromal cells express novel isoforms of the transcriptional modulator CREM and up-regulate ICER in the course of decidualization. Mol Endocrinol. 1997;11:97-113.
100. Telgmann R, Maronde E, Tasken K, Gellersen B. Activated protein kinase A is required for differentiation-dependent transcription of the decidual prolactin gene in human endometrial stromal cells. Endocrinology. 1997;138:929-37.
101. Brar AK, Frank GR, Kessler CA, Cedars MI, Handwerger S. Progesterone-dependent decidualization of the human endometrium is mediated by cAMP. Endocrine. 1997;6:301-7.
102. Pohnke Y, Kempf R, Gellersen B. CCAAT/enhancer-binding proteins are mediators in the protein kinase A-dependent activation of the decidual prolactin promoter. Journal of Biological Chemistry. 1999; 274:24808-18.
103. Brar AK, Kessler CA, Meyer AJ, Cedars MI, Jikihara H. Retinoic acid suppresses in-vitro decidualization of human endometrial stromal cells. Mol Hum Reprod. 1996;2:185-93.
104. Frank GR, Brar AK, Jikihara H, Cedars MI, Handwerger S. Interleukin-1 beta and the endometrium: an inhibitor of stromal cell differentiation and possible autoregulator of decidualization in humans. Biol.Reprod. 1995; 52:184-191.
105. Brar AK, Frank GR, Richards RG, Meyer AJ, Kessler CA, Cedars MI, Klein DJ, Handwerger S. Laminin decreases PRL and IGFBP-1 expression during *in vitro* decidualization of human endometrial stromal cells. Journal of Cellular Physiology. 1995; 163:30-37.
106. Handwerger S, Brar A. Placental lactogen, placental growth hormone, and decidual prolactin. Seminars in Reproductive Endocrinology; 1992: 106-115.

Chapter 10

Structure-Function Relationships in Prolactin

Mary Y. Lorenson and Ameae M. Walker

University of California, Riverside,
Riverside, CA

INTRODUCTION

In this chapter, it is the authors' intent to provide an update on structure-function relationships for PRL. We have only covered those PRL varieties for which additional information has been gained since the excellent and comprehensive review by Sinha in 1995 (1). Focus is on forms for which a different biological activity is becoming evident, although attention is also drawn to reports of some new PRLs, the existence of which may help to explain old controversies. Also included, are mutational studies which have contributed greatly to our understanding of hormone-receptor interactions and, therefore, to the potential impact of specific posttranslational modifications on biological activity. We have attempted to meld old and new data on each form of PRL into a comprehensive view whenever we felt that was possible. When such was not possible, we have suggested lines of investigation which may resolve current controversies.

PROLACTIN

PRL is a protein hormone which, since it is secreted, is translated by ribosomes of the rough endoplasmic reticulum as pre-PRL. The pre-piece, or signal peptide, is cleaved cotranslationally (2) in the cisternal lumen to produce an approximately 23 kDa, mature protein. Although the majority of circulating PRL is released by the pituitary, many other tissues have been shown to synthesize PRL (for review see 3). It is currently unclear how much these alternate sources contribute to circulating PRL under different physiological or pathological circumstances (4). In some cases, it is now evident that peripheral tissue PRL is at least used locally, at or near, its source (reviewed in 3). Also unclear at present is whether some posttranslationally modified forms, produced in the pituitary (e.g., glycosylated, phosphorylated), can be similarly produced in peripheral tissues. Some posttranslational modifications, on the other hand, appear not to occur to any great extent in the pituitary, but occur mainly in peripheral tissues (e.g., 16kDa PRL cleavage) (5). Whether posttranslational modification in peripheral tissues is normally of pituitary-derived or locally-derived PRL, or both, is unknown. The PRL is identical in sequence whether synthesized in peripheral tissues or in the pituitary (reviewed in 3). Different promoter regions on the gene, however, regulate the production of peripheral tissue PRL somewhat differently (6).

MUTANT PROLACTINS

The three-dimensional structure of unmodified PRL, defined as the plain polypeptide sequence minus the pre-piece, has not been determined. In addition, unmodified PRL has not yet been crystallized with the extracellular domains of the PRL receptor to determine which amino acids of the hormone actually interact with those of the receptor. Ideas concerning interacting residues have therefore been based on thoughtful analyses of the similarities in structure among PRL, PLs and GH and their receptors, on mutational studies, or a combination thereof (7-15). Early mutational studies focused on residues conserved among lactogenic hormones (7-10,14,15). More recent studies have largely focused on residues thought, by modeling analogy to other family members, likely to interact with the receptor (11-13). Since activation of the PRL receptor is believed to require receptor dimerization by a single PRL molecule and there is no bilateral symmetry in PRL, two different putative-receptor-interacting regions of PRL have been extensively studied.

Before considering the details of particular mutational studies, it is useful to reflect on the advantages and disadvantages of various expression systems that can be utilized. When utilizing mammalian cell expression systems, the advantages include the cell's own quality control mechanisms which prevent the release of misfolded hormone from the endoplasmic reticulum (16-18). Poor expression levels therefore tell you which mutations are inconsistent with appropriate folding and conversely, one knows that proteins expressed at near-wildtype levels are well folded. The disadvantage of mammalian cell expression systems is that they can perform all of the potential posttranslational modifications of the hormone, leading to expression of a mixture of forms with potentially different biological activities. This may significantly confuse analysis of the mutation. Only when a mutation has 1) no significant effect on expression level, and hence the rate of secretion and time in contact with modifying enzymes, and 2) no likely effect on a posttranslational site (by mutating the derivatized residue or a consensus sequence), can the effects of mutations be reasonably clear.

When utilizing bacterial expression systems, the advantage is the elimination of most posttranslational modifications, but the disadvantages include the usual need to solubilize inclusion bodies with denaturing agents. The success of a mutation analysis therefore depends on the investigator's ability to accurately refold the molecule. Although secretion into the periplasmic space theoretically overcomes this problem, investigators have found the lower yields, increased degradation, increased need for purification and increased contamination by endotoxins to be more of a challenge than problems with refolding.

One major interpretive flaw in mutational analyses is the tendency of investigators to assume that a loss-of-function mutation means that that specific residue directly interacts with the function-mediating molecule. In this case, that would be a mutation in PRL that affects its activation of the PRL receptor. When one considers the complex interactions among amino acids in a three-dimensional, folded protein such as PRL, it should be clear that mutations of amino acids far away from the receptor binding regions, could drastically alter overall conformation and therefore receptor activation. Some investigators have been careful to bear this issue in mind when interpreting their data, while others have not.

When making mutations, conservative substitutions which have a marked effect on receptor activation are most likely to identify crucial residues. In other words, if a minor change in side chain size has a major impact, then that residue in particular is crucial either for maintenance of exact structure, or for interaction with the receptor. Less dramatic effects of conservative substitutions may reflect only the need, for example, to maintain helical structure. One amino acid which exemplifies a crucial residue is R177 in Figure 10-1 (following 2 pages). Aligned sequences of 22 PRLs. Sequences were obtained from the Swiss protein database using Seqweb software of the Genetics Computer Group, Inc. (GCG). Residues discussed in the text are underlined for ease of reference. Putative helices are derived from Goffin *et al.* (145).

Another issue to be considered when analyzing mutational studies is whether the author reported effects on receptor activation (i.e., Nb2 bioactivity) or receptor binding, or both. If interpretations are to be based on binding, or more accurately, competition with iodinated wildtype or standard hormone, then it is crucial to demonstrate that the iodinated hormone has full biological activity. Otherwise the competition that is being measured may not be for the site of true interest and may not be reflected in measures of bioactivity. The importance of this level of attention to detail is illustrated by noting that most tyrosines in PRL are in highly conserved regions of the molecule where an iodine atom could easily cause a distortion (1 and Figure 10-1). Also, if bioactivity is to be compared with binding, the same receptor type should be used since there is evidence of different affinities for the Nb2 *versus* long form (19). The same receptor type has been used in some studies (11) but not others (13). For practical reasons, especially for bioassay, most investigators use a different species of receptor than hormone. Given the fact that binding is different in homologous versus heterologous ligand-receptor systems (20), this should always be borne in mind when interpreting data. Forms of PRL which bind but do not activate the receptor, will act as antagonists to the wildtype hormone. Analysis of antagonism provides an alternate mechanism for analyzing binding under more physiological conditions than are usually used.

RECEPTOR BINDING

Based on all studies to date, PRL is thought to bind to its first receptor via regions of helix 1, loop 1 and helix 4 (reviewed in 21). Binding to the second receptor involves regions of helix 1 (at the interface with helix 3) and helix 3 (reviewed in 21). Because of the ability of excess PRL to undo PRL-mediated dimerization, binding to the second receptor has to be a lower affinity interaction than that to the first receptor (21). We will begin by considering binding to the first receptor and then move on to the second. This work has recently been expertly reviewed in detail elsewhere (21) and hence only major highlights will be referred to here.

One clear conclusion from mutational analyses of the site which binds to the first receptor is that despite the high homology among PRL, GH and PL, the residues are not the same on each hormone for effective interaction with the equally homologous receptors (reviewed in 21). In fact, different residues in PRL and GH are of greatest importance even when binding for both is to the PRL receptor. In part, this may have been anticipated by the need for zinc coordination for GH (but not PRL) to bind to the PRL receptor (22). As discussed in Kinet *et al.* (11), K69, Y169 and H180 play a major role in human PRL binding to the PRL receptor, whereas the aligned equivalent residues in hGH (R64, Y160 and D171) can be mutated without significantly affecting binding to the same receptor. I58, S62, E65 and R183 in hGH are required for zinc-mediated interaction with the PRL receptor, while the aligned equivalent residues (L63, E67, E70 and R192) in PRL can be mutated without affecting receptor binding. Even when equivalent residues in both hormones are used in receptor binding, they do not seem to be equally important

1	10	20	30	40	50	60	70	80	90	100
human	LPI**C**PGGAAR	**C**QVTLRDLFD	R**AV**VL**S**HYI**H**	**N**LSSEM**F**SEF	DKRYTHGRGF	ITKAINS**CHT**	SSLA**TP**ED**K**E	QAQQMNQKDF	LSLIVSILRS	WNEPLYHLVT
monkey	LPVCPGGAAR	CQVTLRDLFD	RAVVLSHYIH	**N**LSSEMFSEF	DKRYTHGRGF	ITRAINSCHT	SSLPTPEDKE	QAQQMNQKDF	LSLIVSILRS	WNEPLYHLVT
			helix 1				loop 1		helix 2	

	8	18	28	38	48	58	68	78	88	98
rat	LPVCSGG..D	CQTPLPELFD	RVVMLSHYIH	TLYTDMFIEF	DKQYVQDREF	IAKAINDCP**T**	SSLA**T**PEDKE	QAQKVPPEVL	LNLILSLVHS	WNDPLFQLIT
mouse	LPICSAG..D	CQTSLRELFD	RVVILSHYIH	TLYTDMFIEF	DKQYVQDREF	MVKVINDCPT	SSLATPEDKE	QALKVPPEVL	LNLILSLVQS	SSDPLFQLIT
hamster	LPICPGG..N	CQMPLQELFD	RVIMLSHYIY	MLSADMFIEL	DKQYAQDHEF	IAKAISDCPT	SSLATPEGKE	EAQQVPPEVL	LNLILSLVHS	WNDPLFQLVT

	1 10	20	30	40	50	60	70	80	90	100
goat	TPVCPNGPGN	CQVSLRDLFD	RAVMVSHYIH	**N**LSSEMFNEF	DKRYAQGKGY	ITMALNSCHT	SSLPTPEDKE	QAQQTHHEVL	MSLILGLLRS	WNDPLYHLVT
sheep	TPVCPNGPGN	CQVSLRDLFD	RAVMVSHYIH	**N**LSSEMFNEF	DKRYAQGKGF	ITMALNSCHT	SSLPTPEDKE	QAQQTHHEVL	MSLILGLLRS	WNDPLYHLVT
bovine	TPVCPNGPGN	CQVSLRDLFD	RAVMV**S**HYIH	DLS**S**EMFNEF	DKRYAQGKGF	ITMALNSCHT	SSLPTPEDKE	QAQQTHHEVL	MSLILGLLR**S**	WNDPLYHLVT
allig. 1	LPICPSGSVN	CQVSLGELFD	RAVKLSHYIH	FLSSEMFNEF	DERYAQGRGF	ITKAVNGCH**N**	ASLTTPEDKE	QAQQIHHEDL	LNLVLGVLRS	WNDPLLHLVT
allig. 2	LPICPSGSVN	CQVSLGELFD	RAVKLSHYIH	FLSSEMFNEF	DERYAQGRGF	ITKAVNGCHT	ASLTTPEDKE	QAQQIHHEDL	LNLVLGVLRS	WNDPLLHLVT
croc. 1	LPICPSGSVN	CQVSLGELFD	RAVKLSHYIH	FLSSEMFNEF	DERYAQGRGF	ITKAVNGCH**N**	ASLTTPEDKE	QAQQIHHEDL	LNLVLGVLRS	WNDPLLHLVT
croc. 2	LPICPSGSVN	CQVSLGELFD	RAVKLSHYIH	FLSSEMFNEF	DERYAQGRGF	ITKAVNGCHT	ASLTTPEDKE	QAQQIHHEDL	LNLVLGVLRS	WNDPLLHLVT
chicken	LPICPIGSVN	CQVSLGELFD	RAVKLSHYIH	YLSSEIFNEF	DERYAQGRGF	ITKAVNGCHT	SSLTTPEDKE	QAQQIHHEDL	LNLVVGVLRS	WNDPLIHLAS
turkey	LPICSSGSVN	CQVSLGELFD	RAVRLSHYIH	FLSSEIFNEF	DERYAQGRGF	ITKAVNGCHT	SSLTTPEDKE	QTQQIHHEEL	LNLILGVLRS	WNDPLIHLAS
turtle	LPICPSGSVG	CQVSLENLFD	RAVKLSHYIH	SLSSEMFNEF	DERYAQGRGF	LTKAINGCHT	SSLTTPEDKE	QAQQIHHEDL	LNLVLGVLRS	WNDPLLHLVS
camel	LPICPSGAVN	CQVSLRDLFD	RAVILSHYIH	**N**LSSEMFNEF	DKRYAQGRGF	MTKAINSCHT	SSLSTPEDKE	QAQQIHHEDL	LNLVLRVLRS	WNDPLYHLVT
pig	LPICPSGAVN	CQVSLRDLFD	RAVILSHYIH	**N**LSSEMFNEF	DKRYAQGRGF	ITKAINSCHT	SSLSTPEDKE	QAQQIHHEVL	LNLILRVLRS	WNDPLYHLVT
rabbit	LPICPSGAVN	CQVSLRDLFD	RAVILSHHIH	KLSSEMFNEF	DKRYTQGRGF	ITKAINSCHT	SSLSTPEDKE	QAQQIHHEDL	LNMVLRVLPS	WNDPLYHLVT
horse	LPICPSGAVN	CQVSLRELFD	RAVILSHYIH	**N**LSSEMFNEF	DKRYAQGRGF	VTKAINSCHT	SSLSTPEDKE	QAQQIHHEDL	LNLILRVLKS	WNDPLYHLVS
cat	LPICPSGAVN	CQVSLRDLFD	RAVILSHYIH	**N**LSSEMFNEF	DKRYAQGRGF	ITKAINSCHT	SSLPTPEDKE	QAQQIHHEDL	LNVILRVLRS	WNDPLYHLVT
whale	LPICPSGAVN	CQVSLRDLFD	RAVILSHYIH	**N**LSSEMFNEF	DKRYAQGRGF	ITKAIDSCHT	SSLQTPEDKE	QAQZIHHEVL	VSLILGVLRS	WNBPLYHLVT
elephant	IPVCPRGSVR	CQVSLPDLFD	RAVMLSHYIH	SLSSDMFHEF	NKQYALGRGF	IPRAINSCHT	SSISTPEDKD	QAQQTHHEVL	MDLILGLLRS	WNDPLDHLAS

Figure 10-1.

	101 110	120	130	140	150	160	170	180	190	199
human	EVRGMQEAPE	AILSKAVEIE	EQ**TKR**LLE**G**M	E**LI**VSQVH**P**E	TKENEIYPV**W**	**S**GLPSLQMAD	EESRLSAY**Y**N	LL**HC**L**RR**D**SH**	**K**IDN**Y**LKLLK	**C**RIIHNNN**C**
monkey	EVRGMEEAPE	AILSKAVEIE	EQTKRLLEGM	E**LI**VSQVHPE	TKENEIYPVW	TGLPSLQMAD	EESRLSAYYN	LLHCLRRDSH	KIDNYLKLLK	CRIIHNNNC
		helix 3						helix 4		

	99	108	118	128	138	148	158	168	178	188 197
rat	GLGGIHEAPD	AIISRAKEIE	EQNKRLLEGI	EKIISQAYPE	AKGNEI**YLVW**	SQLPSLQGVD	EESKDLAFYN	NIRCL**RR**D**S**H	KVDNYL**KF**L**R**	**C**QIVHKNNC
mouse	GVGGIQEAPE	YILSRAKEIE	EQNKQLLEGV	EKIISQAYPE	AKGNGIYFVW	SQLPSLQGVD	EESKILSLRN	TIRCLRRDSH	KVDNFLKVLR	CQIAHQNNC
hamster	EVDGIHEASD	AIISRAKEIG	EQNKRLLEGI	EKILGQAYPE	AKGNEIYSVW	SQFPSLQGVD	EESRDLAIYN	KVRCLRRDSH	KVDNYLKLLR	CRVVHNNNC

	101 110	120	130	140	150	160	170	180	190	199
goat	EVRGMKGVPD	AILSRAIEIE	EENKRLLEGM	EMILGQVIPG	AKETEPYPVW	SGLPSLQTKD	EEARHSAFYN	LLHCLRRDSS	KIDTYLKLLN	CRIIYNNNC
sheep	EVRGMKGVPD	AILSRAIEIE	EENKRLLEGM	EMIFGQVIPG	AKETEPYPVW	SGLPSLQTKD	EDARHSAFYN	LLHCLRRDSS	KIDTYLKLLN	CRIIYNNNC
bovine	EVRGMKGAPD	AILSRAIEIE	EENKRLLEGM	EMIFGQVIPG	AKETEPYPVW	SGLPSLQTKD	EDARYSAFYN	LLHCLR**R**D**SS**	KIDTYLKLLN	CRIIYNNNC
allig. 1	EVQRIKEAPD	TILWKAVEIE	EQNKRLLEGM	EKIVGRVQPG	DTGNEVYSRW	SGLPSLQLAD	EDSRLFAFYN	LLHCGRRDSH	KIDNYLKLLK	CRLIHDSNC
allig. 2	EVQRIKEAPD	TILWKAVEIE	EQNKRLLEGM	EKIVGRVQPG	DTGNEVYSRW	SGLPSLQLAD	EDSRLFAFYN	LLHCGRRDSH	KIDNYLKLLK	CRLIHDSNC
croc. 1	EVQRIKEAPD	TILWKAVEIE	EQNKRLLEGM	EKIIGRVQPG	DTGNEVYSRW	SGLPSLQLAD	EDSRLFAFYN	LLHCGRRDSH	KIDNYLKLLK	CRLIHDSNC
croc. 2	EVQRIKEAPD	TILWKAVEIE	EQNKRLLEGM	EKIIGRVQPG	DTGNEVYSRW	SGLPSLQLAD	EDSRLFAFYN	LLHCGRRDSH	KIDNYLKLLK	CRLIHDSNC
chicken	EVQRIKEAPD	TILWKAVEIE	EQNKRLLEGM	EKIVGRVHSG	HAGNEIYSHS	DGLPSLQLAD	EDSRLFAFYN	LLHCHRRDSH	KIDNYLKVLK	CRLIHDSNC
turkey	EVQRIKEAPD	TILWKAVEIE	EQNKRLLEGM	EKIVGRIHSG	DAGNEVFSQW	DGLPSLQLAD	EDSRLFAFYN	LLHCLRRDSH	KIDNYLKVLK	CRLIHDNNC
turtle	EVQSIKEAPD	TIL.KAVEIE	EQDKRLLEGM	EKIVGQVHPG	EIENELYSPW	SGLPSLQQVD	EDSRLFAFYN	LLHCLRRDSH	KIDNYLKLLK	CRLIHDNDC
camel	EVRGMQEAPD	AILSRAIEIE	EQNKRLLEGM	EKIVGQVHPG	VKENEIYSVW	SGLPSLQMAD	EDTRLFAFYN	LLHCLRRDSH	KIDNYLKLLK	CRIIYDSNC
pig	EVRGMQEAPD	AILSRAIEIE	EQNKRLLEGM	EKIVGQVHPG	IKENEVYSVW	SGLPSLQMAD	EDTRLFAFYN	LLHCLRRDSH	KIDNYLKLLK	CRIIYDSNC
rabbit	EVRGMQEAPD	AILSKAIEIE	EQNRRLLEGM	EKIVGQVHPG	IKENEIYSVW	SGLPSLQMAD	EDARLFAFYN	LLHCLRRDSH	KIDNYLKLLK	CRIIYDSNC
horse	EVRGMQEAPE	AILSKAIEIE	EQNRRLLEGM	EKIVGQVQPR	IKENEVYSVW	SGLPSLQMAD	EDSRLFAFYN	LLHCLRRDSH	KIDNYLKLLK	CRIVYDSNC
cat	EVRGLHEAPD	AILSRAIEIE	EQNRRLLEGM	EKIVHQVHPG	VRENEVYSVW	SGLPSLQMAD	EDSRLFAFYN	LLHCLRRDSH	KIDSYLKLLK	CRIVYDSNC
whale	EVRGMQEAPD	AILSRAIQIE	EENKRLLEGM	EKIVGQVHPG	VKENEVYSVW	SGLPSLQMAD	EDTRLFAFYD	LLHCLRRDSH	KIDSYLKLLK	CRIIYNSNC
elephant	EVHSLPKAPS	ALLTKATEVK	EENQRLLEGI	EKIVDQVHPG	AKENKAYSVW	SGLPSLQTTD	EDARLFAFYN	LFRCLRRDSH	KIDSYLKLLK	CRIVYNNNC

Figure 10-1 continued.

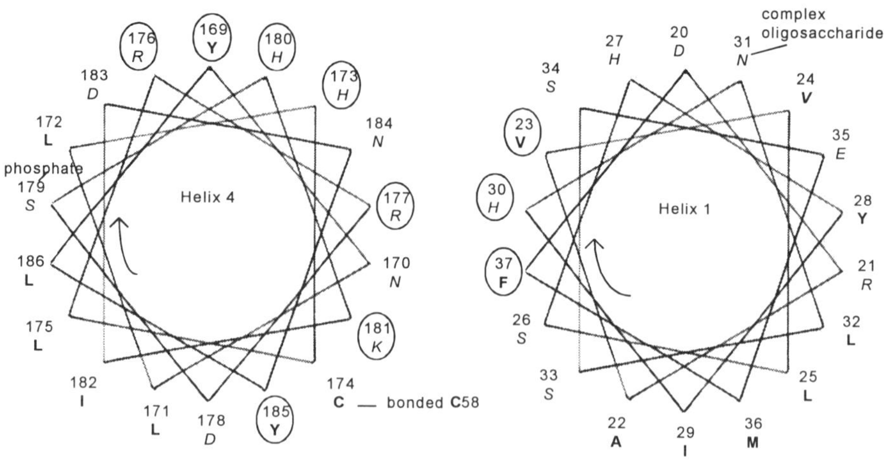

Figure 10-2. Parts of PRL involved in binding the first receptor. Bold letters are hydrophobic amino acids. Italicized letters are hydrophilic amino acids. Crucial residues are circled. Arrows show the direction of the helix

(11). The important conclusions of these studies, showing that small differences can have large effects, serve to remind us that although we are frequently constrained by the availability of exact tools, we must not forget that GH is not PRL (likewise for the receptors), and that a cow is not a human is not a rat. In addition, it illustrates the potential of various posttranslational modifications (natural such as phosphorylation or investigator-produced such as iodination) to alter receptor binding.

When the residues important to PRL conformational stability/binding/receptor activation (10,11) are labeled on helical wheel representations of those regions of the molecule, one can see that for helix 1, two are hydrophobic residues (V23 and F37) and one is hydrophilic (H30) (Figure 10-2). These residues are, however, all located on the charged side of the amphipathic helix. In the three-dimensional model of Kinet *et al.* (11), F37 and H30 are pictured as close to the receptor binding site, whereas V23 faces the interior of the molecule. N31 in this region is the site of glycosylation of human, ovine, horse, monkey, goat, baboon, camel and pig PRL, and the likely site of glycosylation of cat and whale PRL (see later in chapter). Glycosylation at this site involves the addition of a complex, approximately 2 kDa sugar (see later in chapter). Without modeling the three-dimensional structure of the receptor alongside the three-dimensional structure of PRL, it is hard to predict the full impact of glycosylation at this site on receptor activation. Addition of such a bulky residue in this critical region, however, seems likely to negatively affect receptor binding and activation and, as discussed later in the chapter, this seems to be the case.

For helix 4 (Figure 10-2), the residues, identified by Kinet *et al.* (11) or Luck (10) as critical to receptor activation, are on the hydrophilic side; two are hydrophobic (Y169 and Y185) and five are hydrophilic (R176, H180, H173, R177, K181). Three-dimensional modeling predicts that K181 and Y185 stabilize loop 1 which is also involved in binding the first receptor, while the other residues are directly involved in receptor binding. It is in this critical region that rat, bovine and human PRL are phosphorylated (see later in chapter) and this phosphorylation site (S179) is illustrated on the helical wheel. Phosphorylation at this site would be predicted to cause a twist in the helix such that the phosphate group was moved towards the receptor binding face and away from the hydrophobic interior of the molecule. S179 phosphorylated PRL binds to PRL receptors causing 1) greater activation of casein gene expression than unmodified PRL (23), and 2) antagonism of Nb2 cell proliferation stimulated by unmodified PRL (24). It is noteworthy that this antagonism is produced by changing a residue in a region involved in binding to the first receptor (see later in chapter for further discussion). Loop 1 includes residues 58 to 76 on Figure 10-1. Critical residues include H59, P66 and K69. Rat PRL has a minor phosphorylation site at T58 or T63 (25) which is equivalent to T60 or T65 on the human sequence. Phosphorylation in this region may well affect receptor binding; whether positively or negatively is unknown at present. One form of crocodile (croc. 1) and alligator (allig. 1) prolactin has an N-linked glycosylation site at position 60 (substitution of N for T at 60) (Figure 10-1). This also would be predicted to have a major impact on receptor binding.

Binding to the second receptor is thought to be mediated very differently. The residues important to conformation/binding are located on the hydrophobic faces of helices 1 and 3 and form a hydrophobic pocket in this area (Figure 10-3). The crucial residues, based on the negative impact on biological activity of their mutation to tryptophan, are A22 and S26 in helix 1 (12). In helix 3, mutation of G129 (although in this case to an R) also reduced biological activity. Goffin *et al.* (12) have proposed that binding to the receptor is via this hydrophobic pocket. The N31 which becomes glycosylated in human PRL is on the opposite side of the helix and yet still has a very negative impact on receptor activation (reviewed in 1. Also see later section). Whether any of the negative impact on receptor activation is due to interference with second receptor binding is unclear at present.

Because signal transduction is thought to be initiated by receptor dimerization, mutations which are thought likely to prevent dimerization, by altering binding to the second receptor, should act as antagonists. The G129R mutant is a PRL antagonist (26).

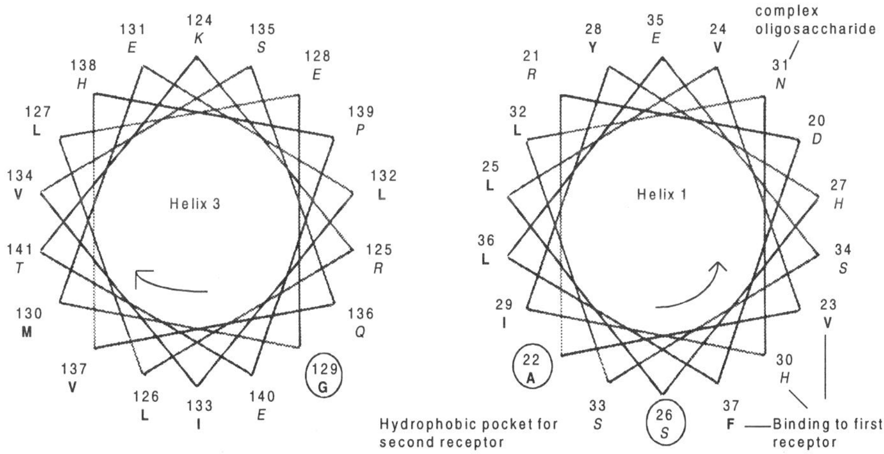

Figure 10-3. Parts of PRL involved in binding the second receptor. Bold letters are hydrophobic amino acids. Italicizedletters are hydrophilic amino acids. Crucial residues are circled. Arrows show the direction of the helix.

However, the most potent antagonist yet described is S179 phosphate (24), a posttranslational modification of a residue on helix 4 that normally faces into the interior of the molecule. If the mechanism of action is prevention of dimerization, this could be by interference with this hydrophobic pocket, either by causing a twist in helix 3 or by opening up the interior of the molecule. However, although inhibition of receptor dimerization was thought to be the mechanism of action of the hGH antagonist, G120R (equivalent to the hPRL G129R), it is now known that dimerization of the receptor is not prevented (27). Moreover, other receptors in the PRL/GH family have been shown to preexist as dimers without ligand binding (28). It appears more likely therefore that it is not dimerization *per se* that causes signal transduction, but the conformational changes in the receptors which are brought about by interaction with the ligand. Worth noting in this regard is the proximity of the two regions of helix 1 involved in interactions with both receptors Figure 10-3).

POSTTRANSLATIONAL MODIFICATIONS

This section of the chapter is organized according to the frequency with which a posttranslational modification is likely to occur (i.e., the conservation among PRLs of the known modified residue/consensus sequence).

DISULFIDE BOND FORMATION

All PRLs contain three disulfide bonds. These occur between C4 and C11, C58 and C174 and C191 and C199, thereby producing small N- and C-terminal loops and a large loop in the middle. Substitution of each C with an S demonstrated that the C^{4}-C^{11} and C^{191}-C^{199} bridges were not required for full biological activity (29). These mutational studies were in agreement with earlier biochemical analyses using specific reduction of each bond (30). The amino acids within these loops are still required for activity (for example see later under kallikrein cleavage), but not the loop structure itself. By contrast, the central disulfide loop is an absolute requirement for biological activity (29,30). Although disulfides are usually made co-translationally and often in association with heat shock proteins, the disulfides in PRL are reorganized during the production of dense secretory granules (see later under big PRL) and need to be resorted to their proper positions during exocytosis. It is currently unknown which cysteine residues are involved in oligomerization in the granules although intragranular cysteine thiols in the C-terminal region have been reported (31). Thus, it is possible that the large C^{50}-C^{179} is intact and only the N-terminal and/or C-terminal cysteines are involved in hormone oligomerization.

PHOSPHORYLATION

Phosphorylation at serine 179

Serine 179 (177 in rodents) is found in all PRLs in the Swiss protein database and, with the surrounding amino acids CLRRDS^{179} HKID, comprises the most conserved motif among all PRLs listed. Work from the authors' laboratory has demonstrated that S177 is the primary site of phosphorylation in rat PRL (25), that bovine PRL is phosphorylated on both S179 and S180 (32) and that molecular mimicry of phosphorylation at S177/179 in both the rat and human hormone produces molecules with the same biological activities as the phosphorylated hormone (33). This phosphorylation site is not present in GH or rat PL-I, but is present in rat PL-II (Swiss protein database). Thus the potential for phosphorylation at this site is associated with PRL-like and not GH-like activities.

Naturally phosphorylated rat PRL acts as an antagonist to Nb2 cell proliferation promoted by unmodified PRL (24), serves as an autocrine regulator of GH_3 cell proliferation (34) and as an autocrine regulator of normal mammotroph secretion (35). Its secretion is also physiologically regulated with different proportions of unmodified to phosphorylated PRL being released from the pituitary at different stages of the estrous cycle (36) and at different times during pregnancy and pseudopregnancy (37). The phosphorylated form is very stable in the circulation, requiring hours of exposure to high concentrations of phosphatase to achieve dephosphorylation (24). An S179D molecular mimic of phosphorylated PRL also antagonizes the stimulation of Nb2 cell proliferation brought about by unmodified PRL (33). In general, a large amount of recently accumulated data using S179D PRL (23,38-44) is consistent with the general hypothesis that phosphorylated PRL acts as an antagonist in tissues where/when unmodified PRL promotes growth, but that it acts as a superagonist when the effects of PRL are not related to growth. This latter activity is exemplified by the much larger stimulation of casein gene expression by the S179D molecular mimic *versus* that of the unmodified PRL (23). The S179D mimic has been shown to compete for JAK 2 activation by unmodified PRL (45). This is one way it may act as a growth antagonist. In addition, the S179D mimic and, by its presence in NIDDK PRL, the phosphorylated hormone, have been shown to signal differently in Nb2 cells (45) and, more recently, in HC11 mammary cells (unpublished data from the author's laboratory). Different signalling is consistent with the different biological activities observed.

Phosphorylation of threonine 58/60 or 63/65

Second, a more minor site of phosphorylation (this time of a threonine) was identified in rat PRL within the tryptic peptide AINDCPT^{58}SSLAT^{63}PEDK (25). This phosphorylation occurs in the doubly phosphorylated variety (i.e., in a molecule already phosphorylated on S177). It is included at this point in the chapter because it is part of the second most frequently conserved motif in PRL: CHT^{58}SSLXT^{63}PEDKEQA and, according to the model of Kinet *et al.* (11), would be likely to have an effect on binding to receptor 1. It is unknown which of these Ts is used. The level of secretion of this doubly phosphorylated form is fairly consistent when expressed as a percentage of the total PRL being released, except in the afternoon of proestrus when its absence is conspicuous (36).

Phosphorylation at serine 90

The only other analysis of phosphorylation sites to date has been for bovine PRL (46). These authors conclude that bovine PRL is primarily phosphorylated on serine 90, with minor phosphorylation sites on serines 26 and 34 (46). The methodology employed would have missed the serine at 179: S179 would be in the tryptic peptide, D^{178}SSK, which is highly charged, especially when phosphorylated, thereby running through a C18 column. Nevertheless, it appears that bovine PRL may also be phosphorylated at S90. S90 is not present in most fish PRLs (fish PRLs not presented in Figure 10-1), but is present in reptilian and mammalian PRLs. It is also present in GHs and PLs and is in the middle of helix 2 for each hormone. S90 is therefore a residue, conserved in this family (except for most fishPRLs), but one that is not specific to PRL. Physiological studies have been performed using isolated preparations of non-phosphorylated and phosphorylated hormone and, more recently using a molecular mimic of S90 phosphorylated bovine PRL, S90E: Interpretation of the results with isolated preparations is complicated by the presence of more than one major species of phosphorylated hormone

(S179/180 plus S90). Nevertheless, dephosphorylation of the phosphorylated species caused an increase in Nb2 cell proliferative activity (47), just as we had previously shown for rat PRL phosphorylated on S177 (24). Also complicating analysis of the bovine preparations is the fact that they show some signs of damage during isolation. The reported Nb2 bioactivity of the isolated, non-phosphorylated material is only 75% of NIDDK bovine PRL (47). Given that, according to the authors, between 20 and 80% of NIDDK bovine PRL should be phosphorylated (46) and the author's contention that phosphorylated PRL does not bind to the Nb2 PRL receptor (47), the non-phosphorylated isolate should have at least 120% the activity of NIDDK PRL. If the non-phosphorylated isolate is damaged, it is likely that the phosphorylated isolate is also. When tested in rats, a phosphorylated isolate made by an improved procedure (48) was found to be without luteotropic activity (49), as one would expect for a preparation that does not bind to PRL receptors. When the S90E mimic was tested for Nb2 bioactivity, it was found to have very little (50). When the importance of S90 to wildtype biological activity was tested, a S90A mutation had full activity (51). Thus, on the one hand phosphorylation or pseudophosphorylation of S90 seems to produce a biologically inert molecule but, on the other, a large change in the side chain of the residue at position 90 has no effect on wildtype bioactivity which argues against major importance for this amino acid. By contrast, similar mutation of human S179 to A in the authors' laboratory eliminated all Nb2 activity (33), a finding consistent with that observed in another laboratory using a mammalian expression system which showed rat S177A to have only 3.9% of the activity of the wildtype hormone (13).

Based on these data, it appears that phosphorylation on S90 in the bovine hormone renders it inactive. The purpose of this is unclear since it is not dephosphorylated in the circulation (49). By contrast, phosphorylation on S177/179 produces a molecule with markedly different biological activities to the unmodified hormone. The reported minor sites of phosphorylation in the bovine hormone, however, seem located in a region with the potential to have a major impact on binding to the first receptor. S26 and S34 sit on either side of the three residues on helix 1 thought to be involved in interactions with the first receptor (Figure 10-2). However, no molecular mimics of these have been tested thus far.

The prolactin kinases

The activity of the S177/179 kinase is reduced after estrogen treatment of either pituitary cells in culture or intact female rats (36,52). Thus estrogen treatment of animals results not only in an increased amount of PRL release from the pituitary (53), but an increase in the proliferative activity of that PRL (24). Thyrotropin releasing hormone and dopamine, by contrast, have no effect on kinase activity (52). We have shown that phosphorylation of S177/179 can be duplicated by protein kinase A (PKA) (25,54) and PKA also phosphorylates chicken, turkey, and sheep PRL (55). Duplication of phosphorylation does not, however, identify PKA as the PRL kinase. Recent studies from the authors' laboratory have identified p21-activated kinase within rat and bovine PRL secretory granules and have shown that this kinase also phosphorylates S177/179 (32). It remains to be determined, however, whether this is the true physiological kinase. One major problem in studies of the kinase is the development of a method to distinguish between kinases present on the cytoplasmic face versus those present on the granule core face of the granule membranes. Many kinases have membrane-anchoring post-translational modifications (56) and hence can associate with cytoplasmic faces, a situation in which their presence may contribute, for example, to granule fusion events, but not to PRL phosphorylation. Studies in which exogenous PRL is added to granules (or other subcellular fractions) to achieve significant phosphorylation (57) do not distinguish between cytoplasmic-face and granule-core-face enzymes and may produce misleading results. Competition between intragranular PRL phosphorylation and the phosphorylation of a specific substrate for the suspected kinase is reasonable proof of the likely identity of PRL kinase, but a form of that kinase, capable of being placed in the secretory pathway then needs to be identified.

Phosphorylated PRL in other species

Since the demonstration of rat PRL phosphorylation in the authors' laboratory in 1986 (54), PRL has been directly shown to be phosphorylated in bovine pituitaries (46) (plus see earlier section on phosphorylation sites) and ovine and turkey pituitaries (55). In addition, human pituitary PRL has been indirectly shown to be phosphorylated by the increased Nb2 bioactivity produced by acid phosphatase treatment (33). Ashing studies have shown ovine, rat and turkey prolactin to contain 24, 21 and 68% phosphorylated hormone, respectively (55), but to date the only studies on physiological regulation of phosphorylation have been conducted on rats. These studies show that the proportion of phosphorylated material being released varies under different physiological circumstances (36,37). It is assumed therefore that the same will be eventually demonstrated for ovine and turkey PRL and hence that these figures should not be considered absolute. Given the conservation of at least one phosphorylation site among all PRLs, it also seems likely that all PRLs will eventually be shown to be phosphorylated to some degree.

Phosphorylated PRL in milk

PRL has been detected in the milk of cows (58), goats (59), pigs (60), sheep (61), rats and humans (62,63). This PRL can pass across the immature intestinal epithelium into the circulation of the newborn where it can influence later hypothalamic, reproductive and immune functions (64-66).

Studies largely presented in reviews by Grosvenor (67) have suggested that the majority of PRL in milk is phosphorylated. Phosphorylated PRL was only identified as such, however, by its ability to bind to metal ions. More recent studies by Ellis and Picciano (68) have suggested that a variety of PRL forms are present in milk. These include glycosylated PRL, phosphorylated PRL, glycosylated and phosphorylated PRL and, what appear to be complexes between these and cleaved forms of PRL. However, the methodologies employed in this study cannot exclude the possibility that material identified in their study as glycosylated or phosphorylated PRL was not simply PRL bound to another glycosylated or phosphorylated substance. In this context, it is important to note the proportion of PRL bioactivity at molecular weights higher than 40, or even, 100 kDa (see later under big PRLs). Nevertheless, this study contains some evidence of the existence of both phosphorylated and perhaps phosphorylated and glycosylated PRL in milk, produced by demonstrating increased Nb2 activity after dephosphorylation and deglycosylation of the milk fractions.

Given the mammary gland's capacity to phosphorylate casein, one might reasonably expect that cells of this gland could also phosphorylate PRL. This would either be PRL made in the mammary gland (69) or PRL in the process of transcytosis from circulation to milk (70). This issue has yet to be resolved.

PROTEOLYTIC CLEAVAGE

A variety of cleaved PRLs have been described in pituitary, target tissues and serum (for review see 1). Many descriptions of these fragments have been based on Western blots and so one needs to consider the known cross-reactivity of many anti-PRL antisera with IgG light chains (71) (also see later under glycosylated PRL). In other words some reported fragments may be fragments of light chains and not PRL. Studies of hormone processing are also severely complicated by the ever-present need to distinguish between processing and degradation. At the moment, evidence is strongest for two processing events. The first is a cleavage to produce an "open loop", 25 kDa PRL which, after disulfide reduction, yields 14-16 kDa and 6-8 kDa fragments. The second is a kallikrein cleavage to produce a 21 kDa PRL and three small C-terminal peptides. In the case of the first cleavage, the evidence is convincing because one of the generated fragments has distinct biological activities (see below). In the case of the second cleavage, the evidence is convincing because the processing enzyme is physiologically regulated, as is release of the fragments produced (72-74).

25 kDa and 16 kDa PRL

If PRL is cleaved in the large disufide loop, a so-called open loop variant is produced. Although such a variant has been reported in the pituitary, it represents only 0.6-1% of pituitary PRL (5). It is more likely therefore that biologically important cleavage in the large loop occurs at target tissues. A number of target tissues, including the liver and lactating mammary gland, have been shown to contain enzymes capable of a suitable cleavage (75). Cleavage in these tissues occurs under acidic conditions (75), a circumstance consistent with proteolysis inside an endosomal compartment after internalization of the ligand-receptor complex. To produce a free N-terminal 16 kDa fragment, it is also necessary for the dilsulfide between C58 and C174 to be reduced. How this is achieved is unknown, but again seems to be an event which could more easily be orchestrated intracellularly.

It is presently unclear what enzyme is responsible for the cleavage. Cathepsin D was a good candidate because it mimicked the cleavage of rat PRL by mammary tissue (76). It is also an enzyme which requires glutathione (77) and hence would function in a reducing environment. Cleavage of rat PRL by cathepsin D excises a tripeptide, LVW, by cleaving between Y145 and L146 and W148 and S149 (76). Human PRL, however, is not cleaved by cathepsin D despite the presence of the $W^{150}S^{151}$ pair (78) (Figure 10-1). It seems, therefore, that the YL cleavage must precede the WS cleavage in rat PRL. Human, sheep, goat, cow, monkey and fish PRLs have YP in place of YL and so none of these PRLs would be cleaved by cathepsin D. Most other species have YS rather than YL, but there is no information for any of them about cathepsin cleavage.

Thrombin will cleave both human and rat PRL to produce a 16 kDa fragment, but this is a C-terminal fragment with no known biological activity (78). Thrombin cleavage, however, may be responsible for observations of 16 kDa PRL in serum and hence future studies of 16 kDa in serum must utilize a specific antibody to distinguish between the two possible forms.

When primate recombinant PRLs were expressed in C127 mammary cells, Cole *et al.* reported production of a 16 kDa fragment (79). This cleavage was between L132 and I133. An L and I pair at this position only exists in primate PRL (Figure 10-1).

It appears therefore that the correct processing enzyme has yet to be identified. Clapp *et al.* predicted that cleavage would most likely occur between the dibasic residues at 124-125 (80). Cleavage between dibasic residues is typical in polypeptide processing (81). A recombinant version of 16 kDa PRL ending at residue 123 has the appropriate biological activity (see below). This potential site is present in most, but not all PRLs, but so far no appropriate dibasic cleaving enzyme has been identified in target tissues.

In a later section, we suggest that glycosylation of an N in an NXC sequence might be useful in the production of 16 kDa PRL because it might prevent the large disulfide loop from forming. There is no obvious correlation in the literature, however, between the degree of glycosylation of PRL and the ease with which it can be cleaved (75,76,82). It is nevertheless important to bear in mind in this context that the natural enzyme activity has yet to be identified.

16 kDa (16K) PRL has generated a lot of interest because it is antiangiogenic (80). The form of 16K PRL being used in studies of antiangiogenesis, however, is different in different laboratories: Some studies use recombinant 16K hPRL terminating at residue T123 (80) while others use recombinant 16K hPRL terminating at P139 (83) or an implanted P142 (84). These recombinant 16K PRLs also have a C58S mutation so that no intermolecular disulfides will form during folding of the reduced, solubilized inclusion bodies (80,83,84). Although this is a conservative mutation, it is at the beginning of loop 1 and next to H59 which is critical to binding to the first receptor. As emphasized earlier, even conservative mutations in critical regions may alter activity. In addition, various labs compare results with natural rat PRL cleaved with target tissue enzymes or cathepsin D. In the latter case, the terminating residue on the 16K fragment is Y145 (76). There seems to be some flexibility therefore in the length of the so-called 16K PRL in terms of its antiangiogenic activity.

Antiangiogenic activity has been demonstrated by inhibition of bFGF-stimulated endothelial cell proliferation *in vitro* (80,83,84) and by inhibition of angiogenesis in chick chorioallantoic membranes (80,84) or bFGF-stimulated rat cornea (85). Human PRL and the 16K fragment bind to heparin (86) and so part of this antiangiogenic activity can be attributed to competition with bFGF or other heparin-binding growth factors for binding to heparin. Rat PRL, however, does not bind to heparin (86) and so none of its antiangiogenic properties can be ascribed to this competition phenomenon. In a recent study, full-length PRL and 16K PRL were found to have opposite effects on angiogenesis, but at different stages of chorioallantoic membrane development (84). At early stages, 16K PRL was antiangiogenic whereas 23 kDa PRL was without effect. At later stages, 16K PRL was without effect whereas 23 kDa PRL promoted angiogenesis. In brain endothelial cells this translated to opposite effects of 23 kDa and 16K PRL on MAP kinase activity (84).

Endothelial cells possess a specific 16K PRL receptor although it has yet to be cloned (87). What is meant by specific is the lack of binding of 23 kDa PRL. The work of Struman *et al.* (above) implied the additional presence of a classical PRL receptor on endothelial cells, but until a recent report by Merkle *et al.* (88) no one had succeeded in demonstrating the presence of classical PRL receptor mRNA. These two receptors (classical and 16K) presumably are responsible for mediating opposite effects on MAP kinase activity (84). 16K PRL has also been shown to stimulate expression of endothelial cell plasminogen activator inhibitor (83) and the inducible form of nitric oxide synthase in pulmonary fibroblasts and alveolar type II pneumocytes (89). Endothelial cells and corneal epithelial cells express PRL (85,90) and locally-produced PRL could be used for either angiogenic or antiangiogenic activity in the case of the former, or for continual antiangiogenic activity in the case of the latter.

Kallikrein cleavage of PRL

The kallikreins belong to a family of enzymes involved in prohormone processing (91). A glandular kallikrein is expressed in both the intermediate and anterior lobes of the pituitary where its expression is 20-fold higher in females (72-74,92,93). Expression of kallikrein parallels that of PRL during development and during pregnancy (94,95). Estrogen stimulates and dopamine represses expression (96,97). Such physiological regulation suggests an important biological function. Kallikrein cleavage of PRL occurs in secretory granules where its activity is regulated by granule condensation (98), thiol availability in the form of glutathione (98,99) and divalent cations (100). Direct PRL-zinc coordination in granules also protects two cleavage sites (100). The sites of cleavage of rat PRL were determined by Powers and Hatala (99) and shown to be between R174 and R175, K185 and F186 and, R188 and C189. These cleavage sites are conserved in most PRLs. Ovine and bovine PRL do not have the most C-terminal site and substitutions at position 173/175 in crocodile, alligator, and chicken PRL may affect the most N-terminal site (Figure 10-1). Recent work from the authors' laboratory has demonstrated that kallikrein does not cleave bovine PRL despite the presence of two out of the three normal sites (unpublished data). This implies that kallikrein cleavage is a sequential process which works its way in from the C-terminus and/or that other minor modifications in the sequence (compared to rat PRL), such as the H180S substitution, sufficiently alters conformation to prevent cleavage. In this regard, however, it is important to note that phosphorylation of S177 in rat PRL does not influence kallikrein cleavage (99) thereby making the first explanation more likely.

Although kallikrein cleavage occurs in the pituitary, most of the kallikrein present is in the inactive pro-form (96). This may be a mechanism allowing the cell to regulate activity just before PRL release. In other words, the cell may package sufficient kallikrein with the PRL to cleave a large proportion, but the amount of enzyme activated in the secretory granule would be according to immediate physiological need. The co-packaged pro- kallikrein is released from the cell (101) and could still be activated in the circulation. If it remained complexed with PRL, some processing could then potentially still occur. If it does not, it seems unlikely that PRL processing would occur because of the much larger quantities of alternate substrates present in the circulation.

Kallikreins are also present in a number of PRL target tissues and hence a similar processing of full-length PRL could occur there (91).

The 21 kDa fragment produced has no Nb2 activity (101) and to date no other reproducible, unique activity. There is, however, some indication that the released peptides may affect tyrosine hydroxylase activity in the hypothalamus (L. Arbogast, personal communication).

GLYCOSYLATION

Glycosylated PRL (G-PRL) was reported as early as 1984 in ovine pituitaries; this had an asparagine N-linked oligosaccharide at the single consensus sequence NXS (positions 31-33) (102). Of the PRL's shown in Figure 10-1, this sequence is conserved in human, monkey, goat, sheep, camel, pig, horse, cat and whale. It is also found in recombinant baboon PRL whose structure is only different from monkey PRL at positions 3 (I for V) and 43 (Q for R) (79). As reviewed by Sinha (1), glycosylated PRL has been isolated and confirmed at this site in human, monkey, baboon, goat, sheep, camel and pig; more recently, glycosylated horse (equine) PRL has also been isolated showing

N-glycosylation at N31 (103). Pituitary equine PRL and its glycosylated form were purified by acetone extraction/precipitation, NaCl and isoelectric precipitation, gel filtration, and concanavalin A sepharose (ConA) affinity chromatography (103). Butnev *et al.* (103) demonstrated by mass spectrometry analysis and SDS-PAGE mobilities that, like the G-PRLs from the above species, the equine G-PRL had a molecular weight of 25 kDa compared to 23 kDa for non-glycosylated PRL. Due to the similarity of the carbohydrate composition in other species, they proposed that the predominant unit in N-glycosylated PRLs is a fucosylated mono- and/or biantennary complex with a sulfated GalNAc at a Man-α1,3 branch (103). A NeuAc-Gal as a terminal monosaccharide in one of the branches is also possible (103).

Is there glycosylation in PRLs that do not have the NXS sequence, or are there glycosylation sites in addition to NXS in some PRLs? Substitutions for the crucial N include T in mouse and rat, M in hamster, D in bovine, Y in chicken, K in rabbit, S in turtle and elephant, and F in turkey, alligator and crocodile PRLs (Figure 10-1). However, glycosylation has been reported in rat, bovine, turkey, alligator and crocodile PRLs (reviewed in 1), despite the fact that each is missing an NX(S/T) In addition, in ovine PRL, a second site for glycosylation has been postulated (104). There are several possibilities to explain the results. First is the possibility that what is being reported as glycosylation may, in fact, be an artifact of the procedure. Bovine PRL, for example, was thought to have glycosylated forms (105), but during cell free biosynthesis no glycosylation occurred (104), suggesting that the reported bovine G-PRL may have been a different protein. In some studies, glycosylation is suggested from positive bands on immunoblots (using anti-PRL antisera) with molecular weights approximately 2 kDa greater than native, unmodified PRL. These proteins also may bind to ConA. In an important paper by Heffner *et al.* (71), the authors demonstrated that several anti-human prolactin sera potently crossreacted with immunoglobulin light chains that migrate at 25 kDa under reducing conditions. IgG light chains can arise from blood within tissue samples when isolated, from the circulation if serum levels are measured, or from fetal calf serum used in cell culture medium. These light chains would also bind to ConA. PRL binding to immunoglobulin G will be discussed in greater detail in a later section. The possibility that one is observing changes in IgG rather than G-PRL is extremely important, since serum and placenta IgG levels have been shown to vary with the estrous cycle and pregnancy in cows, rats, and humans (106,107). The work of Heffner *et al.* (71), then, emphasizes the need to run controls for this possibility. (Does the anti-PRL one is using bind to IgG light chain alone?)

In studies on rat glycosylated PRL (108,109), the above potential artifact does not appear to be a significant issue. Non-reducing conditions for electrophoresis were used so, unless there is some free IgG light chain under certain physiologic conditions, migration of IgG would not be in the vicinity of 25-27 kDa. Additionally, rat IgG light chains may not be glycosylated (personal communication, Herve Bazin). Light chains could be present, however, in cell culture studies due to fetal calf serum. It is of interest that only one band of the reported doublet (25 kDa and 26 kDa) described as glycosylated rat PRL appears to be bound to ConA (specific for α-D-Glu and α-D-Man), and appears to be affected by trifluoromethanesulfonic acid (TMSF) deglycosylation (108). Additionally, these two bands can not be distinguished from native PRL on reducing gels, supposedly due to the change in conformation of monomeric, non-glycosylated hormone (108). The glycosylation that is seen in the rat may be O-glycosylation at serine or threonine and some evidence points in this direction. Although O-glycosylation usually involves mono- or disaccharide additions, larger oligosaccharide additions can be O-linked and could explain the 2 kDa increase in size. Some shift in electrophoretic mobility occurs with both TMSF (108) and O-sialoendopeptidase digestion (109). One question is, however, that if O-glycosylation occurs in rat PRL and results in a high proportion of glycosylated hormone, why is O-glycosylation not a major modification of other PRLs as well? There is the possibility that the protein glycosylated is a *related* protein but not PRL. For example, somatolactins that are found in the pituitary intermediate lobe have approximately 210 amino acids and N-glycosylation sites have been identified (110). To date, somatolactins have not been identified in the rat; work has been primarily in advanced teleosts and other fish. Answers to the question of if and where rat PRL is glycosylated must await sequencing of the putative glycosylated rat PRL peptides.

The sequence of turkey PRL has F substituted for N at position 31, and no other NXS exists. However, like the rat PRL, glycosylated isoforms have been reported (111). Although it was initially believed that turkey G-PRL was O-glycosylated, O-glycosidase treatment had no effect (112). Neuraminidase treatment resulted in a partial shift from 27 to 26 kDa, like the partial shift seen with O-glycosidase treatment in rat PRL (from 26 to 25 kDa). On the other hand, N-glycosidase treatment of turkey PRL resulted in the complete disappearance of the 27 kDa band and concomitant increase in the 23/24 kDa isoform (112). These investigators found that a 27 kDa band was dominant during incubation behavior periods whereas the non-glycosylated PRL was most prevalent during out of lay stages of the reproductive cycle (113). Since it appears that the electrophoretic studies with turkey PRL were performed under reducing conditions, it seems imperative that controls for the presence of IgG light chains be carried out to be confident that the 27 kDa band is, indeed, glycosylated PRL.

N-glycosylation at an alternate site

Another possibility to explain the above results is that N-glycosylation is occurring at an alternate site on the hormone. Either there is another sequence that is recognized by the enzyme for N-glycosylation, or there is a previously unrecognized hormone variant that contains an NX(S/T). As for the first possibility, it has been shown that glycosylation can occur at NXC (e.g., human von Willebrand factor (114) and CD69 (115)). NXC has been suggested as a second site possibility in ovine PRL where two N-glycosylated isoforms were identified (104). NXC at positions 56 to 58 (54 to 56 in rodents) is present in all PRLs except hamster (SDC) and whale (DSC) (Figure 10-1), and fishes (not shown in Figure 10-1). NNC is also present at the C-terminus in human, monkey, mouse, rat, hamster, turkey, goat, sheep, cow, and elephant and NDC terminates camel PRL (Figure 10-1). Since the cysteine has to be in the sulfhydryl form, glycosylation would have to take place *before* hormone disulfide bond formation. This would be possible since proximal glycosylation takes place in the endoplasmic reticulum. However, depending on whether this affected the large loop or the C-terminal loop, one might expect profound effects on storage, secretion, and function. One would also question why it has only been seen in sheep PRL and not as an additional glycosylation site in other PRLs containing the NXC sequence. If glycosylation is possible at NXC, however, one could speculate that glycosylation at this point might prevent the large disulfide loop from forming, and therefore be important for the formation of 16K PRL (see earlier section).

Considering the possibility of variants of PRL that may have a previously unrecognized alternate N-glycosylation NX(S/T) consensus sequence, Noso *et al.* (116) have reported variants in both the alligator and crocodile hormone. An N has been substituted for a T, the result of a genetic code change from AC(C/T) to AA(C/T). Thus, in a highly conserved region of PRLs (#57-#63), a new glycosylation site, NAS (#60-#62), was identified. Although conservative alterations in the pre-PRL region have been identified in mouse PRL (117), it is unknown whether PRL modifications exist elsewhere like those in the reptiles studied. For example, differences due to strain of rat (and thus the potential variant) could account for glycosylated rat PRL being observed in some laboratories (Wistar rats are used in the laboratories of Bollengier *et al.* (108,109)) and not others. If this site, near C57 were glycosylated in other species, it would be expected to have profound effects on formation of the large disulfide loop and, in turn, on structure and biological activity.

Physiological regulation

In the human, when the proportion of glycosylated and non-glycosylated PRL is estimated by using specific monoclonal antibodies, the amount of G-PRL is approximately 16-24% of the total plasma and pituitary PRL. Transfection of human PRL cDNA into pituitary tissue, Chinese hamster ovary cells, and Cos-1 cells also generated the same degree of glycosylation (118). Using assays specific to NG- or G-PRL, opioid administration (e.g., phenoperidine) was shown to stimulate PRL secretion without changing the ratio of forms (119) whereas marked increases in non-glycosylated PRL (NG-PRL) compared to G-PRL were observed when release of PRL was stimulated by thyrotropin releasing hormone or metoclopropamide (120). Thus, the type of secretagogue may influence the mammotrophs' secretory response. In contrast to these results, Gambino *et al.* (121) found that TRH had no effect on B/I ratios (Nb2 proliferation activity/immunoactivity). Since glycosylation lowers bioactivity (reviewed in 1) but is recognized by the antibody, one would expect that decreases in G-PRL compared to NG-PRL as reported by Brue *et al.* (120) would increase the B/I ratio. Although this discrepancy in results has not been resolved, Gambino *et al.* (121) also found that PRL from serum had to be immunopurified prior to immunoassay. This procedure removed non-lactogenic cytokine-like factors that interfered with the Nb2 assay (non parallel to standard). These authors also found that both the B/I and G-PRL/NG-PRL ratios did not change in maternal or fetal circulation during pregnancy, or in patients with prolactinomas (121). Two individual patients with adenomas, however, showed high immunoactive PRL, mild clinical features, and lowered B/I ratios. Although this would be consistent with elevated G-PRL forms, G-PRL/NG-PRL were within the normal range, suggesting that mechanisms other than altered glycosylation were contributing to the lowered bioactivity (121).

What is the physiological significance of glycosylated PRL?

Glycosylation at N31 would likely result in interference of receptor binding/activation and thus, it is not surprising that human G-PRL has diminished activity in pigeon crop sac, Nb2 lymphoma cell proliferation, and casein synthesis assays (reviewed in 1). One would not expect the mammotrophs to glycosylate the hormone merely to lower biological activity; it would be more efficient to lower the rate of PRL synthesis. Other purposes for glycosylating PRL could include: 1) The glycosylated PRL could affect down regulation of some target tissue receptors. Alternatively, it could bind to a different receptor than non-glycosylated PRL. Direct evidence for these possibilities, however, has not been observed. 2) Glycosylated PRL could act as a circulating reservoir of PRL, having decreased biological activity until deglycosylation. There is no evidence at present, however, for total de-glycosylation occurring in the circulation. 3) Glycosylated hormone could have a different half life from non-glycosylated hormone in the circulation. Brue *et al.* (120), however, saw no change in G-PRL/NG-PRL ratios in patients with chronic renal failure compared to normal individuals suggesting that any clearance by the kidney was unaffected by glycosylation. Moreover, a change in half life would only be relevant if either total deglycosylation could occur or G-PRL had a distinct function. . 4) Glycosylated PRL could be differentially taken up by the liver. It is possible that the liver is responsible for much of the G-PRL clearance. Perhaps the role of the oligosaccharide on PRL is to target the hormone to the asialoglycoprotein receptor on the liver. In the liver, potentially bioactive cleaved forms of the hormone could be produced. 5) Finally, one can speculate that since differences in G-PRL/NG-PRL have been unremarkable under a variety of conditions, glycosylation may play a role in the pituitary either in secretory granule storage or in the hormone's susceptibility to posttranslational modifications (e.g., phosphorylation, kallikrein cleavage, 16K formation). Secretory granules store PRL, in part, as high molecular weight intermolecular disulfide-bonded hormone oligomers (122,123). The specific PRL cysteine residues involved in intermolecular disulfides are unknown, but addition of the bulky carbohydrate moiety (especially if it occurred at NXC) could influence these reactions. Susceptibility to specific proteases or kinases could also be increased/decreased, although the site(s) for 16K production, kallikrein cleavage, and phosphorylation do not appear to occur near the probable glycosylation site(s). However, if the conformational change was marked upon carbohydrate addition, or if the carbohydrate interfered with disulfide formation, protease or kinase action could be affected. These speculations must await further study, of course, before conclusions can be drawn.

"MULTIMERIC" FORMS OF PROLACTIN

Molecular heterogeneity of PRL in the circulation is well established (reviewed in 124,125), and high molecular weight forms in human plasma were reported as early as 1974 by Suh and Frantz (126). Circulating forms of PRL are referred to as "little" (23 kDa, monomeric), "big" (~50 kDa, dimeric), and "big big" (excluded from Sephadex G100). What comprises the multimeric forms can be divided into two general categories. The first includes monomeric PRL linked to other PRL monomers, forming higher molecular weight complexes; these are found primarily in pituitary secretory granules and, under certain conditions, may also be in the circulation. The second category includes PRL that is associated with immunoglobulins in the circulation. The latter category includes: 1) PRL covalently bound to IgG; 2) PRL associated with IgG but not as a PRL-anti-PRL complex; and 3) PRL bound to an IgG autoantibody to PRL. The literature in this area is complicated since differentiation among these possibilities is not always addressed, and many times PRL binding to IgG is *assumed* to be an autoantibody reaction. PRL may also bind to other serum components in the human, such as heparin (86) but these will not be discussed in this section.

High molecular weight forms of PRL in pituitary secretory granules

Both PRL and GH are stored in secretory granules, in part, as relatively osmotically inactive, oligomeric forms (granular hormone is excluded from G200 Sephadex (127)) (Figure 10-4). Like the storage of hexameric insulin coordinated with zinc in insulin granules, oligomeric PRL has the advantage that little metabolic energy is expended by the cell to maintain a large amount of readily available hormone (123). Also, upon stimulation of secretion, oligomer depolymerizes to lower molecular weight forms with a resulting increase in intragranular osmotic pressure before or concomitant with exocytosis. Oligomeric hormone is poorly immunoreactive against antibody raised to monomeric PRL (122,127-129). Data suggest structural stability is maintained in oligomers by intermolecular disulfide bridges and/or intermolecular ionic interactions. Thiols, EDTA, and urea increase monomerization (and immunoactivity) while divalent cations and disulfides diminish conversion (and immunoactivity) (122,123,127-131). Zinc in particular may be crucial to pituitary PRL storage forms. In addition to inhibiting monomerization and posttranslational modifications (e.g., kallikrein cleavage) (100), studies with rat PRL suggest zinc binds at D181 with coordination from intragranular cysteine thiol(s) at C189 and/or C197 (31). After monomerization, disulfide interchange and disulfide bond formation, residues from helix 1 may replace the contributions of the two cysteines.

Big PRL

During exocytosis, the depolymerization of oligomeric PRL probably occurs with the aid of reduced glutathione in granules (123) and an increase in pH from the acidic intragranular environment to the extracellular milieu. Since dissolution of the dense granule hormone is often observed in the extracellular space on electronmicroscopy (132), it is reasonable that dimers and higher multimers of PRL can be seen in the circulation, especially with potently stimulated secretion (Figure 10-4). At least some of the size variants of PRL are due to this phenomenon. Big PRL is not as bioactive and occurs asymptomatically in 0.4% of Japanese women and 0.02% of Japanese men (133).

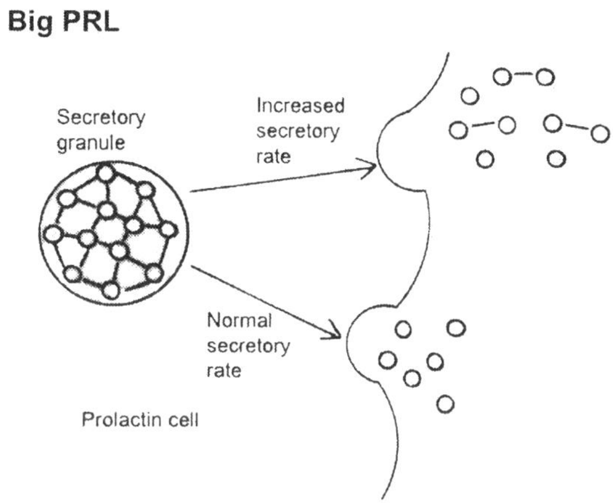

Figure 10-4. Release of Big PRL. Increased secretory rate could be caused by a defect in hypothalamic control or the secretory machinery. o——o represents a disulfide-linked PRL dimer. o represents PRL monomer. o——o Stable in circulation. Not as bioactive. Incidence: 0.4% of Japanese women, 0.02% of Japanese men (133).

Big big PRL: Covalent binding of PRL to immunoglobulin G. Covalent binding of PRL to IgG was first reported in 1992 (134). A 50 kDa binding protein was demonstrated by two dimensional gel electrophoresis of anti-PRL-precipitated sera. Since the anti-PRL did not directly recognize either 50 kDa IgG heavy chains or the 25 kDa IgG light chains, and since the levels of IgG (or subclasses) were not different from controls, it was suggested that PRL was binding to IgG heavy chains (134). In 1992, Cohen *et al.* (135) also showed that ^{125}I-PRL bound to IgG and its 50 kDa subunit. It was later demonstrated in the authors' laboratory that in normal serum the attachment of PRL is to the Fab region of IgG heavy chain (subclasses 1-4) (136). The bonding is catalyzed by a transglutaminase and the complex contains nearly 1 mole N^6-(γ-glutamyl)lysine crosslinks per mole of the 175 kDa complex (136). On non-reducing gels, bands positive for both IgG and PRL were found at ~192 kDa and ~219 kDa (IgG migrated as 167 kDa).

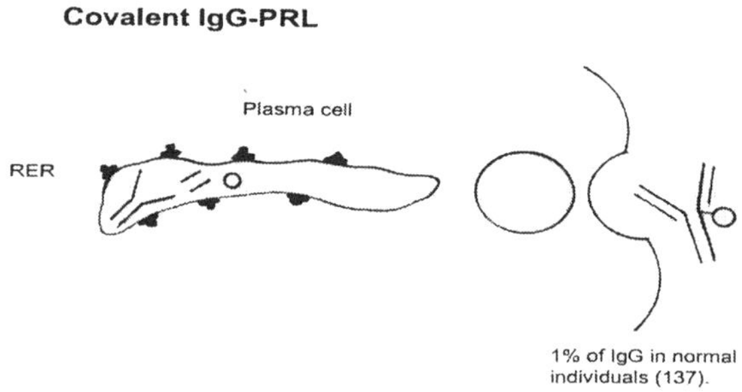

Figure 10-5. Formation of covalent IgG-PRL in a plasma cell. The rough endoplasmic reticulum (RER) synthesizes heavy chains, light chains and PRL (o). Heavy and light chains are linked by disulfide bonds. PRL is linked to the heavy chain by γ-glutamyl-lysine bond (136)

On reducing gels, doubly positive bands were seen at 98 and 75 kDa. Surprisingly, up to 1% of the total IgG (measured by immunoradiometric assay) was in the form of this complex (136). Thus, there is much more PRL in serum covalently bound to IgG than is present as free PRL. Biosynthetic labeling studies showed that the covalent complex was synthesized in some B cells (136) (Figure 10-5).

Although the PRL-IgG complex could bind to the PRL receptor on Nb2 cells, it was inactive in the Nb2 proliferation assay (136). Similarly, the complex did not cause proliferation of normal peripheral blood lymphocytes. A critical finding of this work was that the PRL-IgG complex acted as a stimulant for the proliferation of immature B lymphocytes from patients with chronic lymphocytic leukemia. This response required the involvement of both the PRL and the Fc receptor (136); neither IgG nor PRL alone resulted in a proliferative response. On the other hand, the presence of PRL inhibited the PRL-IgG proliferative effect. Consistent with these findings is that high PRL levels during lactation were shown to be inversely correlative with B-cell (CD19+) and serum IgG concentrations (137). Thus, a unique cascade of events may occur when both Fc and PRL receptors are occupied by the PRL-IgG complex. Further investigations are necessary to correlate levels of this specific PRL-IgG complex with physiological and pathological states.

Big big PRL: Non-covalent binding of PRL with IgG. The work of Heffner *et al.* (71) and Cohen *et al.* (135) clearly demonstrated that IgG is, in essence, a binding protein for PRL and that this interaction is not an antigen-antibody complex (Figure 10-6). It is understandable that investigators who observe PRL binding to IgG would assume these were auto-antibody complexes and refer to them as such but, to date, this has not been rigorously demonstrated. That said, however, it seems likely that autoantibodies to PRL will be found in some individuals. IgG-PRL complexes therefore can be covalent, non-covalent or antigen-antibody complexes (Figures 5 & 6). In the following discussion of the observed changes in multimeric forms of PRL in the circulation (primarily IgG-PRL), the exact physical protein-protein interactions are not defined.

Big big PRL: Changes in physiological and pathological states. One must be cautious in trying to quantify the amount of big big PRL since available anti-PRL antibodies recognize big big PRL to varying degrees (138). This is especially true of Elisa assays requiring PRL recognition by two distinct antibodies. With this in mind, PRL has been reported to be primarily monomeric in normal and hyperprolactinemic human sera. Significant amounts of big PRL with low monomer and no big big PRL have been noted in several male hyperprolactinemic patients with pituitary adenomas, but in these cases, there is no abnormal sexual function (139) (i.e., the big PRL is not very bioactive). These cases may be rare since approximately one fourth of the patients with hyperprolactinemia have macro-prolactinemia, defined as a sustained high PRL level in which the predominant circulating form of PRL is big big PRL (140). Although big big PRL is also often the predominant form in patients with prolactinomas (140), macroprolactinemia is especially likely in hyperprolactinemic patients with normal ovarian function, with up to 70-80% as the big big form (142). Interestingly, Guay *et al.* (142) studied six men who had erectile dysfunction. Hyperprolactinemia was considered the cause of the dysfunction until potency returned after counseling, suggesting a psychological cause. On further study, there was found to be a predominance of big big PRL (>85%), and that the big big PRL was relatively inactive. Hattori and Inagaki (143) have recently studied four hyperprolactinemic women (mean PRL of 326 $\pm$ 216 μg/L *versus* <30 μg/L in normals) with normal pituitary function, no clinical symptoms of hyperprolactinemia such as amenorrhea and galactorrhea, and no generalized autoimmune disorders. One, in fact, had a normal pregnancy without bromocriptine after hyperprolactinemia was confirmed. On gel chromatography, 84 $\pm$ 5.5% of PRL was big big PRL. By protein G affinity chromatography, it was confirmed that this was a PRL-IgG complex (~60% of the total PRL was bound). Further experimentation suggested that the IgG-PRL complex in this instance may be an antibody-antigen interaction since its clearance properties were different from a mixture of PRL and IgG (143).

INFLUENCE OF ONE POSTTRANSLATIONAL MODIFICATION ON ANOTHER

Although it is an attractive hypothesis that one posttranslational modification may influence another, there is at present no evidence to support this. None of the demonstrated sites of phosphorylation or glycosylation occur in close proximity to each other or in close proximity to known proteolytic cleavage sites. The one exception to this is phosphorylation on rat S177 which is two amino acids C terminal to a kallikrein cleavage site. Specific investigation of this issue showed no influence of phosphorylation on kallikrein cleavage (100).

ACKNOWLEDGEMENTS

Recent work in the authors' laboratory was supported by USDA grant 97-35208-4861, CCRP grant IPF0052, USAMRMC grant BC 990711 and a neuroimmunology award from the Cancer Federation. The authors would like to thank all past and present lab members whose efforts contributed to work described. A special thanks goes to Ms. Nancy Price for graphics and word processing and Min Lu for many trips to the library.

REFERENCES

1. Sinha YN. Structural variants of prolactin: occurrence and physiological significance. Endocr Rev 1995;16:354-369.
2. Jackson RC, Blobel G. Post-translational cleavage of presecretory proteins with an extract of rough microsomes from dog pancreas containing signal peptidase activity. Proc Natl Acad Sci USA 1977;74:5598-5602.
3. Ben-Jonathan N, Mershon JL, Allen DL, Steinmetz RW. Extrapituitary prolactin: distribution, regulation, functions, and clinical aspects. Endocr Rev 1996;17: 639-669.
4. Nagy E, Berczi I. Hypophysectomized rats depend on residual prolactin for survival. Endocrinology 1991;128: 2776-2784.
5. Andries M, Tilemans D, Denef C. Isolation of cleaved prolactin variants that stimulate DNA synthesis in specific cell types in rat pituitary cell aggregates in culture. Biochem J 1992;281: 393-400.

6. Gellersen B, Kempf R, Telgmann R, DiMattia GE. Nonpituitary human prolactin gene transcription is independent of Pit-1 and differentially controlled in lymphocytes and in endometrial stroma. Mol Endocrinol 1994;8: 356-373.
7. Goffin V, Norman M, Martial JA. Alanine-scanning mutagenesis of human prolactin: importance of the 58-74 region for bioactivity. Mol Endocrinol 1992;6:1381-1392.
8. Luck DN, Gout PW, Beer CT, Smith M. Bioactive recombinant methionyl bovine prolactin: structure-function studies using site-specific mutagenesis. Mol Endocrinol 1989;3:822-831.
9. Luck DN, Gout PW, Kelsay K, Atkinson T, Beer CT, Smith M. Recombinant methionyl bovine prolactin: loss of bioactivity after single amino acid deletions from putative helical regions. Mol Endocrinol 1990; 4:1011-1016.
10. Luck DN, Huyer M, Gout PW, Beer CT, Smith M. Single amino acid substitutions in recombinant bovine prolactin that markedly reduce its mitogenic activity in Nb2 cell cultures. Mol Endocrinol 1991;5: 1800-1886.
11. Kinet S, Goffin V, Mainfroid V, Martial JA. Characterization of lactogen receptor-binding site 1 of human prolactin. J Biol Chem 1996;271:14353-14360.
12. Goffin V, Struman I, Mainfroid V, Kinet S, Martial JA. Evidence for a second receptor binding site on human prolactin. J Biol Chem 1994;269:32598-32606.
13. Kato Y, Maruyama, O, Chung H-O, Tomizawa K, Kato T. Amino acids in highly conserved regions near the C-terminus of rat prolactin (PRL) play critical roles similar to those in binding of human GH to the PRL receptor. Biochem Biophys Res Commun 1996;222:547-552.
14. Davis JA, Linzer DI. Mutational analysis of a lactogenic hormone reveals a role for lactogen-specific amino acid residues in receptor binding and mitogenic activity. Mol Endocrinol 1989: 3: 1987-1995.
15. Davis JA, Linzer DI. A mutant lactogenic hormone binds, but does not activate, the prolactin receptor. Mol Endocrinol 1989;3:949-953.
16. Sidrauski C, Chapman R, Walker P. The unfolded protein response: an intracellular signaling pathway with many surprising features. Trends Cell Biol 1998;8:245-249.
17. Sommer T, Wolf D. Endoplasmic reticulum degradation: reverse protein flow of no return. FASEB 1997;11:1227-1233.
18. Suzuki T, Lennarz W. Complex, two-way traffic across the membrane of the endoplasmic reticulum. J Biol Chem 1998;273:10083-10086.
19. Ali S, Pellegrini T, Kelly PA. A prolactin-dependent immune cell line (Nb2) expresses a mutant form of prolactin receptor. J Biol Chem 1991;266:20110-10117.
20. Gertler A, Grosclaude J, Strasburger CJ, Nir S, Djiane J. Real-time kinetic measurements of the interactions between lactogenic hormones and prolactin-receptor extracellular domains from several species support the model of hormone-induced transient receptor dimerization. J Biol Chem 1996;271:24482-24491.
21. Goffin V, Shiverick KT, Kelly PA, Martial JA. Sequence-function relationships within the expanding family of prolactin, growth hormone, placental lactogen, and related proteins in mammals. Endocrine Rev 1996;17:385-410.
22. Fuh G, Colosi P, Wood WI, Wells JA. Mechanism-based design of prolactin receptor antagonists. J Biol chem 1993;268:5376-5381.
23. Kuo CB, Chen C, Xu X, Yang L, Liu Y, Coss D, Walker AM. Unmodified prolactin promotes ductal and lubolo-alveolar growth, while a molecular mimic of phosphorylated prolactin promotes lobuloalveolus formation and casein gene expression in the pregnant rat mammary gland. The Endocrine Society 82nd Annual Meeting, June 2000, Toronto, Canada.
24. Wang YF, Walker AM. Dephosphorylation of standard prolactin produces a more biologically active molecule: evidence for antagonism between nonphosphorylated and phosphorylated prolactin in the stimulation of Nb2 cell proliferation. Endocrinology 1993;133:2156-2160.

25. Wang YF, Liu JW, Mamidi M, Walker AM. Identification of the major site of rat prolactin phosphorylation as serine 177. J Biol Chem 1996;271:2462-2469.
26. Goffin V, Kinet S, Ferrag F, Binart N, Martial JA, Kelly PA. Antagonistic properties of human prolactin analogs that show paradoxical agonistic activity in the Nb2 bioassay. J Biol Chem 1996;28:16573-16579.
27. Harding PA, Wang X, Okada S, Chen WY, Wan W, Kopchick JJ. Growth hormone (GH) and a GH antagonist promote GH receptor dimerization and internalization. J Biol Chem 1996;271:6708-6712.
28. Livnah O, Stura EH, Middleton SA, Johnson DL, Jolliffe LK, Wilson IA. Crystallographic evidence for preformed dimers of erythropoietin receptor before ligand activation. Science 1999;283:987-990.
29. Luck DN, Gout PW, Sutherland ER, Fox K, Huyer M, Smith M. Analysis of disulphide bridge function in recombinant bovine prolactin using site-specific mutagenesis and renaturation under mild alkaline conditions: a crucial role for the central disulphide bridge in the mitogenic activity of the hormone. Protein Eng 1992;5:559-567.
30. Doneen BA, Bewley TA, Li CH. Studies on prolactin. Selective reduction of the disulfide bonds of the ovine hormone. Biochemistry 1979;22:4851-4869.
31. Lorenson MY, Patel TJ, Liu J-W, Walker AM. Prolactin (PRL) is a zinc-binding protein: I. Zinc interactions with monomeric PRL and divalent cation protection of intragranular PRL cysteine thiols. Endocrinology 1996;137:809-816.
32. Lorenson MY, Tuazon PT, Traugh JA, Walker AM. p21-activated protein kinase (γ-PAK) is present in rat and bovine prolactin (PRL) secretory granules and can phosphorylate PRL on the serine crucial to biological activity. The Endocrine Society 82nd Annual Meeting, June 2000, Toronto, Canada.
33. Chen T-J, Kuo CB, Tsai KF, Liu J-W, Chen D-Y, Walker AM. Development of recombinant human prolactin receptor antagonists by molecular mimicry of the phosphorylated hormone. Endocrinology 1998;139:609-616.
34. Krown KA, Wang YF, Ho TW, Kelly PA, Walker AM. Prolactin isoform 2 as an autocrine growth factor for GH3 cells. Endocrinology 1992;131:595-602.
35. Ho TW, Greenan JR, Walker AM. Mammotroph autoregulation: the differential roles of the 24K isoforms of prolactin. Endocrinology 1989;124:1507-1414.
36. Ho TW, Leong FS, Olaso CH, Walker AM. Secretion of specific nonphosphorylated and phosphorylated rat prolactin isoforms at different stages of the estrous cycle. Neuroendocrinol 1993; 58:160-165.
37. Ho TWC, Kawaminami M, Walker AM. Secretion of phosphorylated and nonphosphorylated monomer prolactin isoforms during rat pregnancy and pseudopregnancy. Endocrine J 1993;1:435-439.
38. Walker SE, Yang L, Kuo CB, Walker AM. A pilot study suggests that prolactin shortens, and a prolactin receptor antagonist lengthens, the life span of NZB/NZW mice with spontaneous lupus erythematosus. Abstract. 81st Annual Meeting of the Endocrine Society, San Diego, CA, 1999.
39. Yang L, Kuo CB, Liu Y, Coss D, Xu X, Chen C, Bagdasarian G, Oster-Granite M-L, Walker AM. Multiple abnormalities associated with increased amounts of phosphorylated prolactin during rat pregnancy. Abstract. 81st Annual Meeting of the Endocrine Society, San Diego, CA, 1999.
40. Yang L, Kuo CB, Xu X, Liu Y, Coss D, Chen C, Walker AM. A molecular mimic of phosphorylated prolactin (PRL) is luteotropic, but inhibits mammary gland development during pregnancy. Abstract. 81st Annual Meeting of the Endocrine Society, San Diego, CA, 1999.
41. Xu X, Kuo CB, Yang L, Walker AM. An autocrine prolactin (PRL) growth loop in the human prostate cancer cell line, DU145: Inhibition of cell proliferation by a prolactin receptor antagonist. Abstract. 81st Annual Meeting of the Endocrine Society, San Diego, CA, 1999.

42. Coss C, Yang L, Kuo CB, Xu X, Luben RA, Walker AM. Effects of unmodified PRL and pseudophosphorylated PRL, administered during pregnancy, on the development of pup bone. The Endocrine Society 82nd Annual Meeting, June 2000, Toronto, Canada.
43. Yang L, Kuo CB, Xu X, Coss D, Walker AM. Administration of a prolactin receptor antagonist to pregnant rats resulted in increased fetal intrathymic apoptosis and no gamma delta T cell seeding of the epidermis. The Endocrine Society 82nd Annual Meeting, June 2000, Toronto, Canada.
44. Xu X, Kreye E, Kuo CB, Yang L, Walker AM. Administration of the human PRL mutant, S179D, markedly reduces the incidence of tumors when DU145 human prostate cancer cells are grown in nude mice. The Endocrine Society 82nd Annual Meeting, June 2000, Toronto, Canada.
45. Coss D, Kuo CB, Yang L, Ingleton P, Luben R, Walker AM. Dissociation of Janus kinase 2 and signal transducer and activator of transcription 5 activation after treatment of Nb2 cells with a molecular mimic of phosphorylated prolactin. Endocrinology 1999:140:5087-5094.
46. Kim BG, Brooks CL. Isolation and characterization of phosphorylated bovine prolactin. Biochem J 1993;296:41-47.
47. Wicks JR, Brooks CL. Biological activity of phosphorylated and dephosphorylated bovine prolactin. Mol Cell Endocrinol 1995;112:223-229.
48. Brooks CL, Isaacs LA, Wicks JR. Preparative purification of phosphorylated and nonphosphorylated bovine prolactins. Mol Cell Endocrinol 1994;99:301-305.
49. Brooks CL, Saiduddin S. Phosphorylation of bovine prolactin eliminates luteotropic activity in the rat. Life Sci 1998;14:1281-1287.
50. Maciejewski PM, Peterson FC, Anderson PJ, Brooks CL. Mutation of serine 90 to glutamic acid mimics phosphorylation of bovine prolactin. J Biol Chem 1995;270:27661-27665.
51. Rhee HK, Sun Z, Kim SS, Goffin V, Martial JA, Dannies PS. Biological activity and immunological reactivity of human prolactin mutants. Endocrinology 1995;136:4990-4995.
52. Liu JW, Walker AM. Long term effect of estradiol and thyrotropin releasing hormone on the release of non-phosphorylated and phosphorylated monomer prolactin *in vitro*. Abstract 1363, 76th Annual Meeting of the Endocrine Society, Anaheim, CA,1994.
53. Maurer RA, Kim KE, Day RN, Notides AC. Regulation of prolactin gene expression by estradiol. Prog Clin Biol Res 1990;322:159-169.
54. Oetting WS, Tuazon PT, Traugh JA, Walker AM. Phosphorylation of prolactin. J Biol Chem 1986;261:1649-1652.
55. Arámburo C, Montiel JL, Proudman JA, Berghman LR, Scanes CG. Phosphorylation of prolactin and growth hormone. J Mol Endocrinol 1992;8:183-191.
56. Sowadski JM Ellis CA, Madhusudan. Detergent binding to unmyristylated protein kinase A: structural implications for the role of myristate. J Bioenerg Biomem 1996;28:7-12.
57. Wicks JR, Brooks CL. Prolactin kinase activity in bovine anterior pituitary sub-cellular fractions. Mol Cell Endocrinol 1999;147:125-132.
58. Erb RE, Chew BP, Keller HF, Malven PV. Effect of hormonal treatments prior to lactation on hormones in blood plasma, milk, and urine during early lactation. J Dairy Sci 1977;60:557-565.
59. McMurtry JP, Malven PV. Experimental alterations in prolactin levels in goat milk and blood plasma. Endocrinology 1974;95:559-564.
60. Mulloy AL, Malven PV. Relationships between concentrations of porcine prolactin in blood serum and milk of lactating sows. J Anim Sci 1979;48:876-881.
61. Erb RE, Sitarz NE, Malven PV. Blood plasma and milk prolactin, and effects of sampling techniques on composition of milk from suckled ewes. J Dairy Sci 1977;60:197-203.

62. McMurtry JP, Malven PV. Radioimmunoassay of endogenous and exogenous prolactin in milk of rats. J Endocrinol 1973;61:211-217.
63. Gala RR, Singhakowinta A, Brennan M. Studies on prolactin in human serum, urine and milk. Horm Res 1975;6:310-320.
64. Sinha YN, Vanderlaan WP. Effect on growth of prolactin deficiency induced in infant mice. Endocrinology 1982;110:1871-1876.
65. Grove DS, Bour B, Kacsoh B, Mastro AM. Effect of neonatal milk-prolactin deprivation on the ontogeny of the immune system of the rat. Endocr Regul 1991;25:111-119.
66. Shah GV, Shyr GW, Grosvenor CE, Crowley WR. Hyper prolactinemia after neonatal prolactin (PRL) deficiency in rats: evidence for altered anterior pituitary regulation of PRL secretion. Endocrinology 1988;122:1883-1889.
67. Grosvenor CE, Picciano MF, Baumrucker CR. Hormones and growth factors in milk. Endocr Rev 1992;14:710-728.
68. Ellis LA, Picciano MF. Bioactive and immunoreactive prolactin variants in human milk. Endocrinology 1995;136:2711-2720.
69. Kurtz A, Bristol LA, Tâoth BE, Lazar-Wesley E, Takâacs L, Kacsâoh B. Mammary epithelial cells of lactating rats express prolactin messenger ribonucleic acid. Biol Reprod 1993;48:1095-1103.
70. Ollivier-Bousquet M, Kann G, Durand G. Prolactin transit through mammary epithelial cells and appearance in milk. Endocr Reg 1993;27:115-124.
71. Heffner, LJ, Markoff E, Yuan RW, Lee DW. Glycosylated human prolactin: Detection by immunoblotting of biological fluids is confounded by antisera crossreactivity with immunoglobulins. Biochem Biophys Res Commun 1995;212:48-55.
72. Chao J, Chao L, Swain CC, Tsai J, Margolius HS. Tissue kallikrein in the rat brain and pituitary: regional distribution and estrogen induction in the anterior pituitary. Endocrinology 1987;120:475-482.
73. Hatala MA, Powers CA. Dynamics of estrogen induction of glandular kallikrein in the rat anterior pituitary. Biochim Biophys Acta 1987;926:258-263.
74. Oetting WS, Ho TWC, Greenan JR, Walker AM. Prolactin (PRL) processing by kallikrein: production of the 21-23.5K PRL-like molecule and inferences about PRL storage in mature secretory granules. Endocrinology 1991;129:184-192.
75. Baldocchi RA, Tan L, Nicoll CS. Processing of rat prolactin by rat tissue explants and serum *in vitro*. Endocrinology 1992;130:1653-1659.
76. Baldocchi RA, Tan L, King DS, Nicoll CS. Mass spectrometric analysis of the fragments produced by cleavage and reduction of rat prolactin: Evidence that the cleaving enzyme is cathepsin D. Endocrinology 1993;133:935-938.
77. Barrett AJ. Cathepsin D in proteinases in mammalian cells and tissues. In: AJ Barrett (ed) North-Holland Publishing 1979; pp. 209-248.
78. Khurana S, Liby K, Buckley AR, Ben-Jonathan N. Proteolysis of human prolactin: resistance to cathepsin D and formation of a nonangiostatic, C-terminal 16K fragment by thrombin. Endocrinology 1999;140:4127-4132.
79. Cole ES, Nichols EH, Lauziere K, Edmunds T, McPherson JM. Characterization of the microheterogeneity of recombinant primate prolactin: implications for posttranslational modifications of the hormone *in vivo*. Endocrinology 1991;129:2639-2646.
80. Clapp C, Martial JA, Guzman RC, Rentier-Delrue F, Weiner RI. The 16-kilodalton N-terminal fragment of human prolactin is a potent inhibitor of angiogenesis. Endocrinology 1993;1292-1299.
81. Darby NJ, Smyth DG. Endopeptidases and prohormone processing. Biosci Rep 1990;10:1-13.
82. Clapp C. Analysis of the proteolytic cleavage of prolactin by the mammary gland and liver of the rat: characterization of the cleaved and 16K forms. Endocrinology 1987;121:2055-2064.

83. Lee H, Struman I, Clapp C, Martial J, Weiner RI. Inhibition of urokinase activity by the antiangiogenic factor 16K prolactin: activation of plasminogen activator inhibitor 1 expression. Endocrinology 1998;139:3696-3703.
84. Struman I, Bentzient F, Lee H, Mainfroid V, D'Angelot G, Goffin V, Weiner RI, Martial JA. Opposing actions of intact and N-terminal fragments of the human prolactin/growth hormone family members on angiogenesis: An efficient mechanism for the regulation of angiogenesis. Proc Natl Acad Sci USA 1999;96:1246-1251.
85. Dueñas Z, Torner L, Corbacho AM, Ochoa A, Gutierrez-Ospina G, López-Barrera F, Barrios FA, Berger P, Gonzalo Martínez de la Escalera G, Clapp C. Inhibition of rat corneal angiogenesis by 16kDa prolactin and by endogenous prolactin-like molecules. Invest Ophthalmol Vis Sci 1999;40:2498-2503.
86. Khurana S, Kuns R, Ben-Jonathan N. Heparin-binding property of human prolactin: a novel aspect of prolactin biology. Endocrinology 1999;140:1026-1029.
87. Clapp C, Weiner RI. A specific, high affinity, saturable binding site for the 16-kilodalton fragment of prolactin on capillary endothelial cells. Endocrinology 1992;130:1380-1386.
88. Merkle CJ, Schuler LA, Schaeffer RC, Gribbon JM, Montgomery DW. Structural and functional effects of high prolactin levels on injured endothelial cells. Evidence for an endothelial prolactin receptor. Endocrine 2000, submitted.
89. Corbacho AM, Nava G, Eiserich JP, Noris G, Macotela Y, Struman I, Martínez de la Escalera G, Freeman BA, Clapp C. Proteolytic cleavage confers nitric oxide synthase activity upon prolactin. J Biol Chem 2000, in press.
90. Corbacho AM, Macotela Y, Nava G, Torner L, Dueñas Z, Noris G, Moralis MA, Martínez de la Escalera G, Clapp C. Human umbilical vein endothelial cells express multiple prolactin isoforms. J Endocrinol 2000, in press.
91. Schachter M. Kallikreins and kinins, an overview: some thoughts old and new. Adv Exp Med Biol 1983;156:13-27.
92. Powers CA, Nasjletti A. A kininogenase resembling glandular kallikrein in the rat pituitary pars intermedia. Endocrinology 1983;112:1194-1200.
93. Powers CA, Nasjletti A. A major sex difference in kallikrein-like activity in the rat anterior pituitary. Endocrinology 1984;114:1841-1844.
94. Powers CA, Westlin WF. Glandular kallikrein levels in the rat anterior pituitary during the rat estrous cycle and pregnancy. Biol Reprod 1987;37:306-310.
95. Hatala MA, Powers CA. Development of the sex difference in glandular kallikrein and prolactin levels in the rat anterior pituitary. Biol Reprod 1988;38:846-852.
96. Powers CA. Anterior pituitary glandular kallikrein: trypsin activation and estrogen regulation. Mol Cell Endocrinol 1986;46:163-174.
97. Pritchett DB, Roberts JL. Dopamine regulates expression of the glandular type kallikrein gene at the transcriptional level in the pituitary. Proc Natl Acad Sci USA 1987;84:5545-5549.
98. Oetting WS, Greenan JR, Walker AM. Production and secretion of the 21-23.5 kDa prolactin-like molecules. Mol Cell Endocrinol 1989;61:189-199.
99. Powers CA, Hatala MA. Prolactin proteolysis by glandular kallikrein: *in vitro* reaction requirements and cleavage sites, and detection of processed prolactin *in vivo*. Endocrinology 1990;127:1916-1927.
100. Lorenson MY, Liu J-W, Patel T, Walker AM. Intragranular prolactin phosphorylation and kallikrein cleavage are regulated by zinc and other divalent cations. Endocrine 1996;4:249-257.
101. Anthony PK, Stoltz RA, Pucci ML, Powers CA. The 22K variant of rat prolactin: evidence for identity to prolactin (1-173), storage in secretory granules, and regulated release. Endocrinology 1993;132:806-814.
102. Lewis UJ, Singh RN, Lewis LJ, Seavey BK, Sinha YN. Glycosylated ovine prolactin. Proc Natl Acad Sci USA 1984;81:385-389.

103. Butnev VY, Gotschall RR, Baker VL, Moore WT, Gout PW, Bousfield GR. Glycosylated equine prolactin and its carbohydrate moiety. J Prot Chem 1996;15:413-426.
104. Strickland TW, Pierce JG. Glycosylation of ovine prolactin during cell-free biosynthesis. Endocrinology 1985;116:1295-1298.
105. Lewis UJ, Singh RNP, Lewis LJ, Seavey BK, Sinha YN. Glycosylated ovine prolactin. Proc Natl Acad Sci 1984;USA 81: 385-389.
106. Gentile T, Llambias P, Dokmetjian J, Margni RA. Effect of pregnancy and placental factors on the quality of humoral immune response. Immunol Lett 1998;62: 151-157.
107. Brenner J, Shemesh M, Shore LS, Friedman S, Bider Z, Moalem U, Trainin Z. A possible linkage between gonadal hormones, serum and uterine levels of IgG of dairy cows. Vet Immunol Immunopath 1995;47:179-184.
108. Bollengier F, Velkeniers B, Hooghe-Peters E, Mahler A, Vanhaelst. Multiple forms of rat prolactin and growth hormone in pituitary cell subpopulations separated using a Percoll gradient system: Disulphide-bridged dimers and glycosylated variants. J Endocrinology 1989;120: 201-206.
109. Bollengier F, Mahler A, Matton A, Vanhaelst. Molecular heterogeneity and glycosylation modulation of rat pituitary prolactin isoforms synthesized and secreted *in vitro* in postnatal ontogeny, gestation, lactation and weaning. J Neuroendocrinology 1996;8: 721-730.
110. Ayson FG, de Jesus EG, Amemiya Y, Moriyama S, Hiramo T, Kawauchi H. Isolation and cDNA cloning of somatolactin in rabbitfish (Siganus guttatus). Gen Comp Endocrinology 1999;115:292-300.
111. Corcoran DH, Proudman JA. Isoforms of turkey prolactin: Evidence for differences in glycosylation and in tryptic peptide mapping. Comp Biochem Physiol 1991;99B:563-570.
112. Bedecarrats G, Guemene D, Kuhnlein U, Zadworny D. Changes in levels of immunoreactive prolactin isoforms during a reproductive cycle in turkey hens. Gen and Comp Endocrinology 1999;113:96-104.
113. Bedecarrats G, Guemene D, Morvan C, Crisostomo-Pinto S, Kuhnlein U, Zadworny D. *In vitro* release of isoforms of prolactin from pituitary glands of turkey hens at different physiological stages. Gen and Comp Endocrinology 1999;113:105-111.
114. Titani K, Kumar S, Takio K, Ericsson LH, Wade RD, Ashida K, Walsh KA, Chopek MW, Sadler JE, Fujikawa K. Amino acid sequence of human von Willebrand factor. Biochemistry 1986;25:3171-3184.
115. Vance BA, Wu W, Ribaudo RK, Segal DM, Kearse KP. Multiple dimeric forms of human CD69 result from differential addition of N-glycans to typical (Asn-X-Ser/Thr) and atypical (Asn-X-Cys) Glycosylation motifs. J Biol Chem 1997;272:23117-23122.
116. Noso T, Swanson P, Lance VA, Kawauchi H. Isolation and characterization of glycosylated and non-glycosylated prolactins from alligator and crocodile. Intl J Pept Prot Res 1992;39:250-257.
117. Harigaya T, Nakayama K, Ohkubo H, Nakanishi S, Seo H, Hoshino K. Cloning and sequence analysis of cDNA for mouse prolactin. Biochim Biophys Acta 1986;868:30-38.
118. Hirooka T, Suganuma N, Furuhashi M, Kitagawa T, Ando H, Kikkawa F, Tomoda Y, Seo H. Synthesis and release of glycosylated prolactin in transfected cells with human prolactin complementary deoxyribonucleic acid. Endocrine J 1996;43:423-428.
119. Priou A, Bruder N, Begou D, Morange I, Graziani N, Grisoll F, Brue T. Glycosylated and non-glycosylated prolactin forms are increased after opioid administration as part of surgical anaesthesia. Clin Endocrinology 1995;43:213-217
120. Brue T, Caruso E, Morange I, Hoffman T, Evrin M, Gunz G, Benkirane M, Jaquet P. Immunoradiometric analysis of circulating human glycosylated and nonglycosylated prolactin forms: spontaneous and stimulated secretions. J Clin Endocrinol Metab 1992;75:1338-1344.

121. Gambino GM, Beck-Peccoz P, Borgato S, Faglia G, Spada A, Persani L. Bioactivity and glycosylation of circulating prolactin in various physiological and pathological conditions. Pituitary 1999;2:225-231.
122. Lorenson MY. *In vitro* conditions modify immunoassayability of bovine pituitary prolactin and growth hormone: Insights into their secretory granule storage forms. Endocrinology 1985;116:1399-1407.
123. Greenan J, Lorenson MY, Conconi MV, Walker AM. Alterations in *in situ* prolactin secretory granule morphology and immunoactivity by thiols and divalent cations. Endocrinology 1990;126:512-528.
124. Sinha YA. Prolactin variants. Trends Endocrinol Metab 1992;3:100-106.
125. Smith CR, Norman MR. Prolactin and growth hormone: molecular heterogeneity and measurement in serum. An Clin Biochem 1990;27:542-550.
126. Suh HK, Frantz AG. Size heterogeneity of human prolactin in plasma and pituitary extracts. J Clin Endocrinol Metab 1974;39: 928-935.
127. Lorenson MY, Miska SP, Jacobs LS. Molecular mechanism of prolactin release from pituitary secretory granules. In: Mena F, Valverde CM eds. Prolactin Secretion: A Multidisciplinary Approach. New York, Academic Press, pp. 141-160, 1984.
128. Lorenson MY, Jacobs LS. Thiol regulation of protein, growth hormone, and prolactin release from isolated adenohypophysial secretory granules. Endocrinology 1982;110:1164-1172.
129. Lorenson MY, Robson DL, Jacobs LS. Detectability of pituitary PRL and GH by immunoassay is increased by thiols and suppressed by divalent cations. Endocrinology 1983;12:1880-1882.
130. Lorenson MY, Robson DL, Jacobs LS. Divalent cation inhibition of hormone release from isolated adenohypophysial secretory granules. J Biol Chem 1983;258:8618-8622.
131. Prada MI, Pasolli HA, Torres AI, Aoki A. Bioactivity and immunoreactivity of rat pituitary prolactin submitted to various extraction procedures. J Endocrinol Invest 1995;18:341-345.
132. Pelletier G, Labrie F. Anterior pituitary secretory granules. In: The Secretory Granule. Poisner Am, Trifaro JM (eds) Elsevier Biomedical Press, Amsterdam, pp. 173-209, 1982.
133. Miyai K, Ichihara K, Kondo K, Mori S. Asymptomatic, hyperprolactinemia and prolactinoma in the general population: mass screening by paired assays of serum prolactin. Clin Endocrinol 1986;25:549-554.
134. Walker AM, Peabody CA, Ho TWC, Warner MD. 50 kD prolactin binding protein in schizophrenics on neuroleptic medication. J Psychiatr Neurosci 1992;17:61-67.
135. Cohen H, Coeh O, Gagnon J. Serum prolactin-binding protein (PRL-BP) of human and rat are identified as IgG. C R Acad Sci Paris 1994;317:293-298.
136. Walker AM, Montgomery DW, Saraiya S, Ho TWC, Garewal HS, Wilson J, Lorand L. Prolactin-immunoglobulin G complexes from human serum act as costimulatory ligands causing proliferation of malignant B lymphocytes. Proc Natl Acad Sci USA 1995;92:3278-3282.
137. Zimmer JP, Garza C, Butte NF, Goldman AS. Maternal blood B-cell (CD19+) percentages and serum immunoglobulin concentrations correlate with breast-feeding behavior and serum prolactin concentration. Amer J Reprod Immunol 1998;40:57-62.
138. Cavaco B, Prazeres S, Santos MA, Sobrinho LG, Leite V. Hyperprolactinemia due to big big prolactin is differently detected by commercially availably immunoassays. J Endocrinol Invest 1999;22:203-208.
139. Tritos NA, Guay AT, Malarkey WB. Asymptomatic "big" hyperprolactinemia in two men with pituitary adenomas. Eur J Endocrinol 1998;138:82-85.

140. Cavaco B, Leite V, Santos MA, Arranhado E, Sobrinho LG. Some forms of (big, big) prolactin behave as a complex of monomeric prolactin with an immunoglobulin G in patients with macroprolactinemia or prolactinoma. J Clin Endocrinol Metab 1995;80:2342-2349.
141. Larrea F, Escorza A, Valero A, Hernandez L, Craviolo Mc, Diaz-Sanchez V. Heterogeneity of serum prolactin throughout the menstrual cycle and pregnancy in hyperprolactinemic women with normal ovarian function. J Clin Endocrinol Metab 1989;68:982-987.
142. Guay GT, Sabharwal P, Varma S, Malarkey WB. Delayed diagnosis of psychological erectile dysfnction because of the presence of macroprolactinemia. J Clin Endocrinol Metab 1996;81:2512-2514.
143. Hattori N, Inagaki C. Anti-prolactin (PRL) autoantibodies cause asymptomatic hyperprolactinemia: Bioassay and clearance studies of PRL-immunoglobulin G complex. J Clin Endocrinol Metab 1997;82:3107-3110.
144. Hattori N. The frequency of macroprolactinomia in pregnant women and the heterogeneity of its etiologies. J Clin Endo Metab 1996;81:586-590.
145. Goffin V, Martial JA, Summers NL. Use of a model to understand prolactin and growth hormone specificities. Protein Eng 1995;8:1215-1231.

Chapter 11

Mammary Gland Development

Christopher J. Ormandy, Nelson D. Horseman, Matthew J. Naylor, Jessica Harris, Fiona Robertson, Nadine Binart and Paul A. Kelly

Cancer Research Program, Garvan Institute of Medical Research, Darlinghurst, AUSTRALIA, Department of Molecular and Cellular Physiology, University of Cincinnati, OH and INSERM Unité 344, Faculté de Médecine Hopital Necker-Enfants Malades, Paris, FRANCE

INTRODUCTION

Knockout mice have been made which carry null mutations of genes encoding molecules essential for prolactin release, prolactin, the prolactin receptor and various members of the receptor's signalling pathway, allowing *in vivo* genetic analysis of the role of the lactogenic hormones in mammary gland development. In prolactin and prolactin receptor knockout mice mammary ductal side branching was absent, terminal end bud like structures persisted at the ductal termini well into maturity and no alveolar buds formed along the ductal tree. Transplants of recombined mammary glands formed from stromal and epithelial elements with and without the prolactin receptor showed that prolactin plays an indirect role in these processes and supplementation of progesterone levels in prolactin knockout animals restored ductal side branching but not alveolar bud formation. Up to mid pregnancy prolactin receptor heterozygous animals showed normal ductal and alveolar development, but alveolar development stalled during late pregnancy preventing successful lactation. This phenotype was also seen in knockouts of galanin, a peptide controlling pituitary prolactin release, and Stat5a a member of the receptor's signalling pathway. Transplants of recombined glands containing knockout epithelium and wild type stroma showed no lobulo-alveolar development during pregnancy, while recombinations of wild type epithelium and PRLR knockout stroma showed normal development, demonstrating that a direct action of lactogenic hormones is confined to the epithelium to promote lobuloalveolar development. The role of prolactin in mammary tumor progression was analysed by introduction of the polyoma middle T antigen under the control of the MMTV promoter. Tumors occurred later and grew more slowly in the absence of prolactin, but exhibited identical histopathology to those found in wild type animals and were ultimately no less injurious, demonstrating that prolactin acts as a tumor promoter without differentiative activity, in addition to its established role as a factor enhancing the initiation of tumors by chemical carcinogens.

BACKGROUND

The mouse mammary gland develops in four discrete stages; *in utero*, where a rudimentary ductal structure is first produced, during puberty, when ducts elongate and bifurcate to fill the mammary fat pad, during each estrus cycle, where in a strain dependent manner the density of ductal side branches and alveolar buds increase with each cycle, and during pregnancy, where the alveolar buds which formed on the ductal tree give rise to large lobulo-alveolar structures capable of milk production. Following pregnancy and each estrus the gland undergoes involution, with the loss of most of the epithelial component gained during the preceding event. In humans the gland involutes further with declining ovarian function in later life.

A number of hormonal factors controlling these developmental stages have been described. Embryonic mammary epithelium appears to develop independently of ovarian and pituitary influence, although responsive to hormonal stimuli (1). Hormonal replacement in hypophysectomised, ovariectomised and adrenalectomised mice showed that development of the mammary ducts (resembling pubertal development) was produced by a combination of estrogen and growth hormone while alveolar development (resembling pregnancy) required additional progesterone and prolactin (PRL), (2). These hormonal combinations were shown to produce similar results in serum-free *in vitro* culture of whole mammary glands, although mammary development did not achieve the extent seen in normal animals (3,4).

In rodents, lobulo-alveolar development during pregnancy is initially dependent upon increased PRL production by the pituitary, maintained by the medial preoptic area of the brain in response to cervical stimulation during copulation (5). Development becomes independent of the pituatary at mid gestation (6), due to placental lactogen production by the placenta's giant trophoblast cells (7). These lactogenic hormones may act indirectly via the modulation of endocrine organs capable of producing mammotrophic factors, such as the ovary where lactogenic hormones provide trophic support of the corpora luteum, maintaining estrogen and progesterone production (8). And in the liver PRL and placental lactogens increase the output of IGF-1 (9). PRL may also direct mammary development directly via interaction with PRLRs in the mammary gland, which are expressed by the epithelium and possibly the stroma. The mammary prolactin receptor (PRLR) signals via the Jak2 kinase, to activate the Stat5a, Map kinase and other signalling pathways. Details of these pathways are described in detail in other sections of this volume.

Knockout mice now allow the dissection of these hormonal pathways by hormone replacement and tissue recombination in endocrinologically intact animals (10). These techniques have shown that a mammary stroma estrogen receptor is essential for ductal development (11,12) and that a mammary epithelial progesterone receptor is required for alveolar development (13-15). Genetic analysis of PRL action has been undertaken by the creation of knockout mice carrying null mutations of genes which act at multiple points in the PRL pathway. For example, genes have been knocked out which encode regulators of PRL secretion (16), PRL (17) and its receptor (18), and members of its signalling pathway, such as Stat5a (19). These models have provided new insights on PRL action in the mammary gland, and form the subjects of this review.

DEVELOPMENT IN UTERO – FORMATION OF THE DUCTAL RUDIMENT

At birth, $PRLR^{-/-}$ possessed a mammary rudiment of identical ductal architecture to wild type animals (20). In males the nipples of $PRLR^{-/-}$ animals were destroyed as normal and a rudimentary ductal system, embedded in the mammary fat pad, persisted in two thirds of both PRLR$^{-/-}$ and wild type animals. Both ductal systems underwent normal isometric growth prior to puberty. These observations show that PRL plays no essential role during mammary organogenesis.

DEVELOPMENT DURING PUBERTY AND WITH EACH ESTRUS CYCLE - DUCTAL BRANCHING AND ALVEOLAR BUD FORMATION

FAILURE OF DUCTAL SIDE BRANCHING

At the onset of puberty terminal end buds formed normally in both PRL$^{-/-}$ and PRLR$^{-/-}$ females and ductal elongation and bifurcation began. Examination of the mammary gland at sexual maturity (Figure 11-1) showed that the major ducts appear at the same density in mammary glands of all genotypes, suggesting no abrogation of estrogen/growth hormone action in these glands, but that ductal side failed in PRLR$^{-/-}$ and PRL$^{-/-}$ animals (panel C and E) in comparison to wild type animals (panel A). In virgin wild type glands the degree of ductal side branching, increased with age (compare Figure 11-1 panel A and B), but not in the glands of PRLR$^{-/-}$ animals (compare Figure 11-1 panel C and D) or PRL$^{-/-}$ animals (compare Figure 11-1 panel E and F), where the ductal complexity achieved by 14 weeks remained virtually unchanged for the life of the animal. A similar effect is seen in the Stat5a knockout (Stat5a$^{-/-}$) (21) and the galanin knockout (GALN$^{-/-}$) (16). Thus a distinction needs to be drawn between the processes which cause bifurcation and those which cause side branching; epithelial cells of the side branches may be phenotypically distinct from those of the major ducts.

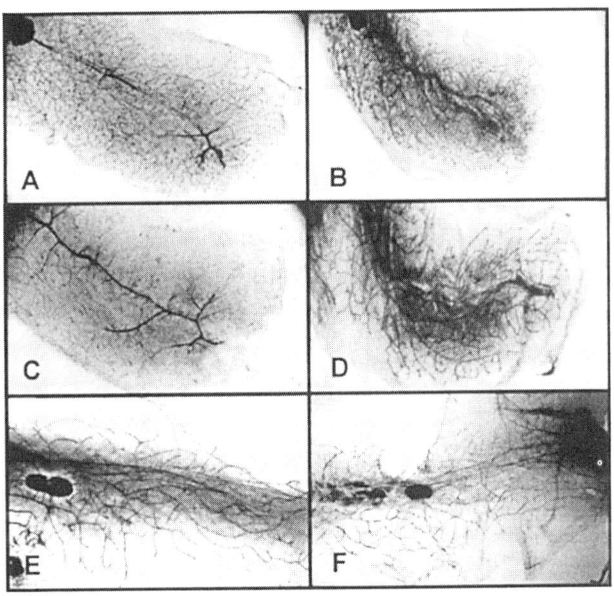

Figure 11-1. PRL causes ductal side branching. Whole mounts at 14 weeks (panels ACE) and 1 year of age (panels BDF) of the 4th inguinal gland from virgin wild type animals (A, B); PRLR$^{-/-}$ (C, D) and PRL$^{-/-}$ (E, F).

PERSISTENCE OF TERMINAL END BUD LIKE STRUCTURES IN PRL$^{-/-}$ AND PRLR$^{-/-}$ ANIMALS

By 14 weeks of age the terminal end buds of the major ducts and side branches in PRLR$^{+/+}$ animals had differentiated to alveolar buds, however, in PRL$^{-/-}$ and PRLR$^{-/-}$ animals a terminal end bud (TEB) like structure persisted at the termini of most ducts

(Figure 11-1, panel C and E). In old animals most of the major ducts had lost the TEB-like structures and the ducts now end without an apparent terminal structure (Figure 11-1, panel D and F). Microdissection of these structures allowed viewing of whole mount preparations at 200 x original magnification in addition to H&E stained sections. The persistent TEB-like structures (Figure 11-2 Panel C and D) showed no resemblance to the alveolar buds seen at the ductal termini of PRLR$^{+/+}$ animals (Figure 11-2 Panel A and B). Comparison with PRLR$^{-/-}$ TEBs observed at 8 weeks of age (which displayed normal histology-data not shown), showed that although the typical direct contact between apical cells and fat cells was maintained, the TEB-like structures were much smaller and contained fewer apical cell layers with no distinct cap cell layer. These histological observations reflect their dormant behavior and indicate that the persistent TEB-like structures were not typical terminal end buds. Alveolar buds were never seen in PRL$^{-/-}$ or PRLR$^{-/-}$ animals.

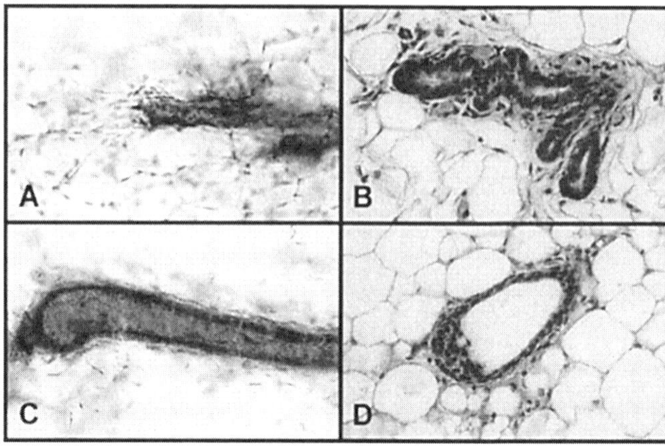

Figure 11-2. Abnormal ductal termini in PRLR$^{-/-}$ animals. Whole mounts (panels A and C) and H&E histology (panels B and D) of terminal ductal structures from 14 week old wild type (panel A and B) and PRLR$^{-/-}$ (panel C and D) virgin females

These aberrant structures are probably the result of the failure of the terminal end buds to differentiate into alveolar buds under the indirect influence of PRL, and may represent an intermediate structure in which mitogenesis and ductal elongation has been suspended, but differentiation to form an alveolar bud has not occurred. That TEB mitogenesis can cease without alveolar differentiation demonstrates that the local factors which signal the edge of the fat pad or the proximity of other ducts must operate independently of alveolar differentiation under the influence of PRL.

PRLRS OUTSIDE THE MAMMARY EPITHELIUM OR STROMA ARE NEEDED FOR NORMAL PUBERTAL DEVELOPMENT OF THE MAMMARY GLANDS

Patterning of the mammary epithelium is strongly influenced by its stroma, for which is strongly inductive. (22). This suggests that the failure of ductal side branching is exerted by the mammary stroma. The PRLR was not detected in the mammary stroma of the mouse or rat (23-26), but has been reported at low levels by immunohistochemistry in human breast cancer stroma using an anti rat PRLR monoclonal (27), and by in situ hybridisation (28). To determine whether PRL acts directly on the mammary epithelial or stromal cells, or indirectly via PRLRs outside the mammary epithelium to influence ductal side branching and alveolar bud formation, we transplanted recombined mam-

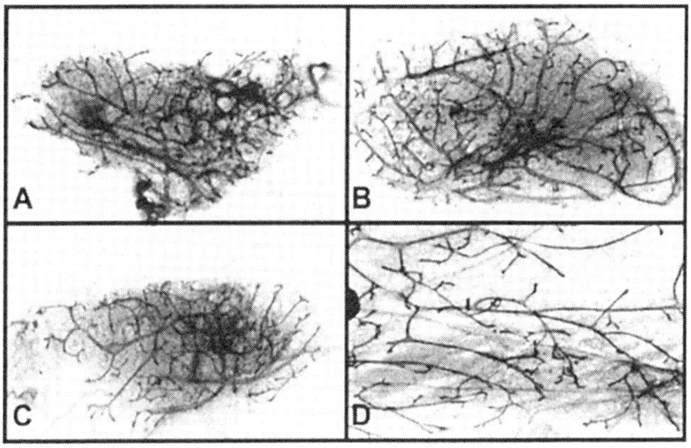

Figure 11-3. PRL directs side branching via indirect mechanisms. Mammary glands formed by tissue recombination and transplantation to virgin RAG1$^{-/-}$ hosts. A; PRLR$^{+/+}$ stroma with PRLR$^{+/+}$ epithelium. B; PRLR$^{-/-}$ stroma with PRLR$^{+/+}$ epithelium. C; PRLR$^{+/+}$ stroma with PRLR$^{-/-}$ epithelium. D; endogenous C57BL/6 RAG1$^{-/-}$ gland.

mary glands formed from epithelium and stroma of both PRLR genotypes, of 129SV/Ola background, into RAG1$^{-/-}$ recipients on the C57BL/6 genetic background. Mice homozygous for the inactivated RAG1 allele are immunocompromised and therefore able to accept allografts (29) and their mammary glands show typical C57BL/6 morphology of low side branching and little alveolar bud formation (data not shown). Recombined glands consisted of a fourth mammary fat pad from a 4 week old animal (that had been cleared of the undeveloped epithelial rudiment) into which was placed a 1 mm^3 portion of mammary gland from a mature animal. The transplant was placed on the muscle wall, under the skin between the 4th and 2nd/3rd mammary fat pad, via a small mid line incision. Ten weeks after surgery the transplanted epithelium had filled the fat pad, and the transplanted mammary glands as well as an endogenous gland were analysed by whole mount histology (Figure 11-3). Whole mount analysis showed no differences in ductal side branching between any of the various combinations of epithelium and stroma. This indicates that a mammary PRLR, whether in the epithelium or stroma, is not required for normal ductal side branching to occur. Thus the side branching phenotype seen in PRL$^{-/-}$ and PRLR$^{-/-}$ animals is due to PRL action outside the mammary gland.

All transplants reproduced the highly side branched ductal pattern of the 129 mouse, in contrast to the endogenous glands showing the typical absence of sidebranching seen in the C57BL/6 strain. When C57BL/6 stroma is used with 129 epithelium the C57Bl/6 pattern is reproduced. This indicates that factors within the stroma that are absent from C57BL/6 but present in 129 are responsible for the strain dependent differences in ductal side branching between 129 and C57BL/6. Like the C57BL/6 endogenous gland none of the transplants produced alveolar buds, suggesting that endocrine factors absent from virgin C57BL/6 but present in virgin 129 are required for alveolar bud formation.

ROLE OF PROGESTERONE IN SIDE BRANCHING

Transplanted progesterone receptor knockout mammary glands show an absence of ductal side branching (15), and progesterone levels are reduced in PRL$^{-/-}$ and PRLR$^{-/-}$ animals (30), suggesting that progesterone is the mediator and that the ovary is the site of PRL action which controls ductal side branching. Treatment of PRL$^{-/-}$ females at 6 weeks of age with progesterone pellets (25mg / 21 day release) for 19 days resulted in the formation of ductal side branches, but not alveolar buds (31) (Figure 11-4). Supplementation with PRL by pituitary implantation resulted in normal development, both side branches and alveolar buds formed (not shown). These experiments demonstrate that

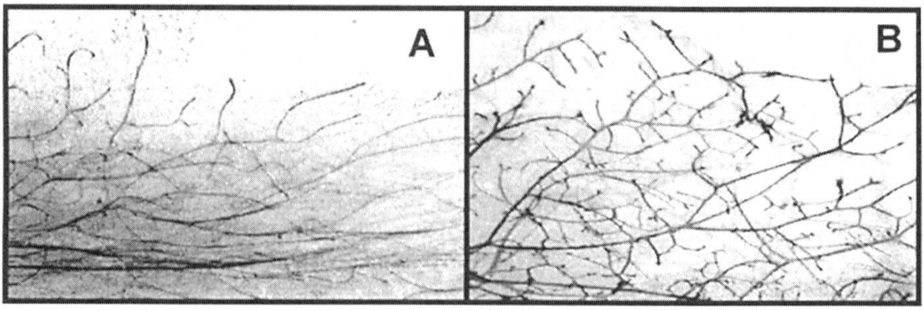

Figure 11-4. PRL indirectly influences ductal side branching via progesterone. Effect of progesterone treatment on ductal side branching in $PRL^{-/-}$ animals. Panel A, untreated. Panel B treated for 18 days with progesterone pellet (25mg).

progesterone mediates the side branching deficit in $PRL^{-/-}$ mice. Progesterone is not the endocrine factor causing alveolar bud formation, which is consistent with the ability of transplanted progesterone receptor knockout glands to form alveolar buds, but not extensive side branches during pregnancy (15). Whether progesterone acts via the stroma or epithelium to induce side branch formation remains an open question that can answered by tissue recombination using 129 mammary tissue elements.

PREGNANCY – LOBULOALVEOLAR DEVELOPMENT

FAILED FIRST LACTATION IN $PRLR^{+/-}$ ANIMALS

Heterozygous animals on the 129 SvPas/Ola or 129Sv/C57BL/6 background mated at 6 weeks of age were unable to lactate at their first pregnancy when mated at 6 weeks. Aging to 20 weeks prior to mating reduced the severity of the lactational deficit. The subsequent pregnancy resulted in lactational capacity sufficient for pup survival and normal growth rate but with a developmental lag of approximately 3 days due to slow onset of growth (20).

Analysis of mammary development showed that prior to day 15 of pregnancy, ductal elongation, branching and the number of lobules formed were similar between PRLR genotypes in response to the hormonal environment of pregnancy (data not shown). From day 15 development of $PRLR^{+/-}$ glands stalled, and greater development of the lobulo-alveoli became increasingly apparent in wild type animals. At one day post partum $PRLR^{+/-}$ animals which were incapable of lactation showed mostly lobules of stage 2 and 3, with a few lobules of stage 4 at the periphery of the fat pad, in comparison to $PRLR^{+/+}$ animals which showed a fat pad densely packed with stage 4 lobules (Figure 11-5 Panels A and D). Heterozygous animals exhibiting partial lactation showed many more stage 4 lobules than animals unable to lactate, but far fewer than seen in $PRLR^{+/+}$ animals (data not shown). Microdissection of lobules from animals 1 day post partum demonstrated that the lobules from $PRLR^{+/-}$ animals showed the formation of multiple alveoli that had failed to engorge with milk post partum (Figure 11-5 Panels B and E). H&E stained serial sections (Figure 11-5 panel C and F) confirmed that the alveoli were not engorged with milk and showed that although the diameter of the alveoli in $PRLR^{+/-}$ animals were smaller, they contained a similar number of epithelial cells as wild type animals. Thus these $PRLR^{+/-}$ lobules showed identical architecture to wild type lobules but were not expanded by milk secretion, suggesting a failure of the final stage of functional differentiation as the cause of failed lactation. Interestingly it is the alveoli at

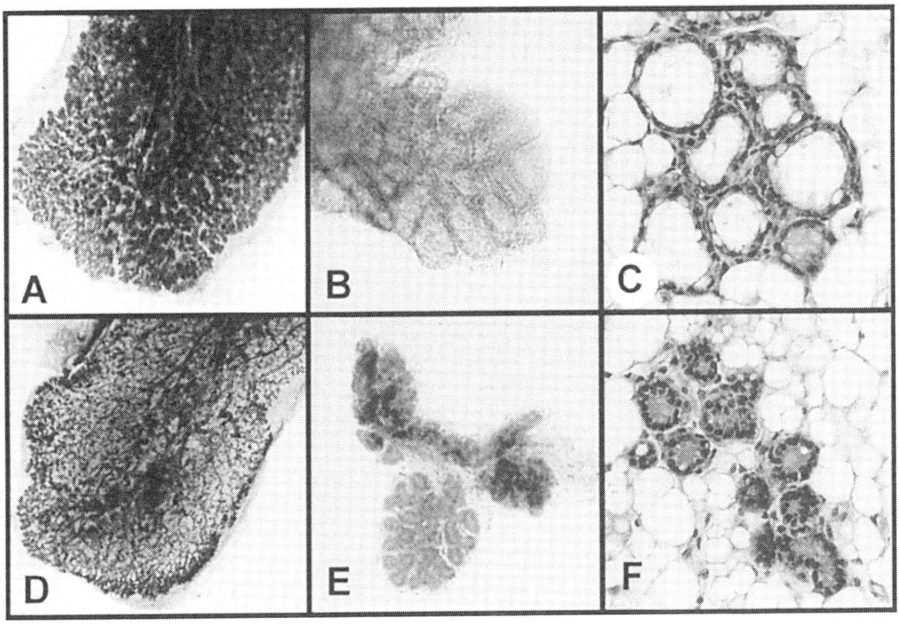

Figure 11-5. Mammary development during pregnancy in PRLR+/- animals. Wild type (panels A, B, C) and PRLR+/- (panels D, E, F) at 1 day postpartum analysed by low power whole mount (panels A and D). high power whole mount (panels B and E) and H&E histology (panels C and F).

the periphery of the fat pad, those which formed last, which were most developed. This is a consistent finding in both the 4th and 2nd/3rd glands, and suggests that the initiation of lactation proceeds in a wave from the periphery towards the nipple in PRLR+/- animals. No evidence for this was found in wild type animals, but it may occur too quickly to be detected

These defects are also seen in GALN-/- females (16), which show dramatically reduced PRL levels during pregnancy, and Stat5a-/- females (19), which lacks the signalling component essential to the differentiative activity of PRL. Whole mounts of glands from a failed lactation from PRLR+/- (Figure 11-5D), Stat5a-/- (Figure 11-6A) or GALN-/- animals (Figure 11-6B) all display formed lobuloalveoli that have failed to engorge with milk.

Figure 11-6. Stat5a-/-, *GALN-/- and* PRLR-/- *mice show similar phenotypes. Mammary development 1 day post partum of* Stat5a-/- *(panel A) and GALN-/- (panel B) glands.*

THE PRLR IS REQUIRED IN THE MAMMARY EPITHELIUM BUT NOT THE MAMMARY STROMA FOR LOBULOALVEOLAR DEVELOPMENT.

PRLR$^{-/-}$ females are infertile, preventing an analysis of the effects of pregnancy on mammary development in these animals. To circumvent this problem we made recombined mammary glands from epithelial and stromal elements of both PRLR$^{-/-}$ and PRLR$^{+/+}$ genotypes (from 129 Sv background) prior to transplant to RAG1$^{-/-}$ recipients of C57Bl/6 background. The engrafted animals were mated 8 weeks after surgery and the transplanted and endogenous glands analysed 1 day post partum. Glands formed from wild type epithelium and stroma (Figure 11-7A), showed development identical to endogenous glands (Figure 11-7D). Glands formed from PRLR$^{-/-}$ stroma and PRLR$^{+/+}$ epithelium also developed normally (Figure 11-7B), demonstrating that a stromal PRLR is not required for normal development. Glands formed from PRLR$^{-/-}$ epithelium and wild type stroma (Figure 11-7C) showed no alveolar development, but did form side branches and alveolar buds. Identical results for PRLR$^{-/-}$ epithelium were obtained using the cleared fat pad technique (32), where no beta casein expression was seen in PRLR$^{-/-}$ glands. These results demonstrate that an epithelial PRLR is not required for alveolar bud formation but is required for lobuloalveolar development, the next stage of mammary development. The formation of alveolar buds in PRLR$^{-/-}$ epithelium at pregnancy, or in PRL$^{-/-}$ glands in response to a pituitary transplant, where none form during the virgin estrus cycle, indicates that PRL induces a systemic factor other than progesterone which is permissive for this event. This represents a second indirect effect of PRL on mammary development.

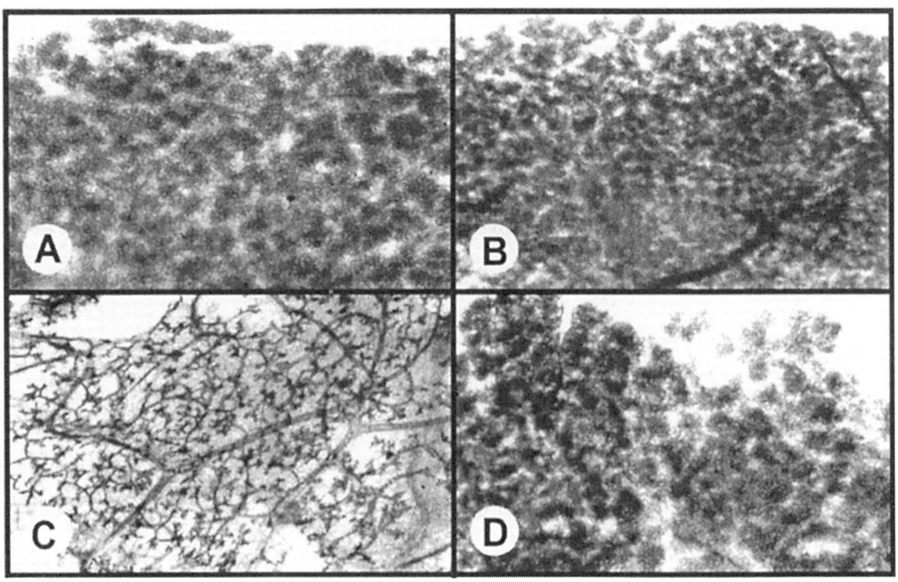

Figure 11-7. PRL acts exclusively via the epithelium to direct lobuloalveolar development. Mammary glands formed by tissue recombination and transplantation to RAG1$^{-/-}$ hosts, analysed 1 day post partum. A; PRLR$^{+/+}$ stroma with PRLR$^{+/+}$ epithelium. B; PRLR$^{-/-}$ stroma with PRLR$^{+/+}$ epithelium. C; PRLR$^{+/+}$ stroma with PRLR$^{-/-}$ epithelium. D; endogenous C57BL/6 RAG1$^{-/-}$ gland.

One of the most striking features of these experiments is the close similarity between the mammary glands of PRL$^{-/-}$, PRLR$^{-/-}$ and progesterone receptor KO mice. These models display failed ductal side branching, persistent TEB-like structures and lobuloalveolar failure during pregnancy. Both hormones are essential for development of the alveoli at pregnancy and PRL and progesterone cooperate in an indirect mechanism to promote side branching from the ductal epithelium post puberty. Investigation has shown that these hormones also interact. In mouse mammary cells (33,34) and human breast cancer cells (35) PRL and progesterone upregulate each other's receptors, providing a mechanism for their synergistic interaction during alveolar formation. There is also interaction between the progesterone receptor and Stat5a, to redirect transcriptional activity (36). The nature of the interaction changes in late pregnancy, progesterone now holds PRLR levels in check (37,38), preventing lactation before parturition, suggesting a modulation of the interaction between these hormones with changing mammary epithelial cell phenotype (39).

Progesterone treatment of PRLR$^{-/-}$ females following mating is able to fully restore the deficits of preimplantation embryo development and implantation, but cannot sustain fetal growth past mid term, with the result that only 20% of implantations survive long enough to be delivered by C section. These animals, when successfully fostered are normal, indicating a maternal or placental defect (40). In PRL$^{-/-}$ animals progesterone supplementation is able to fully restore fertility and normal numbers of pups are delivered. The mammary glands from pregnancies maintained to term by progesterone show normal development in PRL$^{-/-}$ animals (31) and failed lobuloalveolar development in PRLR$^{-/-}$ animals (40), indicating that the action of placental lactogen, which is can act in PRL$^{-/-}$ but not PRLR$^{-/-}$ animals, is able to fully compensate for PRL, both for mammary development and possibly for maintenance of the placenta. Placental lactogens (PL) cannot act in the absence of the PRLR. The simplest explanation is that the PRLR is the PL receptor, or is an essential component of the PL receptor.

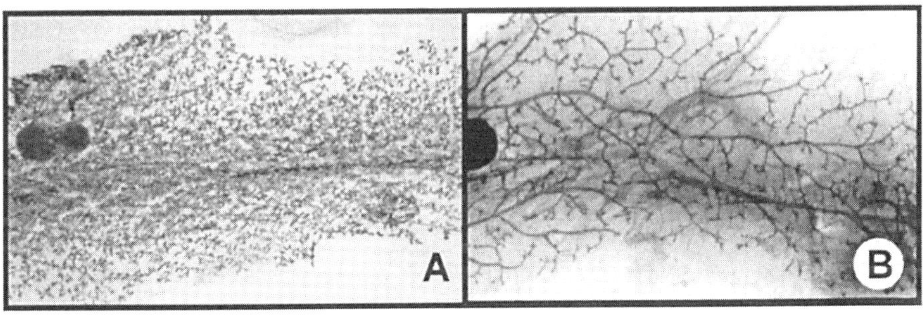

Figure 11-8. Mammary development during progesterone rescue of pregnancy. Effect of pregnancy maintained by progesterone treatment on ductal side branching and alveolar development in PRL$^{-/-}$ animals (panel A) and PRLR$^{-/-}$ animals (panel B).

INVOLUTION – LOSS OF LOBULOALVEOLI AND TISSUE REMODELING

Involution of the mammary gland is induced by weaning and proceeds in two stages; The initial phase, reversible by renewed suckling, is characterised by engorgement followed by programmed alveolar cell death, but not tissue remodeling. The second phase is irreversible and is characterised by disruption of the basement membrane and destruction of alveolar architecture. Teat sealing experiments, where one gland is forced into involution by engorgement while pups continue to remove milk from the others, have established that suckling-induced maintenance of the systemic hormonal environment of lactation can prevent the onset of the second phase of involution in the sealed gland (41). This environment includes PRL, in addition to progesterone and oxytocin, which fall following weaning concomitant with the loss of mammary phosphorylated Stat5a (42), Supplemental PRL can promote cell survival during involution (43). The failure of PRL$^{-/-}$ and PRLR$^{-/-}$ epithelium to produce lobuloalveoli prevents the use of these models in the study of PRL as a survival factor during involution, however the Stat5a knockout has been investigated. When crossed with a mouse transgenic for alpha transforming growth factor (αTGF), which shows delayed involution at weaning, the ability of the gland to involute is partially restored. The αTGF transgenic shows persistent Stat5a phosphorylation during failed involution, indicating that EGFR signalling can phosphorylate Stat5a (42). In the teat sealed glands the loss of Stat5a phosphorylation occurs despite the maintenance of serum PRL (41). Thus Stat5a phosphorylation integrates PRLR, EGFR and locally generated engorgement signals to make alveolar cell survival decisions.

TUMORIGENESIS – ROLE OF THE PRL PATHWAY IN TRANSGENIC MODELS OF BREAST CANCER

Chemical carcinogenesis in rodents has established PRL an important hormone driving the development and growth of these rodent mammary tumors (44). In humans high serum PRL carries a 2-3 fold increased relative risk of breast cancer, independently of steroid hormones (45). A previous pregnancy, and thus PRL-driven lobuloalveolar development, is protective against rodent chemical carcinogenesis and human breast cancer (46). The mechanisms behind these observations remain obscure. How is PRL involved in the initiation of tumors - does it act simply because it forces cells into cycle (as a mitogen) or prevents apoptosis (as a survival factor), to allow the replication of damaged DNA, or does PRL act to induce a state of differentiation (gene expression) more amenable to transformation. How does PRL act to promote tumor progression – again as a survival factor or mitogen allowing accumulation of the subsequent DNA damage essential to phenotypic change, or as a factor influencing the differentiation of tumor cells? Most intriguingly how is previous exposure to PRL protective for mammary cancer? Does exposure result in differentiation that renders the epithelium refractory to transformation? Knockout and transgenic mice now make possible a genetic analysis of this problem, and some initial results have recently been reported.

The PRL knockout mouse has been crossed with transgenic mice expressing the polyoma virus middle T antigen under the control of the MMTV promoter (MMTV-PyVT) (31). This model shows extensive development of mammary tumors at 3 weeks of age, prior to puberty. From the developmental studies outlined above we know that PRL is not required for mammary development prior to puberty. Thus this model tests the role of PRL in tumor progression, since tumors are initiated prior to prolactin action. On average tumors first became palpable at 51.7 days in MMTV-PyVT females and 60.7 days in PRL$^{-/-}$, MMTV-PyVT animals ($p<0.02$) and reached 1 cm at 69.5 and 81.2 days respectfully ($p<0.05$). Comparison of the growth curves indicates a 30% reduction in growth rate in the absence of prolactin. There was no difference in their histological appearance and the tumors were ultimately no less injurious in PRLR$^{-/-}$ mice. In this model the absence of PRL reduces the growth rate of a pre-initiated tumor without altering the state of tumor differentiation.

Tumor formation has also been examined in the WAP-αTGF, Stat5a$^{-/-}$ mice (42). WAP-αTGF mice form well differentiated adenomas and adenocarcinomas in mice bred repeatedly, which is required to induce αTGF expression via the WAP promoter. This is the time when PRL exerts its greatest developmental effect on the mammary gland and so this model tests the role of prolactin in tumor initiation during lobuloalveolar development. When crossed with the Stat5a$^{-/-}$ animals the incidence of hyperplasia was reduced, tumorigenesis was delayed by 1.5 months ($p<0.003$) and tumors were confined to a single gland, rather than multiple glands in the presence of Stat5a. Stat5a genotype made no difference to the survival of these animals. Although the presence of Stat5a in the EGFR signalling pathway makes the direct role of PRL in this model difficult to assess, it provides clear evidence that PRL's major signalling molecule can influence both the rate of tumor initiation and latency to detection. Similar survival rates between genotypes suggests no Stat5a effect on differentiation.

Much remains to be investigated using these and other transgenic crosses under construction, but the similarities between the PyVT and αTGF models, and what is known about the normal role of PRL, allows some initial conclusions to be drawn. The role of PRL as a survival factor during involution may account for its modest but significant effects on tumor initiation and growth. A more robust effect would be expected if PRL was a major mitogen for these cells. Although the failure of these tumors to show differences in their state of differentiation is unexpected, given the strong differentiative role attributed to PRL, we know from the PRLR$^{+/-}$, Stat5a$^{-/-}$ and GALN$^{-/-}$ animals that the final stage of mammary differentiation fails before prolactin signalling is lost completely, and so in non pregnant animals we should not expect to see a differentiative action of prolactin. We do not yet know if these differences in tumor initiation and growth are due to direct or indirect actions of prolactin.

MODEL OF PRL ACTION IN THE MAMMARY GLAND

PRL acts indirectly to control ductal side branching via an action in the ovary to control progesterone secretion. It acts directly on the mammary epithelium to drive lobuloalveolar development at pregnancy, and at this point alveolar buds form on PRLR$^{-/-}$ ducts. During involution PRL has a cell survival action which prevents the second stage of involution. In the transgenic models of mammary cancer examined to date the absence of PRL or Stat5a reduces the rate of tumor formation.

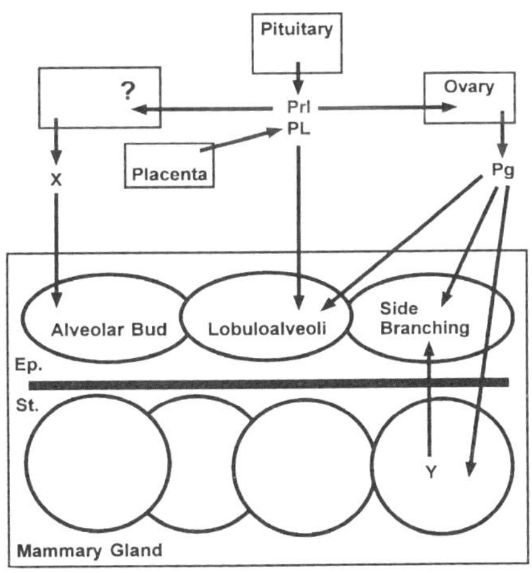

Figure 11-9. Model of PRL or placental lactogen (PL) action on the mammary epithelium (Ep.) and stroma (St.) to influence ductal patterning and terminal ductal structure of the mammary epithelium. An unknown site (?) secretes an unknown factor X to cause alveolar bud development under prolactin control. Side branching is influenced by stromal factor Y. Factors X and Y show mouse strain dependent activity. Progesterone (Pg), under prolactin control, acts via the epithelium or stroma to influence side branching.

REFERENCES

1. Ceriani RL. Fetal mammary gland differentiation *in vitro* in response to hormones II. Biochemical findings. Dev. Biol. 1970;21:530-546.
2. Nandi S. Endocrine control of mammary gland development and function in the C3H/He Crg1 mouse. J. Natl. Cancer Inst. 1958;21:1039-1063.
3. Vonderhaar B. Regulation of development of the normal mammary gland by hormones and growth factors. In: Lippman M, Dickson R, eds. Breast cancer: cellular and molecular biology. Boston USA: Kluwer Academic Publishers; 1988:252-266.
4. Ichinose R, Nandi S. Lobuloalveolar differentiation in mouse mammary tissue *in vitro*. Science. 1964;145:496-497.
5. Jakubowski M, Terkel J. Female reproductive function and sexually dimorphic prolactin secretion in rats with lesions in the medial preoptic-anterior hypothalamic continuum. Neuroendocrinology. 1986;43:696-705.
6. Collip J, Selye H, Thomson D. Further observations on the effect of hyphophysectomy on lactation. Proc. Soc. Exp. Biol. Med. 1933;30:913.
7. Thordarson G, Talamantes F. Role of the placenta in mammary gland development and function. In: Neville MC, Daniel CW, eds. The Mammary Gland. New York: Plenum Press; 1987:459-498.
8. Galosy S, Talamantes F. Luteotropic actions of placental lactogens at midpregnancy in the mouse. Endocrinology. 1995;136:3993-4003.
9. Wennbo H, Kindblom J, Isaksson OG, Tornell J. Transgenic mice over expressing the prolactin gene develop dramatic enlargement of the prostate gland. Endocrinology. 1997;138:4410-4415.
10. Hennighausen L, Robinson G. Think globally, act locally: the making of a mouse mammary gland. Genes Dev. 1998;12:449-455.
11. Korach KS, Couse JF, Curtis SW, Washburn TF, Lindzey J, Kimbro KS, Eddy EM, Migliaccio S, Snedeker SM, Lubahn DB, Schomberg DW, Smith EP. Estrogen receptor gene disruption: molecular characterization and experimental and clinical phenotypes. Rec. Prog. Horm. Res. 1996;51:159-186.
12. Cunha G, Young P, Hom Y, Cooke P, Taylor J, Lubahn D. Elucidation of a role for stromal steroid hormone receptors in mammary gland growth and development using tissue recombinations. J. Mammary Gland Biol. Neoplasia. 1997;2:393-402.
13. Lyndon JP, DeMayo FJ, Funk CR, Mani SK, Hughes AR, Montgomery CA, Shyamala G, Conneely OM, O'Malley BW. Mice lacking progesterone receptor exhibit pleiotropic reproductive abnormalities. Genes Dev. 1995;9:2266-2278.
14. Humphreys R, Lyndon J, O'Malley B, Rosen J. Mammary gland development is mediated by both stromal and epithelial progesterone receptors. Mol. Endocrinol. 1997;11:801-811.
15. Brisken C, Park S, Vass T, Lyndon JP, O'Malley B, Weinberg RA. A paracrine role for the epithelial progesterone receptor in mammary gland development. Proc. Natl. Acad. Sci. USA. 1998;95:5076-5081.
16. Wynick D, Small CJ, Bacon A, Holmes FE, Norman M, Ormandy CJ, Kilic E, Kerr NCH, Ghatei M, Talamantes F, Bloom SR, Pachnis V. Galanin regulates prolactin release and lactotroph proliferation. Proc. Natl. Acad. Sci. (USA). 1998;95:12671-12676.
17. Horseman N, Zhao W, Montecino-Rodriquez E, Tanaka M, Nakashima K, Engle S, Smith F, Markoff E, Dorshkind K. Defective mammopoiesis, but normal hematipoiesis, in mice with targeted disruption of the prolactin gene. EMBO J. 1997;16:101-110.
18. Ormandy CJ, Camus A, Barra J, Damotte D, Lucas B, Buteau H, Brousse N, Babinet C, Binart N, Kelly PA. Null mutation of the prolactin receptor gene produces multiple reproductive defects in the mouse. Genes Dev. 1997;11:167-178.

19. Liu X, Robertson G, Wagner K-U, Garret L, Wynshaw-Boris A, Hennighausen L. Stat5a is mandatory for adult mammary gland development and lactogenesis. Genes Dev. 1997;11:179-186.
20. Ormandy CJ, Binart N, Kelly PA. Mammary gland development in prolactin receptor knockout mice. J. Mamm. Gland Biol. Neoplasia. 1997;2:355-364.
21. Teglund S, McKay C, Schuetz E, van Deursen J, Stravopodis D, Wang D, Brown M, Bodner S, Grosveld G, Ihle J. Stat5a and Stat5b Proteins Have Essential and Nonessential, or Redundant, Roles in Cytokine Responses. Cell. 1998;93:841-850.
22. Sakakura T, Nishizuka Y, Dawe CJ. Mesenchyme-dependent morphogenesis and epithelium-specific cytodifferentiation in mouse mammary gland. Science. 1976;194:1439-1441.
23. Ouhtit A, Morel G, Kelly PA. Visualization of gene expression of short and long forms of prolactin receptor in rat reproductive tissues. Biol Reprod. 1993;49:528-536.
24. Ouhtit A, Morel G, Kelly PA. Visualization of gene expression of short and long forms of prolactin receptor in the rat. Endocrinology. 1993;133:135-144.
25. Meister B, Jacobsson G, Elde R. Observations on the localization of prolactin receptor mRNA in rat tissues as revealed by in situ hybridization. Acta Physiol Scand. 1992;146:533-534.
26. Shirota M, Kurohmaru M, Hayashi Y, Shirota K, Kelly PA. Detection of in situ localization of long form prolactin receptor messenger RNA in lactating rats by biotin-labelled riboprobe. Endocr J. 1995;42:69-76.
27. Reynolds C, Montone KT, Powell CM, Tomaszewski JE, Clevenger CV. Expression of prolactin and its receptor in human breast carcinoma. Endocrinology. 1997;138:5555-5560.
28. Mertani H, Garcia-Caballero T, Lambert A, Gerard F, Palayer C, Boutin J-M, Vonderhaar B, Waters M, Lobie P, Morel G. Cellular expression of growth hormone and prolactin receptors in human breast disorders. Int J Cancer. 1998;79:202-211.
29. Mombaerts P, Iacomini J, Johnson R, Herrup K, Tonegawa S, Papaioannou V. Rag-1-deficient mice have no mature B and T lymphocytes. Cell. 1992;68:869-877.
30. Clement-Lacroix P, Ormandy CJ, Lepescheux L, Ammann P, Damotte D, Goffin V, Bouchard B, Gaillard-Kelly M, Binart N, Baron R, Kelly PA. Reduced bone formation in prolactin receptor deficient mice. Endocrinology. 1999;140:96-105.
31. Vomachka A, Pratt S, Lockeffeer J, Horsemann N. Prolactin gene-disruption arrests mammary gland development and retards T-antigen-induced tumor growth. Oncogene. 2000;19:1071-1084.
32. Brisken C, Kaur S, Park S, Binart N, Sutherland RL, Weinberg RA, Kelly PA, Ormandy CJ. Prolactin controls mammary gland development via direct and indirect mechanisms. Dev Biol. 1999;210:96-106.
33. Sakai S, Bowman P, Yang J, McCormick K, Nandi S. Glucocorticoid regulation of prolactin receptors on mammary cells in culture. Endocrinology. 1979;104 (5):1447-1449.
34. Edery M, Imagawa W, Larson L, Nandi S. Regulation of estrogen and progesterone receptor levels in mouse mammary epithelial cells grown in serum-free collagen gel culture. Endocrinology. 1985;116:105-112.
35. Ormandy CJ, Hall RE, Manning DL, Robertson JFR, Blamey RG, Kelly PA, Nicholson RI, Sutherland RL. Coexpression and cross-regulation of the prolactin receptor and sex steroid hormone receptors in breast cancer. J. Clin. Endocrinol. Metab. 1997;82:3692-3699.
36. Richer JK, Lange CA, Manning NG, Owen G, Powell R, Horwitz KB. Convergence of progesterone with growth factor and cytokine signalling in breast cancer. Progesterone receptors regulate signal transducers and activators of transcription expression and activity. J Biol Chem. 1998;273:31317-31326.
37. Jahn GA, Edery M, Belair L, Kelly PA, Djiane J. Prolactin receptor gene expression in rat mammary gland and liver during pregnancy and lactation. Endocrinology. 1991;128:2976-2984.

38. Djiane J, Durand P. Prolactin-progesterone antagonism in self regulation of prolactin receptors in the mammary gland. Nature. 1977;266:641-643.
39. Vonderhaar BK, Biswas R. Prolactin effects and regulation of its receptors in human mammary tumour cells. In: Medina D, Heppner G, Anderson E, eds. Cellular and Molecular Biology of Experimental Mammary Cancer. New York: Plenum Publishing Co.; 1987:205-219.
40. Binart N, Helloco C, Ormandy C, Barra J, Clement-Lacroix P, Baran N, Kelly P. Rescue of preimplantory egg development and embryo implantation in prolactin receptor deficient mice after progesterone administration. Endocrinology. 2000;In Press.
41. Li M, Liu X, Robinson G, Bar-Peled U, Wagner K, Young W, Hennighausen L, Furth P. Mammary-derived signals activate programmed cell death during the first stage of mammary gland involution. Proc Natl Acad Sci U S A. 1997;94:3425-3430.
42. Humphreys R, Hennighausen L. Signal transducer and activator of transcription 5a influences mammary epithelial cell survival and tumorigenesis. Cell Growth Diff. 1999;10:685-694.
43. Travers M, Barber M, Tonne RE, Quarrie L, Wilde C, Flint D. The role of prolactin and growth hormone in the regulation of casein gene expression and mammary cell survival: relationships to milk synthesis and secretion. Endocrinology. 1996;137:1530-1539.
44. Welsch CW. Host factors affecting the growth of carcinogen-induced rat mammary carcinomas: a review and tribute to Charles Brenton Huggins. Cancer Res. 1985;45:3415-3443.
45. Hankinson S, Willet W, Michaud D, Manson J, Colditz G, Longcope C, Rosner B, Speizer F. Plasma prolactin levels and subsequent risk of breast cancer in postmenopausal women. J. Natl. Cancer Inst. 1999;91:629-634.
46. Medina D, Smith G. Chemical carcinogen-induced tumorigenesis in parous, involuted mouse mammary glands. J. Natl. Cancer Inst. 1999;91:967-969.

Chapter 12

Actions of Prolactin in the Prostate Gland

J. Kindblom, K. Dillner,
J. Törnell and H. Wennbo

Research Center for Endocrinology and Metabolism,
Department of Physiology, Göteborg University and
AstraZeneca Transgenic Center, Mölndal, SWEDEN

INTRODUCTION

The study of normal prostate development and prostate pathobiology has historically been mainly focused on actions of androgens. Androgens regulate proliferation and differentiation of prostatic cells. Although indispensable, androgens alone are not sufficient for the normal development and function of the prostate gland (1). The physiological importance of non-androgenic hormones and growth factors is well established. Prolactin (PRL) is one factor with the potential to act on normal prostate, alone or synergistically, with androgens (2). The possibility of PRL involvement in the progression of prostate cancer was first indicated in clinical studies more than 25 years ago. Additional clinical observations are presented along with findings in experimental work, providing possible explanations for PRL action in normal and pathophysiological conditions of the prostate gland.

PROSTATE ANATOMY

The prostate gland is an exocrine gland found only in mammals. It secretes enzymes, amines, lipids and metal ions, essential for the normal function of spermatozoa. One unique function of the prostate gland is the capacity to produce and secrete high levels of citrate. There are fundamental differences between the prostate anatomy of human and other primates and non-primates, e.g. dog, rat and mouse. In humans and other primates, the prostate is associated with the urethra contiguously below the urinary bladder and prostatic ducts emanate from the urethra radiating towards the periphery completely surrounding the urethra. The adult human gland can be divided into four zones based on morphology; the anterior fibromuscular stroma, the central zone, the peripheral zone and the transition zone. The two latter are of more clinical interest because prostatic carcinoma arises nearly exclusively from the peripheral zone and benign prostate hyperplasia (BPH) from the transition zone (3).

Primate prostate anatomy is in contrast to the anatomy of rodents, where organized and encapsulated individual lobes are located in specific positions around the urethra, but not completely circumscribing it (4). The prostate of rodents can be divided into four lobes; the dorsal, lateral, ventral and anterior (coagulating gland). The dorsolateral lobes are the parts of murine prostate that are considered the most homologous to the human prostate (5). The proportion between prostate epithelial and stromal compartments differs between species. In the adult rat the epithelial:stromal ratio is approximately 5:1. In human and other primates, stromal and epithelial cells are present in approximately equal numbers (6,7).

PROSTATE CANCER AND BENIGN PROSTATE HYPERPLASIA IN HUMAN AND IN ANIMAL MODELS

Prostate gland disorders are age-related diseases affecting a majority of elderly men in the western world. The increasing incidence of prostatic disorders with age coincides in man with changes in hormonal patterns. Testosterone and GH levels decrease while estrogen levels increase. Conflicting data exists on changes in PRL levels with increasing age (8-11). Among mammals with a prostate gland, humans and dogs are the only species known to suffer from benign prostatic hyperplasia (BPH) and prostate carcinoma (12). However, recent data indicate that chimpanzees may also develop spontaneous BPH (13). Experimental models for *in vivo* studies of these disorders in other species have been established, mainly in rodents. Examples of models include transgenic mice, such as the transgenic mouse model for prostate carcinoma, TRAMP (14,15), and the PRL transgenic mouse with prostate hyperplasia resembling human BPH (16). One important species difference in prostate is the lack of prostate specific antigen (PSA) expression in all species except for man and dog. Cell lines derived from human prostate cancers have also been important tools for *in vitro* studies of prostate malignancy. Studies in these cell lines have provided interesting data concerning PRL action and progression of prostate cancer.

PRL IN PROSTATE PATHOPHYSIOLOGY

The role of PRL in the regulation of normal development and growth of the prostate gland has been well studied (2,17). The PRL receptor (PRLR) is expressed in both human and rodent prostate (18,19). Interestingly, expression of the PRL ligand has also been demonstrated locally in both human and rodent prostate (see section 6). Several functions in the differentiated gland have been proposed, regulation of citrate metabolism being the most thoroughly investigated (2). The role of PRL action in promoting pathological conditions of the prostate has been extensively investigated. There is compelling evidence that PRL, independently of androgens, can exert a trophic effect on malignant prostate cells both *in vitro* and *in vivo*. Furthermore, clinical studies have indicated a correlation between circulating PRL levels and prognosis of prostate cancer (see below). The clinical importance of these findings remains to be proven. To interpret findings in rodent versus human studies, one needs to be aware of the important differences in influence of PRL on circulating androgen levels. In man, PRL is known to decrease circulating androgen levels through depression of gonadotropin release from the pituitary gland (20), whereas in rodents, PRL can elevate circulating androgen levels by increasing the response to luteinizing hormone in the testis (21).

PRL IN HUMAN PROSTATE CANCER AND BPH

More than 20 years ago, a beneficial effect of hypophysectomy in combination with castration compared with castration alone was observed in patients with disseminated prostate cancer (22,23). This indicated a role for one or more pituitary hormones, such as PRL, in advanced prostate cancer.

Prolactin serum levels in prostate pathology have been studied, but no clear correlation has been established in the clinical setting. Significantly higher prolactin serum levels have been reported in patients with prostate cancer (24,25) and patients with BPH (26). However, other studies report no differences in serum prolactin levels between prostatic carcinoma patients and age-matched controls (10,27). There is evidence that elevated prolactin serum levels correlate with poorer prognosis in patients with advanced prostate cancer (28,29). In addition, some clinical trials of advanced prostate cancer treatment have indicated a significant improvement in clinical response when combining conventional treatment with prolactin suppression (25,30-33).

Several antipsychotic drugs are known to produce dose-related increases in plasma prolactin levels, presumably because of a dopamine receptor blockade (34). However, no clear data concerning relative incidence of prostate cancer or BPH in treated patients is available.

Interestingly, prolactin serum levels have been reported to transiently decrease following prostatectomy or transurethral resection of the prostate, TURP, (26, 35,36). The same finding has been made in rat (37). Another study reported no difference in prolactin serum levels following prostate surgery (38), however, postoperative serum samples in this study were obtained at a later time point compared to the aforementioned. The importance of this finding is unclear, but could suggest a role of the prostate in feedback regulation of PRL, or may indicate that significant levels of PRL are expressed locally in the gland. The latter was indicated in one study were a significant increase in serum prolactin levels was noted during retropubic prostatectomy, just after removal of the prostate, as compared with levels just prior to its removal (39). This difference might be explained by high concentrations of PRL in the prostatic tissue, part of which could release into the bloodstream during the traumatic squeezing of the prostate.

EXPERIMENTAL ANIMAL DATA

Enhanced growth of rodent prostate lobes after pituitary grafting under the renal capsule (40,41), or local grafting to a specific lobe (42,43) has been reported. In rat, anterior pituitary grafting to the lateral prostate results in significant growth specifically in the lateral lobe compared to controls (42). These animals were castrated and subcutaneously implanted with a testosterone tubing at the same time. These results indicate a local direct effect of PRL on the lateral prostate lobe. In mice, implantation of a single anterior pituitary into the ventral prostate of intact mice results in a significant increase in weight and the area occupied by the ventral prostate. Prostate growth was associated with the elevation of circulating PRL. In addition, hyperplastic lesions were also noted in the grafted prostate lobes of these animals (43).

In other work, hyperprolactinemia has been reported to induce prostatic dysplasia *in vivo.* Noble rats, treated with testosterone and estradiol-17β2 for a prolonged time period, develop dorsolateral lobe dysplasia, a pre-neoplastic lesion. In these rats, the dysplasia is mediated via estradiol-induced hyperprolactinemia, as evidenced by effective inhibition of dysplastic evolution through bromocriptine co-treatment (44).

In rats, a transient increase in PRL secretion prior to puberty can result in lateral prostate inflammation. Prepubertal exposure to compounds that increase PRL secretion also increase the incidence of lateral prostate inflammation in the adult rat (45). A recent study reports that early lactational exposure to atrazine, a toxic agent that suppresses suckling-induced PRL release, leads to altered PRL regulation and subsequent prostatitis in the male offspring. Without early lactational exposure to PRL (postnatal day 1-9), tuberoinfundibular neuronal growth is impaired and as a consequence prepubertal PRL levels become elevated. This results in higher incidence and severity of lateral prostate inflammation in the offspring, evident at 120 days of age (46).

NULL MUTATED AND TRANSGENIC ANIMAL MODELS

Null mutated mice have been generated both for the PRL ligand and the PRL receptor (47,48). At present, there is no publication that describes the phenotypes of the prostate gland in any detail. PRL mutant males are fertile (47), whereas in the male PRLR null mutated mice a delay in full fertility occurs in some animals (unpubl.). Accessory sexual organs exhibit no obvious morphological or histological abnormalities in either model. Preliminary studies of the prostate gland in PRLR knockout males did not reveal any major histological alterations (unpublished). The data from the knockout mice indicate that PRL action is not of major importance for male fertility and normal anatomical development of the prostate gland. However, studies of more functional aspects of the gland need to be carried out in these animals.

Transgenic mice expressing the rat PRL gene develop a dramatic hyperplasia of the prostate gland. All prostate lobes are affected, resulting in an organ wet weight increase of more than ten times by 12 months of age. Histologically, the hyperplasia is characterized by epithelial proliferation with hypersecretory activity and a marked increase in interductal stroma (16). Castration and testosterone substitution studies performed on these animals indicate that the elevation in serum testosterone levels present in the transgenic animals is not necessary for development of the prostatic phenotype (unpublished). In rats, long-term treatment with androgens does not induce growth of the prostate gland above the maximum size obtained in normal adult controls (49). We have treated adult mice with high doses of testosterone for several months and no significant increase in prostate size could be observed (unpublished). This would support the hypothesis that increased testosterone levels in the PRL transgenic mice are not responsible for the prostate hyperplasia.

BIOLOGICAL EFFECTS OF PRL IN THE PROSTATE

PROLIFERATION

Regarding prostate carcinoma, there is evidence that PRL exerts a stimulatory effect on malignant cells *in vitro*. PRL stimulates proliferation in well-established androgen-insensitive cell lines, such as DU145 and PC3 (50). In human BPH organ cultures, PRL can significantly increase the cell proliferation rate of BPH specimens (51). Dihydrotestosterone, estrogen and progesterone were also assessed and all exerted weaker proliferating effects than PRL. Associated with the growth-promoting effect of prolactin is its effect on ornithine decarboxylase (ODC) in the lateral prostate of rat. ODC is a rate-limiting enzyme in polyamine biosynthesis, and polyamines have been classified as growth mediators due to their effects on DNA- and RNA-synthesis in somatic cells (52,53).

APOPTOSIS

In a recent *in vitro* study, PRL was shown to significantly inhibit apoptosis in androgen deprived dorsal and lateral prostate cultures, as assessed by nuclear morphology and in situ DNA fragmentation analysis. This indicates a possible physiological role for PRL as a survival factor for prostate epithelium (54). In earlier *in vivo* work, a significant delay of castration-induced regression of the lateral rat prostate was noted in pituitary graft-bearing animals (55-57). In addition, these studies also indicated that androgen receptors (AR) did not mediate PRL actions on the prostate gland, as evidenced by the failure of flutamide, an androgen receptor antagonist, to inhibit the delay in prostatic regression. These results also reveal a lobe-specific response to PRL in the androgen-deprived prostate.

CITRATE PRODUCTION

The major function of the prostate gland is to accumulate and secrete extraordinarily high levels of citrate. In addition to citrate, the normal and BPH prostate also accumulate the highest levels of zinc in the body. These specialized metabolic processes are the result of unique metabolic capabilities of the secretory epithelial cells. In prostate cancer the capability for citrate production is lost and the ability for high zinc accumulation is diminished (58).

In rats, PRL has been shown to cause an androgen-independent stimulation of citrate production exclusively in the lateral lobe of the prostate (59). The stimulatory effect of PRL on citrate production has also been confirmed in the monkey (60). *In vitro* studies in prostate epithelial cells and organ cultures, obtained from pig and rat lateral lobes, as well as in the human cancer cell lines LNCaP and PC-3, demonstrated that the increased citrate level was due to androgen-independent PRL-induced transcription of the precursor of mitochondrial aspartate aminotransferase (m-AAT)(61,62). M-AAT is the key enzyme in the metabolic pathway of prostate citrate production. This transcription has been shown to be mediated via the protein kinase C (PKC) pathway (62,63). Furthermore, *in vitro* studies of rat lateral prostate epithelial cells have shown that PRL stimulates the biosynthesis of pyruvate dehydrogenase (PDH E1α), an enzyme involved in the supply of acetyl-CoA for the citrate synthesis (64).

Recent studies of mitochondrial (m)-aconitase, the key enzyme in the citrate oxidation, have provided direct evidence that PRL, androgen-independently, regulates the m-aconitase gene in citrate-producing mammalian cells, including human (65). Earlier *in vivo* and *in vitro* studies of rat and pig prostate demonstrated that the regulation of m-aconitase is cell-specific and can be stimulatory, as in the case of rat ventral prostate, or inhibitory, as in rat lateral prostate. In non-citrate producing cells, including rat dorsal prostate, the hormone has none of these effects (66-68).

In vivo and *in vitro* studies of rats have revealed that the accumulation of zinc in the prostate also is regulated by PRL, independently of androgens. PRL increases both cellular and mitochondrial zinc levels of citrate-producing lateral prostate cells, decreases the zinc levels in ventral prostate cells and has no effect on dorsal prostate cells (69). Since a high level of zinc has an inhibitory effect on m-aconitase and thereby citrate oxidation, this provides an additional way of regulating the citrate level.

Regarding transport molecules, PRL has been reported to have a stimulatory effect on the aspartate transporter (2). Aspartate is used in citrate production.

In vitro studies of human malignant prostate cell lines, LNCaP and PC-3 cells, recently confirmed that these cells also possess a ZIP-type plasma membrane zinc uptake transporter which is regulated by PRL (70). The ZIP gene family of eukaryotic metal ion transporters includes several human and rodent genes. It is suggested that this ZIP-type zinc transporter is responsible for the ability to accumulate and transport high amounts of zinc in prostate cells.

INTERACTIONS OF PRL WITH ANDROGENS IN THE PROSTATE.

PRL has been shown to potentiate the action of androgens in the support and stimulation of prostatic growth and metabolism (71-73). This has been hypothezised to be accomplished through increasing prostate receptivity to androgens, mainly by affecting androgen receptor (AR) levels and 5α-reductase activity. Results suggest that PRL is involved in regulating AR synthesis, at least partially by direct action on the prostate gland. In immature, hypophysectomized male rats, PRL treatment can significantly increase AR mRNA levels (74). Findings in adult castrated and pituitary-grafted rats suggest that PRL promotes lateral prostatic growth via an increase in nuclear AR levels, and thus optimizes tissue response to circulating testosterone (75). Furthermore, pituitary grafting in immature rats can produce a significant increase in the weight of the seminal vesicles and the ventral and anterior prostate lobes (76). In the ventral prostate, nuclear AR content increased, whereas the cytosolic AR content decreased, suggesting increased translocation of the AR to the nucleus. In a study on human BPH patients, cytosolic and nuclear levels of AR were shown to be proportional to plasma PRL levels (77). These findings indicate plasma prolactin involvement in the regulation of AR content also in the benign human prostate.

In rats, hyperprolactinemia by pituitary grafting can lead to increased 5α-reductase activity in the testis (78), but indications of a PRL-induced increase in 5α-reductase activity in the prostate are limited to enhancement of androgen effects (79). PRL mediation of steroid uptake through alterations of the plasma membrane permeability in benign hyperplastic human prostate tissue has also been reported (80).

POSSIBLE INTERACTIONS BETWEEN PROLACTIN AND OTHER GROWTH FACTORS.

PRL can interact with several growth factors shown to influence the prostate/ prostate cells. Insulin-like growth factor 1 (IGF-I) is important for the normal development of the gland in mice (81) probably mediating the effect of growth hormone (GH). It is also shown that PSA can cleave IGF binding protein 3 (IGFBP3) and decrease the affinity of IGF-I to IGFBP3. Dissociation will render IGF-I available to the receptor to stimulate proliferation (82). Serum levels of IGF-I were increased in PRL transgenic mice. However, bovine GH transgenic mice also displayed elevated IGF-I levels, without showing any significant effects on prostate growth (16). PRL can also induce expression of IGF-I and IGF-I receptor in rat prostate (74). This could suggest a role for locally produced IGF-I in the prostate.

Fibroblast growth factor 10 (FGF10) is expressed in mesenchymal prostate tissue and may serve as a paracrine growth factor for epithelial growth (83). FGF10 is also expressed in rat prostate tumors (84). The PRL promoter contains elements responsive to FGF (85) and PRL is known to be produced locally in the prostate (see section 6).

Transforming growth factor alpha (TGF-α) and epidermal growth factor (EGF) are expressed in normal and pathological conditions in the basal epithelial cells in the prostate. The EGF receptor is expressed on the same sort of cells making an autocrine action possible (86). PRL may regulate this action since PRL can prevent EGF from inducing EGF-R association with Grb2 (87).

Transforming growth factor beta (TGF-β) is expressed in human prostate (88) and by introducing a dominant negative type II TGF-β into a rat prostate cell line the number of tumors increased when the cells were injected into nude mice (89). TGF-β can suppress PRL secretion from lactotrophic pituitary cells (90) and similar mechanism is possible for locally produced PRL in the prostate.

Taken together, these data suggest that there are multiple possible interactions between PRL and other growth factors active in the prostate, although most of them are speculative in the context of prostate physiology/pathology.

OTHER TRANSCRIPTS REGULATED BY PRL

PRL increases the steady-state levels of transcripts coding for several prostate specific proteins in rodents: the ventral-lobe specific C3 subunit of the prostatic binding protein (prostatein), the dorsolateral lobe specific secretory proteins, prostatic basic protein (probasin) and Royal Winnipeg Ballet (RWB) (91). These PRL effects are, at least in part, androgen-independent.

TABLE 12-1. Summary of PRL regulated transcripts in the prostate

Transcript	↑↓	Description	Species
ODC	↑	Enzyme involved the synthesis of polyamines. Site of regulation: lateral lobe	Rat (52)
AR	↑	Member of the steroid receptor family	Rat (74)
IGF-I	↑	Peptide hormone	Rat (74)
IGF-I receptor	↑	Member of the insulin receptor family	Rat (74)
PRLR	↑	Member of thecytokine receptor family. Site of regulation: dorsal lobe	Rat (18)
m-asconitase		Enzyme which catalyzes the oxidation of citrate	Rat, pig, human cancer cells (65)
- Ventral rat prostate:	↑		
- Lateral ratprostatre:	↓		
m-AAT	↑	Enzyme involved the synthesis of citrate. Site of regulation in rat: lateral lobe	Rat, pig, human cancer cells (61,62)
PDH E1$_\alpha$	↑	Enzyme which catalyzes the oxidation of pyruvate	Rat (64)
Aspartate transporter	↑	Molecule which transports aspartate from the blood into prostate cells	Rat (68)
C3 subunit of prostatein	↑	Prostate-specific protein primarily expressed in the ventral lobe	Rat (91)
Probasin	↑	Prostate-specific protein primarily expressed in the dorsolateral lobes	Rat (91)
RWB	↑	Prostate-specific protein primarily expressed in the dorsolateral lobes	Rat (91)

POSSIBLE AUTO/PARACRINE ACTION OF PRL IN THE PROSTATE GLAND

Ligand expression in human and rat prostate tissue has been reported (19,20). The expression of PRL ligand in rat dorsal and lateral prostate was found to be androgen dependent, both *in vivo* and *in vitro* (92). These results indicate a role of PRL as an autocrine/paracrine growth factor that is regulated by androgen and could mediate androgen effects in the rat prostate. Enhanced prostate growth observed after pituitary grafting locally to the prostate in rodents further supports the existence of local direct effects of PRL in the prostate (42,43).

In PRL transgenic mice, PRL serum levels varied between the three separate transgenic founder lines, but the degree of prostate hyperplasia were equal in the transgenic lines (16). As both the endogenous PRL gene and the rPRL transgene was expressed in the prostate, the phenotype could possibly be explained by auto/paracrine PRL action.

Recently recognized, the existence of crosstalk between the signal transduction systems of steroid hormones and peptide hormones/growth factors (93-95) provides a mechanism for locally produced growth factor influence on AR activation. In the progression of prostate cancer to an androgen-independent state, local growth factors, such as PRL, may prove instrumental in regulation of cell growth.

As reviewed earlier, the presence of prolactin receptors in both human and rodent prostate gland is well documented (18,19, 96-98). The possibility for regulation of PRLR expression by its own ligand has been reported in several tissues, including the prostate gland. In rat prostate, both testosterone and estrogen was shown to regulate the level of the long PRL receptor mRNAs in a tissue-specific manner (18). Increased PRLR expression levels have been reported in dysplastic lesions, whereas in lower grade carcinomas the receptor expression levels approximated those found in normal prostatic epithelium (98). Results from this study suggest that PRL may participate in early neoplastic transformation of the gland. Furthermore, elevated tissue levels of PRL in latent moderately or poorly differentiated type prostatic carcinoma have been reported (99).

FUTURE PROSPECTS

The potential importance of PRL for normal prostate function and disease needs to be investigated in more detail. To understand the physiological role of PRL in the prostate, a more functional analysis of the PRL and PRLR knockout mice will be very important. Furthermore, a more detailed characterization of PRL effects and the potential interaction with androgens on a molecular level in the prostate is necessary. The potential auto/paracrine effects of PRL in the prostate must also be taken in to consideration when new clinical studies are initiated.

REFERENCES

1. Wilding G. Endocrine control of prostate cancer. Review. Cancer Surv 1995;23:43-62
2. Costello LC, Franklin RB. Effect of prolactin on the prostate. Prostate 1994;24:162-6
3. McNeal JE. The prostate gland-morphology and pathobiology. Monogr Urol 1983;4:3
4. Cunha GR, *et al.* The endocrinology and developmental biology of the prostate. Endocr Rev 1987;8:338-62.
5. Price D. Comparative aspects of development and structure in the prostate. Natl Cancer Inst Monogr 1963;12:351-369.
6. DeKlerk DP, Coffey DS. Quantitative determination of prostatic epithelial and stromal hyperplasia by a new technique: biomorphometrics. Invest Urol 1978;16:240
7. Bartsch G, Rohr HP. Comparative light and electron microscopy study of human, dog and rat prostate. Urol Int 1980;35:91.
8. Davidson JM, Chen JJ, Crapo L, Gray GD, Greenleaf WJ, Catania JA Hormonal changes and sexual function in aging men. J Clin Endocrinol Metab 1983;57:71-7
9. Nankin HR, Calkins JH. Decreased bioavailable testosterone in aging normal and impotent men. J Clin Endocrinol Metab 1986;63:1418-20
10. Hammond GL, Kontturi M, Maattala P, Puukka M, Vihko R Serum FSH, LH and prolactin in normal males and patients with prostatic diseases. Clin Endocrinol 1977;7:129-35
11. Vekemans M, Robyn C. Influence of age on serum prolactin levels in women and men. Br Med J 1975;4:738-9
12. Maini A, Archer C, Wang CY, Haas GP. Comparative pathology of benign prostatic hyperplasia and prostate cancer. *In vivo* 1997;11:293-9
13. Steiner MS, Couch RC, Raghow S, Stauffer D. The chimpanzee as a model of human benign prostatic hyperplasia. J Urol 1999;162:1454-61

14. Greenberg NM, DeMayo F, Finegold MJ, Medina D, Tilley WD, Aspinall JO, Cunha GR, Donjacour AA, Matusik RJ, Rosen JM. Prostate cancer in a transgenic mouse. Proc Natl Acad Sci 1995;92:3439-43.
15. Gingrich JR, Barrios RJ, Morton RA, Boyce BF, DeMayo FJ, Finegold MJ, Angelopoulou R, Rosen JM, Greenberg NM. Metastatic prostate cancer in a transgenic mouse. Cancer Res 1996;56:4096-102
16. Wennbo H, Kindblom J, Isaksson OG, Tornell J. Transgenic mice overexpressing the prolactin gene develop dramatic enlargement of the prostate gland. Endocrinology 1997;138:4410-5
17. Reiter E, Hennuy B, Bruyninx M, Cornet A, Klug M, McNamara M, Closset J, Hennen G. Effects of pituitary hormones on the prostate. Prostate 1999;38:159-65
18. Nevalainen MT, Valve EM, Ingleton PM, Harkonen PL. Expression and hormone regulation of prolactin receptors in rat dorsal and lateral prostate. Endocrinology 1996;137:3078-88.
19. Nevalainen MT, Valve EM, Ingleton PM, Nurmi M, Martikainen PM, Harkonen PL. Prolactin and prolactin receptors are expressed and functioning in human prostate J Clin Invest 1997;99:618-27
20. Segal S, Yaffe H, Laufer N, Ben-David M. Male hyperprolactinemia:effects on fertility. Fertil Steril 1979;32:556-61
21. Dombrowicz D, Sente B, Closset J, Hennen G. Dose-dependent effects of human prolactin on the immature hypophysectomized rat testis. Endocrinology 1992;130:695-700
22. Brendler H. Adrenalectomy and hypophysectomy for prostatic cancer. Urology 1973;2:99-102
23. Tindall GT, Payne NS, Nixon DW. Transsphenoidal hypophysectomy for disseminated carcinoma of the prostate gland. Results in 53 patients. J Neurosurg 1979;50:275-82
24. Farnsworth WE, Gonder MJ. Prolactin and prostate cancer. Urology 1977;10:33-4
25. Tasar C, Akdas A, Kirkali Z. The effect of ergobromocriptine on serum testosterone and prolactin levels in patients with carcinoma of the prostate. Int Urol Nephrol 1986;18:71-4
26. Turkolmez K, Bozlu M, Sarica K, Gemalmaz H, Ozdiler E, Gogus O. Effects of transurethral prostate resection and transurethral laser prostatectomy on plasma hormone levels. Urol Int 1998;61:162-7
27. Jacobi GH, Rathgen GH, Altwein JE. Serum prolactin and tumors of the prostate: unchanged basal levels and lack of correlation to serum testosterone. J Endocrinol Invest 1980;3:15-8
28. Mee AD, Khan O, Mashiter K. High serum prolactin associated with poor prognosis in carcinoma of the prostate. Br J Urol 1984;56:698-701
29. Morales A, Nickel JC. Clinical relevance of plasma testosterone and prolactin changes in advanced cancer of prostate treated with diethylstilbestrol or estramustine phosphate. Urology 1985;26:477-81
30. Jeromin L. The serum levels of testosterone and prolactin in patients with prostatic carcinoma treated with various doses of Fostrolin and bromocriptin. Int Urol Nephrol 1982;14:51-6
31. Rana A, Habib FK, Halliday P, Ross M, Wild R, Elton RA, Chisholm GD. A case for synchronous reduction of testicular androgen, adrenal androgen and prolactin for the treatment of advanced carcinoma of the prostate. Eur J Cancer 1995;31:871-5
32. Jeromin L, Wisniewski J, Rozniecki M, Janiak Z, Borkiewicz J. Five-year clinical follow-up of prostatic cancer patients treated with fosfestrol and bromocriptine. Int Urol Nephrol 1987;19:81-5
33. Jacobi GH, Altwein JE, Hohenfellner R. Adjunct bromocriptine treatment as palliation for prostate cancer: experimental and clinical evaluation. Scand J Urol Nephrol Suppl 1980;55:107-12

34. Kleinberg DL, Davis JM, de Coster R, Van Baelen B, Brecher M. Prolactin levels and adverse events in patients treated with risperidone. J Clin Psychopharmacol 1999;19:57-61
35. Sasagawa I, Nakada T, Suzuki H, Adachi Y, Adachi M. Effect of transurethral resection of prostate on plasma hormone levels in benign prostatic hyperplasia Br J Urol 1993;72:611-4
36. Phadke MA, Vanage GR, Sheth AR. Circulating levels of inhibin, prolactin, TSH, LH, and FSH in benign prostatic hypertrophy before and after tumor resection. Prostate 1987;10: 115-22
37. Hurkadli KS, Joseph R, Kadam MS, Sheth AR. Involvement of prostate in the regulation of serum levels of FSH, LH and prolactin in male rats. Prostate 1986;9:411-6
38. Miller LR, Partin AW, Chan DW, Bruzek DJ, Dobs AS, Epstein JI, Walsh PC. Influence of radical prostatectomy on serum hormone levels. J Urol 1998;160:449-53
39. Ron M, Shapiro A, Caine M, Ben-David M, Palti Z. Serum prolactin levels in men during retropubic prostatectomy. Urology 1980;15:150-1
40. Negro-Vilar A, Saad WA, McCann SM. Evidence for a role of prolactin in prostate and seminal vesicle growth in immature male rats. Endocrinology 1977;100:729-37
41. Lee C, Hopkins D, Holland JM. Reduction in prostatic concentration of endogenous dihydrotestosterone in rats by hyperprolactinemia. Prostate 1985;6:361-7
42. Schacht MJ, Niederberger CS, Garnett JE, Sensibar JA, Lee C, Grayhack JT. A local direct effect of pituitary graft on growth of the lateral prostate in rats. Prostate 1992;20:51-8
43. Mori T, Nagasawa H. Effects of pituitary grafting and CB-154 treatment on the growth of prostates in mice. Acta Anat 1984;120:180-4.
44. Lane KE, Leav I, Ziar J, Bridges RS, Rand WM, Ho SM. Suppression of testosterone and estradiol-17beta-induced dysplasia in the dorsolateral prostate of Noble rats by bromocriptine. Carcinogenesis 1997;18:1505-10
45. Stoker TE, Robinette CL, Britt BH, Laws SC, Cooper RL. Prepubertal exposure to compounds that increase prolactin secretion in the male rat: effects on the adult prostate. Biol Reprod 1999;61:1636-43
46. Stoker TE, Robinette CL, Cooper RL. Maternal exposure to atrazine during lactation suppresses suckling-induced PRL release and results in prostatitis in the adult offspring. Toxicol Sci 1999;52:68-79
47. Horseman ND, Zhao W, Montecino-Rodriguez E, Tanaka M, Nakashima K, Engle SJ, Smith F, Markoff E, Dorshkind K. Defective mammopoiesis, but normal hematopoiesis, in mice with a targeted disruption of the prolactin gene. EMBO J 1997;16:6926-35
48. Ormandy CJ, Camus A, Barra J, Damotte D, Lucas B, Buteau H, Edery M, Brousse N, Babinet C, Binart N, Kelly PA. Null mutation of the prolactin receptor gene produces multiple reproductive defects in the mouse. Genes Dev 1997;11:167-78
49. Berry SJ, Isaacs JT. Comparative aspects of prostatic growth and androgen metabolism with ageing in the dog versus the rat. Endocrinology 1984 114:511-520.
50. Janssen T, Darro F, Petein M, Raviv G, Pasteels JL, Kiss R, Schulman CC. *In vitro* characterization of prolactin-induced effects on proliferation in the neoplastic LNCaP, DU145, and PC3 models of the human prostate. Cancer 1996;77:144-9
51. de Launoit Y, Kiss R, Jossa V, Coibion M, Paridaens RJ, De Backer E, Danguy AJ, Pasteels JL. Influences of dihydrotestosterone, testosterone, estradiol, progesterone, or prolactin on the cell kinetics of human hyperplastic prostatic tissue in organ culture. Prostate 1988;13:143-53
52. Rui H, Purvis K. Independent control of citrate production and ornithine decarboxylase by prolactin in the lateral lobe of the rat prostate. Mol Cell Endocrinol 1987;52:91-5
53. Russell DH. Ornithine decarboxylase as a biological and pharmacological tool. Pharmacology 1980;20:117-29

54. Ahonen TJ, Harkonen PL, Laine J, Rui H, Martikainen PM, Nevalainen MT. Prolactin is a survival factor for androgen-deprived rat dorsal and lateral prostate epithelium in organ culture. Endocrinology 1999;140:5412-21
55. Kolbusz WE, Lee C, Grayhack JT. Delay of castration induced regression in rat prostate by pituitary homografts. J Urol 1982;127:581-4
56. Assimos D, Smith C, Lee C, Grayhack JT. Action of prolactin in regressing prostate: independent of action mediated by androgen receptors. Prostate 1984;5:589-95
57. Smith C, Assimos D, Lee C, Grayhack JT. Metabolic action of prolactin in regressing prostate: independent of androgen action. Prostate 1985;6:49-59
58. Costello LC, Franklin RB. Novel role of zinc in the regulation of prostate citrate metabolism and its implications in prostate cancer. Prostate 1998;35:285-96
59. Grayhack JT, Lebowitz A. Effects of prolactin on citric acid of lateral lobe of prostate of Sprague-Dawley rats. Invest Urol 1967;5:87-94
60. Arunakaran J, Aruldhas MM, Govindarajulu P. Effect of prolactin and androgens on the prostate of bonnet monkeys, acaca radiata: I. Nucleic acids, phosphatases, and citric acid. Prostate 1987;10:265-73
61. Franklin RB, Costello LC. Prolactin directly stimulates citrate production and mitochondrial aspartate aminotransferase of prostate epithelial cells. Prostate 1990;17:13-8
62. Franklin RB, Zou J, Gorski E, Yang YH, Costello LC. Prolactin regulation of mitochondrial aspartate aminotransferase and protein kinase C in human prostate cancer cells. Mol Cell Endocrinol 1997;127:19-25
63. Franklin RB, Ekiko DB, Costello LC. Prolactin stimulates transcription of aspartate aminotransferase in prostate cells Mol Cell Endocrinol 1992;90:27-32
64. Costello LC, Liu Y, Franklin RB. Prolactin specifically increases pyruvate dehydrogenase E1 alpha in rat lateral prostate epithelial cells. Prostate 1995;26:189-93
65. Costello LC, Liu Y, Zou J, Franklin RB. Mitochondrial aconitase gene expression is regulated by testosterone and prolactin in prostate epithelial cells. Prostate 2000;42:196-202
66. Liu Y, Costello LC, Franklin RB. Prolactin specifically regulates citrate oxidation and m-aconitase of rat prostate epithelial cells. Metabolism 1996;45:442-9
67. Costello LC, Liu Y, Franklin RB. Testosterone and prolactin stimulation of mitochondrial aconitase in pig prostate epithelial cells. Urology 1996;48:654-9
68. Costello LC, Franklin RB. Citrate metabolism of normal and malignant prostate epithelial cells. Urology 1997;50:3-12
69. Liu Y, Franklin RB, Costello LC. Prolactin and testosterone regulation of mitochondrial zinc in prostate epithelial cells. Prostate 1997;30:26-32
70. Costello LC, Liu Y, Zou J, Franklin RB. Evidence for a zinc uptake transporter in human prostate cancer cells which is regulated by prolactin and testosterone. J Biol Chem 1999;274:17499-504
71. Thomas JA, Manandhar MS, Keenan EJ, Edwards WD, Klase PA. Effects of prolactin and dihydrotestosterone upon the rat prostate gland. Urol Int 1976;31:265-71
72. Keenan EJ, Kemp ED, Ramsey EE, Garrison LB, Pearse HD, Hodges CV. Specific binding of prolactin by the prostate gland of the rat and man. J Urol 1979;122:43-6
73. Jones R, Riding PR, Parker MG. Effects of prolactin on testosterone-induced growth and protein synthesis in rat accessory sex glands. J Endocrinol 1983;96:407-16
74. Reiter E, Bonnet P, Sente B, Dombrowicz D, de Leval J, Closset J, Hennen G. Growth hormone and prolactin stimulate androgen receptor, insulin-like growth factor-I (IGF-I) and IGF-I receptor levels in the prostate of immature rats. Mol Cell Endocrinol. 1992;88:77-87.
75. Prins GS. Prolactin influence on cytosol and nuclear androgen receptors in the ventral, dorsal, and lateral lobes of the rat prostate. Endocrinology 1987;120:1457-64
76. Coert A, Nievelstein H, Kloosterboer HJ, Loonen P, van der Vies J. Effects of hyperprolactinemia on the accessory sexual organs of the male rat. Prostate 1985;6:269-76

77. Odoma S, Chisholm GD, Nicol K, Habib FK Evidence for the association between blood prolactin and androgen receptors in BPH. J Urol 1985;133:717-20
78. Takeyama M, Nagareda T, Takatsuka D, Namiki M, Koizumi K, Aono T, Matsumoto K. Stimulatory effect of prolactin on luteinizing hormone-induced testicular 5 alpha-reductase activity in hypophysectomized adult rats. Endocrinology 1986;118:2268-75
79. Yamanaka H, Kirdani RY, Saroff J, Murphy GP, Sandberg AA. Effects of testosterone and prolactin on rat prostatic weight, 5alpha-reductase, and arginase. Am J Physiol 1975;229:1102-9
80. Farnsworth WE. Prolactin effect on the permeability of human benign hyperplastic prostate to testosterone. Prostate 1988;12:221-9
81. Ruan W. Powell-Braxton L. Kopchick JJ. Kleinberg DL. Evidence that insulin-like growth factor I and growth hormone are required for prostate gland development. Endocrinology 1999;140:1984-9,
82. Sutkowski DM. Goode RL. Baniel J. Teater C. Cohen P. McNulty AM. Hsiung HM. Becker GW. Neubauer BL. Growth regulation of prostatic stromal cells by prostate-specific antigen. Journal of the National Cancer Institute 1999;91:1663-9
83. Thomson AA. Cunha GR.. Prostatic growth and development are regulated by FGF10. Development 1999;126:3693-701
84. Lu W. Luo Y. Kan M. McKeehan WL. Fibroblast growth factor-10. A second candidate stromal to epithelial cell andromedin in prostate. Journal of Biological Chemistry 1999;274:12827-34
85. Schweppe RE. Frazer-Abel AA. Gutierrez-Hartmann A. Bradford AP. Functional components of fibroblast growth factor (FGF) signal transduction in pituitary cells. Identification of FGF response elements in the prolactin gene. Journal of Biological Chemistry 1997;272:30852-9
86. De Miguel P, Royuela, Bethencourt R, Ruiz A, Fraile B, Paniagua R. Immunohistochemical comparative analysis of transforming growth factor alpha, epidermal growth factor, and epidermal growth factor receptor in normal, hyperplastic and neoplastic human prostates. Cytokine 1999;11:722-7
87. Johnson JL. Fenton S. Sheffield LG. Prolactin inhibits epidermal growth factor-induced Ras-MAPK signaling in mammary epithelial cells. Journal of Biological Chemistry 1996;271:21574-8
88. Eklov S. Funa K. Nordgren H. Olofsson A. Kanzaki T. Miyazono K. Nilsson S. Lack of the latent transforming growth factor beta binding protein in malignant, but not benign prostatic tissue. Cancer Research 1993;53:3193-7
89. Tang B. de Castro K. Barnes HE. Parks WT. Stewart L. Bottinger EP. Danielpour D. Wakefield LM. Loss of responsiveness to transforming growth factor beta induces malignant transformation of nontumorigenic rat prostate epithelial cells. Cancer Research 1999;59:4834-42
90. Abraham EJ. Faught WJ. Frawley LS. Transforming growth factor beta1 is a paracrine inhibitor of prolactin gene expression. Endocrinology 1998;139:5174-81
91. Reiter E, Lardinois S, Klug M, Sente B, Hennuy B, Bruyninx M, Closset J, Hennen G. Androgen-independent effects of prolactin on the different lobes of the immature rat prostate. Mol Cell Endocrinol 1995;112:113-22
92. Nevalainen MT, Valve EM, Ahonen T, Yagi A, Paranko J, Harkonen PL Androgen-dependent expression of prolactin in rat prostate epithelium *in vivo* and in organ culture. FASEB J 1997;11:1297-307.
93. Griffiths K, Morton MS, Nicholson RI. Androgens, androgen receptors, antiandrogens and the treatment of prostate cancer. Eur Urol 1997;32:24-40
94. Reinikainen P, Palvimo JJ, Janne OA. Effects of mitogens on androgen receptor-mediated transactivation. Endocrinology 1996;137:4351-7

95. Stöcklin E, Wissler M, Gouilleux F, Groner B. Functional interactions between Stat5 and the glucocorticoid receptor. Nature 1996;383:726-8
96. Keenan EJ, Klase PA, Thomas JA. Effects of prolactin on DNA synthesis and growth of the accessory sex organs in male mice. Endocrinology 1981;109:170-5
97. Aragona C, Friesen HG. Specific prolactin binding sites in the prostate and testis of rats. Endocrinology 1975;97:677-84
98. Leav I, Merk FB, Lee KF, Loda M, Mandoki M, McNeal JE, Ho SM. Prolactin receptor expression in the developing human prostate and in hyperplastic, dysplastic, and neoplastic lesions. Am J Pathol 1999;154:863-70
99. Yatani R, Kusano I, Shiraishi T, Miura S, Takanari H, Liu PI. Elevated prolactin level in prostates with latent carcinoma. Ann Clin Lab Sci 1987;17:178-82

Chapter 13

Prolactin Regulation of Cell Proliferation and Apoptosis

Arthur R. Buckley

University of Cincinnati Medical Center, Cincinnati, OH

INTRODUCTION

The idea that PRL may function as a stimulator of a trophic response in a broad spectrum of tissues is one that has evolved slowly since its initial discovery nearly 70 years ago. In large part, the comparative reluctance to embrace a role for PRL as a regulator of growth in tissues beyond those involved in reproduction and nurturing of the young, is due to its well characterized and extensively studied effects in the mammary gland in which it is essential for lactation and for which it was originally named (1). Indeed, its effect to stimulate proliferation in the crop-sac, together with the formation of crop-milk in pigeons and doves as well as to induce lactation in mammals led to identification of PRL as a new anterior pituitary hormone (2, 3).

These early observations resulted in development of widely employed bioassays for PRL based on the crop-sac and mammary gland proliferative responses in pigeons and mice, respectively. However, over the last two decades, these animal-intensive bioassays have been replaced, to a large extent, by the rat Nb2 T-lymphoma cultured cell system (4). Cells of the various Nb2 lines, like the pigeon crop-sac and mammary gland, undergo proliferation in response to PRL stimulation. Due to their exquisite sensitivity to physiological concentrations of lactogens and specificity, Nb2 cells are now widely utilized as a PRL bioassay. Most importantly, these lines have also proven valuable as a model system to investigate PRL-receptor interactions, signal transduction pathways, hormonal regulation of transcriptional machinery, and expression specific gene products linked to mitogenesis. In addition to the Nb2 model, this chapter will also address other cell systems ranging from lymphocytes to pancreatic β-cells in which PRL has been reported to stimulate proliferation.

Recent evidence also indicates that PRL functions as a survival factor for some cell types, and in others, promotes apoptotic cell death. That PRL regulates tissue size by controlling death as well as proliferation is a relatively new area of investigation for the hormone. Moreover, it represents an exciting avenue for future study of molecular targets that underlie this process. Therefore, a second major focus of this chapter will be the actions of PRL on apoptosis.

REGULATION OF PROLIFERATION BY PRL

The role of PRL to stimulate growth in certain tissues, in part, lead to its discovery and continues to form the basis for its bioassay. Work provided by numerous investigators has extended and refined our knowledge of this action which is discussed below.

PIGEON CROP-SAC

The capacity of the pigeon and dove crop-sac to grow in response to administration of anterior pituitary gland extracts paved the way for identification of PRL as a mammotropic and lactogenic hormone (1,2). When injected systemically into columbid birds, PRL induced proliferation as well as differentiation in crop-sac mucosal epithelium (5). The product of the crop-sac, termed "crop-milk" , consists of aggregates of epithelial cells ultimately sloughed from the mucosal lining of the organ. Intradermal administration of PRL directly produced proliferation in mucosal epithelial cells (6). This response led to the development of a bioassay for PRL preparations due to its specificity, dose-dependency, and ease of quantitation (7,8). This bioassay was instrumental to the evaluation of several variants of PRL including glycosylated (9) and truncated forms (10).

In addition to its utility as a PRL bioassay, the pigeon crop-sac response also represented an early model system used for investigation of mechanistic aspects of PRL-stimulated mitogenesis. Administration of PRL reproducibly induced a 3-5 fold increase in gland weight (8) which was found to reflect significantly increased RNA and DNA content (11), total protein (12), and a marked augmentation of [^{3}H]-thymidine and [^{3}H]-uridine incorporation into newly synthesized DNA and RNA, respectively (13). Moreover, PRL-stimulated signaling mechanisms leading to hypertrophy and hyperplasia as well as its effects on specific expression of protein products were extensively studied in this tissue. The crop-sac PRL receptor was cloned (14) and based upon its amino acid sequence and structural motifs, it was found to represent a new member of the cytokine super-family of receptors which included receptors for growth hormone and numerous interleukins. Early studies of PRL signaling (15) indicated that cyclic nucleotides slightly affected PRL-provoked crop-sac growth. However, it was clear that these substances only represented a small component of the PRL-generated signal. More recent studies demonstrated that intercellular mechanisms observed in the mammary gland and other model systems are also operative in the crop-sac.

Horseman and colleagues employed the lactogen-stimulated crop-sac response to investigate PRL signaling. Using differential hybridization, these researchers identified a PRL-inducible mRNA that encoded a protein of 35.5 kDa (16) that was subsequently cloned and identified as a unique member of the annexin gene family (Icp 35, 17). Other studies revealed that PRL-signaling to Icp 35 transcription required an interaction between a γ-interferon activation sequence (GAS) within its 5'-promoter and a p91 transcription factor which was subsequently found to be a member of the signal transducers and activators of transcription (Stat) family of transcription factors (18, 19). Thus, signaling from the activated PRL receptor leading to initiation of growth in the crop sac appeared to be similar to that observed in mammary gland and Nb2 cells (see below).

The proliferative response of the pigeon crop-sac elicited by PRL as a model system played a pivotal role in our current understanding of the actions of the hormone. It facilitated the discovery of PRL as an anterior pituitary hormone, served as the first PRL bioassay, allowed identification and characterization of PRL variants, and provided researchers with a powerful tool to study molecular mechanisms contributing to PRL-regulated proliferation and differentiation.

MAMMARY GLAND

The mechanism by which PRL regulates mammary gland growth and differentiated function has been area of intensive investigation for decades. Since identification of the hormone as "the hitherto undefined pituitary principle which is essential for lactation in mammals" (1), considerable effort by numerous laboratories has been expended on defining the precise function(s) and underlying mechanisms of PRL's actions in this target tissue.

The mammary gland lobuloductal network continues to develop into adulthood as a function of reproductive status of the female. Its morphogenesis includes branching, budding, and extension stages, each of which, reflects complex cell-cell interactions between stromal and epithelial cells and the extracellular matrix (20). This process is governed by the coordinated actions of multiple hormones and growth factors. In early puberty, estrogen, growth hormone, and growth factors regulate rapid growth of the epithelium resulting in formation of terminal end buds (21-25). Prolactin and progesterone subsequently stimulate branching of the lobular buds later in puberty which ultimately form rudimentary lobules (26). Upon stimulation by PRL, placental lactogens, and progesterone, the lobular epithelium proliferates during pregnancy. With parturition, serum concentrations of progesterone and placental lactogens fall which is followed by lactation (27,28). In humans, lactation is maintained under the influence of PRL but also requires corticoids, thyroid and growth hormones, insulin and gonadal steroids (29). Prolactin is essential for maintenance of lactation since administration of bromocryptine, a dopamine agonist that inhibits hormone secretion by the anterior pituitary gland, halts lactogenesis (30). Moreover, studies in mice with targeted disruption of PRL itself or its receptor have provided unequivocal evidence that PRL is essential for mammary gland organogenesis and differentiated function associated with lactation (31,32). Thus, for each stage of mammary gland development, growth, and differentiation, PRL appears to be obligatory.

The actions of PRL in mammary epithelium have been studied *in vivo* (21), in gland explants (22), in primary mammary cells (33), and in cultured cell lines (34,35). Utilizing a novel model of normal mammary epithelial cells, isolated as non-milk producing organoids, Darcy *et al.* (33), showed that PRL was required for induction of growth, alveolar and multilobular branching, and accumulation of casein. Signaling mechanisms underlying regulation of gene expression in response to PRL have been extensively investigated in rodent and human mammary epithelial cells. Prolac3nogenesis in rodents (45); however, whether it participates in analogous manner in human breast cancer remains controversial. Prolactin receptors have been reported to be present in 70 - 95% of human breast biopsies (47,48). In one study, 95% of breast tumors were found to be positive for intracellular PRL (48). In addition, PRL was shown to serve as a chemoattractant for several human breast cancer cell lines *in vitro* (49) suggesting that it may promote growth and/or cell migration in these cells via an autocrine/paracrine manner. Despite these observations, a definitive contribution of PRL in the pathogenesis of human breast cancer remains to be demonstrated.

HEMATOPOIETIC CELLS

The hypothesis that prolactin (PRL), acting as a cytokine, functions as an immunomodulator was originally suggested in the early 1980s and has been intensively investigated during the twenty years that have transpired since its formulation. During this period, numerous laboratories, employing multiple experimental immune cell paradigms have identified, cloned, and characterized PRL receptors expressed on immune cells, studied intracellular signaling mechanisms activated by ligation of its receptor, investigated transcription of genes induced by PRL in lymphocytes, observed nuclear translocation of the hormone, as well as evaluated the consequences of its (or receptor) ablation (by hypophysectomy, bromocryptine administration, or targeted gene disruption) on the immune response *in vivo*. However, despite a plethora of biochemical and molecular biological evidence linking PRL to immune cell regulation, a consensus regarding the precise function of the hormone in the immune system has remained elusive.

To date, a large body of evidence has accumulated which suggests that PRL regulates proliferation of immune system cells. Its administration to hypophysectomized rats restores immunocompetence (50-52) and in combination with interleukin-2 (IL-2), phytohemagglutinin, or *Staphylococcus aurieus cowan*, it enhances mitogenesis in natural killer and T and B lymphocytes (53,54). Prolactin serves as a cofactor for IL-2 stimulated T-cell proliferation (55). Moreover, its immuno-neutralization blocked mitogenesis stimulated by IL-2 and IL-4 in cytokine-responsive cell lines (56). Rodent and human lymphocytes synthesize and secrete PRL (57-61) and express cell surface PRL receptors (62,63). In addition, PRL is directly mitogenic in the rat pre-T cell line, Nb2 lymphoma (3). Due to the robust proliferative response stimulated by PRL in this lactogen-dependent lymphoma, it has become a widely employed paradigm for investigation of molecular mechanisms coupled to PRL receptor stimulation in T lineage lymphocytes. The PRL receptor expressed in these cells has been cloned (64) and, together with its counterparts in other cells and tissues, has been found to be a truncated variant of that found in other hormonal targets. However, notwithstanding the volume of accumulated data in support of an important function for PRL in regulation of the *in vivo* immune response, recent work utilizing genetic models of PRL and PRL receptor deficiency have raised questions about whether PRL is obligatory for immune function.

Recently, two laboratories reported results from studies in which the ontogeny and functional activity of the immune system was investigated using mouse models of targeted disruption of the PRL gene (31) and its receptor (65). Knockout of either the hormone or its receptor produced marked deficits in development of the mammary gland and in reproductive function. However, the animals exhibited normal primary and secondary lymphopoiesis as well as innate immunity. In each case, thymic and splenic cellularity, as well as subset distributions of lymphocytes were not altered, nor was the capacity to produce antibodies, or proliferate in response to mitogenic stimuli *in vitro*. Furthermore, elimination of PRL or its receptor did not appear to suppress the function of the immune system in response to several well characterized antigenic challenges. Based upon these observations, both groups concluded that PRL most likely is not obligatory for the normal development and functioning of the immune response *in vivo* (31,65,66).

Clearly, the results obtained in PRL and PRL receptor knockout models demonstrated normally functioning immune cells *in vitro* and *in vivo* as evaluated by well-established laboratory methods. Nevertheless, they do not exclude the possibility that aberrations in immune function may have occurred that were not detected by the assays chosen nor the concept of functional redundancy by alternative cytokines, hormones, or growth factors which might subserve the role of PRL. For example, growth hormone, which shares a high degree of sequence homology with PRL and binds to cytokine-family receptors, also has been reported to stimulate T-lymphocyte proliferation (67). It is possible that this, or an alternative, hormone may support the function of PRL. Moreover, these studies did not address whether PRL plays a functional role in the immune response under stressful conditions (66). Since PRL and adrenal corticoids are both components of a physiological stress response, the latter is particularly important since recent evidence from several laboratories suggests that PRL may regulate lymphocyte population density by suppressing cell death mechanisms in the face of elevated glucocorticoids (see below).

OTHER CELLS AND TISSUES

In addition to regulating proliferation in the crop-sac, mammary gland, and immune cells, PRL has been reported to provoke cell multiplication in pancreatic islets, cells resident within the CNS, and a subset of cells present within skin. Of these diverse systems, a functional role for PRL as a modulator of growth and functional differentiation of pancreatic β-cells is supported to the greatest extent by the available experimental evidence.

The dramatically altered hormonal milieu that occurs during pregnancy has been long recognized as an important contributor to adaptive changes that occur in the pancreas in response to a substantially increased demand for insulin (68). In the late 1980s (69), it was demonstrated that neonatal and adult rat islets exposed to PRL *in vitro* produced more insulin and that the former cells exhibited increased proliferation. Growth hormone appeared to have similar effects albeit with lower efficacy (70). In other studies, the effects of homologous lactogenic hormones were evaluated in neonatal rat and mouse, and adult rat and human islets in comparison to their respective growth hormones (71). The results revealed that placental lactogens and PRL stimulated insulin secretion and islet cell proliferation to the greatest extent. When primary pancreatic β-cells were incubated with PRL, slight proliferation was observed within 24 hrs; however, the maximal effect was delayed until 6-10 days. These results were interpreted to indicate a direct effect of the hormone on induction of cell cycle progression in a significant population of previously quiescent cells (72). Results from experiments conducted in the rat insulinoma cell line, INS-1, corroborated the effects of lactogens observed in the primary β-cell cultures and extended the functional consequences of these hormones to positive regulation of glut-2 mRNA expression (73). Thus, results from multiple *in vitro* studies demonstrate that PRL and/or placental lactogens, both of which are elevated during pregnancy, are key components of the adaptive response of the pancreas to this physiological condition.

A role for PRL in regulation of β-cell function and proliferation was also supported by observations made in experiments conducted *in vivo*. Results obtained from analysis of pregnant rats indicated that increased pancreatic function correlated with augmented levels of the PRL receptor (74). In subsequent studies, increased expression of the receptor was found to be the result of hormone-signaling to activation of Stat5 (75), which has also been linked to the effects of PRL on insulin secretion and induction of β-cell proliferation (76). In another study, PRL administration to mice significantly reduced hyperglycemia and the early inflammatory response provoked by streptozotocin administration presumably reflecting either an immunologic mechanism of the hormone or its capacity to induce β-proliferation (77). Finally, an effect of PRL on function of the endocrine pancreas that has potential clinical utility was suggested by the observation that co-transplantation of islets in mice accompanied with pituitary grafts, which release large amounts of PRL into serum, significantly increased insulin graft viability, their performance, and vascularization (78). Verification of these positive effects of PRL on transplanted islets may lead to a previously unsuspected clinical role for it in the therapy of diabetes.

Cells within the CNS and skin represent two additional targets in which PRL has been reported to stimulate proliferation. Treatment of cultured rat astrocytes with PRL triggered mitogenesis coupled to activation of protein kinase C (79,80). Results obtained in a CNS wounding model of astrogliosis *in vivo* also provided evidence of astrocyte activation by PRL (80). Similar to astrocytes, treatment of keratinocytes obtained from newborns with PRL evoked proliferation (81). However, a recent report indicated that while expression of the PRL receptor was elevated with induction of differentiation of these cells, PRL was unable to induce proliferation or activate Stat5 (82). Whether PRL regulates physiological proliferation in these cells remains, at best, controversial.

REGULATION OF APOPTOSIS BY PRL

The "flip-side" of proliferative cell renewal is cell loss by apoptosis. The latter process is one in which PRL regulation recently has been implicated in several target tissues. Prolactin has been reported to suppress apoptosis in hematopoietic cells, particularly in Nb2-T cells, in mammary and prostate glands, and to induce cell death in the corpus luteum. Interestingly, PRL has been long known to inhibit thyroid hormone-induced amphibian metamorphosis, a physiological response that reflects extensive cell death (83,84). Similarly, the immunosuppressive effects of glucocorticoids in mammals have been recognized for decades. That an agonist/antagonist relationship existed between glucocorticoids and PRL, analogous to it and thyroid hormones in amphibians, was first reported in the late 1980s (85). However, it was not appreciated until the last decade that PRL functions as a suppressor of hormone-activated apoptosis in each of these seemingly disparate systems (86,87). The remainder of this chapter will focus on this newly described action of PRL.

NB2 LYMPHOMA CELLS

When studies of glucocorticoid immunosuppression were extended to PRL-dependent Nb2 cells it was observed that addition of dexamethasone (DEX), a potent glucocorticoid, suppressed proliferation in the presence of serum or PRL without altering cell viability (85,87). However, when the cells were cultured with DEX, in the absence of PRL, concentration-dependent cytolysis occurred characterized by fragmentation of genomic DNA (87). Since the appearance of a ladder-pattern of fragmented DNA as a consequence of endonuclease digestion is widely viewed as hallmark of apoptosis, this observation suggested that PRL functioned in this system as suppressor of the cell death program.

In addition, to PRL-dependent Nb2 cell lines, several autonomous sublines were developed by lactogen starvation and cloning of surviving cells (88). While retaining PRL sensitivity, the subline designated Nb2-SFJCD1, proliferated in the absence of hormone and resisted apoptosis in response to glucocorticoids (89). Thus, PRL-dependent and autonomous Nb2 cell lines represented a useful system to investigate mechanisms which contributed to activation of apoptosis by adrenal steroids or other stimuli, suppression by PRL, and molecular events underlying apoptosis-resistance. In this context, it was demonstrated that resistance to DEX-activated apoptosis in Nb2-SFJCD1 cells could be reversed by exposure of the cells to butyrate (90), a fatty acid differentiating agent. Butyrate transiently reverted PRL-independence for proliferation to a hormone-dependent phenotype (88,91). Furthermore, it also reversed resistance to apoptosis induced by DEX and hormonal deprivation (90). Using this experimental system, we and others have investigated PRL-stimulated signaling mechanisms and candidate genes that may mediate suppression of apoptosis.

The protein product of the founding member of the *bcl-2* gene family of apoptosis regulators is a 26-kDa membrane associated protein that can interact with itself, or dimerize with other homologous family members (92-95). Originally identified as a protooncogene expressed in lymphoid neoplasia, Bcl-2 represents a well-characterized suppressor of apoptosis (96,97). The accumulated evidence to date indicates that the relative levels of Bcl-2 and a second family member, Bax, are key determinants of whether cells succumb to apoptotic stimuli or survive. Thus, when Bax is expressed at elevated levels compared to Bcl-2, apoptosis is favored; if Bcl-2 predominates, cells are afforded a survival advantage (92).

Three *bcl-2* transcripts were found to be responsive to PRL in Nb2 cells (98). Hormone addition increased each in a time- and concentration-dependent manner with their maximal accumulation within 6 hrs. Prolactin-induced *bcl-2* expression appeared to be a direct effect of the hormone. A 1.0 kb *bax* transcript and its protein product were also detected in Nb2 cells (98). Prolactin stimulated its expression with maximal levels observed after 8-12 hrs. In contrast, the Bax protein, although apparent in quiescent cells, decreased to undetectable levels 4-6 hrs following PRL treatment.

In quiescent Nb2 cells, a condition that favors a modest induction of apoptosis, Bcl-2 was depressed while Bax was expressed at higher levels. In contrast, PRL treatment, rapidly augmented the Bcl-2:Bax ratio. As indicated, treatment of Nb2 cells with DEX caused apoptosis which was abrogated by PRL treatment possibly through elevated *bcl-2*. These observations suggested that a component of the mechanism by which PRL inhibited apoptosis likely reflected activation of the *bcl-2* gene and attenuation of Bax expression.

The expression of other Bcl-2 family members including *mcl-1*, *bad*, and *bcl-x_L* were also evaluated in the Nb2 paradigm. Of these, only *bcl-x_L* was found to be responsive to PRL. Similar to the anti-apoptotic actions of *bcl-2*, *bcl-x_L* is a suppressor of this process that interacts with other family members. It has been reported to suppress redistribution of cytochrome c, a pro-apoptotic event (99-101) and may interfere with the activation of certain caspases (102). Treatment with PRL significantly increased *bcl-x_L* and its protein product again, within 6 hrs (Kochendoerffer and Buckley, unpublished observations). The stimulatory effect of PRL on *bcl-x_L* was not altered by DEX. Therefore, at least two protein members of the *bcl-2* family, eg. Bcl-2 and Bcl-x_L, may participate in PRL-mediated apoptosis suppression under the conditions of lactogen deprivation or following exposure to glucocorticoids.

The protooncogene, *pim-1*, encodes a conserved cytosolic serine-threonine kinase that is expressed in myeloid and lymphoid tissues and is thought to be required for lymphocyte activation (103). Originally identified in malignant lymphomas, *pim-1* has been linked to proliferation (104) as well as to suppression of apoptosis in hematopoietic cells (105). The addition of PRL to quiescent Nb2 cultures stimulated *pim-1* expression to levels nearly 60-fold above that found in control cultures within 4 hrs. Results from [^{3}H]-thymidine pulse-labeling experiments revealed that the PRL-stimulated mRNA peak corresponded to early-mid G1 cell cycle progression. Results from other experiments revealed that PRL-induced *pim-1* expression was concentration-dependent and a direct effect of the hormone. These results suggested that, similar to interferon refulatory factor-1 and *c-myc*, (106,107), *pim-1* represented a PRL-regulated immediate-early gene.

Pim-1 expression was also studied in Nb2 cells exposed to DEX (107). Here, PRL also markedly increased its expression. In addition, significantly elevated levels of the Pim-1 protein accumulated in parallel to the observed alterations in its mRNA in cells treated with DEX in combination with PRL. Expression of *pim-1* was also investigated in the PRL-independent and DEX resistant Nb2-SFJCD1 subline. In control Nb2-SFJCD1 cultures, *pim-1* expression was found to be constitutive; it was augmented by the addition of PRL with kinetics similar to those observed in its PRL-dependent counterpart (109).

The expression of the Bcl-2 family genes, *bcl-2* and *bax*, as well as *pim-1* were studied to determine whether the alterations in their expression subsequent to the addition of PRL to DEX-treated Nb2 cultures were also apparent in butyrate-pretreated Nb2-SFJCD1 cells under similar conditions (90). The expression of *bax* at both the mRNA and protein levels was found to be unaltered by the addition of DEX-alone or it in combination with PRL. A reduction in *bcl-2* mRNA was observed in DEX-treated cultures; nevertheless, it did not appear to produce any deviation in the level of its protein product. The addition of PRL also failed to increase the Bcl-2 protein. Thus, in butyrate-pretreated Nb2-SFJCD1 cells challenged with DEX alone or with DEX plus PRL the Bcl-2:Bax ratio remained unchanged compared to control cultures. These results lead to the conclusion that these two apoptosis-regulatory proteins most likely did not contribute to the anti-apoptotic effect afforded by PRL in the butyrate-pretreated cells.

Compared to the expression of the Bcl-2 family members, *pim-1* was markedly altered by PRL in the autonomous line following butyrate treatment, then challenged with DEX. Significantly increased *pim-1* mRNA was observed following the addition of PRL to the DEX treated cultures. More importantly, the combined hormone treatment triggered a marked increase in Pim-1 protein synthesis. Thus, PRL increased transcription/translation of *pim-1* in the presence of a potent apoptotic stimulus suggesting a role for this kinase as a mediator of lactogen-provoked suppression of the death program.

As discussed above, PRL stimulated rapid expression of several anti-apoptotic genes. In each case, increased transcription was detected within a few hours and maximal protein product accumulation was observed within 6 hrs after PRL in quiescent Nb2 cells. In addition, recent evidence indicated that expression of the recently described X-linked inhibitor of apoptosis (XIAP) gene is also regulated by PRL and that its accumulation occurred substantially later during cell cycle progression (Krishnan and Buckley, unpublished observations). Mammalian XIAP is a member of a gene family that blocks apoptosis in cells exposed to viral infection and growth factor deprivation (110). It has been shown to exhibit extensive sequence homology to baculovirus IAP which encodes a protein that inhibits at least two members of the caspase family of cell-death proteases (110,111). Stimulation of quiescent cultures with PRL significantly increased expression of XIAP with peak levels observed by 12 hrs which were sustained through 24 hrs, time points which corresponded with entry of cells into S and progression through G2/M phases (Krishnan and Buckley, unpublished observations). It was also detected in apoptosis-resistant Nb2-SFJCD1 cells but substantially reduced by butyrate exposure. In this setting, PRL addition stimulated its expression with a similar kinetic pattern to that observed in the hormone-dependent line (Krishnan and Buckley, unpublished observations). A possible conclusion from these studies is that XIAP may protect PRL-treated cells from apoptosis during the later phases of cell cycle.

In numerous previous studies, lactogen-stimulated Nb2 cells have been utilized to investigate PRLR activation of signal transduction pathways linked to regulation of specific gene transcription. Of the pathways evaluated, activation of Jak2/Stat has been most commonly implicated in expression of PRL-responsive genes. In several hematopoietic cell systems, including Nb2 cells, *pim-1* has been reported to be regulated by this signaling mechanism (112,113). Since the 5'-promoter region of the *pim-1* gene contains several GAS-related elements, it appeared likely that its PRL-induced expression most likely reflected activation of Jak/Stat.

However, results from promoter/reporter experiments, which employed a 1268 bp segment of the 5'-*pim-1* promoter, and a series of deletion mutants indicated that two elements [distal (DE), -427 to -336 bp and proximal (PE), -104 to -1] but not several GAS or GAS-like sequences were required for promoter activation by PRL (114). In other experiments, the effect of PRL to stimulate Nb2 cell transcription factor binding to the DE and PE was evaluated. The results indicated that factors within PRL-treated cells rapidly and specifically bound to these promoter sequences; however, no binding was observed to GAS-like elements also present within the promoter. In addition, DNAse I protection experiments revealed a nuclear factor-1 (NF-1)-like motif within the promoter that appeared to regulate *pim-1* expression in unstimulated Nb2 cells (114).

The accumulated evidence suggests that at least four gene products, *bcl-2*, *bcl-xL*, *pim-1*, and *XIAP* may regulate suppression of apoptosis in T-lineage lymphocytes treated with PRL (Figure 13-1). Each of these was induced by PRL in quiescent Nb2 T-cells under conditions in which apoptosis was activated. In addition, recent evidence (115) has suggested that yet another protein, Bag-1, a Bcl-2- associated protein that has been linked to signaling by the serine kinase, Raf-1, may also participate in the survival effects of PRL in Nb2 cells. Taken together, PRL appears to regulate an spectrum of apoptosis-associated intermediates to facilitate survival in cells in cell death has been activated ranging from direct mediators of apoptosis itself (Bcl-2, Bcl-x_L, XIAP), to associated proteins (Bag-1), and those coupled to signaling mechanisms (Pim-1).

CELL CYCLE ASSOCIATED APOPTOSIS GENE EXPRESSION IN Nb2 CELLS

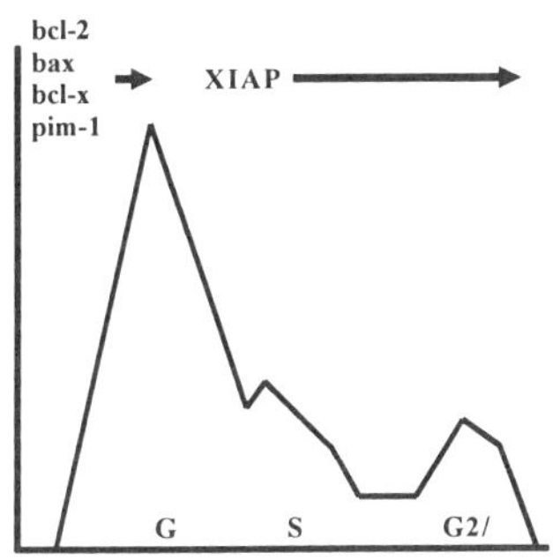

Figure 13-1. Schematic representation indicating the pattern of PRL-stimulated expression of apoptosis-associated genes in quiescent Nb2 cells. Expression of bcl-2, bcl-xL and pim-1 is increased and Bax decreased by PRL during early-mid G1 (2-6 hrs). XIAP is maximal during progression from G1/S to G2/M (12-24 hrs).

MAMMARY GLAND AND PROSTATE

The observation of an anti-apoptotic action of PRL in Nb2 cells has recently been extended to mammary gland and prostate. Secretion of milk can be terminated by removal of suckling young, which produces rapid accumulation of milk and cessation of its requisite protein synthesis, or by hormone deprivation with continuation of suckling. Deficiency of PRL, together with growth hormone, decreased milk components and mammary DNA content (116,117). Administration of bromocryptine and anti-growth hormone antisera to rats in mid-lactation significantly decreased mammary gland weight and reduced gland DNA content, an effect blocked by PRL re-administration (118). When endonuclease-mediated DNA fragmentation was determined, hormone deprivation caused significant hydrolysis which was also inhibited by PRL but not growth hormone. Further evidence supporting an anti-apoptotic role for PRL was provided by studies in which the effects of a human PRL antagonist (hPRL-G129R) was evaluated in several PRL-receptor positive human breast cancer cell lines (119). In T-47D cells, a line in which PRL stimulates proliferation, treatment with hPRL-G129R inhibited cell growth. Importantly, when the inhibitory effects of hPRL-G129R were further evaluated, the antagonist was found to induce apoptosis in a concentration-dependent manner in four PRL-receptor positive cell lines (119). Taken together, these results indicate that one function of PRL in mammary epithelial cells may involve maintenance of cell viability.

Prostate epithelium has been reported to produce PRL (120) similar to cells of the immune system and in the mammary gland. Since the prostate expresses PRL receptors (121), these observations suggested that an autocrine/paracrine regulatory loop may exist for PRL in this tissue. Culturing rat prostate explants in androgen-free medium provoked rapid apoptosis in the gland as determined by morphological analysis of nuclei and assessment of DNA fragmentation. The addition of PRL suppressed apoptosis in the dorsal and lateral, but not ventral, prostate cultures (122). Together with observations of apoptosis-suppression in the immune system and mammary gland, these results suggest that in addition to its trophic actions in these tissues, PRL may regulate cell population density by prolonging survival in these hormonal targets.

CORPUS LUTEUM

The actions of PRL in the rat corpus luteum (CL) are complex. On one hand, PRL is luteotrophic, eg., the functioning organ can be maintained for prolonged periods in hypophysectomized animals if PRL is administered (123,124) or added to CL cells in culture (125). Paradoxically, PRL can also be luteolytic (126).

Regression of the CL in humans and rats occurs in two distinct phases (127). Initially, the organ loses its capacity to synthesize and secrete progesterone. This stage is followed by CL involution. During the rat estrous cycle, a surge in circulating PRL occurs on the afternoon of proestrus which is thought to initiate structural luteolysis (126,128,129). It was recently observed that fragmentation of genomic DNA was associated with this event, an effect that was significantly inhibited by bromocryptine administration (130,131). Moreover, PRL injection in bromocryptine-treated rats restored the destruction of the CL. In dispersed luteal tissue, similar effects were found subsequent to the addition of PRL to the culture medium (125). These results suggested that increased PRL levels at proestrus activate apoptosis as a component of structural luteolysis. However, since *in vivo* evidence of apoptosis was also observed at other time points in addition to those associated with increased serum PRL levels it was also concluded that other hormones may participate in this process (130).

Accumulation of macrophages, presumably to phagocytize debris and cellular components, accompanies CL regression (132). Administration of PRL to rats similarly produced recruitment of macrophages (133) and enhanced expression of monocyte chemoattractant protein-1 (134). Inhibition of PRL secretion, not only blocked the increase in apoptotic CL cells but also reduced macrophage infiltration (135). Administration of progesterone receptor antagonists similarly inhibited apoptosis but not accumulation of the phagocytic cells suggesting that both PRL and progesterone may be required for induction of apoptosis associated with CL regression (135).

The mechanism by which PRL stimulates apoptosis in the CL is as yet unknown. Since PRL appears to trigger macrophage infiltration during luteolysis (133,134) it is possible that these cells represent cellular targets that mediate some of the actions of the hormone. Support for a PRL-regulated immunological component in CL regression was provided by the observation that cultured cells prepared during proestrus expressed high levels of Fas ligand (FasL), a protein that stimulates apoptosis-inducing signals, in response to PRL addition (136). The FasL expressing cells were found to co-localize with CD-3 positive T-lymphocytes (136) suggesting the effects of PRL on CL may reflect, at least in part, an action of the hormone on certain hematopoietic cell populations.

ACKNOWLEDGEMENTS

Work presented from the author's laboratory was supported by grants from the National Institutes of Health, the American Institute for Cancer Research, and the Ohio Cancer Research Associates.

REFERENCES

1. Riddle O, Bates RW, Dykshorn SW. A new hormone of the anterior pituitary. Proc Soc Exper Biol Med 1932; 29:1211-1212.
2. Riddle O, Bates RW, Dykshorn SW. The preparation, identification and assay of prolactin - a hormone of the anterior pituitary. Am J Physiol 1933;105:191-216.
3. Riddle O. Prolactin in vertebrate function and organization. J Natl Cancer Inst 1968;31:1039-1110.
4. Gout PW, Beer CT, Noble RL. Prolactin-stimulated growth of cell cultures established from malignant Nb rat lymphomas. Cancer Res 1980;40:2433-2436.
5. Dumont JN. Prolactin-induced cytologic changes in the mucosa of the pigeon crop during crop 'milk' formation. Z Zellforsch 1965;68:755-782.
6. Anderson TR, Pitts DS, Nicoll CS. Prolactin's mitogenic action on the pigeon crop-sac mucosal epithelium involves direct and indirect mechanisms. Gen Comp Endocrinol 1984;54:236-246.
7. Ben-David M. A sensitive bioassay for prolactin based on [3H]methylthymidine uptake by the pigeon crop mucous epithelium. Proc Soc Exp Biol Med 1967;125:705-708.
8. Nicoll CS. Bioassay of prolactin: analysis of the pigeon crop sac response to local prolactin injection by an objective and quantitative method. Endocrinology 1967;80:641-655.
9. Lewis UJ, Singh RNP, Lewis LL, Seavey BK, Sinha YN. Glycosylated ovine prolactin. Proc Natl Acad Sci USA 1984;81:385-389.
10. Clapp C, Sears PS, Russell DH, Richards J, Levay-Young BK, Nicoll CS. Biological and immunological characterization of cleaved and 16K forms of rat prolactin. Endocrinology 1988;122:2892-2898.
11. Nicoll C, Bern HA. Further analysis of the occurrence of pigeon crop sac-stimulating activity (prolactin) in the vertebrate adenohypophysis. Gen Comp Endocrinol 1968;11:5-16.
12. Pukac LA, Horseman ND. Regulation of pigeon crop gene expression by prolactin. Endocrinology 1984;114:1718-1724.
13. Bani G, Sacchi TB, Bigazzi M. Response of the pigeon crop sac to mammotrophic hormones: comparison between relaxin and prolactin. Gen Comp Endocrinol 1990;80:16-23.

14. Chen X, Horseman ND. Cloning, expression, and mutational analysis of the pigeon prolactin receptor. Endocrinology 1994;135:269-276.
15. Anderson TR, Mayer GL, Nicoll CS. Cyclic nucleotides and the control of epithelial cell proliferation: cyclic CMP may be a partial mediator of the response of the pigeon crop-sac to prolactin. J Cyclic Nucl Res 1981;7:225-233.
16. Pukac LA, Horseman ND. Regulation of cloned prolactin-inducible genes in pigeon crop. Mol Endocrinol 1987;1:188-194.
17. Horseman ND. A prolactin-inducible gene product which is a member of the calpactin/lipocortin family. Mol Endocrinol 1989;3:773-779.
18. Xu YH, Horseman ND. Nuclear proteins and prolactin-induced annexin Icp35 gene transcription. Mol Endocrinol 1992;6:375-383.
19. Sidis Y, Horseman ND. Prolactin induces rapid p95/p70 tyrosine phosphorylation, and protein binding to GAS-like sites in the anx Icp35 and c-fos genes. Endocrinology 1994;134:1979-1985.
20. Emerman JT, Bissell MJ. Cultures of mammary epithelial cells: extracellular matrix and functional differentiation. Adv Cell Culture 1988;6:137-159.
21. Lyons WR, Li Ch, Johnson RE. The hormonal control of mammary growth and lactation. Recent Prog Horm Res 1958;14:219-250.
22. Dilley WG, Nandi S. Rat mammary gland differentiation *in vitro* in the absence of steroids. Science 1968;161:59-60.
23. Topper YJ, Freeman CS. Multiple hormone interactions in the developmental biology of the mammary gland. Physiol Rev 1980;60:1049-1106.
24. Oka T, Yoshimura M, Lavandero S, Wada K, Ohba Y. Control of growth and differentiation of the mammary gland by growth factors. J Dairy Sci 1991;74:2788-2800.
25. Kleinberg DL. Early mammary gland development: growth hormone and IGF-1. J Mam Gland Biol Neoplasia 1997;2:49-57.
26. Horseman ND. Prolactin and mammary gland development. J Mam Gland Biol Neoplasia 1999;4:79-88.
27. Vonderhaar BK. Studies on the mechanisms by which thyroid hormones enhance a-lactalbumin activity in explants from mouse mammary glands. Endocrinology 1977;100:1423-1431.
28. Martin R, Glass M, Wilson G, Woods K. Human lactalbumin and hormonal factors in pregnancy and lactation. Clin Endocrinol 1973;13:323-330.
29. Hearn J. Pituitary inhibition of pregnancy. Nature 1973;241:207-8.
30. Brun F, del Re R, del Pozo E. Prolactin inhibition and suppression of puerperal lactation by a Br-ergocriptine (CD 154): a comparison with estrogen. Obstet Gynecol 1973;41:884-892.
31. Horseman ND, Zhao W, Montecino-Rodriguez E, Tanaka M, Nakashima K, Engle SJ, Smith F, Markoff E, Dorshkind K. Defective mammopoiesis, but normal hematopoiesis, in mice with a targeted disruption of the prolactin gene. EMBO J 1997;16:6926-6935.
32. Ormandy C, Camus JA, Barra J, Damotte D,, Lucas B, Buteau H, Edery M, Brousse N, Babomet C, Binart N, Kelly PA. Null mutation of the prolactin receptor gene produces multiple reproductive defects in the mouse. Genes Dev 1997;11:167-178.
33. Darcy KM, Shoemaker SF, Lee P-P H, Vaughan MW, Black JD, Ip MM. Prolactin and epidermal growth factor regulation of proliferation, morphogenesis, and functional differentiation of normal rat mammary epithelial cells in three dimensional primary culture. J Cell Physiol 1995;163:346-364.
34. Moriggl R, Berchtold S, Friedrich K, Standke GJR, Kammer W, Heim M, Wissler M, Stocklin E, Gouilleux F, Groner B. Comparison of the transactivation domains of Stat5 and Stat6 in lymphoid cells and mammary epithelial cells. Mol Cell Biol 1997;17:3663-3678.
35. Ali S, Ali S. Prolactin receptor regulates Stat5 tyrosine phosphorylation and nuclear translocation by two separate pathways. J Biol Chem 1998;273:7709-7716.

36. Lebrun J-J, Ali S, Ullrich A, Kelly PA. Proline-rich sequence mediated Jak2 association to the prolactin receptor is required but not sufficient for signal transduction. J Biol Chem 1995;270:10664-10670.
37. DaSilva L, Rui H, Erwin RA, Howard OM, Kirken RA, Malabarba M, Hackett RH, Larner AC, Farrar WL. Prolactin recruits STAT1, STAT3 and STAT5 independent of conserved receptor tyrosines TYR402, TYR479, TYR515 and TYR580.Mol Cell Endocrinol 1996;117:131-140.
38. Wakao H, Gouilleux F, Groner B. Mammary gland factor (MGF) is a novel member of the cytokine regulated transcription factor gene family and confers the prolactin response. EMBO J 1994;13:2182-2191.
39. Liu X, Robinson GW, Gouilleux F, Groner B, Hennighausen L. Cloning and expression of Stat5 and an additional homologue (Stat5b) involved in prolactin signal transduction in mouse mammary tissue. Proc Natl Acad Sci USA 1995;92:8831-8835.
40. Gouilleux F, Wakao H, Mundt M, Groner B. Prolactin induces phosphorylation of Tyr694 of Stat5 (MGF), a prerequisite for DNA binding and induction of transcription. EMBO J 1994;13:4361-4369.
41. Kane IM, Leder P. NDF/heregulin induces persistence of terminal end buds and adenocarcinomas in the mammary glands of transgenic mice. Oncogene 1996;12:1781-1788.
42. Rothman P, Kreider B, Azam M, Levy D, Wegenka U, Eilers A, Decker T, Forn F, Kashieva H, Ihle J, Schindler C. Cytokines and growth factors signal through tyrosine phosphorylation of a family of related transcription factors. Immunity 1994;1:457-468.
43. Decker T, Lew DJ, Mirkovitch J, Darnell JE. Cytoplasmic activation of GAF, and IFN-gamma-regulated DNA-binding factor. EMBO J 1991;10:927-932.
44. Barash I. Prolactin and insulin synergize to regulate the translation modulator PHAS-I via mitogen-activated protein kinase-independent by wortmannin- and rapamycin-sensitive pathway. Mol Cell Endocrinol 1999;155:37-49.
45. Vonderhaar BK. Prolactin and its receptors in human breast cancer. In: Manni A, ed. Endocrinology of Breast Cancer. Totowa, New Jersey: Humana Press, 1999:261-279.
46. Das R, Vonderhaar BK. Involvement of SHC, Grb2, Sos, and Ras in prolactin signal transduction in mammary epithelial cells. Oncogene 1996;13:1139-1145.
47. Das R, Vonderhaar BK. Prolactin as a mitogen in mammary cells. J Mam Gland Biol Neoplasia 1997;2:29-39.
48. Reynolds C, Montone KT, Powell CM, Tomaszewski JE, Clevenger CV. Expression of prolactin and its receptor in human breast carcinoma. Endocrinology 1997;138:5555-5560.
49. Maus MV, Reilly SC, Clevenger CV. Prolactin as a chemoattractant for human breast carcinoma. Endocrinology 1999;140:5447-5450.
50. Berczi, I, Nagy, E, Kovacs, K, Horvath, E. Regulation of humoral immunity in rats by pituitary hormones. Acta Endocrinol (Copenh) 1981;98:506-513.
51. Nagy, E, Berczi, I. Prolactin and contact sensitivity. Allergy 1981;36:429-432.
52. Berczi, I, Nagy, E, Asa, SL, Kovacs. The influence of pituitary hormones on adjuvant arthritis Rheum 1984;27:682-688.
53. Matera, L, Casano, A, Bellone, G, Oberholtzer, E. Modulatory effect of prolactin on the resting and mitogen-induced activity of T, B, and NK lymphocytes. Brain Behav Immun 1992; 6:409-417.
54. Athreya, BH, Pletcher, J, Zulian, F, Weiner, DB, Williams, WV. Subset-specific effects of sex hormones and pituitary gonadotropins on human lymphocyte proliferation *in vitro*. Clin Immunol Immunopathol 1993;66:201-211.
55. Clevenger CV, Russell DH, Appasamy PM, Prystowsky MB. Regulation of IL2-driven T-lymphocyte proliferation by prolactin. Proc Natl Acad Sci USA 1990;87:6460-6464.
56. Hartmann DP, Holoday JW, Bernton EW. Inhibition of lymphocyte proliferation by antibodies to prolactin. FASEB J 1989:3:2194-2202.

57. Montgomery, DW, Zukoski, CF, Shah, GN, Buckley, AR, Pacholecyzk, T, Russell, DH. Concanavalin A stimulated murine splenocytes produce a factor with prolactin-like bioactivity and immunoreactivity. Biochem Biophys Res Commun 1987;145:692-698.
58. O'Neal, KD, Montgomery, DW, Truong, TM, Yu-Lee, L-Y. Prolactin gene expression in human thymocytes. Mol Cell Endocrinol 1992;87:R19-R23.
59. Montgomery, DW, Shen, GK, Ulrich, ED, Steiner, LL, Parish, PR, Zukoski, CF. Human thymocytes express a prolactin-like messenger ribonucleic acid and synthesize bioactive prolactin-like proteins. Endocrinology 1992;131:3019-3026.
60. Sabharwal, P, Glaser, R, Lafuse, W, Varma, S, Lia, Q, Arkins, S, Kooijman, R, Kutz, L, Kelley, KW, Malarkey, WB. Prolactin synthesized and secreted by human peripheral blood mononuclear cells: an autocrine growth factor for lymphoproliferation. Proc Natl Acad Sci USA 1992;89:7713-7716.
61. Pellegrini, I, Labrun, JJ, Ali, S, Kelly, PA. Expression of prolactin and its receptor in human lymphoid cells. Mol Endocrinol 1992;6:1023-1031.
62. Russell, DH, Kibler, R, Matrisian, L, Larson, DF, Poulos, B, Magun, BE. Prolactin receptors on human T and B lymphocytes: antagonism of prolactin binding by cyclosporine. J Immunol 1985;134:3027-3031.
63. Gagnerault, M-C, Tourine, P, Savino, W, Kelly, PA, Dardenne, M. Expression of prolactin receptors in murine lymphoid cells in normal and autoimmune situations. J Immunol 1993;150:5673-5681.
64. Ali, S, Pellegrini, I, Kelly, PA. A prolactin-dependent immune cell line (Nb2) expresses a mutant form of prolactin receptor. J Biol Chem 1991;266:20110-20117.
65. Bouchard B, Ormandy CJ, DiSanto JP, Kelly PA. Immune system development and function in prolactin receptor-deficient mice. J Immunol 1999;163:576-582.
66. Dorshkind K, Horseman N. The roles of prolactin (PRL), growth hormone (GH), insulin-like growth factor-I (IGF-I), and thyroid hormones in lymphocyte development and function: insights from genetic models of hormone and hormone receptor deficiency. Endocrine Rev, in press.
67. Postel-Vinay M-C, de Mello Coelho V, Gagnerault M-C, Dardenne M. Growth hormone stimulates the proliferation of activated mouse T-lymphocytes. Endocrinology 1997;138:1816-1820.
68. Sorenson RL, Brelje TC. Adaptation of islets of Langerhans to pregnancy: beta-cell growth, enhanced insulin secretion and the role of lactogenic hormones. Horm Metab Res 1997:29:301-307.
69. Brelje TC, Allaire P, Hegre O, Sorenson RL. Effect of prolactin versus growth hormone on islet function and the importance of using homologous mammsosmatotropic hormones. Endocrinology 1989;125:2392-2399.
70. Brelje TC, Sorenson RL. Role of prolactin versus growth hormone on islet B-cell proliferation *in vitro*: implications for pregnancy. Endocrinology 1991;128:45-57.
71. Brelje TC, Scharp DW, Lacy PE, Ogren L, Talamantes F, Robertson M, Friesen HG, Sorenson RL. Effect of homologous placental lactogens, prolactins, and growth hormones on islet B-cell division and insulin secretion in rat, mouse, and human islets: implication for placental lactogen regulation of islet function during pregnancy. Endocrinology 1993;132:879-887.
72. Brelje TC, Parsons JA, Sorenson RL. Regulation of islet B-cell proliferation by prolactin in rat islets. Diabetes 1994;43:263-273.
73. Fleenor D, Petryk A, Driscoll P, Freemark M. Constitutive expression of placental lactogen in pancreatic beta cells: effects on cell morphology, growth, and gene expression. Pediatr Res 2000;47:136-142.
74. Sorenson RL, Stout LE. Prolactin receptors and Jak2 in Islets of Langerhans: an immunohistochemical analysis. Endocrinology 1995;136:4092-4098.

75. Galsgaard ED, Nielson JH, Moldrup A. Regulation of prolactin receptor (PRLR) gene expression in insulin-producing cells. J Biol Chem 1999;274:18686-18692.
76. Stout LE, Svensson AM, Sorenson RL. Prolactin regulation of islet-derived INS-1 cells: characteristics and immunocytochemical analysis of STAT5 translocation. Endocrinology 1997;138:1592-1603.
77. Holstad M, Sandler S. Prolactin protects against diabetes induced by multiple low doses of streptozotocin in mice. J Endocrinol 1999;163:229-234.
78. Davalli AM, Scaglia L, Brevi M, Sanvito F, Freschi M, Cavallaro U, Parlow AF, Pontiroli AE. Pituitary cotransplantation significantly improves the performance, insulin content, and vascularization of renal subcapsular islet grafts. Diabetes 1999;48:59-65.
79. DeVito WJ, Avakian C, Stone S, Okulicz WC. Prolactin-stimulated mitogenesis of cultured astrocytes is mediated by a protein kinase C-dependent mechanism. J Neurochem 1993;60:835-842.
80. DeVito WJ, Stone S, Shamgochian M. Prolactin induced expression of glial fibrillary acidic protein and tumor necrosis factor-alpha at a wound site in the rat brain. Mol Cell Endocrinol 1995;108:125-130.
81. Girolomoni G, Phillips JT, Bergstresser PR. Prolactin stimulates proliferation of cultured human keratinocytes. J Invest Dermatol 1993;101:275-279.
82 Poumay Y, Jolivet G, Pittelkow MR, Herphelin F, DePotter IY, Mitev V, Houdebine LM. Human epidermal keratinocytes upregulate expression of the prolactin receptor after the onset of terminal differentiation, but do not respond to prolactin. Arch Biochem Biophys 1999;364:247-253.
83. White BA, Nicoll CS. Hormonal control of amphibian metamorphosis. In: Gilbert LI, Frieden E, eds. Metamorphosis. A problem in developmental biology. New York: Plenum Press, 1981:363-396.
84. Ray LB, Dent JN. Observations on the interaction of prolactin and thyroxine in the tail of the bullfrog tadpole. Gen Comp Endocrinol 1986;64:36-43.
85. Russell DH, Buckley AR, Montgomery DW, Putnam CW, Zukoski CF. Antiproliferative properties of corticosteroids and the cyclopeptides, cyclosporine and didemnin B, involve inhibition of prolactin receptor binding in rat Nb2 node lymphoma cells. J Cell Biochem 1987;11A:39 (abst).
86. Tata JR. Hormonal regulation of programmed cell death during amphibian metamorphosis. Biochem Cell Biol 1994;72:581-588.
87. Fletcher-Chiappini SE, Comptom MM, LaVoie HA, Day EB, Witorsch RJ. Glucocorticoid-prolactin interactions in Nb2 lymphoma cells: antiproliferative versus anticytolytic effects. Proc Soc Exp Biol Med 1993;202:345-352
88. Gout PW. Transient requirement for prolactin as a growth initiator following treatment of autonomous Nb2 node rat lymphoma cell cultures with butyrate. Cancer Res 1987;47:1751-1755.
89. Witorsch RJ, Day EB, LaVoie HA, Hashemi N, Taylor JK. Comparison of glucocorticoid-induced effects in prolactin-dependent and autonomous rat Nb2 lymphoma cells. Proc Soc Exp Biol Med 1993;203:454-460.
90. Buckley AR, Krumenacker JS, Buckley DJ, Leff MA, Magnuson NS, Reed JC, Miyashita T, de Jong G, Gout PW. Butyrate-induced reversal of dexamethasone resistance in autonomous rat Nb2 lymphoma cells. Apoptosis 1997;2:518-528.
91. Buckley AR, Leff MA, Buckley DJ, Magnuson NS, de Jong G, Gout PW. Alterations in pim-1 and c-myc expression associated with sodium butyrate-induced growth factor dependency in autonomous rat Nb2 lymphoma cells. Cell Growth Diff 1996;7:1713-1721.
92. Oltvai ZN, Milliman CL, Korsmeyer SJ. Bcl-2 heterodimerizes *in vivo* with a conserved homolog, Bax, that accelerates programmed cell death. Cell 1993;74:609-619.

93. Ohta K, Iwai K, Kasahara Y, Taniguchi N, Krajeeski S, Reed JC, Miyawaki, T. Immunoblot analysis of cellular expression of Bcl-2 family proteins, Bcl-2, Bax, Bcl-X and Mcl-1, in human peripheral blood and lymphoid tissues. Int Immunol 1995;7:1817-1825.
94. Yin XM, Oltvai ZN, Korsmeyer SJ. BH1 and BH2 domains of Bcl-2 are required for inhibition of apoptosis and heterodimerization with Bax. Nature 1994;369:321-323.
95. Zha H, Aime-sempe C, Sato T, Reed J.C. Proapoptotic protein Bax heterodimerizes with Bcl-2 and homodimerizes with Bax via a novel domain (BH3) distinct from BH1 and BH2. J Biol Chem 1996;271:7440-7444.
96. Bakhshi A, Jensen JP, Goldman P, Wright JJ, Mcbride OW, Epstein AL, Korsmeyer SJ. Cloning the chromosomal breakpoint of t(14;18) human lymphomas: clustering around JH on chromosome 14 and near a transcriptional unit on 18. Cell 1985;41 899-906.
97. Cleary ML, Sklar J. Nucleotide sequence of a t(14;18) chromosomal breakpoint in follicular lymphoma and demonstration of a breakpoint-cluster region near a transcriptionally active locus on chromosome 18. Proc Natl Acad Sci USA 1985;82:7439-7443.
98. Leff MA, Buckley DJ, Krumenacker JS, Reed JC, Miyashita T, Buckley AR. Rapid modulation of the apoptosis regulatory genes, bcl-2 and bax by prolactin in rat Nb2 lymphoma cells. Endocrinology 1996;137:5456-5462.
99. Minn AJ, Velez P, Schendel SL, Liang H, Muchmore SW, Feslk SW, Fill M, Thompson CB. Bcl-x(L) forms an ion channel in synthetic lipid membranes. Nature 1997;385:353-357.
100. Vanderheiden MG, Chandel NS, Williamson EK, Schumaker PT, Thompson CB. Bcl-xL regulates the membrane potential and volume homeostasis of mitochondria. Cell 1997;91:627-637.
101. Li F, Srinivasan A, Wang Y, Armstrong RC, Tomaselli KJ, Fritz LC. Cell-specific induction of apoptosis by microinjection of cytochrome c. Bcl-xL has activity independent of cytochrome c release. J Biol Chem 1997;272:30299-30305.
102. Chinnalyan AM, O'Rourke K, Lane BR, Dixit VM. Interaction of CED-4 with CED-3 and CED-9: a molecular framework for cell death. Science 1997;275:1122-1126.
103. Hoover D, Friedman M, Reeves R, Magnuson NS. Recombinant human pim-1 protein exhibits serine/threonine kinase activity. J Biol Chem 1991;266:14018-14022.
104. Dautry F, Weil D, Yu J, Dautry-Versat A. Regulation of pim-1 and myb mRNA accumulation by interleukin-2 and interleukin-3 in murine hematopoietic cell lines. J Biol Chem 1988; 263:17615-17620.
105. Lilly M, Kraft A. Enforced expression of the Mr 33,000 Pim-1 kinase enhances factor-independent survival and inhibits apoptosis in murine myeloid cells. Cancer Res 1997;57:5348-5355.
106. Yu-Lee L-Y, Hrachovy JA, Stevens AM, Schwarz LA. Interferon-regulatory factor 1 is an immediate-early gene under transcriptional regulation by prolactin in Nb2 T cells. Mol Cell Biol 1990;10:3087-3084.
107. Murphy PR, Dimattia GE, Freisen HG. Role of calcium in prolactinstimulated c-myc gene expression and mitogenesis in Nb2 lymphoma cells. Endocrinology 1988;122:2476-2485.
108. Krumenacker JS, Buckley DJ, Leff MA, Mccormack JT, De Jong G, Gout PW, Reed JC, Miyashita T, Magnuson NS, Buckley AR. Prolactin-regulated apoptosis of Nb2 lymphoma cells: pim-1, bcl-2, and bax expression. Endocrine 1998;9:163-170.
109. Buckley AR, Buckley DJ, Leff MA, Hoover DS, Magnuson NS. Rapid induction of pim-1 expression by prolactin and interleukin-2 in rat Nb2 lymphoma cells. Endocrinology 1995;136:5252-5259.
110. Duckett CS, Nava VE, Gedrich RW, Clem RJ, Van Dongen JL, Gilfillan MC, Shiels H, Hardwick JM, Thompson CB. A conserved family of cellular genes related to the baculovirus iap gene and encoding apoptosis inhibitors. EMBO J 1996:15:2685-2694.
111. Deveraux QL, Takahashi R, Salvesen GS, Reed JC. X-linked IAP is a direct inhibitor of cell-death proteases. Nature 1997;388:300-304.

112. Mui AL, Wakao H, O'Farrell AM, Harada N, Miyajima A. Interleukin-3, granulocyte-macrophage colony-stimulating factor, and interleukin-5 transduce signals through two STAT5 homologs. J Leuk Biol 1995;57: 799-803.
113. Rui H, Xu J, Mehta S, fang H, Williams J, Dong F, Grimley PM. Activation of the Jak2-Stat5 signaling pathway in Nb2 lymphoma cells by an anti-apoptotic agent, aurintricarboxylic acid. J Biol Chem 1998;273:28-32.
114. Borg KE, Zhang M, Hegge D, Stephen RL, Buckley DJ, Magnuson NS, Buckley AR. Prolactin regulation of pim-1 expression: positive and negative promoter elements. Endocrinology 1999:140:5659-5668.
115. Clevenger CV, Thickman K, Ngo W, Chang W-P, Takayama S, Reed JC. Role of Bag-1 in the survival and proliferation of the cytokine-dependent lymphocyte lines, Ba/F3 and Nb2. Mol Endocrinol 1997;11: 608-618.
116. Barber MC, Travers MT, Finley E, Flint DJ, Vernon RG. Growth-hormone-prolactin interactions in the regulation of mammary and adipose tissue acetyl-CoA carboxylase activity and gene expression in lactating rats. Biochem J 1992;285:469-475.
117. Flint DJ, Gardner M. Evidence that growth hormone stimulates milk synthesis by direct action on the mammary gland and that prolactin exerts effects on milk secretion by maintenance of mammary deoxyribonucleic acid content and tight junction status. Endocrinology 1994;135:1119-1124.
118. Travers MT, Barber MC, Tonner E, Quarrie L, Wilde CJ, Flint DJ. The role of prolactin and growth hormone in the regulation of casein gene expression and mammary cell survival: relationships to milk synthesis and secretion. Endocrinology 1996;137:1530-1539.
119. Chen WY, Ramamoorthy P, Chen N-Y, Sticca R, Wagner TE. A human prolactin antagonist, hPRL-G129R, inhibits breast cancer cell proliferation through induction of apoptosis. Clin Cancer Res 1999;5:3583-3593.
120. Nevalainen MT, Valve EM, Ahonen TA, Yagi A, Paranko J, Harkonen PL. Androgen-dependent expression of prolactin in rat prostate epithelium *in vivo* and in organ culture. FASEB J 1997;11:1297-1307.
121. Nevalainen MT, Martikainen P, Valve EM, Ingleton PM, Nurmi M, Harkonen PL. Prolactin receptors are expressed and functioning in himan prostate. J Clin Invest 1997;99:618-627.
122. Ahonen TJ, Harkonen PL, Laine J, Rui H, Martikainen PM, Navalainen MT. Prolactin is a survival factor for androgen-deprived rat dorsal and lateral prostate epithelium in organ culture. Endocrinology 1999;140:5412-5421.
123. Evans HM, Simpson ME, Lyons MR, Turpeinen K. Anterior pituitary hormones which favor the production of traumatic uterine placentomata. Endocrinology 1941;28:933-940.
124. Greenwald GS, Rothchild I. Formation and maintenance of corpora lutea in laboratory animals. J Anim Sci 1968;27: 139-162.
125. Kanuka H, Matsuyama S, Ohnishi M, Matsumoto Y, Nishihara M, Takahashi M. Prolactin expresses differential effects on apoptotic cell death of luteal cells *in vivo* and *in vitro*. Endocr J 1997;44:11-22.
126. Rothchild I. The regulation of the mammalian corpus luteum. Recent Prog Horm Res 1981;37:183-298.
127. Malven PV. Hypophyseal regulation of luteolysis in the rat. In McKerns KW ed. Hypophyseal Regulation of Luteolysis in the Rat. New York: Meredith, 1969:367-371.
128. Gibori G, Khan I, Warsaw ML, McLean MP, Puryear TK, Nelson S, Durkee TJ, Azhar S, Steinschneider A, Rao MC. Placental-derived regulators and the complex control of luteal cell function. Recent Prog Horm Res 1988;44:377-429.
129. Malven PV. Luteotrophic and luteolytic responses to prolactin in hypophysectomized rats. Endocrinology 1969;84:1224-1229.

130. Matsuyama S, Chang K-T, Kanuka H, Ohnishi M, Ikeda A, Nishihara M, Takahashi M. Occurrence of deoxyribonucleic acid fragmentation during prolactin-induced structural luteolysis in cycling rats. Biol Reprod 1996;54:1245-1251.
131. Kiya T, Endo T, Goto T, Yamamoto H, Ito E, Kudo R, Behrman HR. Apoptosis ond PCNA expression induced by prolactin in structural involution of the rat corpus luteum. J Endocrinol Invest 1998;21:276-283.
132. Bagavandoss P, Kunkel SL, Wiggins RC, Keyes PL. Tumor necrosis factor-a (TNF-a) production and localization of macrophages and T lymphocytes in the rabbit corpus luteum. Endocrinology 1988;122:1185-1187.
133. Bowen JM, Keyes PL, Warren JS, Townson DH, Prolactin-induced regression of the rat corpus luteum: expression of monocyte chemoattractant protein-1 and invasion of macrophages. Biol Reprod 1996;54:1120-1127.
134. Bowen JM, Towns R, Warren JS, Keyes PL. Luteal regression in the normally cycling rat: apoptosis, monocyte chemoattractant protein-1, and inflammatory cell involvement. Biol Reprod 1999;60:740-746.
135. Gaytan F, Bellido C, Morales C, Sanchez-Criado JE. Both prolactin and progesterone in proestrus are necessary for the induction of apoptosis in the regressing corpus luteum of the rat. Biol Reprod 1998;59:1200-1206.
136. Kuranaga E, Kanuka H, Bannai M, Suzuki M, Nichihara M, Takahashi M. Fas/Fas ligand system in prolactin-induced apoptosis in rat corpus luteum: possible role of luteal immune cells. Biochem Biophys Res Commun 1999;260:167-173.

Chapter 14

Mechanisms of Luteal Cell Regulation by Prolactin

Michael Risk and Geula Gibori

University of Illinois at Chicago College of Medicine, Chicago, IL

INTRODUCTION

The corpus luteum is an endocrine gland of limited lifespan formed from the remaining granulosa and theca cells of the follicle following ovulation. The main function of this gland is the synthesis and secretion of progesterone. Progesterone is a steroid hormone required for preparing the endometrium for implantation, regulating pituitary gonadotropin secretion and, importantly, maintaining a quiescent uterus by inhibiting the contractile activity of the myometrium, thus preventing the onset of parturition. In some species, such as the rat and mouse, the corpus luteum is responsible for the secretion of progesterone throughout pregnancy and is necessary for the entire length of gestation. In other species, such as human, the placenta assumes the majority of progesterone secretion at some point in gestation. In all mammals, however, progesterone is required for pregnancy and during a portion of gestation it must come from the corpus luteum.

In rodents, prolactin (PRL) is a crucial hormone, necessary for luteal function and thus the maintenance of pregnancy. The luteotropic action of PRL was first suggested by Dresel in 1935 (1) when he found that administration of PRL delayed the appearance of estrous in mice. Astwood in 1941 (2) demonstrated, with the use of hypophysectomized rats and vaginal cytology, that this delay was due to stimulation of the corpus luteum. A similar study in 1941 established that PRL allows the corpus luteum to support uterine deciduoma (3). Although the main stimulatory effects of PRL on the rodent corpus luteum were shown to be due to prolongation of luteal survival (2,4), it was later discovered that PRL can also cause the disintegration of rat corpora lutea which are no longer functionally active (5). In some species, such as marsupials, PRL does not appear to have a luteotropic function, but rather induces luteolysis (6). In species which undergo embryonic diapause, some are maintained in diapause through tonic inhibition of the corpus luteum by PRL (7) while in others reactivation of the corpus luteum and implantation are initiated by PRL (8). Thus, the effects of PRL on the corpus luteum appear to vary widely across species. In this chapter, we will focus on the PRL regulation of the rodent corpus luteum, with parallels in other species mentioned where applicable.

In rodents, PRL is required for the normal function of the corpus luteum throughout pregnancy, however the source of PRL support switches from the pituitary to the uterus/conceptus with the progression of gestation. Following ovulation on the morning of estrous, the corpus luteum secretes a limited amount of progesterone, lasts two days and then involutes if mating does not occur. Mating on estrous induces the release of PRL from the anterior pituitary via activation of a neural network that begins at the cervix and ends in the hypothalamus (reviewed in (9)). This causes two daily surges of PRL to be released for ten to twelve days regardless of implantation of the embryo (10,11). If fertilization does not occur, the corpus luteum will survive for these 10-12 days leading the animal to remain in a state of pseudopregnancy. The corpus luteum of pseudopregnancy, maintained functional by pituitary PRL, produces enough progesterone to affect the uterine milieu and allows for decidualization of the endometrial stroma in response to artificial stimuli. In the pseudopregnant rat, as in the human, the decidua expresses the PRL gene (12) and secretes a sufficient amount of PRL to sustain corpus luteum function for an additional few days (13). The physiological role of decidual PRL in the normal function of the corpus luteum during pregnancy is not clear since at the time pituitary PRL ceases to be secreted the placenta starts outpouring PRL-like hormones known as rat placental lactogens (rPLs) 1 and 2. These placental hormones bind to the PRL receptor (PRL-R) and sustain corpus luteum function until parturition. (for review see (14-16)).

While PRL has long been known to affect luteal function, the development of transgenic technology has made the importance of this hormone in the physiology of the rodent corpus luteum indisputable. The generation of mice lacking either the PRL or the PRL-R genes made immediately apparent the need for this hormone (17,18). Female mice lacking the PRL or PRL-R genes are sterile, and fail to become pregnant or pseudopregnant. The embryonic development, implantation and decidualization of the uterine stroma are defective in PRL-R$^{-/-}$ mice due to very low levels of progesterone (18,19). All these parameters could be corrected with progesterone treatment (19), indicating that the main action of PRL is to stimulate ovarian production of progesterone. However, progesterone treatment of PRL-R$^{-/-}$ mice could not prevent fetal death in utero(20), most probably because of the absence of decidual PRL-R. Indeed the rodent decidua was shown recently to express both PRL (12) and the PRL-R (21,22) genes. The main role of decidual PRL appears to involve the local stimulation of the estradiol receptor (α and β) (23,24) and also the inhibition of decidual IL-6 (25) and 20α-hydroxysteroid dehydrogenase (20αHSD) (26), both of which are detrimental to fetal survival.

PRL SIGNAL TRANSDUCTION IN THE CORPUS LUTEUM

The PRL-R has been found in the corpora lutea of a number of mammalian species, including rat (27-30) and mouse (31). The characteristics of the PRL-R and its signaling pathways will be extensively discussed elsewhere in this book, so we will largely focus on findings specific to the corpus luteum.

It is by now well established that in the rat, the PRL-R exists in two isoforms, termed long (PRL-RL) and short (PRL-RS) which are 591 and 291 amino acids in length, respectively (28,32). These isoforms are products of the same gene with differences generated through alternative splicing of the mRNA transcript. They are identical in sequence in their N-terminal 261 amino acids, which includes the extracellular and transmembrane domains as well as a portion of their intracellular regions. Their differences lie in the remainder of their intracellular domains, leading one to speculate that they are utilized to transduce distinct signals within PRL target tissues. Expression of PRL-RL and PRL-RS has been shown throughout pregnancy in the corpus luteum of the rat, although the relative amounts of the two appear to change over the course of gestation (33). Luteinization causes a large increase in mRNA levels of both isoforms of the PRL-R (34), particularly PRL-RS. Mature granulosa cells contain PRL-R, almost entirely in the form of PRL-RL, yet are not able to activate Stat5 (34). This is interesting, since activation of Stat is thought to be due to PRL-RL mediated events (35-37), whereas PRL-RS has been found to act as a dominant negative receptor, inhibiting activation of the JAK/Stat pathway and stimulation of promoters through PRL-RL in some experimental models (38,39). In the corpus luteum, acquisition of PRL signaling capacity, at least to Stat5, is associated with an increase in PRL-RS levels relative to PRL-RL, so PRL-RS does not appear to have a dominant negative effect in this system. Whether this is due to a difference in Stat5 isoform utilization (Stat 5a versus Stat 5b) or other signaling component differences remains to be determined. At the end of pregnancy, both PRL-R isoforms decrease precipitously (Figure 14-1) (33,34), reducing the responsiveness of the corpus luteum to rat placental lactogen and PRL secreted at the end of pregnancy. The mechanisms controlling the dynamic expression of both forms of the PRL-R have not yet been totally elucidated, although PRL appears to stimulate the expression of its own receptor (33).

The PRL-R isoforms are members of the cytokine receptor superfamily, due to the presence of specific motifs in their primary structure (40). One important characteristic of this family of receptors is their lack of intrinsic kinase domains. PRL signal transduction, therefore, occurs through activation of intracellular kinases. Many kinase pathways are activated through the PRL-R in various tissues, including JAK2, MAPK (41-43), c-*src* (43) and PI3 kinase/PKB (44-46). Most of the work on signaling in the corpus luteum has focused on the JAK2/Stat 5 system, which appears to be activated in signals propagated through the long form on the PRL-R. This signal transduction pathway has been characterized in a number of other tissues, and was thoroughly described in a recent review (47). JAK2 is a cytoplasmic tyrosine kinase that binds to the intracellular domain of the PRL-R. PRL binding to the PRL-R induces PRL-R dimerization, leading to

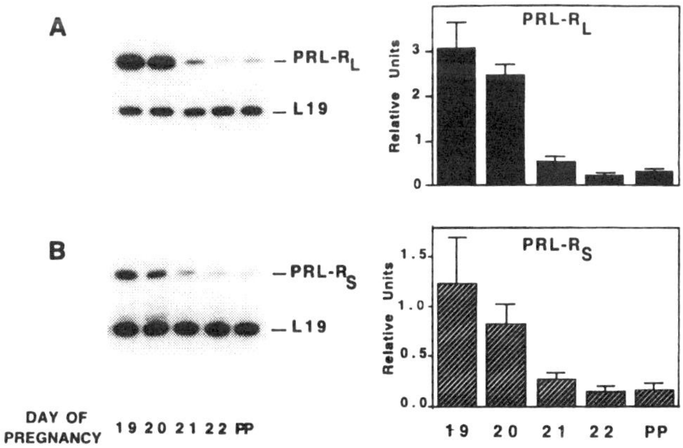

Figure 14-1. PRL-R expression decreases at the end of pregnancy. The mRNA expression of PRL-RL and PRL-RS as determined by semi-quantitative RT-PCR from day 18 of gestation to post-partum (PP). Expression levels of both isoforms of the PRL-R decrease to very low levels beginning on day 20 of pregnancy, and remain low following day 22, the day of parturition. (From Telleria, *et al.* (33)).

autophosphorylation and activation of JAK2 as well as phosphorylation of the PRL-R. These phosphotyrosines then serve as docking sites for Stat 5, which can bind these residues via its SH-2 domain. This recruitment of Stat 5 to the receptor-kinase complex allows JAK2 to phosphorylate Stat 5, and Stat 5 can then dimerize and translocate to the nucleus to affect transcription.

In the rat corpus luteum, JAK2 is expressed constitutively in granulosa cells as well as the corpus luteum (48), while Stat 5, especially Stat 5b, increases markedly upon luteinization and remains elevated throughout gestation (34,48,49). PRL rapidly induces tyrosine phosphorylation of both JAK2 (50) and Stat 5 (49-51). PRL has also been shown to increase Stat 5 DNA binding to its response elements in the rat β-casein (49) and α_2-macroglobulin genes(34,48,51,52) and to activate transcription of a reporter gene through the Stat 5 response element in the α_2-macroglobulin promoter(51,52). Interestingly, this Stat response element in the α_2-macroglobulin promoter was found to bind mainly Stat 5b (51), and stimulation of luteal Stat 5 DNA binding activity *in vivo* with exogenous PRL diminishes after midgestation when the corpus luteum contains a high basal Stat 5 DNA binding activity (48). This led Richards, et al (48), to suggest that the high levels of rPLs may have led to a maximal stimulation of the PRL signaling pathway, thus exogenous PRL had no effect.

Stat 5 actually exists as two isoforms, namely Stat 5a and Stat 5b. Both Stat 5 genes have been knocked out, and the phenotypes testify to the importance of these transcription factors in luteal function. It appears that in the corpus luteum these two proteins are interchangeable, as mice with disruption of either gene alone can become pregnant (53,54), whereas the double knockout of Stats 5a and 5b renders the females infertile (55). In Stat 5a/5b null mice, follicle development and ovulation appear normal, but there are few structures resembling corpora lutea. These structures highly expressed 20αHSD, an enzyme whose expression is normally inhibited by PRL. Of interest was the finding that although pregnancy could take place in either Stat 5a or Stat 5b null mice, fetal death occurs only in Stat 5b knockout mice (54). This phenotype is reminiscent of that seen in PRL-$R^{-/-}$ mice which had been administered progesterone, suggesting that whereas in the ovary Stats 5a and 5b are interchangeable, only Stat 5b may mediate the decidual PRL signal.

In addition to JAK2 and Stat 5, other modulators of this pathway have been found in the rat corpus luteum. *Src* homology 2 domain phosphotyrosine phosphatases 1 and 2 (SHP-1 and SHP-2) are two tyrosine phosphatases which contain SH-2 domains that mediate binding to phosphotyrosine residues. Whereas SHP-1 inhibits cytokine signaling through dephosphorylation of JAK2 (56) and Stat 5b (57), SHP-2 appears to positively regulate PRL signaling through the JAK2/Stat 5 pathway (58,59). The mechanism of SHP-2 action is not yet clear, although it can dephosphorylate Stat 5 (60) and possibly keep a certain level of Stat 5 available for response to ligand. SHP-2 is found in granulosa cells and luteal cells throughout pregnancy(34), where it may assist in PRL signaling. SHP-1 expression is more intriguing, not present in healthy follicles, yet found in the cytoplasm of atretic follicles and the nuclei of luteal cells (34). This led Richards and collaborators (34) to hypothesize that nuclear SHP-1 and SHP-2 do dephosphorylate Stat 5 after activation of transcription, whereas cytosolic SHP-1 may dephosphorylate Stat 5 before it even reaches the nucleus, leading to atresia.

Recent evidence from Hunzicker-Dunn, *et al.* (61,62), has revealed a novel signaling mechanism of PRL in the rat corpus luteum in the form of protein kinase C delta (PKCδ). This kinase, a member of the PKC family of serine/ threonine kinases, is activated by phosphatidyl serine and diacylglycerol (DAG) but does not require calcium for activation. Luteal expression of PKCδ is dramatically increased around the time of midpregnancy in the rat, and appears to be stimulated by estradiol (63), another important regulator of luteal function. *In vitro* studies have shown that rPL1 treatment of luteinized granulosa cells leads to translocation of PKCδ to the membrane fraction and activation of its kinase activity. The mechanism by which PRL could activate PKCδ is not yet known. PRL was shown to induce production of diacylglycerol by rat granulosa cells (64) and PI(3,4,5)P3 after activation of PI3 kinase (44,45). PI(3,4,5)P3 activates several PKC isoforms and activation of PKCδ may involve the PI3 kinase (65). Further studies by the same laboratory demonstrated that rPLI could induce phosphorylation of Stat 3 on both serine (S727) and tyrosine (Y705) (61,62). Both the tyrosine and serine phosphorylations could be blocked with the PKCδ specific inhibitor, rottlerin (62), suggesting PKCδ is upstream of both events. Although tyrosine phosphorylation of the Stats is required for their dimerization, translocation to the nucleus and transcriptional activity, further phosphorylation on serine 727 appears to optimize Stat 3 transcriptional activity (66-71) and PKCδ is a Stat 3 serine kinase. PRL and rPL1 stimulation of relaxin expression appears to involve not only JAK2, but also PKCδ activation and its association with Stat 3 leading to Stat 3 phosphorylation.

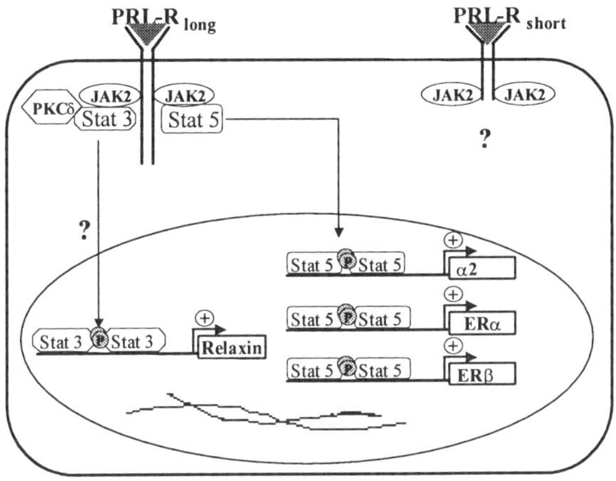

Figure 14-2. PRL signal transduction in the rat corpus luteum. Although signal transduction involving PRL-RS remains to be determined, some of the molecules involved in signaling through PRL-RL in luteal cells have been identified. One pathway through PRL-RL and JAK2 leads to Stat 5 tyrosine phosphorylation, dimerization and translocation to the nucleus to affect gene transcription. Three genes identified as targets of this pathway include α_2-macroglobulin (α2MG), estrogen receptor α (ERα) and estrogen receptor β (ERβ). Another pathway involves JAK2, PKCδ and Stat 3. PRL activates both PKCδ and JAK2 in luteal cells leading to phosphorylation on both serine and tyrosine residues of Stat 3 and eventually causes activation of relaxin expression.

ROLE OF PRL IN LUTEAL DIFFERENTIATION AND HYPERTROPHY

Following ovulation, the follicular cells undergo luteinization, a differentiation process which entails a number of changes in the steroidogenic cells as well as in the vascularization of the gland (for review see (14,72-75)). The granulosa cells exit the cell cycle and terminally differentiate, the molecular evidence for this halt in cell proliferation being the loss of expression of cyclin D2 and the increased expression of p21 and p27 (76). The importance of cell cycle exit in luteinization can be observed in the p27 knockout mouse, which fails to luteinize properly as follicular cells remain proliferative for a period following ovulation, and this leads to infertility (77). In addition, granulosa cells begin to express highly some genes, such as the PRL-R (34), while other genes, such as the follicle stimulating hormone receptor (78), disappear and still others change in their regulation, such as the P450 aromatase (79). Once formed, the corpus luteum expresses elevated levels of PRL-R and Stat 5, and gains the capacity to transduce a PRL signal to Stat 5 (34). The survival of the corpus luteum beyond diestrous depends on PRL stimulation, and associated with PRL stimulation is an increase in luteal size and vascularization. Since the luteal cell is terminally differentiated, changes in corpus luteum size related to these cells are due to hypertrophy (30). The $PRL^{-/-}$ mice (17), $PRL\text{-}R^{-/-}$ mice (18), and Stat 5 a/b knockout mice (55) are all infertile. In the $PRL\text{-}R^{-/-}$ mice (19,20) and Stat 5 a/b knockout mice (55) this appears to be largely due to a lack of functional corpora lutea, indicating a role for PRL in luteal formation and differentiation. The mechanism by which PRL regulates luteal cell survival and hypertrophy are not completely understood, but some potential intermediaries have been identified.

PRL together with estradiol causes an increase in the size of the corpus luteum (80). This is due, in part, to these homones' action on translation, namely on the protein elongation factor 2 (EF2). EF2 catalyzes the translocation of the ribosome along the mRNA transcript during translation. Eukaryotic elongation factor 2 kinase (eEF-2 kinase), previously known as calcium/calmodulin-dependent protein kinase III, phosphorylates EF2 on two threonine residues, leading to its deactivation (reviewed in (81)). PRL has been shown to cause dephosphorylation and thus activation of this enzyme in the corpus luteum *in vivo* (82) whereas estradiol stimulates its expression (83). Thus, EF2 serves as a point at which estradiol and PRL synergize to stimulate overall protein synthesis (82,83) leading to luteal hypertrophy (80).

Another potential mechanism of PRL-induced luteal hypertrophy is through insulin-like growth factor 1 (IGF-1). IGF-1 is known to regulate protein synthesis and growth in a number of tissues (84). The rat corpus luteum has been shown to express the IGF-1 receptor (85,86), and this receptor is expressed mainly on the large luteal cells, suggesting it as a regulator of cell size in this gland (86). PRL stimulates IGF-1 expression through the long form of the PRL-R (87) in a cell line (GG-CL) generated from the large luteal cells of the rat (88). The signaling pathway used by PRL to stimulate IGF-1 in the corpus luteum is still not known.

A third candidate is not so much a tropic mediator as a survival factor. Oxygen free radicals are generated via normal cellular processes (oxidative phosphorylation), in response to pathologic stimuli (drugs, radiation) and by immune cells in response to inflammatory stimuli. Oxygen free radicals can produce irreversible damage to proteins, membrane lipids and DNA (89). Elevated levels are associated with apoptosis (90) as well as luteolysis (91). Oxygen free radicals have been shown to be detrimental to luteal function, as evidenced by their ability to inhibit steroidogenesis (92-95). Superoxide dismutases (SODs) are a group of enzymes which are involved in scavenging free radicals by inactivating them and can protect cells from the damaging effects of free radicals (90). SODs are found in the corpora lutea of rat and human (96,97), and in the rat Cu/Zn SOD and Mg SOD are expressed throughout pregnancy (98). The names of these isoforms are based on their preferred metal cofactor, and the former is located in the cytoplasm while the later is a mitochondrial enzyme. PRL and rat placental lactogens have been shown to increase mRNA levels of both of these SOD isoforms *in vivo* and *in vitro* (99). Through the induction of these enzymes, PRL can prevent the cellular damage and decrease in progesterone secretion brought on by oxygen free radicals.

Vascularization is another crucial event in the formation and hypertrophy of the corpus luteum. The biphasic secretion of PRL induced by mating leads to increased endothelial cell proliferation in the corpus luteum of pregnancy (100). The size and angiogenesis of the corpus luteum increases rapidly from midpregnancy at a time when the placenta starts secreting high levels of PRL-like hormones in the form of rPL1 (101). Part of this effect may be due to the enhanced expression of the estrogen receptor by PRL (102,103). Estradiol has been shown to mediate the proliferation of endothelial cells in the corpus luteum (101,104), and in this endeavor estradiol appears to employ basic fibroblast growth factor (bFGF) (104,105). bFGF is a well known angiogenic factor which has been shown to induce proliferation in endothelial cells of the corpus luteum (101,106), and FGF receptors are expressed in endothelial cells of the ovine corpus luteum (107). PRL increases luteal expression of another mediator of vascularization, transforming growth factor beta (TGF-β) (108). TGF-β can also stimulate endothelial cell proliferation, although under some conditions it inhibits proliferation (109). In addition, both bFGF and TGF-β are involved in matrix degradation, which plays an important part in vascularization. Proteolysis appears to be stimulated by bFGF, while TGF-β leads to less protease activity (110), thus balancing out the angiogenesis process. Vascular endothelial growth factor (VEGF) is a third molecule implicated in the luteal vascularization process. VEGF is a growth factor known to be vital to angiogenesis in a number of physiological and pathological situations, including corpus formation (for review see (111)). VEGF is found in the corpus luteum of the rat (112), mouse (113), sheep (114), cow (115), primate (116,117), and human (118-120). Prevention of VEGF signaling during luteinization, through the injection of either VEGF antibodies (121) or soluble forms of the VEGF receptor (122), leads to decreased vascularization of the corpus luteum, diminished serum progesterone, and stunted growth of the corpus luteum. Thus, the formation of new blood vessels appears to be a vital event needed for the proper formation and development of the corpus luteum. In bovine (115), primate (117) and human (123) luteinized granulosa cells VEGF is stimulated by luteinizing hormone (LH). PRL has not been shown to regulate VEGF in the corpus luteum, but PRL does stimulate luteal LH-R expression (124). This suggests that vascularization is managed by both estradiol, through bFGF, and PRL, through TGF-β and these pathways may interact with LH and VEGF to appropriately regulate this process.

Another PRL regulated protein that may play a role in hypertrophy and vascularization is α_2-macroglobulin. PRL has been shown to enhance α_2-macroglobulin mRNA and protein levels in the corpus luteum (125). This increase is mediated by Stat 5b binding to its response element in the promoter of this gene (48,51,52). α_2-Macroglobulin is a member of the thiol ester plasma protein family, so named because of the unique cyclic thiol ester bond (126). This 720 kD tetramer acts as a protease inactivator as well as a carrier and targeting protein for cytokines. It can either inhibit cytokine action by preventing ligand-receptor interactions or promote cytokine responses by preventing cytokine degradation and assisting in their targeting to certain tissues by binding to the α_2-macroglobulin receptor. IGF-1 (127), TGF-β (128) and bFGF (129) are three of the numerous cytokines which have been demonstrated to bind to α_2-macroglobulin. By modulating expression of α_2-macroglobulin in the corpus luteum, PRL may regulate the actions of a number of factors and thus such diverse processes as hypertrophy and vascularization leading to tissue remodeling.

PRL-MEDIATED EFFECTS ON STEROIDOGENESIS

For many years, PRL was thought to act solely on the corpus luteum as a permissive agent, allowing it to maintain its function and enabling it to respond to other luteotropic stimuli, such as luteinizing hormone (LH) and estradiol (130). While these effects are true, increasing evidence has demonstrated that PRL and placental lactogens alone can increase *in vitro* progesterone production by mouse (131-133), hamster (134), rat (135-137), pig (138-140), mink (141), ferret (142) and human (143,144) luteal cells. Thus, it appears that the effects of PRL on luteal progesterone secretion can be classified as either indirect effects (e.g., permitting other luteotropic hormones to act) or direct effects which lead to increased progesterone production.

PERMISSIVE EFFECT OF PRL ON LH STIMULATION OF STEROIDOGENESIS

PRL is known to affect a number of genes whose expression is key to the function of the corpus luteum. One 73 kD protein known to be regulated by PRL is the luteinizing hormone receptor (LH-R), a member of the super-family of G-protein coupled receptors. Signaling through this receptor mainly involves activation of adenylate cyclase and production of cAMP (for review see (145)), although LH can activate other kinases (146-148). LH is an important stimulator of luteal steroidogenesis in a number of species, including the rat. It upregulates acutely progesterone production through the stimulation of cholesterol transport to P450 side chain cleavage on the inner mitochondrial membrane. This event is thought to be the rate-limiting step in progesterone synthesis, and appears to involve steroidogenic acute regulatory protein (149,150). Initial evidence suggesting an effect of PRL on LH-R levels were provided by the finding that LH stimulation of progesterone secretion requires a previous exposure of the corpus luteum to PRL (151). Subsequently, LH binding sites in rat corpora lutea were shown to be increased in response to PRL (152-154). More recently, PRL has been shown to stimulate LH-R mRNA expression in rat granulosa cells (155) and corpora lutea (124,156,157), however the signaling mechanism by which PRL brings about higher levels of LH-R expression is still not known.

PERMISSIVE EFFECT OF PRL ON ESTRADIOL PRODUCTION AND ACTION

Although LH can stimulate luteal progesterone secretion directly, its long-term effect, at least in the rat, is to promote ovarian estradiol production through the stimulation of P450 17α, an enzyme necessary for the conversion of progesterone to androstenedione (158-161). In fact, it has been shown that estradiol can maintain the corpus luteum in the absence of LH (162-164). Thus, the main luteotropic factors in the rat are PRL and estradiol, and these two hormones synergize to elicit maximal progesterone secretion by the rat corpus luteum (165). This synergy is not totally understood, but appears to result from interactions between these two hormones on a number of levels. We have already mentioned two proteins recognized to be regulated by both of these hormones, PKCδ and EF2. One of the earliest proteins found to be involved in this synergy, however, was the estrogen receptor. Early studies demonstrated that PRL is a prerequisite for estradiol stimulation of progesterone synthesis (165) and further demonstrated that PRL elevates the number of estrogen binding sites and allows estradiol to regulate luteal function (102). More recently two major isoforms of the estrogen receptor have been cloned, and termed estrogen receptor alpha (ERα) (166,167) and beta (ERβ) (168). These receptors are both members of the nuclear receptor superfamily and possess a high degree of homology, yet are the products of separate genes and have some differences in expression and regulation. In the rat corpus luteum, studies (103) have shown that both forms of the estrogen receptor are expressed, albeit with differing patterns over the course of pregnancy. PRL and placental lactogens have been shown to stimulate ERα and ERβ protein and mRNA both *in vivo* and *in vitro*, and recent evidence demonstrates that this is due to regulation at the transcriptional level. Promoter studies have revealed that PRL stimulates both ERα and ERβ through the JAK2/Stat 5 pathway, and that Stat 5 regulates their transcription through specific Stat 5 response elements present in each of their promoters (169,170).

In addition to stimulating levels of estradiol receptors in the corpus luteum, there is evidence that PRL plays a role in the biosynthesis of estradiol. The production of androstenedione substrate for estradiol synthesis in the ovary is stimulated by LH (158,162). At midpregnancy, LH levels drop (10), and the placenta becomes the source of androstenedione used by the corpus luteum to generate estradiol (171-175). The conversion of androstenedione to estradiol involves two enzymes, P450 aromatase and 17βhydroxysteroid dehydrogenase (17βHSD). Aromatase is expressed in the corpus luteum throughout pregnancy, however levels increase dramatically at midpregnancy (176,177). It appears that this increase in aromatase expression is stimulated by PRL and placental lactogens (79,176,178). The 17βHSD of the rodent corpus luteum termed 17βHSD7 was previously known in the rat as Prolactin Receptor-Associated Protein (PRAP) (179,180). This enzyme is expressed during mid to late pregnancy in rodent corpus luteum (179,181), associates specifically with the short form of the PRL-R (182), is phosphorylated on tyrosine residues, and contains a putative site for phosphorylation by JAK2 (182). Whether PRL, through the short form of the PRL-R, affects 17βHSD activity in the corpus luteum remains a subject of investigation.

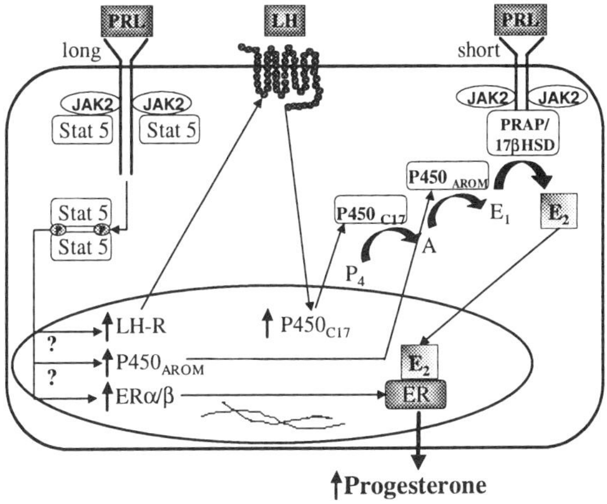

Figure 14-3. Permissive effects of PRL on estradiol synthesis and action. PRL plays a role in estradiol production by luteal cells. PRL has been shown to enhance the expression of LH-R allowing LH to stimulate P450 17α which helps generate the androstenedione substrate for estradiol biosynthesis. PRL also increases P450 aromatase expression allowing for the aromatization of androgens. Finally, PRAP/17βHSD7, responsible for the last step in estradiol biosynthesis, associates with PRL-RS. In addition to allowing for estradiol biosynthesis, PRL renders luteal cells responsve to estradiol by increasing ER α/β expression. The regulation of ER occurs at the transcriptional level through JAK2/Stat 5. However, the signaling mechanism by which PRL stimulates LH-R and P450 aromatase is still unknown.

PRL-MEDIATED REGULATION OF THE RELAXIN GENE IN THE CORPUS LUTEUM

Relaxin is a hormone secreted by the corpus luteum whose synthesis is regulated by both PRL and estradiol. Relaxin is structurally related to insulin, and the corpus luteum appears to be the main source of its production (for review see (183,184)). Although relaxin does not appear to be involved in progesterone secretion, it does play a role in the maintenance of gestation and in preparing the mother for parturition and lactation. Its purported effects include the lengthening of the interpubic ligament and cervical ripening, inhibiting contractions of the uterine myometrium and enhancing development of the mammary gland, and some of these effects were recently verified with the development of a relaxin knockout mouse (185). In the pig, PRL has been shown to increase relaxin secretion from corpora lutea at the end of gestation and following parturition (186,187). More extensive studies in the rat have shown that estradiol and PRL synergize to induce the expression of relaxin. Originally, reports established that relaxin expression depends on signals derived from the placental units and luteal estradiol (188-191). Recently, Hunzicker-Dunn, *et al.* (61,62), established the synergism between estradiol and rat placental lactogen on relaxin expression and further defined a novel signal transduction pathway of PRL-R agonists in the corpus luteum involving PKCδ. Estradiol was shown to increase PKCδ protein expression (63,192) whereas PRL causes its activation (61,62). In final analysis, the synergism between estradiol and PRL on relaxin appears to converge on PKCδ, with estradiol upregulating the expression of this kinase and PRL causing its activation.

PRL EFFECTS ON LUTEAL PROGESTERONE PRODUCTION

Two basic mechanisms by which PRL may affect progesterone secretion by the corpus luteum are substrate availability and enzyme activity. Cholesterol is the substrate for the biosynthesis of all steroid hormones and, following the identification of PRL as a luteotropic hormone, studies found that PRL could affect cholesterol metabolism in the corpus luteum. One early study noted that the ability of estradiol or LH to cause cholesterol accumulation in the cycling corpus luteum required previous exposure to lactogens (193). PRL in this case may have acted through elevation of LH or estradiol receptor expression (see above), but later reports found that PRL alone could increase luteal levels of cholesterol substrate for progesterone biosynthesis (151,194,195).

Cholesterol can be made available for steroidogenesis via *de novo* synthesis, uptake from circulating lipoproteins and liberation from cholesterol esters in cytoplasmic lipid droplets. The first, *de novo* synthesis, involves the generation of cholesterol from acetyl coenzyme A by the luteal cell itself. The rate limiting step in this pathway is catalyzed by 3-hydroxy-3-methylglutaryl coenzyme A reductase (HMG CoA reductase), an acutely regulated enzyme responsible for the conversion of HMG CoA to mevalonic acid. This enzyme is controlled by expression and is activated through dephosphorylation. This protein is present in the corpus luteum throughout gestation and its expression appears to be regulated by luteotropic factors, such as estradiol (196,197). Although PRL has been found to increase the amount of acetate incorporated into cholesterol, cholesterol esters and progestins in cultured luteal cells (151,195), no investigations to date implicate PRL as a regulator of HMG CoA reductase in the corpus luteum. Evidence suggests, however, that this is not the major pathway by which cholesterol is obtained for luteal steroidogenesis *in vivo*. Instead, the primary source of cholesterol for steroid biosynthesis appears to be circulating lipoproteins.

Cholesterol and cholesterol esters are obtained by the corpus luteum from low density lipoproteins (LDLs) and high density lipoproteins (HDLs), and *in vivo* and *in vitro* evidence suggests that this is the main source of cholesterol for steroidogenesis in most species (for review see (198)). There are some differences between species as to the preference of lipoprotein (i.e., HDL versus LDL). The rat corpus luteum has been shown to prefer HDLs (199), while most other species, including humans, appear to favor LDLs. This preference is likely related to the fact that in rats, most of the circulating cholesterol is contained in HDLs, while in other species, such as humans and pigs it is carried in LDLs. The uptake of LDLs occurs mainly through the binding of apolipoproteins B and E to the LDL receptor. The bound receptor undergoes endocytosis, and free cholesterol is formed following lysosomal digestion. HDL metabolism appears not to involve endocytosis of the particle, but rather selective uptake of cholesterol and cholesterol esters from the receptor-bound HDL particle. Recently, the receptor responsible for this selective uptake pathway for HDL has been identified as the class B scavenger receptor, SR-BI (200).

PRL has been shown to affect lipoprotein utilization in the corpus luteum of the rat, pig and ferret (142,201). In the porcine corpus luteum, PRL increases progesterone secretion from both dispersed luteal cells (138) and luteinized granulosa cells (139,140) in the presence of exogenous lipoproteins. Further studies in porcine luteinized granulosa cells demonstrated increased LDL membrane binding and degradation in response to PRL (202), suggesting that PRL upregulates LDL receptor expression and metabolism. In the rat, PRL appears to have similar effects on HDL uptake and metabolism. *In vitro* studies have shown that PRL can stimulate progesterone secretion in rat luteal cells when they are incubated with HDL or LDL (203,204), but not with 25-OH-cholesterol, which passes freely into the cell. *In vivo* studies in pseudopregnant rats have shown that PRL increases HDL binding sites on luteal membranes as well as uptake of HDL (205,206). Though binding studies have been done, the mechanism of PRL action on lipoprotein receptors is still not clear. Functional SR-B1 receptors have been found in the rat corpus luteum (207-209), but a stimulation of SR-B1 expression by PRL has not yet been reported. Even if PRL does regulate lipoprotein receptors, this may not be the rate-limiting step in progesterone secretion *in vivo*. The HDL receptor, SR-BI, has been knocked out, and female knockout mice, while infertile, secrete normal amounts of progesterone (210). The fertility defect instead appears to be due to abnormal embryo development.

Another method for increasing cholesterol availability lies in its release from intracellular stores of cholesterol esters within lipid droplets, which are abundant luteal cells. Cholesterol esterase activity is accomplished by a number of enzymes, but in the corpus luteum, the main source of this activity appears to be neutral cholesterol esterase activity. Neutral cholesterol esterase is an 84 kD cytoplasmic protein expressed in many tissues, including the ovary (211,212). The "neutral" in its name refers to its optimal pH, which is around 7. It is also referred to as hormone-sensitive lipase (HSL), as the same protein serves both functions in various tissues. Studies have shown that PRL increases luteal cholesterol esterase activity (213,214), and have correlated this with elevations in progesterone secretion (214). Regulation of this enzyme can occur through phosphorylation or through regulation of expression. With the cloning of hormone sensitive lipase/ neutral cholesterol esterase (HSL), it was shown that the effect of PRL on HSL is due to an increase in expression. In superovulated rats, inhibition of PRL secretion by hypophysectomy or bromocryptine treatment leads to a decrease in ovarian HSL mRNA and protein, while PRL replacement restores levels of HSL protein and activity (215).

The second mechanism by which PRL may directly affect progesterone secretion is through regulation of steroidogenic enzymes. Synthesis of progesterone requires two enzymes: the P450 side chain cleavage (P450scc) enzyme and 3 beta hydroxysteroid dehydrogenase (3βHSD). P450scc is located in the inner mitochondrial membrane and, with the assistance of the electron transport proteins adrenodoxin and adrenodoxin reductase, catalyzes the conversion of cholesterol to pregnenolone. 3βHSD is a microsomal enzyme localized in the smooth endoplasmic reticulum which catalyzes the conversion of pregnenolone to progesterone. PRL has been shown to increase levels of

P450scc in pregnant corpora lutea *in vivo* and in luteinized granulosa cells *in vitro* (178,216). In addition, transfection of Chinese hamster ovary (CHO) cells with the rabbit PRL-R and treatment with PRL leads to inhibition of mitogenesis and increased expression of P450scc, which the authors purport is a differentiation of these cells (217). Most studies regarding the effects of PRL on 3βHSD have shown a decrease related to the luteolytic activity of PRL (see below). One study did demonstrate that PRL could increase 3βHSD activity in a luteinized granulosa cell model (218). A more recent paper identified a Stat 5 response element in the promoter of human type II 3βHSD (219) and showed that PRL could cause Stat 5 binding to this response element and increase transcription in reporter gene assays. However, although PRL can stimulate P450scc and 3βHSD levels, these enzymes do not appear to be the rate limiting steps in luteal steroidogenesis, and are generally expressed in excess in the corpus luteum.

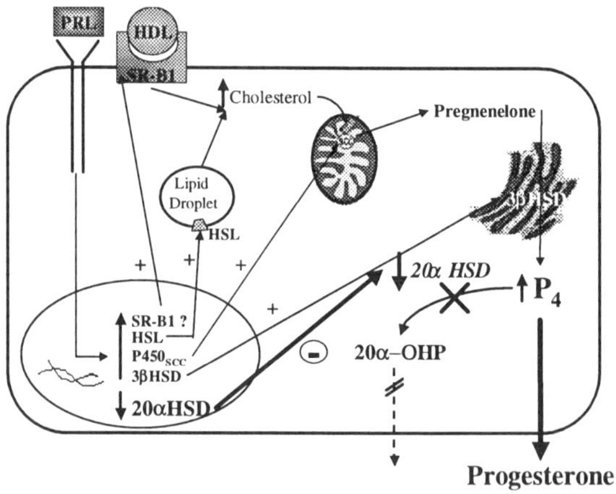

Figure 14-4. Effects of PRL on luteal progesterone production. PRL enhances luteal levels of cholesterol by increasing HSL, which liberates cholesterol from lipid droplets, and possibly by stimulating expression of SR-B1, the HDL receptor. Cholesterol is converted to progesterone by P450scc and 3βHSD, both of which have been shown to be upregulated by PRL. However, these enzymes do not appear to be rate limiting in luteal steroidogenesis, and are generally expressed in excess in the corpus luteum. The best-characterized mechanism for PRL-stimulated progesterone secretion is the inhibition of 20αHSD expression, which prevents catabolism of progesterone.

One of the best-established effects of PRL in the corpus luteum is its ability to inhibit the expression of 20α hydroxysteroid dehydrogenase (20αHSD). The 20αHSD enzyme is responsible for the conversion of progesterone, the major product of the corpus luteum during pregnancy, to 20α-hydroxyprogesterone, an inactive metabolite unable to maintain pregnancy. The inhibition of 20αHSD activity by PRL was shown both *in vivo* (220) and *in vitro* (221). A number of years ago, Albarracin, et al, demonstrated that this enzyme represents a major protein in the rat corpus luteum (222). More recently, the cDNA encoding 20αHSD has been cloned (223,224), allowing for a more discerning investigation as to this inhibitory effect of PRL. This, along with the development of a specific antibody, led to the demonstration that PRL-mediated inhibition of 20αHSD involved a decrease in 20αHSD mRNA and protein levels (225). The mechanism of PRL action on this gene has been investigated, although it is still not clear. The tyrosine kinase inhibitors AG18 and genistein did not block PRL mediated inhibition of 20αHSD in luteinized granulosa cells (50), but these may not be effective in preventing JAK2 activity. The serine/threonine phosphatase inhibitor, calyculin-A, was shown to hinder the PRL-induced decrease in 20αHSD activity in dispersed luteal cells (226), suggesting a role for phosphatases in this pathway. The 20αHSD promoter has been cloned, and sequence analysis identified, among other things, two putative sites for binding of Stat 5 (227). In addition, Stat 5 a/b knockout mice, who are infertile, have many ovarian structures resembling regressing corpora lutea which contain high levels of 20αHSD mRNA (55). Stat 5b was shown to be involved indirectly in the inhibition of 20αHSD expression by PRL (26,50). In decidual cells PRL inhibition of 20αHSD gene transcription appears to be mediated by PRL stimulation of interferon regulatory factor-1 (IRF-1). PRL upregulates IRF-1 expression (26,228,229) and IRF-1 is a potent inhibitor of 20αHSD gene expression (26).

LUTEOLYTIC EFFECTS OF PROLACTIN

While the effects of PRL on steroidogenesis, cell hypertrophy and relaxin production have been extensively studied, the paradoxical ability of PRL to induce structural luteolysis is less understood. Luteolysis was originally divided by Malven (5,230) into two separate events, termed functional luteolysis and structural luteolysis. The former is generally defined as the loss in the capacity of the corpus luteum to secrete progesterone, at least at a level necessary to support pregnancy. Functional luteolysis may be a misnomer, as this event is not accompanied by a cessation in steroidogenic activity by the corpus luteum. Rather, progesterone biosynthesis does take place but progesterone is rapidly metabolized to 20α-OH progesterone by 20αHSD and is not secreted. The lack of PRL causes the induction of 20αHSD gene expression (50,225). Structural luteolysis, on the other hand, refers to the morphological destruction of the corpus luteum, and normally follows functional luteolysis. Whereas PRL plays an essential role in the normal function of the corpus luteum, it is also responsible for structural luteolysis, which involves luteal cell death and the degeneration and disappearance of this gland. Also, in contrast to functional luteolysis, structural luteolysis may be associated with diminished steroidogenesis (231), as PRL causes decreased 3β-HSD expression during structural luteolysis (232,233). Inhibition of proestrous PRL surges in cycling rats causes the accumulation of corpora lutea in their ovaries (234,235) suggesting that PRL secreted at proestrous is important in transforming the corpus luteum into a degenerating corpus albicans. In addition, following hypophysectomy, the loss of PRL leads to functional luteolysis, and 20α-hydroxyprogesterone becomes the major secretory product of these corpora lutea (236). Though nonessential at this point, corpora lutea persist for an indefinite amount of time, with an early study documenting their survival out to 14.5 months (237). When PRL is administered to these rats two to three days after hypophysectomy, the corpora lutea disintegrate and undergo a massive reduction in size (5,230). This indicates that the absence of PRL for several days causes significant changes in the corpus luteum leading PRL to affect a totally different set of molecules. In more recent years, some of the genes regulated during this event have been found, which helps to explain the changes that were initially observed.

Early studies identified structural luteolysis by gross anatomical observations, such as a decrease in ovarian weight, as well as by histological observations, which included an increase in "the total number of connective tissue or stromal elements in the corpus" and a decrease in "recognizable lutein cells" (5). Recent investigations demonstrated that PRL- induced structural luteolysis involves apoptosis, or programmed cell death, and PRL causes DNA degradation in corpora lutea at proestrous (238). This effect is specific to corpora lutea which have undergone functional luteolysis (239). PRL is an anti-apoptotic agent in many tissues (47), including functional corpora lutea (239), so it is perplexing as to why it is different in functionally regressed corpora lutea. The process of apoptosis is sometimes thought of as a failed mitosis, as these two processes share some of their cellular machinery to elicit their effects (240). A recent study found that PRL induces the expression of proliferating cell nuclear antigens (PCNA) in regressing corpora lutea (241), suggesting that this apoptosis might be a failed mitosis. The functional state of the luteal cell, therefore, somehow determines the response of the luteal cell to PRL- survival or apoptosis.

The apoptotic effect of PRL appears, most probably, not to be direct on the luteal cells, but may involve the participation of other cell types, namely cells of the immune system. One cell type identified as a potential regulator is the monocyte/macrophage. These cells are a normal component of cell-mediated immunity and are capable of inducing apoptosis through a number of mechanisms. Monocyte/ macrophage invasion has been associated with luteal involution in a wide variety of species (242-250) and in the rat corpus luteum this invasion has been shown to be initiated by PRL (100,231). Migration of leukocytes from the blood into tissues is a complex process which involves a number of molecules (251). One facet of this process is chemoattraction, where molecules secreted at the site bind to specific receptors on leukocytes and spur their transit into the tissue. Monocyte chemoattractant protein-1 (MCP-1) is a chemokine that serves this purpose; it acts specifically on monocytes to stimulate their migration into tissues and is secreted by a variety of cell types (252). MCP-1 is expressed in regressing corpora lutea from many species, including the rat (253) and human (254). Studies in the rat have shown that regressing corpora lutea begin to express MCP-1 in response to PRL, and this expression precedes the infiltration of macrophages (231,255). In addition to chemoattraction, the migration of monocytes into tissues also requires the vascular bed of the tissue to express the proper adhesion proteins. Monocytes flowing through the bloodstream need to attach to the endothelial cells before they cross it and this involves the interaction of proteins on the leukocyte and endothelial cell (251,256). ICAM-1 is a member of the immunoglobulin superfamily whose expression is induced by cytokines and other inflammatory signals. It is expressed on endothelial cells, although not exclusively, and mediates infiltration of various immune cells by binding to integrin receptors on these cells, namely LFA-1 and Mac-1. PRL was recently found to increase ICAM-1 expression in the regressing corpora lutea of hypophysectomized rats, and this expression appeared to be related to vascularization (257). A study on MCP-1 found that this chemokine can increase the ability of LFA-1 and Mac-1 to bind to ICAM-1 (258), therefore PRL may be increasing binding activity in both the endothelial cell and the monocyte through MCP-1 and ICAM-1. It is interesting to note also that although MCP-1 was initially discovered and named based on its ability to specifically activate monocyte migration (259), more recently it was found to also mediate T-cell migration (260). In addition, the integrins which bind ICAM-1 (LFA-1, Mac-1) are also found on T-cells (251), so both MCP-1 and ICAM-1 may also be assisting in T-cell invasion during luteal involution.

Indeed, other studies have implicated T-lymphocytes in mediating PRL-induced luteal regression. Although T-cells are capable of inducing apoptosis through a number of mechanisms, investigations in the corpus luteum have been focused on Fas-mediated apoptosis. Fas/APO-1/CD95 is a member of the tumor necrosis factor (TNF) receptor family, and binding of Fas ligand (FasL) to this protein induces apoptosis in cells expressing Fas (for review see (261)). FasL is a membrane-bound protein, related to TNF, expressed in T-cells and natural killer cells. It can be cleaved to produce a soluble form, but both forms are capable of inducing apoptosis. Fas is expressed in the corpus luteum of rodents and in human granulosa/lutein cells (262,263,264). Treatment with an antibody that activates Fas leads to apoptosis in the corpus luteum (262). FasL is also present in the corpus luteum, and its expression is enhanced by PRL. Treatment with a blocking anti-FasL antibody inhibits the ability of PRL to induce apoptosis in culture (265). Separation of the various cell types demonstrated that the large cell population, which included the luteal cells, contained Fas, while the small cell population, which included T-cells, contained FasL. Since T-cells express the PRL-R (266-268), PRL may act on these cells to induce the expression of FasL. Either fraction cultured separately in the presence of PRL did not undergo apoptosis, but when both fractions were cultured together PRL did induce apoptosis (269). These studies suggest that PRL-induced apoptosis in regressing corpora lutea involves Fas on luteal cells interacting with FasL on T-cells. Because progesterone reduces the expression of Fas, but not that of FasL, Takahashi and collaborators (270) proposed that progesterone may suppress PRL-induced luteal cell apoptosis by inhibiting luteal Fas expression and that luteal cells, which express 20α-HSD lose their progesterone secreting capacity and begin to express Fas on their surface. PRL stimulates FasL in immune cells leading to the interaction with Fas in luteal cells and induction of cell death. Whether progesterone indeed plays a role in mediating PRL-induced apoptosis is not widely accepted (271-273) and remains the subject of further investigation.

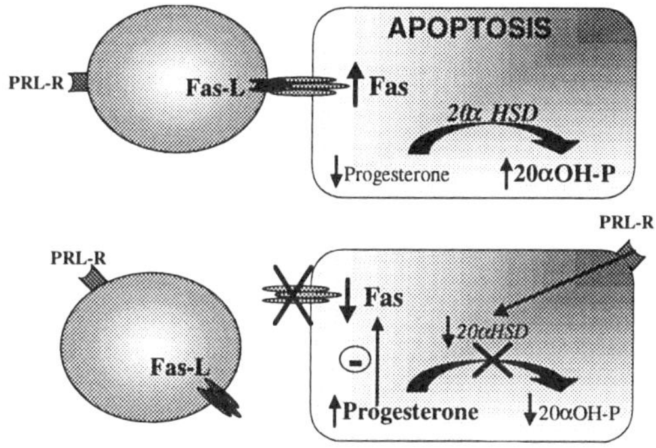

Figure 14-5. PRL and progesterone regulate Fas-mediated apoptosis in the corpus luteum. Corpora lutea maintained in the absence of PRL begin to express 20αHSD which transforms the progesterone synthesized to 20α-OH progesterone. The decrease in progesterone allows for Fas expression in luteal cells. PRL stimulation of these corpora lutea leads to stimulation of FasL expression in T-cells, which interacts with Fas on the luteal cells to induce apoptosis. In the lower panel, during pregnancy, PRL inhibits 20αHSD expression and progesterone is the major product of the corpus luteum. Progesterone inhibits Fas expression in the luteal cell, and thus prevents interaction with FasL on T-cells and apoptosis. (Modified from (269))

A number of other changes have been shown to occur during PRL-induced luteal regression. Regression is associated with tissue remodeling. As luteal cells undergo apoptosis the area becomes a scar which eventually disappears. This remodeling appears to involve activation of matrix-metalloproteinases (MMPs) and tissue inhibitors of metalloproteinases (TIMPs). Studies have demonstrated increased MMP-1 and MMP-2 activity in PRL-induced luteolysis, as well as increases in TIMP-1 (274,275) which may mediate some of the structural changes by regulating collagen synthesis and degradation. As mentioned above, oxygen radicals are associated with luteolysis, and one oxygen radical scavenger that is abundant in the functional corpus luteum is ascorbate (276). PRL induces ascorbate depletion in regressing corpora lutea (274), thus promoting free radical damage. Whether this is a primary effect on ascorbate secretion or secondary to scavenging peroxides generated from macrophages remains to be determined.

CONCLUSION

In summary, PRL not only regulates a wide variety of biological processes in many species, but its action in a single endocrine gland such as the corpus luteum is highly complex and multi-faceted. PRL regulates a great number of genes in both luteal, immune and endothelial cells of the corpus luteum leading to either survival and stimulation of this gland or its demise. These two opposite and contrasting effects of PRL appear to involve not only a multitude of genes and different cell types but may also depend on the progesterone capability of the luteal cell and the expression of 20αHSD. Future studies may shed some light on the mechanism behind the multiplicity of effects PRL has on the corpus luteum.

ACKNOWLEDGMENT

This chapter was supported by NIH grant HD11119.

REFERENCES

1. Dresel I. The effect of prolactin on the estrous cycle of non-parous mice. Science. 1935;82:173.
2. Astwood EB. The regulation of corpus luteum function by hypophysial luteotrophin. Endocrinology. 1941;28:309-320.
3. Evans HM, Simpson ME, Lyons WR, Turpeinen K. Anterior pituitary hormones which favor the production of traumatic uterine placentomata. Endocrinology. 1941;28:933-945.
4. Tobin CE. Effects of lactogen on normal and adrenalectomized female rats. Endocrinology. 1942;31:197-200.
5. Malven PV, Sawyer CH. A luteolytic action of prolactin in hypophysectomized rats. Endocrinology. 1966;79:268-274.
6. Hinds LA. Control of pregnancy, parturition and luteolysis in marsupials. Reprod Fertil Dev. 1990;2:535-552.
7. Tyndale-Biscoe CH, Hinds L. Hormonal control of the corpus luteum and embryonic diapause in macropodid marsupials. J Reprod Fertil Suppl. 1981;29:111-117.
8. Papke RL, Concannon PW, Travis HF, Hansel W. Control of luteal function and implantation in the mink by prolactin. J Anim Sci. 1980;50:1102-1107.
9. Freeman ME. The neuroendocrine control of the ovarian cycle in the rat. In: Knobil E, Neill JD, eds. The Physiology of Reproduction. Second ed. New York: Raven Press; 1994:613-658.
10. Morishige W, Pepe GJ, Rothchild I. Serum luteinizing hormone, prolactin and progesterone levels during pregnancy in the rat. Endocrinology. 1973;92:1527-1530.
11. Freeman ME, Neill JD. The pattern of prolactin secretion during pseudopregnancy in the rat: a daily nocturnal surge. Endocrinology. 1972;90:1292-1294.
12. Prigent-Tessier A, Tessier C, Hirosawa-Takamori M, Boyer C, Ferguson-Gottschall S, Gibori G. Rat decidual prolactin. Identification, molecular cloning, and characterization. J Biol Chem. 1999;274:37982-37989.
13. Gibori G, Rothchild I, Pepe GJ, Morishige WK, Lam P. Luteotrophic action of decidual tissue in the rat. Endocrinology. 1974;95:1113-1118.
14. Gibori G. The corpus luteum of pregnancy. In: Adashi EY, Leung PCK, eds. The Ovary. New York: Raven Press; 1993:261-317.
15. Ogren L, Talamantes F. The placenta as an endocrine organ: polypeptides. In: Knobil E, Neill JD, eds. The Physiology of Reproduction. Second ed. New York: Raven Press; 1994:875-946.
16. Soares M, Faria TN, Roby KF, Deb S. Pregnancy and the prolactin family of hormones: coordination of anterior pituitary, uterine, and placental expression. Endocr Rev. 1991;12:402-423.
17. Horseman ND, Zhao W, Montecino-Rodriguez E, Tanaka M, Nakashima K, Engle SJ, Smith F, Markoff E, Dorshkind K. Defective mammopoiesis, but normal hematopoiesis, in mice with a targeted disruption of the prolactin gene. EMBO J. 1997;16:6926-6935.

18. Ormandy CJ, Camus A, Barra J, Damotte D, Lucas B, Buteau H, Edery M, Brousse N, Babinet C, Binart N, Kelly PA. Null mutation of the prolactin receptor gene produces multiple reproductive defects in the mouse. Genes Dev. 1997;11:167-178.
19. Binart N, Helloco C, Ormandy CJ, Barra J, Clement-Lacroix P, Baran N, Kelly PA. Rescue of preimplantatory egg development and embryo implantation in prolactin receptor-deficient mice after progesterone administration. Endocrinology. 2000;141:2691-2697.
20. Reese J, Binart N, Brown N, Ma WG, Paria BC, Das SK, Kelly PA, Dey SK. Implantation and decidualization defects in prolactin receptor (PRLR)-deficient mice are mediated by ovarian but not uterine PRLR. Endocrinology. 2000;141:1872-1881.
21. Jayatilak Pg GG. Ontogeny of prolactin receptors in rat decidual tissue: binding by a locally produced prolactin-like hormone. J Endocrinol. 1986;110:115-121.
22. Gu Y, Srivastava RK, L. CD, Linzer DI, G. G. The decidual prolactin receptor and its regulation by decidua-derived factors. Endocrinology. 1996;137:4878-4885.
23. Basuray R, Jaffe RC, Gibori G. Role of decidual luteotropin and prolactin in the control of luteal cell receptors of estradiol. Biol Reprod. 1983;28:551-556.
24. Tessier C, Deb S, Prigent-Tessier A, Ferguson-Gottschall S, Gibori GB, Shiu RPC, Gibori G. Estrogen receptors alpha and beta in the rat decidua: cell specific expression and differential regulation by steroid hormones and prolactin. Endocrinology (in press). 2000.
25. Deb S, Tessier C, Prigent-Tessier A, Barkai U, Ferguson-Gottschall S, Srivastava RK, Faliszek J, Gibori G. The expression of interleukin-6 (IL-6), IL-6 receptor, and gp130-kilodalton glycoprotein in the rat decidua and a decidual cell line: regulation by 17beta-estradiol and prolactin. Endocrinology. 1999;140:4442-4450.
26. Tessier C, Prigent-Tessier A, Ferguson-Gottschall S, Djiane J, Gibori G. Decidual expression of 20alpha-hydroxysteroid dehydrogenase and mechanism of its regulation by rat decidual PRL (rdPRL). In: Meeting of the Society for the Study of Reproduction. Madison, WI: Society for the Study of Reproduction; 2000:196.
27. Dunaif AE, Zimmerman EA, Friesen HG, Frantz AG. Intracellular localization of prolactin receptor and prolactin in the rat ovary by immunocytochemistry. Endocrinology. 1982;110:1465-1471.
28. Shirota M, Banville D, Ali S, Jolicoeur C, Boutin JM, Edery M, Djiane J, Kelly PA. Expression of two forms of prolactin receptor in rat ovary and liver. Mol Endocrinol. 1990;4:1136-1143.
29. Hu ZZ, Dufau ML. Multiple and differential regulation of ovarian prolactin receptor messenger RNAs and their expression. Biochem Biophys Res Commun. 1991;181:219-225.
30. Nelson SE, McLean MP, Jayatilak PG, Gibori G. Isolation, characterization, and culture of cell subpopulations forming the pregnant rat corpus luteum. Endocrinology. 1992;130:954-966.
31. Clarke DL, Linzer DI. Changes in prolactin receptor expression during pregnancy in the mouse ovary. Endocrinology. 1993;133:224-232.
32. Boutin JM, Jolicoeur C, Okamura H, Gagnon J, Edery M, Shirota M, Banville D, Dusanter-Fourt I, Djiane J, Kelly PA. Cloning and expression of the rat prolactin receptor, a member of the growth hormone/prolactin receptor gene family. Cell. 1988;53:69-77.
33. Telleria CM, Parmer TG, Zhong L, Clarke DL, Albarracin CT, Duan WR, Linzer DI, Gibori G. The different forms of the prolactin receptor in the rat corpus luteum: developmental expression and hormonal regulation in pregnancy. Endocrinology. 1997;138:4812-4820.
34. Russell DL, Richards JS. Differentiation-dependent prolactin responsiveness and stat (signal transducers and activators of transcription) signaling in rat ovarian cells. Mol Endocrinol. 1999;13:2049-2064.

35. Lebrun JJ, Ali S, Goffin V, Ullrich A, Kelly PA. A single phosphotyrosine residue of the prolactin receptor is responsible for activation of gene transcription. Proc Natl Acad Sci U S A. 1995;92:4031-4035.
36. Chang WP, Clevenger CV. Modulation of growth factor receptor function by isoform heterodimerization. Proc Natl Acad Sci U S A. 1996;93:5947-5952.
37. Gouilleux F, Wakao H, Mundt M, Groner B. Prolactin induces phosphorylation of Tyr694 of Stat5 (MGF), a prerequisite for DNA binding and induction of transcription. EMBO J. 1994;13:4361-4369.
38. Berlanga J, Garcia-Ruiz JP, Perrot-Applanat M, Kelly PA, Edery M. The short form of the prolactin (PRL) receptor silences PRL induction of the beta-casein gene promoter. Mol Endocrinol. 1997;11:1449-1457.
39. Perrot-Applanat M, Gualillo O, Pezet A, Vincent V, Edery M, Kelly PA. Dominant negative and cooperative effects of mutant forms of prolactin receptor. Mol Endocrinol. 1997;11:1020-1032.
40. Silvennoinen O, Saharinen P, Paukku K, Takaluoma K, Kovanen P. Cytokine receptor signal transduction through Jak tyrosine kinases and Stat transcription factors. Apmis. 1997;105:497-509.
41. Buckley AR, Rao YP, Buckley DJ, Gout PW. Prolactin-induced phosphorylation and nuclear translocation of MAP kinase in Nb2 lymphoma cells. Biochem Biophys Res Commun. 1994;204:1158-1164.
42. Das R, Vonderhaar BK. Transduction of prolactin's (PRL) growth signal through both long and short forms of the PRL receptor. Mol Endocrinol. 1995;9:1750-1759.
43. Piccoletti R, Maroni P, Bendinelli P, Bernelli-Zazzera A. Rapid stimulation of mitogen-activated protein kinase of rat liver by prolactin. Biochem J. 1994;303:429-433.
44. Ratovondrahona D, Fournier B, Odessa MF, Dufy B. Prolactin stimulation of phosphoinositide metabolism in CHO cells stably expressing the PRL receptor. Biochem Biophys Res Commun. 1998;243:127-130.
45. Hunter S, Koch BL, Anderson SM. Phosphorylation of cbl after stimulation of Nb2 cells with prolactin and its association with phosphatidylinositol 3-kinase. Mol Endocrinol. 1997;11:1213-1222.
46. al-Sakkaf K, Dobson PR, Brown BL. Activation of phosphatidylinositol 3-kinase by prolactin in Nb2 cells. Biochem Biophys Res Commun. 1996;221:779-784.
47. Bole-Feysot C, Goffin V, Edery M, Binart N, Kelly PA. Prolactin (PRL) and its receptor: actions, signal transduction pathways and phenotypes observed in PRL receptor knockout mice. Endocr Rev. 1998;19:225-268.
48. Russell DL, Norman RL, Dajee M, Liu X, Henninghausen L, Richards JS. Prolactin-induced activation and binding of stat proteins to the IL-6RE of the alpha 2-macroglobulin (alpha 2M) promoter: relation to the expression of alpha 2M in the rat ovary. Biol Reprod. 1996;55:1029-1038.
49. Ruff SJ, Leers-Sucheta S, Melner MH, Cohen S. Induction and activation of Stat 5 in the ovaries of pseudopregnant rats. Endocrinology. 1996;137:4095-4099.
50. Zhong L, Parmer TG, Robertson MC, Gibori G. Prolactin-mediated inhibition of 20alpha-hydroxysteroid dehydrogenase gene expression and the tyrosine kinase system. Biochem Biophys Res Commun. 1997;235:587-592.
51. Dajee M, Fey GH, Richards JS. Stat 5b and the orphan nuclear receptors regulate expression of the alpha2-macroglobulin (alpha2M) gene in rat ovarian granulosa cells. Mol Endocrinol. 1998;12:1393-1409.
52. Dajee M, Kazansky AV, Raught B, Hocke GM, Fey GH, Richards JS. Prolactin induction of the alpha 2-Macroglobulin gene in rat ovarian granulosa cells: stat 5 activation and binding to the interleukin-6 response element. Mol Endocrinol. 1996;10:171-184.

53. Liu X, Robinson GW, Wagner KU, Garrett L, Wynshaw-Boris A, Hennighausen L. Stat5a is mandatory for adult mammary gland development and lactogenesis. Genes Dev. 1997;11:179-186.
54. Udy GB, Towers RP, Snell RG, Wilkins RJ, Park SH, Ram PA, Waxman DJ, Davey HW. Requirement of STAT5b for sexual dimorphism of body growth rates and liver gene expression. Proc Natl Acad Sci U S A. 1997;94:7239-7244.
55. Teglund S, McKay C, Schuetz E, van Deursen JM, Stravopodis D, Wang D, Brown M, Bodner S, Grosveld G, Ihle JN. Stat5a and Stat5b proteins have essential and nonessential, or redundant, roles in cytokine responses. Cell. 1998;93:841-850.
56. Klingmuller U, Lorenz U, Cantley LC, Neel BG, Lodish HF. Specific recruitment of SH-PTP1 to the erythropoietin receptor causes inactivation of JAK2 and termination of proliferative signals. Cell. 1995;80:729-738.
57. Ram PA, Waxman DJ. Interaction of growth hormone-activated STATs with SH2-containing phosphotyrosine phosphatase SHP-1 and nuclear JAK2 tyrosine kinase. J Biol Chem. 1997;272:17694-17702.
58. Ali S, Chen Z, Lebrun JJ, Vogel W, Kharitonenkov A, Kelly PA, Ullrich A. PTP1D is a positive regulator of the prolactin signal leading to beta-casein promoter activation. EMBO J. 1996;15:135-142.
59. Berchtold S, Volarevic S, Moriggl R, Mercep M, Groner B. Dominant negative variants of the SHP-2 tyrosine phosphatase inhibit prolactin activation of Jak2 (janus kinase 2) and induction of Stat5 (signal transducer and activator of transcription 5)-dependent transcription. Mol Endocrinol. 1998;12:556-567.
60. Yu C, Jin YJ, Burakoff SJ. Cytosolic tyrosine dephosphorylation of STAT5. Potential role of SHP-2 in STAT5 regulation. J Biol Chem. 2000;275:599-604.
61. Peters CA, Maizels ET, Hunzicker-Dunn M. Activation of PKC delta in the rat corpus luteum during pregnancy. Potential role of prolactin signaling. J Biol Chem. 1999;274:37499-37505.
62. Peters CA, Maizels ET, Robertson MC, Shiu RPC, Soloff MS, Hunzicker-Dunn M. Induction of relaxin messenger RNA expression in response to prolactin receptor activation requires protein kinase C delta signaling. Mol Endocrinol. 2000;14:576-590.
63. Cutler RJ, Maizels ET, Hunzicker-Dunn M. Delta protein kinase-C in the rat ovary: estrogen regulation and localization. Endocrinology. 1994;135:1669-1678.
64. Fanjul LF, Marrero I, Gonzalez J, Quintana J, Santana P, Estevez F, Mato JM, Ruiz de Galarreta CM. Does oligosaccharide-phosphatidylinositol (glycosyl-phosphatidylinositol) hydrolysis mediate prolactin signal transduction in granulosa cells? Eur J Biochem. 1993;216:747-755.
65. Le Good JA, Ziegler WH, Parekh DB, Alessi DR, Cohen P, Parker PJ. Protein kinase C isotypes controlled by phosphoinositide 3-kinase through the protein kinase PDK1. Science. 1998;281:2042-2045.
66. Wen Z, Zhong Z, Darnell JE, Jr. Maximal activation of transcription by Stat1 and Stat3 requires both tyrosine and serine phosphorylation. Cell. 1995;82:241-250.
67. Wen Z, Darnell JE, Jr. Mapping of Stat3 serine phosphorylation to a single residue (727) and evidence that serine phosphorylation has no influence on DNA binding of Stat1 and Stat3. Nucleic Acids Res. 1997;25:2062-2067.
68. Uddin S, Majchrzak B, Wang PC, Modi S, Khan MK, Fish EN, Platanias LC. Interferon-dependent activation of the serine kinase PI 3'-kinase requires engagement of the IRS pathway but not the Stat pathway. Biochem Biophys Res Commun. 2000;270:158-162.
69. Yokogami K, Wakisaka S, Avruch J, Reeves SA. Serine phosphorylation and maximal activation of STAT3 during CNTF signaling is mediated by the rapamycin target mTOR. Curr Biol. 2000;10:47-50.
70. Zhang X, Blenis J, Li HC, Schindler C, Chen-Kiang S. Requirement of serine phosphorylation for formation of STAT-promoter complexes. Science. 1995;267:1990-1994.

71. Ng J CD. STAT3 is a serine kinase target in T lymphocytes. Interleukin 2 and T cell antigen receptor signals converge upon serine 727. J Biol Chem. 1997;272:24542-24549.
72. Robker RL, Richards JS. Hormonal control of the cell cycle in ovarian cells: proliferation versus differentiation. Biol Reprod. 1998;59:476-482.
73. Richards JS, Fitzpatrick SL, Clemens JW, Morris JK, Alliston T, Sirois J. Ovarian cell differentiation: a cascade of multiple hormones, cellular signals, and regulated genes. Recent Prog Horm Res. 1995;50:223-254.
74. Richards JS. Hormonal control of gene expression in the ovary. Endocr Rev. 1994;15:725-751.
75. Richards JS, Russell DL, Robker RL, Dajee M, Alliston TN. Molecular mechanisms of ovulation and luteinization. Mol Cell Endocrinol. 1998;145:47-54.
76. Robker RL, Richards JS. Hormone-induced proliferation and differentiation of granulosa cells: a coordinated balance of the cell cycle regulators cyclin D2 and p27Kip1. Mol Endocrinol. 1998;12:924-940.
77. Tong W, Kiyokawa H, Soos TJ, Park MS, Soares VC, Manova K, Pollard JW, Koff A. The absence of p27Kip1, an inhibitor of G1 cyclin-dependent kinases, uncouples differentiation and growth arrest during the granulosa-luteal transition. Cell Growth Differ. 1998;9:787-794.
78. Camp TA, Rahal JO, Mayo KE. Cellular localization and hormonal regulation of follicle-stimulating hormone and luteinizing hormone receptor messenger RNAs in the rat ovary. Mol Endocrinol. 1991;5:1405-1417.
79. Krasnow JS, Hickey GJ, Richards JS. Regulation of aromatase mRNA and estradiol biosynthesis in rat ovarian granulosa and luteal cells by prolactin. Mol Endocrinol. 1990;4:13-12.
80. Alloiteau JJ. Hypertrophie du corps jaune chez la ratte a antehypophyse greffee dans le rein et recevant du propionate de testosterone. Secretion d'un oestrogene par le corps jaune. Current Research of Academic Science (Paris). 1959;249:1718-1724.
81. Ryazanov AG, Spirin AS. Phosphorylation of elongation factor 2: a mechanism to shut off protein synthesis for reprogramming gene expression. In: Ilan J, ed. Translational Regulation of Gene Expression. New York: Plenum; 1993:433-455.
82. Albarracin CT, Palfrey HC, Duan WR, Rao MC, Gibori G. Prolactin regulation of the calmodulin-dependent protein kinase III elongation factor-2 system in the rat corpus luteum. J Biol Chem. 1994;269:7772-7776.
83. Rao MC, Palfrey HC, Nash NT, Greisman A, Jayatilak PG, Gibori G. Effects of estradiol on calcium-specific protein phosphorylation in the rat corpus luteum. Endocrinology. 1987;120:1010-1018.
84. LeRoith D, Werner H, Beitner-Johnson D, Roberts CT, Jr. Molecular and cellular aspects of the insulin-like growth factor I receptor. Endocr Rev. 1995;16:143-163.
85. Zhou J, Chin E, Bondy C. Cellular pattern of insulin-like growth factor-I (IGF-I) and IGF-I receptor gene expression in the developing and mature ovarian follicle. Endocrinology. 1991;129:3281-3288.
86. Parmer TG, Roberts CT, Jr., LeRoith D, Adashi EY, Khan I, Solan N, Nelson S, Zilberstein M, Gibori G. Expression, action, and steroidal regulation of insulin-like growth factor-I (IGF-I) and IGF-I receptor in the rat corpus luteum: their differential role in the two cell populations forming the corpus luteum. Endocrinology. 1991;129:2924-2932.
87. Sugino N, Telleria CM, Tessier C, Gibori G. Regulation and role of the insulin-like growth factor I system in rat luteal cells. J Reprod Fertil. 1999;115:349-355.
88. Sugino N, Zilberstein M, Srivastava RK, Telleria CM, Nelson SE, Risk M, Chou JY, Gibori G. Establishment and characterization of a simian virus 40-transformed temperature-sensitive rat luteal cell line. Endocrinology. 1998;139:1936-1942.

89. Fridovich I. Fundamental aspects of reactive oxygen species, or what's the matter with oxygen? Ann N Y Acad Sci. 1999;893:13-18.
90. McCord JM. Superoxide radical: controversies, contradictions, and paradoxes. Proc Soc Exp Biol Med. 1995;209:112-117.
91. Riley JC, Behrman HR. *In vivo* generation of hydrogen peroxide in the rat corpus luteum during luteolysis. Endocrinology. 1991;128:1749-1753.
92. Behrman HR, Aten RF. Evidence that hydrogen peroxide blocks hormone-sensitive cholesterol transport into mitochondria of rat luteal cells. Endocrinology. 1991;128:2958-2966.
93. Kodaman PH, Aten RF, Behrman HR. Lipid hydroperoxides evoke antigonadotropic and antisteroidogenic activity in rat luteal cells. Endocrinology. 1994;135:2723-2730.
94. Musicki B, Aten RF, Behrman HR. Inhibition of protein synthesis and hormone-sensitive steroidogenesis in response to hydrogen peroxide in rat luteal cells. Endocrinology. 1994;134:588-595.
95. Endo T, Aten RF, Leykin L, Behrman HR. Hydrogen peroxide evokes antisteroidogenic and antigonadotropic actions in human granulosa luteal cells. J Clin Endocrinol Metab. 1993;76:337-342.
96. Sugino N, Takiguchi S, Kashida S, Karube A, Nakamura Y, Kato H. Superoxide dismutase expression in the human corpus luteum during the menstrual cycle and in early pregnancy. Mol Hum Reprod. 2000;6:19-25.
97. Sugino N, Nakamura Y, Takeda O, Ishimatsu M, Kato H. Changes in activities of superoxide dismutase and lipid peroxide in corpus luteum during pregnancy in rats. J Reprod Fertil. 1993;97:347-351.
98. Sugino N, Telleria CM, Gibori G. Differential regulation of copper-zinc superoxide dismutase and manganese superoxide dismutase in the rat corpus luteum: induction of manganese superoxide dismutase messenger ribonucleic acid by inflammatory cytokines. Biol Reprod. 1998;59:208-215.
99. Sugino N, Hirosawa-Takamori M, Zhong L, Telleria CM, Shiota K, Gibori G. Hormonal regulation of copper-zinc superoxide dismutase and manganese superoxide dismutase messenger ribonucleic acid in the rat corpus luteum: induction by prolactin and placental lactogens. Biol Reprod. 1998;59:599-605.
100. Gaytan F, Morales C, Bellido C, Aguilar E, Sanchez-Criado JE. Role of prolactin in the regulation of macrophages and in the proliferative activity of vascular cells in newly formed and regressing rat corpora lutea. Biol Reprod. 1997;57:478-486.
101. Tamura H, Greenwald GS. Angiogenesis and its hormonal control in the corpus luteum of the pregnant rat. Biol Reprod. 1987;36:1149-1154.
102. Gibori G, Richards JS, Keyes PL. Synergistic effects of prolactin and estradiol in the luteotropic process in the pregnant rat: regulation of estradiol receptor by prolactin. Biol Reprod. 1979;21:419-423.
103. Telleria CM, Zhong L, Deb S, Srivastava RK, Park KS, Sugino N, Park-Sarge OK, Gibori G. Differential expression of the estrogen receptors alpha and beta in the rat corpus luteum of pregnancy: regulation by prolactin and placental lactogens. Endocrinology. 1998;139:2432-2442.
104. Gibori G, Khan I, Warshaw ML, McLean MP, Puryear TK, Nelson S, Durkee TJ, Azhar S, Steinschneider A, Rao MC. Placental-derived regulators and the complex control of luteal cell function. Recent Prog Horm Res. 1988;44:377-429.
105. Guthridge M, Bertolini J, Cowling J, Hearn MT. Localization of bFGF mRNA in cyclic rat ovary, diethylstilbesterol primed rat ovary, and cultured rat granulosa cells. Growth Factors. 1992;7:15-25.
106. Gospodarowicz D, Massoglia S, Cheng J, Fujii DK. Effect of fibroblast growth factor and lipoproteins on the proliferation of endothelial cells derived from bovine adrenal cortex, brain cortex, and corpus luteum capillaries. J Cell Physiol. 1986;127:121-136.

107. Doraiswamy V, Knutson DL, Grazul-Bilska AT, Redmer DA, Reynolds LP. Fibroblast growth factor receptor (FGFR)-1 and -2 in the ovine corpus luteum throughout the estrous cycle. Growth Factors. 1998;16:125-135.
108. Matsuyama S, Takahashi M. Immunoreactive (ir)-transforming growth factor (TGF)-beta in rat corpus luteum: ir-TGF beta is expressed by luteal macrophages. Endocr J. 1995;42:203-217.
109. RayChaudhury AD, Amore PA. Endothelial cell regulation by transforming growth factor-beta. J Cell Biochem. 1991;47:224-229.
110. Pepper MS, Belin D, Montesano R, Orci L, Vassalli JD. Transforming growth factor-beta 1 modulates basic fibroblast growth factor-induced proteolytic and angiogenic properties of endothelial cells *in vitro*. J Cell Biol. 1990;111:743-755.
111. Reynolds LP, Redmer DA. Growth and development of the corpus luteum. J Reprod Fertil Suppl. 1999;54:181-191.
112. Phillips HS, Hains J, Leung DW, Ferrara N. Vascular endothelial growth factor is expressed in rat corpus luteum. Endocrinology. 1990;127:965-967.
113. Shweiki D, Itin A, Neufeld G, Gitay-Goren H, Keshet E. Patterns of expression of vascular endothelial growth factor (VEGF) and VEGF receptors in mice suggest a role in hormonally regulated angiogenesis. J Clin Invest. 1993;91:2235-2243.
114. Redmer DA, Dai Y, Li J, Charnock-Jones DS, Smith SK, Reynolds LP, Moor RM. Characterization and expression of vascular endothelial growth factor (VEGF) in the ovine corpus luteum. J Reprod Fertil. 1996;108:157-165.
115. Garrido C, Saule S, Gospodarowicz D. Transcriptional regulation of vascular endothelial growth factor gene expression in ovarian bovine granulosa cells. Growth Factors. 1993;8:109-117.
116. Ravindranath N, Little-Ihrig L, Phillips HS, Ferrara N, Zeleznik AJ. Vascular endothelial growth factor messenger ribonucleic acid expression in the primate ovary. Endocrinology. 1992;131:254-260.
117. Christenson LK, Stouffer RL. Follicle-stimulating hormone and luteinizing hormone/chorionic gonadotropin stimulation of vascular endothelial growth factor production by macaque granulosa cells from pre- and periovulatory follicles. J Clin Endocrinol Metab. 1997;82:2135-2142.
118. Yan Z, Weich HA, Bernart W, Breckwoldt M, Neulen J. Vascular endothelial growth factor (VEGF) messenger ribonucleic acid (mRNA) expression in luteinized human granulosa cells *in vitro*. J Clin Endocrinol Metab. 1993;77:1723-1725.
119. Kamat BR, Brown LF, Manseau EJ, Senger DR, Dvorak HF. Expression of vascular permeability factor/vascular endothelial growth factor by human granulosa and theca lutein cells. Role in corpus luteum development. Am J Pathol. 1995;146:157-165.
120. Gordon JD, Mesiano S, Zaloudek CJ, Jaffe RB. Vascular endothelial growth factor localization in human ovary and fallopian tubes: possible role in reproductive function and ovarian cyst formation. J Clin Endocrinol Metab. 1996;81:353-359.
121. Fraser HM, Dickson SE, Lunn SF, Wulff C, Morris KD, Carroll VA, Bicknell R. Suppression of luteal angiogenesis in the primate after neutralization of vascular endothelial growth factor. Endocrinology. 2000;141:995-1000.
122. Ferrara N, Chen H, Davis-Smyth T, Gerber HP, Nguyen TN, Peers D, Chisholm V, Hillan KJ, Schwall RH. Vascular endothelial growth factor is essential for corpus luteum angiogenesis. Nat Med. 1998;4:336-340.
123. Lee A, Christenson LK, Patton PE, Burry KA, Stouffer RL. Vascular endothelial growth factor production by human luteinized granulosa cells *in vitro*. Hum Reprod. 1997;12:2756-2761.
124. Segaloff D, Wang HY, Richards JS. Hormonal regulation of luteinizing hormone/chorionic gonadotropin receptor mRNA in rat ovarian cells during follicular development and luteinization. Mol Endocrinol. 1990;4:1856-1865.

125. Gaddy-Kurten D, Richards JS. Regulation of alpha 2-macroglobulin by luteinizing hormone and prolactin during cell differentiation in the rat ovary. Mol Endocrinol. 1991;5:1280-1291.
126. Borth W. Alpha 2-macroglobulin, a multifunctional binding protein with targeting characteristics. FASEB J. 1992;6:3345-3353.
127. da Silva G, Teixeira N, Bell SC. Major secretory product of the mesometrial decidua in the rat, a variant of alpha-2-macroglobulin, binds insulin-like growth factor I via a protease-dependent mechanism. Mol Reprod Dev. 1996;44:103-110.
128. Huang SO, Grady P, Huang JS. Human transforming growth factor beta.alpha 2-macroglobulin complex is a latent form of transforming growth factor beta. J Biol Chem. 1988;263:1535-1541.
129. Dennis PA, Saksela O, Harpel P, Rifkin DB. Alpha 2-macroglobulin is a binding protein for basic fibroblast growth factor. J Biol Chem. 1989;264:7210-7216.
130. Rothchild I. The regulation of the mammalian corpus luteum. Recent Prog Horm Res. 1981;37:183-298.
131. Galosy SS, Talamantes F. Luteotropic actions of placental lactogens at midpregnancy in the mouse. Endocrinology. 1995;136:3993-4003.
132. Yuan W, Greenwald GS. Progesterone production *in vitro* by mouse luteal cells: response to follicle-stimulating hormone, luteinizing hormone, and prolactin. Proc Soc Exp Biol Med. 1997;214:265-270.
133. Thordarson G, Galosy S, Gudmundsson GO, Newcomer B, Sridaran R, Talamantes F. Interaction of mouse placental lactogens and androgens in regulating progesterone release in cultured mouse luteal cells. Endocrinology. 1997;138:3236-3241.
134. Yuan W, Greenwald GS. Luteotropic effects of follicle-stimulating hormone (FSH): I. FSH has *in vitro* luteotropic and synergistic effects with luteinizing hormone and prolactin on progesterone production by hamster luteal cells during pregnancy. Biol Reprod. 1994;51:43-49.
135. Crisp TM. Hormone requirements for early maintenance of rat granulosa cell cultures. Endocrinology. 1977;101:1286-1297.
136. Centola GM. Hormone requirements for long-term maintenance of rat granulosa cell cultures. Adv Exp Med Biol. 1979;112:225-233.
137. Matsuyama S, Shiota K, Tachi C, Nishihara M, Takahashi M. Splenic macrophages enhance prolactin and luteinizing hormone action in rat luteal cell cultures. Endocrinol Jpn. 1992;39:51-57.
138. Rajkumar K, Malinek J, Murphy BD. Effect of lipoproteins and luteotrophins on progesterone accumulation by luteal cells from the pregnant pig. Steroids. 1985;45:119-134.
139. Rajkumar K, Klingshorn P, Chedrese PJ, Murphy BD. Prolactin, LH, and estradiol-17 beta in utilization of lipoprotein substrate by porcine granulosa cells *in vitro*. Can J Physiol Pharmacol. 1988;66:561-566.
140. Chedrese PJ, Rajkumar K, Ly H, Murphy BD. Dose response of luteinized porcine granulosa cells *in vitro* to prolactin: dependency on pre-exposure to human chorionic gonadotrophin. Can J Physiol Pharmacol. 1988;66:1337-1340.
141. Murphy BD, Rajkumar K, Gonzalez Reyna A, Silversides DW. Control of luteal function in the mink (Mustela vison). J Reprod Fertil Suppl. 1993;47:181-188.
142. McKibbin PE, Rajkumar K, Murphy BD. Role of lipoproteins and prolactin in luteal function in the ferret. Biol Reprod. 1984;30:1160-1166.
143. Alila HW, Rogo KO, Gombe S. Effects of prolactin on steroidogenesis by human luteal cells in culture. Fertil Steril. 1987;47:947-955.
144. McNatty KP, Sawers RS, McNeilly AS. A possible role for prolactin in control of steroid secretion by the human Graafian follicle. Nature. 1974;250:653-655.

145. Ji TH, Ryu KS, Gilchrist R, Ji I. Interaction, signal generation, signal divergence, and signal transduction of LH/CG and the receptor. Recent Prog Horm Res. 1997;52:431-453.
146. Davis JS. Mechanisms of hormone action: luteinizing hormone receptors and second-messenger pathways. Curr Opin Obstet Gynecol. 1994;6:254-261.
147. Davis JS. Modulation of luteinizing hormone-stimulated inositol phosphate accumulation by phorbol esters in bovine luteal cells. Endocrinology. 1992;131:749-757.
148. Morris JK, Richards JS. Luteinizing hormone induces prostaglandin endoperoxide synthase-2 and luteinization *in vitro* by A-kinase and C-kinase pathways. Endocrinology. 1995;136:1549-1558.
149. Clark BJ, Wells J, King SR, Stocco DM. The purification, cloning, and expression of a novel luteinizing hormone-induced mitochondrial protein in MA-10 mouse Leydig tumor cells. Characterization of the steroidogenic acute regulatory protein (StAR). J Biol Chem. 1994;269:28314-28322.
150. Clark BJ, Soo SC, Caron KM, Ikeda Y, Parker KL, Stocco DM. Hormonal and developmental regulation of the steroidogenic acute regulatory protein. Mol Endocrinol. 1995;9:1346-1355.
151. Armstrong DT, Miller LV, Knudsen KA. Regulation of lipid metabolism and progesterone production in rat corpora lutea and ovarian interstitial elements by prolactin and luteinizing hormone. Endocrinology. 1969;85:393-401.
152. Holt JA, Richards JS, Midgley AR, Jr., Reichert LE, Jr. Effect of prolactin on LH receptor in rat luteal cells. Endocrinology. 1976;98:1005-1013.
153. Grinwich DL, Hichens M, Behrman HR. Control of the LH receptor by prolactin and prostaglandin F2alpha in rat corpora lutea. Biol Reprod. 1976;14:212-218.
154. Richards JS, Williams JJ. Luteal cell receptor content for prolactin (PRL) and luteinizing hormone (LH): regulation by LH and PRL. Endocrinology. 1976;99:1571-1581.
155. Piquette GN, LaPolt PS, Oikawa M, Hsueh AJ. Regulation of luteinizing hormone receptor messenger ribonucleic acid levels by gonadotropins, growth factors, and gonadotropin-releasing hormone in cultured rat granulosa cells. Endocrinology. 1991;128:2449-2456.
156. Gafvels M, Bjurulf E, Selstam G. Prolactin stimulates the expression of luteinizing hormone/chorionic gonadotropin receptor messenger ribonucleic acid in the rat corpus luteum and rescues early pregnancy from bromocriptine-induced abortion. Biol Reprod. 1992;47:534-540.
157. Bjurulf E, Selstam G, Olofsson JI. Increased LH receptor mRNA and extended corpus luteum function induced by prolactin and indomethacin treatment *in vivo* in hysterectomized pseudopregnant rats. J Reprod Fertil. 1994;102:139-145.
158. Khan I, Sridaran R, Johnson DC, Gibori G. Selective stimulation of luteal androgen biosynthesis by luteinizing hormone: comparison of hormonal regulation of P45017 alpha activity in corpora lutea and follicles. Endocrinology. 1987;121:1312-1319.
159. Hedin L, Rodgers RJ, Simpson ER, Richards JS. Changes in content of cytochrome P450(17)alpha, cytochrome P450scc, and 3-hydroxy-3-methylglutaryl CoA reductase in developing rat ovarian follicles and corpora lutea: correlation with theca cell steroidogenesis. Biol Reprod. 1987;37:211-223.
160. Bogovich K, Scales LM, Higginbottom E, Ewing LL, Richards JS. Short term androgen production by rat ovarian follicles and long term steroidogenesis by thecal explants in culture. Endocrinology. 1986;118:1379-1386.
161. Bogovich K, Richards JS. Androgen biosynthesis in developing ovarian follicles: evidence that luteinizing hormone regulates thecal 17 alpha-hydroxylase and C17-20-lyase activities. Endocrinology. 1982;111:1201-1208.
162. Gibori G, Keyes PL, Richards JS. A role for intraluteal estrogen in the mediation of luteinizing hormone action on the rat corpus luteum during pregnancy. Endocrinology. 1978;103:162-169.

163. Gibori G, Rodway R, Rothchild I. The luteotrophic effect of estrogen in the rat: prevention by estradiol of the luteolytic effect of an antiserum to luteinizing hormone in the pregnant rat. Endocrinology. 1977;101:1683-1689.
164. MacDonald GJ. Maintenance of pregnancy in the rat in the absence of LH. Proc Soc Exp Biol Med. 1978;159:441-443.
165. Gibori G, Keyes PL. Luteotropic role of estrogen in early pregnancy in the rat. Endocrinology. 1980;106:1584-1588.
166. Koike S, Sakai M, Muramatsu M. Molecular cloning and characterization of rat estrogen receptor cDNA. Nucleic Acids Res. 1987;15:2499-2513.
167. Walter P, Green S, Greene G, Krust A, Bornert JM, Jeltsch JM, Staub A, Jensen E, Scrace G, Waterfield M, *et al.* Cloning of the human estrogen receptor cDNA. Proc Natl Acad Sci U S A. 1985;82:7889-7893.
168. Kuiper GG, Enmark E, Pelto-Huikko M, Nilsson S, Gustafsson JA. Cloning of a novel receptor expressed in rat prostate and ovary. Proc Natl Acad Sci U S A. 1996;93:5925-5930.
169. Frasor J, Zhong Z, Barkai U, Gibori GB, Mao J, Gibori G. Prolactin up-regulation of estrogen receptor alpha gene transcription is tyrosine kinase-dependent. In: Meeting of the Society for the Study of Reproduction. Pullman, WA: Society for the Study of Reproduction; 1999:162.
170. Frasor J, Kitamura T, Park-Sarge OK, Gibori G. Transcriptional regulation of estrogen receptor beta by prolactin. In: Meeting of the Society for the Study of Reproduction. Madison, WI: Society for the Study of Reproduction; 2000:276.
171. Gibori G, Sridaran R. Sites of androgen and estradiol production in the second half of pregnancy in the rat. Biol Reprod. 1981;24:249-256.
172. Gibori G, Chatterton RT, Chien JL. Ovarian and serum concentrations of androgen throughout pregnancy in the rat. Biol Reprod. 1979;21:53-56.
173. Matt DW, MacDonald GJ. *In vitro* progesterone and testosterone production by the rat placenta during pregnancy. Endocrinology. 1984;115:741-747.
174. Gibori G, Kraicer PF. Conversion of testosterone to estrogen by isolated corpora lutea of pregnancy in the rat. Biol Reprod. 1973;9:309-316.
175. Jackson JA, Albrecht ED. The development of placental androstenedione and testosterone production and their utilization by the ovary for aromatization to estrogen during rat pregnancy. Biol Reprod. 1985;33:451-457.
176. Gibori G, Sridaran R, Basuray R. Control of aromatase activity in luteal and ovarian nonluteal tissue of pregnant rats. Endocrinology. 1982;111:781-788.
177. Hickey GJ, Chen SA, Besman MJ, Shively JE, Hall PF, Gaddy-Kurten D, Richards JS. Hormonal regulation, tissue distribution, and content of aromatase cytochrome P450 messenger ribonucleic acid and enzyme in rat ovarian follicles and corpora lutea: relationship to estradiol biosynthesis. Endocrinology. 1988;122:1426-1436.
178. Hickey GJ, Oonk RB, Hall PF, Richards JS. Aromatase cytochrome P450 and cholesterol side-chain cleavage cytochrome P450 in corpora lutea of pregnant rats: diverse regulation by peptide and steroid hormones. Endocrinology. 1989;125:1673-1682.
179. Nokelainen P, Peltoketo H, Vihko R, Vihko P. Expression cloning of a novel estrogenic mouse 17 beta-hydroxysteroid dehydrogenase/17-ketosteroid reductase (m17HSD7), previously described as a prolactin receptor-associated protein (PRAP) in rat. Mol Endocrinol. 1998;12:1048-1059.
180. Risk M, Duan W, Nelson S, Azhar S, Gibori G. Characterization of PRAP as a novel 17-beta hydroxysteroid dehydrogenase enzyme. In: Meeting of the Society for the Study of Reproduction. Pullman, WA: Society for the Study of Reproduction; 1999:162.
181. Duan WR, Parmer TG, Albarracin CT, Zhong L, Gibori G. PRAP, a prolactin receptor associated protein: its gene expression and regulation in the corpus luteum. Endocrinology. 1997;138:3216-3221.

182. Duan WR, Linzer DIH, Gibori G. Cloning and characterization of an ovarian-specific protein that associates with the short form of the prolactin receptor. J Biol Chem. 1996;271:15602-15607.
183. Sherwood OD, Downing SJ, Guico-Lamm ML, Hwang JJ, O Day-Bowman MB, Fields PA. The physiological effects of relaxin during pregnancy: studies in rats and pigs. Oxf Rev Reprod Biol. 1993;15:143-189.
184. Bani D. Relaxin: a pleiotropic hormone. Gen Pharmacol. 1997;28:13-22.
185. Zhao L, Roche PJ, Gunnersen JM, Hammond VE, Tregear GW, Wintour EM, Beck F. Mice without a functional relaxin gene are unable to deliver milk to their pups. Endocrinology. 1999;140:445-453.
186. Felder KJ, Klindt J, Bolt DJ, Anderson LL. Relaxin and progesterone secretion as affected by luteinizing hormone and prolactin after hysterectomy in the pig. Endocrinology. 1988;122:1751-1760.
187. Li Y, Molina JR, Klindt J, Bolt DJ, Anderson LL. Prolactin maintains relaxin and progesterone secretion by aging corpora lutea after hypophysial stalk transection or hypophysectomy in the pig. Endocrinology. 1989;124:1294-1304.
188. Lao Guico MS, Sherwood OD. Effect of oestradiol-17 beta on ovarian and serum concentrations of relaxin during the second half of pregnancy in the rat. J Reprod Fertil. 1985;74:65-70.
189. Goldsmith LT, Grob HS, Weiss G. *In vitro* induction of relaxin secretion in corpora lutea from nonpregnant rats. Ann N Y Acad Sci. 1982;380:60-74.
190. Golos TG, Sherwood OD. Control of corpus luteum function during the second half of pregnancy in the rat: a direct relationship between conceptus number and both serum and ovarian relaxin levels. Endocrinology. 1982;111:872-878.
191. Goldsmith LT, De La Cruz JL, Weiss G, Castracane VD. Steroid effects on relaxin secretion in the rat. Biol Reprod. 1982;27:886-890.
192. Peters CA, Cutler RE, Maizels ET, Robertson MC, Shiu RP, Fields P, Hunzicker-Dunn M. Regulation of PKC delta expression by estrogen and rat placental lactogen-1 in luteinized rat ovarian granulosa cells. Mol Cell Endocrinol. 2000;162:181-191.
193. Everett JW. Hormonal factors responsible for deposition of cholesterol in the corpus luteum of the rat. Endocrinology. 1947;41:364-377.
194. Zarrow MX, Clark JH. Gonadotropin regulation of ovarian cholesterol levels in the rat. Endocrinology. 1969;84:340-346.
195. Armstrong DT, Knudsen KA, Miller LS. Effects of prolactin upon cholesterol metabolism and progesterone biosynthesis in corpora lutea of rats hyophysectomized during pseudopregnancy. Endocrinology. 1970;86:634-641.
196. Azhar S, Khan I, Chen YD, Reaven GM, Gibori G. Regulation of luteal cell 3-hydroxy-3-methylglutaryl coenzyme A reductase activity by estradiol. Biol Reprod. 1985;32:333-341.
197. Puryear TK, McLean MP, Khan I, Gibori G. Mechanism for control of hydroxymethylglutaryl-coenzyme A reductase and cytochrome P-450 side chain cleavage message and enzyme in the corpus luteum. Endocrinology. 1990;126:2910-2918.
198. Gwynne JT, Strauss JF. The role of lipoproteins in steroidogenesis and cholesterol metabolism in steroidogenic glands. Endocr Rev. 1982;3:299-329.
199. Andersen JM, Dietschy JM. Relative importance of high and low density lipoproteins in the regulation of cholesterol synthesis in the adrenal gland, ovary, and testis of the rat. J Biol Chem. 1978;253:9024-9032.
200. Acton S, Rigotti A, Landschulz KT, Xu S, Hobbs HH, Krieger M. Identification of scavenger receptor SR-BI as a high density lipoprotein receptor. Science. 1996;271:518-520.
201. Rajkumar K, Martinuk SD, Agu GO, Murphy BD. *In vitro* binding and utilization of lipoproteins by luteal cells from ferrets treated with dopaminergic drugs during pseudopregnancy. Gen Comp Endocrinol. 1987;67:282-291.

202. Rajkumar K, Ly H, Chedrese PJ, Murphy BD. Effect of prolactin and cyclic AMP on 125I-labelled low density lipoprotein uptake and metabolism by luteinized porcine granulosa cells in culture. Can J Physiol Pharmacol. 1988;66:1450-1454.
203. Rajkumar K, Couture RL, Murphy BD. High density lipoprotein uptake and utilization by rat corpora lutea: the effects of prolactin and PGF 2 alpha. In: Strauss JF, Menon KMJ, eds. Lipoprotein and cholesterol metabolism in steroidogenic tissues. Philadelphia: George F. Stickley Company; 1985:147-154.
204. Menon M, Peegel H, Menon KM. Lipoprotein augmentation of human chorionic gonadotropin and prolactin stimulated progesterone synthesis by rat luteal cells. J Steroid Biochem. 1985;22:79-84.
205. Murphy BD, Rajkumar K, McKibbin PE, Macdonald GJ, Buhr MM, Grinwich DL. The effects of hypophysectomy and administration of pituitary hormones on luteal function and uptake of high density lipoproteins by luteinized ovaries and adrenals of the rat. Endocrinology. 1985;116:1587-1597.
206. Rajkumar K, Couture RL, Murphy BD. Binding of high-density lipoproteins to luteal membranes: the role of prolactin, luteinizing hormone, and circulating lipoproteins. Biol Reprod. 1985;32:546-555.
207. Azhar S, Nomoto A, Leers-Sucheta S, Reaven E. Simultaneous induction of an HDL receptor protein (SR-BI) and the selective uptake of HDL-cholesteryl esters in a physiologically relevant steroidogenic cell model. J Lipid Res. 1998;39:1616-1628.
208. Reaven E, Nomoto A, Leers-Sucheta S, Temel R, Williams DL, Azhar S. Expression and microvillar localization of scavenger receptor, class B, type I (a high density lipoprotein receptor) in luteinized and hormone- desensitized rat ovarian models. Endocrinology. 1998;139:2847-2856.
209. Li X, Peegel H, Menon KM. In situ hybridization of high density lipoprotein (scavenger, type 1) receptor messenger ribonucleic acid (mRNA) during folliculogenesis and luteinization: evidence for mRNA expression and induction by human chorionic gonadotropin specifically in cell types that use cholesterol for steroidogenesis. Endocrinology. 1998;139:3043-3049.
210. Trigatti B, Rayburn H, Vinals M, Braun A, Miettinen H, Penman M, Hertz M, Schrenzel M, Amigo L, Rigotti A, Krieger M. Influence of the high density lipoprotein receptor SR-BI on reproductive and cardiovascular pathophysiology. Proc Natl Acad Sci U S A. 1999;96:9322-9327.
211. Yeaman SJ. Hormone-sensitive lipase—a multipurpose enzyme in lipid metabolism. Biochim Biophys Acta. 1990;1052:128-132.
212. Hui DY. Molecular biology of enzymes involved with cholesterol ester hydrolysis in mammalian tissues. Biochim Biophys Acta. 1996;1303:1-14.
213. Behrman HR, Orczyk GP, Macdonald GJ, Greep RO. Prolactin induction of enzymes controlling luteal cholesterol ester turnover. Endocrinology. 1970;87:1251-1256.
214. Klemcke HG, Brinkley HJ. Effects of bromocriptine and PRL on luteal and adrenal cholesterol ester hydrolase and serum progesterone concentrations in mature pseudopregnant rats. Biol Reprod. 1980;22:1029-1039.
215. Aten RF, Kolodecik TR, Macdonald GJ, Behrman HR. Modulation of cholesteryl ester hydrolase messenger ribonucleic acid levels, protein levels, and activity in the rat corpus luteum. Biol Reprod. 1995;53:1110-1117.
216. Oonk RB, Krasnow JS, Beattie WG, Richards JS. Cyclic AMP-dependent and -independent regulation of cholesterol side chain cleavage cytochrome P-450 (P-450scc) in rat ovarian granulosa cells and corpora lutea. cDNA and deduced amino acid sequence of rat P- 450scc. J Biol Chem. 1989;264:21934-21942.
217. Bignon C, Daniel N, Kermabon AY, Djiane J. Prolactin induces growth inhibition and promotes differentiation of CHO cells stably transfected with prolactin receptor complementary DNA. FEBS Lett. 1995;358:84-88.

218. Jones PB, Valk CA, Hsueh AJ. Regulation of progestin biosynthetic enzymes in cultured rat granulosa cells: effects of prolactin, beta 2-adrenergic agonist, human chorionic gonadotropin and gonadotropin releasing hormone. Biol Reprod. 1983;29:572-585.
219. Feltus FA, Groner B, Melner MH. Stat5-mediated regulation of the human type II 3beta-hydroxysteroid dehydrogenase/delta5-delta4 isomerase gene: activation by prolactin. Mol Endocrinol. 1999;13:1084-1093.
220. Lamprecht SA, Lindner HR, Strauss JF. Induction of 20alpha-hydroxysteroid dehydrogenase in rat corpora lutea by pharmacological blockade of pituitary prolactin secretion. Biochim Biophys Acta. 1969;187:133-143.
221. Lahav M, Lamprecht SA, Amsterdam A, Lindner HR. Suppression of 20 alpha-hydroxysteroid dehydrogenase activity in cultured rat luteal cells by prolactin. Mol Cell Endocrinol. 1977;6:293-302.
222. Albarracin CT, Gibori G. Prolactin action on luteal protein expression in the corpus luteum. Endocrinology. 1991;129:1821-1830.
223. Mao J, Duan WR, Albarracin CT, Parmer TG, Gibori G. Isolation and characterization of a rat luteal cDNA encoding 20 alpha- hydroxysteroid dehydrogenase. Biochem Biophys Res Commun. 1994;201:1289-1295.
224. Miura R, Shiota K, Noda K, Yagi S, Ogawa T, Takahashi M. Molecular cloning of cDNA for rat ovarian 20 alpha-hydroxysteroid dehydrogenase (HSD1). Biochem J. 1994;299:561-567.
225. Albarracin CT, Parmer TG, Duan WR, Nelson SE, Gibori G. Identification of a major prolactin-regulated protein as 20 alpha- hydroxysteroid dehydrogenase: coordinate regulation of its activity, protein content, and messenger ribonucleic acid expression. Endocrinology. 1994;134:2453-2460.
226. Yamanouchi K, Yamanouchi S, Aoki M, Matsuyama S, Nishihara M, Takahashi M. Effect of calyculin-A, phosphatase inhibitor, on 20 alpha- hydroxysteroid dehydrogenase activity in rat luteal cell culture. Endocr J. 1993;40:699-704.
227. Zhong L, Ou J, Barkai U, Mao JF, Frasor J, Gibori G. Molecular cloning and characterization of the rat ovarian 20 alpha-hydroxysteroid dehydrogenase gene. Biochem Biophys Res Commun. 1998;249:797-803.
228. O'Neal KD, Yu-Lee LY. Differential signal transduction of the short, Nb2, and long prolactin receptors. Activation of interferon regulatory factor-1 and cell proliferation. J Biol Chem. 1994;269:26076-26082.
229. Yu-Lee LY, Hrachovy JA, Stevens AM, Schwarz LA. Interferon-regulatory factor 1 is an immediate-early gene under transcriptional regulation by prolactin in Nb2 T cells. Mol Cell Biol. 1990;10:3087-3094.
230. Malven PV. Luteotrophic and luteolytic responses to prolactin in hypophysectomized rats. Endocrinology. 1969;84:1224-1229.
231. Bowen JM, Keyes PL, Warren JS, Townson DH. Prolactin-induced regression of the rat corpus luteum: expression of monocyte chemoattractant protein-1 and invasion of macrophages. Biol Reprod. 1996;54:1120-1127.
232. Martel C, Labrie C, Dupont E, Couet J, Trudel C, Rheaume E, Simard J, Luu-The V, Pelletier G, Labrie F. Regulation of 3 beta-hydroxysteroid dehydrogenase/delta 5-delta 4 isomerase expression and activity in the hypophysectomized rat ovary: interactions between the stimulatory effect of human chorionic gonadotropin and the luteolytic effect of prolactin. Endocrinology. 1990;127:2726-2737.
233. Martel C, Gagne D, Couet J, Labrie Y, Simard J, Labrie F. Rapid modulation of ovarian 3 beta-hydroxysteroid dehydrogenase/delta 5- delta 4 isomerase gene expression by prolactin and human chorionic gonadotropin in the hypophysectomized rat. Mol Cell Endocrinol. 1994;99:63-71.
234. Billeter E, Fluckinger E. Evidence for a luteolytic function of prolactin in the intact cyclic rat using 2-Br-alpha-Ergokryptine (CB 154). Experientia. 1971;27:464-465.

235. Wuttke W, Meites J. Luteolytic role of prolactin during the estrous cycle of the rat. Proc Soc Exp Biol Med. 1971;137:988-991.
236. Taya K, Greenwald GS. *In vivo* and *in vitro* ovarian steroidogenesis in the long term hypophysectomized rat. Endocrinology. 1982;110:390-397.
237. Smith PE. Hypophysectomy and a replacement therapy in the rat. Amer J Anat. 1930;45:205-273.
238. Matsuyama S, Chang KT, Kanuka H, Ohnishi M, Ikeda A, Nishihara M, Takahashi M. Occurrence of deoxyribonucleic acid fragmentation during prolactin-induced structural luteolysis in cycling rats. Biol Reprod. 1996;54:1245-1251.
239. Kanuka H, Matsuyama S, Ohnishi M, Matsumoto Y, Nishihara M, Takahashi M. Prolactin expresses differential effects on apoptotic cell death of luteal cells *in vivo* and *in vitro*. Endocr J. 1997;44:11-22.
240. King KL, Cidlowski JA. Cell cycle regulation and apoptosis. Annu Rev Physiol. 1998;60:601-617.
241. Kiya T, Endo T, Goto T, Yamamoto H, Ito E, Kudo R, Behrman HR. Apoptosis and PCNA expression induced by prolactin in structural involution of the rat corpus luteum. J Endocrinol Invest. 1998;21:276-283.
242. Brannstrom M, Giesecke L, Moore IC, van den Heuvel CJ, Robertson SA. Leukocyte subpopulations in the rat corpus luteum during pregnancy and pseudopregnancy. Biol Reprod. 1994;50:1161-1167.
243. Dore MA. Structural aspects of luteal function and regression in the ovary of the domestic dog. J Reprod Fertil Suppl. 1989;39:41-53.
244. Bagavandoss P, Kunkel SL, Wiggins RC, Keyes PL. Tumor necrosis factor-a (TNF-a) production and localization of macrophages and T lymphocytes in the rabbit corpus luteum. Endocrinology. 1988;122:1185-1187.
245. Hume DA, Halpin D, Charlton H, Gordon S. The mononuclear phagocyte system of the mouse defined by immunohistochemical localization of antigen F4/80: macrophages of endocrine organs. Proc Natl Acad Sci U S A. 1984;81:4174-4177.
246. Paavola LG. The corpus luteum of the guinea pig. IV. Fine structure of macrophages during pregnancy and postpartum luteolysis, and the phagocytosis of luteal cells. Am J Anat. 1979;154:337-364.
247. Hehnke KE, Christenson LK, Ford SP, Taylor M. Macrophage infiltration into the porcine corpus luteum during prostaglandin F2 alpha-induced luteolysis. Biol Reprod. 1994;50:10-15.
248. Bukovsky A, Caudle MR, Keenan JA, Wimalasena J, Upadhyaya NB, Van Meter SE. Is corpus luteum regression an immune-mediated event? Localization of immune system components and luteinizing hormone receptor in human corpora lutea. Biol Reprod. 1995;53:1373-1384.
249. Tsai SJ, Juengel JL, Wiltbank MC. Hormonal regulation of monocyte chemoattractant protein-1 messenger ribonucleic acid expression in corpora lutea. Endocrinology. 1997;138:4517-4520.
250. Takaya R, Fukaya T, Sasano H, Suzuki T, Tamura M, Yajima A. Macrophages in normal cycling human ovaries; immunohistochemical localization and characterization. Hum Reprod. 1997;12:1508-1512.
251. Springer TA. Traffic signals for lymphocyte recirculation and leukocyte emigration: the multistep paradigm. Cell. 1994;76:301-314.
252. Leonard EJ, Yoshimura T. Human monocyte chemoattractant protein-1 (MCP-1). Immunol Today. 1990;11:97-101.
253. Townson DH, Warren JS, Flory CM, Naftalin DM, Keyes PL. Expression of monocyte chemoattractant protein-1 in the corpus luteum of the rat. Biol Reprod. 1996;54:513-520.
254. Senturk LM, Seli E, Gutierrez LS, Mor G, Zeyneloglu HB, Arici A. Monocyte chemotactic protein-1 expression in human corpus luteum. Mol Hum Reprod. 1999;5:697-702.

255. Bowen JM, Towns R, Warren JS, Landis Keyes P. Luteal regression in the normally cycling rat: apoptosis, monocyte chemoattractant protein-1, and inflammatory cell involvement. Biol Reprod. 1999;60:740-746.
256. Springer TA. Adhesion receptors of the immune system. Nature. 1990;346:425-434.
257. Olson KK, Townson DH. Prolactin-induced expression of intercellular adhesion molecule-1 and the accumulation of Monocytes/Macrophages during regression of the rat corpus luteum. Biol Reprod. 2000;62:1571-1578.
258. Weber KS, Klickstein LB, Weber C. Specific activation of leukocyte beta2 integrins lymphocyte function-associated antigen-1 and Mac-1 by chemokines mediated by distinct pathways via the alpha subunit cytoplasmic domains. Mol Biol Cell. 1999;10:861-873.
259. Yoshimura T, Yuhki N, Moore SK, Appella E, Lerman MI, Leonard EJ. Human monocyte chemoattractant protein-1 (MCP-1). Full-length cDNA cloning, expression in mitogen-stimulated blood mononuclear leukocytes, and sequence similarity to mouse competence gene JE. FEBS Lett. 1989;244:487-493.
260. Carr MW, Roth SJ, Luther E, Rose SS, Springer TA. Monocyte chemoattractant protein 1 acts as a T-lymphocyte chemoattractant. Proc Natl Acad Sci U S A. 1994;91:3652-3656.
261. Nagata S. Fas ligand-induced apoptosis. Annu Rev Genet. 1999;33:29-55.
262. Quirk SM, Cowan RG, Joshi SG, Henrikson KP. Fas antigen-mediated apoptosis in human granulosa/luteal cells. Biol Reprod. 1995;52:279-287.
263. Sakamaki K, Yoshida H, Nishimura Y, Nishikawa S, Manabe N, Yonehara S. Involvement of Fas antigen in ovarian follicular atresia and luteolysis. Mol Reprod Dev. 1997;47:11-18.
264. Roughton SA, Lareu RR, Bittles AH, Dharmarajan AM. Fas and Fas ligand messenger ribonucleic acid and protein expression in the rat corpus luteum during apoptosis-mediated luteolysis. Biol Reprod. 1999;60:797-804.
265. Kuranaga E, Kanuka H, Bannai M, Suzuki M, Nishihara M, Takahashi M. Fas/Fas ligand system in prolactin-induced apoptosis in rat corpus luteum: possible role of luteal immune cells. Biochem Biophys Res Commun. 1999;260:167-173.
266. Russell DH, Matrisian L, Kibler R, Larson DF, Poulos B, Magun BE. Prolactin receptors on human lymphocytes and their modulation by cyclosporine. Biochem Biophys Res Commun. 1984;121:899-906.
267. Russell DH, Kibler R, Matrisian L, Larson DF, Poulos B, Magun BE. Prolactin receptors on human T and B lymphocytes: antagonism of prolactin binding by cyclosporine. J Immunol. 1985;134:3027-3031.
268. Pellegrini I, Lebrun JJ, Ali S, Kelly PA. Expression of prolactin and its receptor in human lymphoid cells. Mol Endocrinol. 1992;6:1023-1031.
269. Kuranaga E, Kanuka H, Furuhata Y, Yonezawa T, Suzuki M, Nishihara M, Takahashi R. Requirement of the Fas ligand-expressing luteal immune cells for regression of corpus luteum. FEBS Lett. 2000;472:137-142.
270. Kuranaga E, Kanuka H, Hirabayashi K, Suzuki M, Nishihara M, Takahashi M. Progesterone is a cell death suppressor that downregulates Fas expression in rat corpus luteum. FEBS Lett. 2000;466:279-282.
271. Bowen JM, Keyes PL. The proestrous prolactin surge is not the sole initiator of regressive changes in corpora lutea of normally cycling rats. Biol Reprod. 1999;61:1208-1215.
272. Port CB, Bowen JM, Keyes PL, Townson DH. Effects of a 3 beta-hydroxysteroid dehydrogenase inhibitor on monocyte-macrophage infiltration into rat corpus luteum and on apoptosis: relationship to the luteolytic action of prolactin. Journal of Reproduction & Fertility. 2000;119:93-99.
273. Gaytan F, Bellido C, Morales C, Sanchez-Criado JE. Both prolactin and progesterone in proestrus are necessary for the induction of apoptosis in the regressing corpus luteum of the rat. Biol Reprod. 1998;59:1200-1206.

274. Endo T, Aten RF, Wang F, Behrman HR. Coordinate induction and activation of metalloproteinase and ascorbate depletion in structural luteolysis. Endocrinology. 1993;133:690-698.
275. Hirsch B, Knoke I, Leonhardt S, Pitzel L, Jarry H, Wuttke W. Stimulation of matrix-metalloproteinase-1 and tissue inhibitor of metalloproteinase-1 gene expression in rats by the preovulatory prolactin peak. Eur J Endocrinol. 1999;140:583-589.
276. Musicki B, Kodaman PH, Aten RF, Behrman HR. Endocrine regulation of ascorbic acid transport and secretion in luteal cells. Biol Reprod. 1996;54:399-406.

Chapter 15

Regulation of Pancreatic Islets by Prolactin, Growth Hormone and Placental Lactogen

Robert L. Sorenson and T. Clark Brelje

Department of Genetics, Cell Biology and Development
University of Minnesota, Minneapolis, MN

INTRODUCTION

The role of pituitary hormones in the regulation of glucose homeostasis was first recognized by Bernado Houssay in the 1930's (1-3). Subsequent studies have demonstrated that both prolactin (PRL) and growth hormone (GH) secreted by the pituitary and the closely related placental lactogen (PL) secreted by the placenta during pregnancy are involved in the regulation of carbohydrate and lipid metabolism (4). Although changes in the function of pancreatic islets of Langerhans are reported in many of these conditions, it has been unclear to what extent these reflect a direct effect of the hormones on islets or are a consequence of the changes in other tissues that increase the demand for insulin (e.g., peripheral insulin resistance). In the 1960's, it became possible to examine the direct effect of these hormones on islets *in vitro*. Unfortunately, the species from which islets were easily available (e.g., from rats and mice) were not the same as that of the hormones (e.g., primarily human). The high degree of structural relatedness of these hormones allows a heterologous hormone to have biological properties in tissues from other species that are not shared with the homologous hormone (5-8). Because of these problems with both *in vivo* and *in vitro* studies, no unified picture has emerged to explain the physiological roles of these individual hormones in the regulation of pancreatic function.

To obtain a better understanding of whether these hormones have unique or shared effects on islet function, we examined the effect of the homologous PRL, GH, and PL on islets isolated from several different species (8, 9). In each species, the homologous hormones that bind to the PRL receptor (*i.e.*, PRL and PL) induced profound changes in islet function. Although these include a variety of structural and functional changes, the most important are a lowering of the threshold for glucose-stimulated insulin secretion and a marked increase in islet β-cell proliferation (8, 10-14). These changes allow larger amounts of insulin to be released without the need for a prolonged elevation in the concentration of serum glucose. In contrast, the homologous GHs had little or no effect on these parameters when compared to those induced by PRL receptor activation. These observations were surprising considering the large number of studies demonstrating large effects of GH on islet function (15-19). However, many of these studies are misleading because rat islets were used with heterologous human GH that can readily bind to and activate both the rat PRL and GH receptors.

In this review, we discuss the direct effects of PRL, GH, and PL on pancreatic islet function both *in vivo* and *in vitro*. This new information provides a basis for understanding the direct effects of these hormones on islet function and the difficulties of interpreting the observations from many of the previous studies. The adaptation of islets to meet the increased demand for insulin during pregnancy and the signal transduction mechanisms activated after binding of these hormones to their receptors in the islet will be described. These experiments also suggest an explanation for the difference in response of islets to these closely related hormones. The mechanisms involved in the up-regulation of islet function with a lowering of the threshold for glucose-stimulated insulin secretion will be examined. Finally, the importance of these hormones in the regulation of islet function during pregnancy and other physiological conditions in which the secretion of these hormones is altered will be described. Our recent reviews on the changes in islets during pregnancy and the regulation of β-cell proliferation by these hormones should be consulted for additional information on aspects of islet function not discussed here (20, 21).

EFFECT OF HOMOLOGOUS SPECIES OF PRL/PL/GH ON ISLETS OF LANGERHANS

To characterize the effects of PRL and GH on islets, initial experiments were carried out *in vivo* to examine the hormone effects on insulin secretion (10, 22). In these experiments rats were fitted with osmotic pumps containing either ovine PRL (oPRL) or bovine GH (bGH). Because of insufficient quantities of the homologous hormones, heterologous hormones with well-characterized receptor specificities in rats were used in these experiments. The hormones were infused for four days and resulted in hormone concentrations 15-20 fold greater than the endogenous hormone concentrations. The pancreases were then perfused with a gradient of increasing glucose concentrations and the extent of insulin secretion determined. This protocol allowed determination of the glucose-stimulation threshold as well as the extent of release at higher glucose concentrations. The results showed clear differences between PRL and GH treatment. With PRL there was a marked decrease in the threshold of glucose-stimulated insulin secretion and enhanced secretion at all levels above the threshold (Figure 15-1). In contrast, there was no effect of GH on either the glucose-stimulation threshold or the extent of insulin secretion at above threshold levels (Figure 15-1).

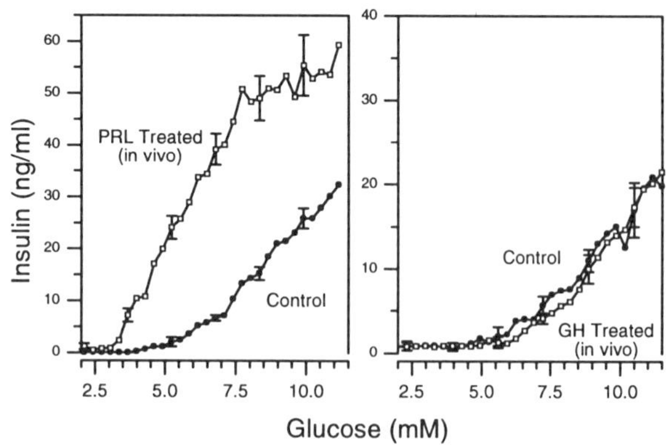

Figure 15-1. Effect of 4 days of in vivo treatment with PRL or GH on insulin secretion from isolated perfused rat pancreases (10). PRL resulted in enhanced glucose induced insulin secretion with a shift in the threshold for glucose stimulation. GH had no effect on glucose-stimulated insulin secretion.

To determine whether these changes are a direct effect of the hormones on islets, studies using the homologous PRL, PL and GH were done with cultured neonatal and adult rat islets, neonatal mouse islets and adult human islets. In these experiments, islets were cultured in the presence of 500 ng/ml of the respective hormones for 7 days and the amount of insulin secreted was determined (9). GH had a small but significant effect on insulin secretion from neonatal rat and mouse islets, but not from the adult rat and human islets. In contrast, PRL and PL had much greater effects on insulin secretion with larger increases from neonatal than adult islets (Figure 15-2). Interestingly, PL was even more potent with a detectable effect at concentrations as low as 10 ng/ml whereas PRL required more than 50 ng/ml (9).

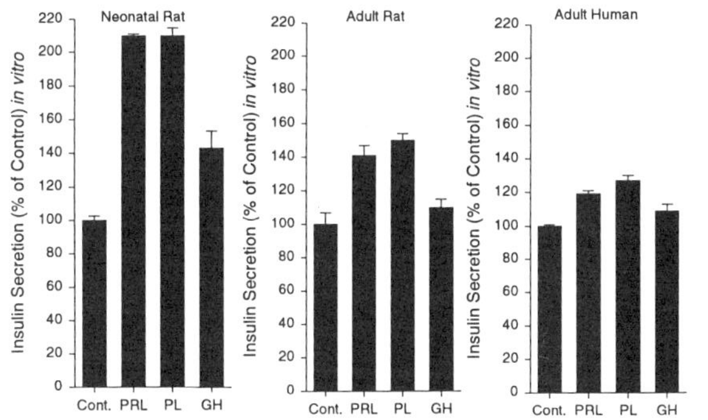

Figure 15-2. Effect of *in vitro* treatment with the homologous PRL, PL and GH on cumulative insulin secretion for 7 days (9). PRL and PL significantly enhanced insulin secretion from neonatal and adult rat islets, and adult human islets. GH was only effective in increasing insulin secretion from the neonatal rat islets. Although rat PL-I was used for these experiments with rat islets, similar results have been reported with the closely related rat PL-II (23).

In addition to insulin secretion, there was a question concerning the ability of these hormones to directly stimulate the proliferation of islet cells. In these experiments, rat islets were cultured in the presence of 500 ng/ml of PRL, PL or GH for 7 days with 5-bromo-2′-deoxyuridine (BrdU) being present for the last 24 hours. The incorporation of BrdU into dividing cells was then determined by immunohistochemistry. GH had a small effect on islet cell proliferation with its effect being larger in neonatal than adult rat islets (Figure 15-3). In contrast, PRL and the PL resulted in a much larger (3 to 4-fold) increase in islet cell division. We only had the opportunity to examine a few samples of human islets for the effects of the hormones on islet cell division. In islets from two of the three subjects, PRL and PL treatment increased cell division, and in one of the three, GH had an effect (21). These results, although limited, are compatible with those observed with rodent islets.

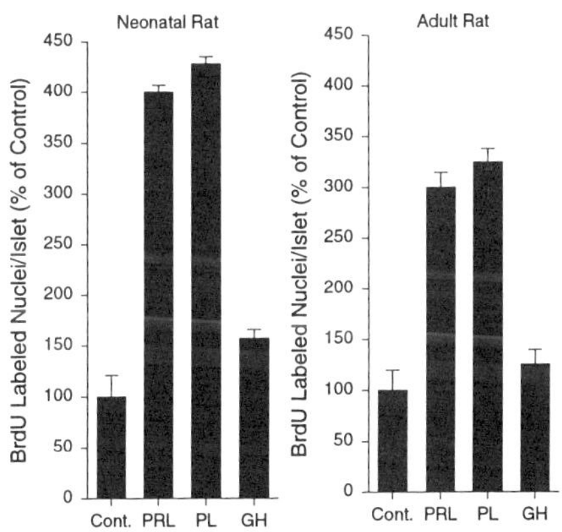

Figure 15-3. Effect of *in vitro* treatment with the homologous PRL, PL and GH on BrdU labeling of islets after 7 days of culture (9). PRL and PL significantly enhanced BrdU labeling of dividing islet cells in neonatal and adult islets. GH only had a small effect in neonatal islets.

An important characteristic of the effects of PRL on islets is that it requires a long period of time to increase insulin secretion and islet cell proliferation. Glucose and most other hormones regulate insulin secretion in a very rapid manner to quickly respond to the influx of nutrients following a meal. In contrast, the effect of PRL on insulin secretion is first observed after 24 hours of treatment and continues to increase over the next 4 to 6 days when it reaches a maximum (Figure 15-4) (11, 22). The effect of PRL on islet cell division similarly requires a long period of time to have its full impact (Figure 15-4) (11). This suggests that members of the PRL, PL, GH family are important to the long-term adaptation of islet function rather than the moment-to-moment regulation of insulin secretion.

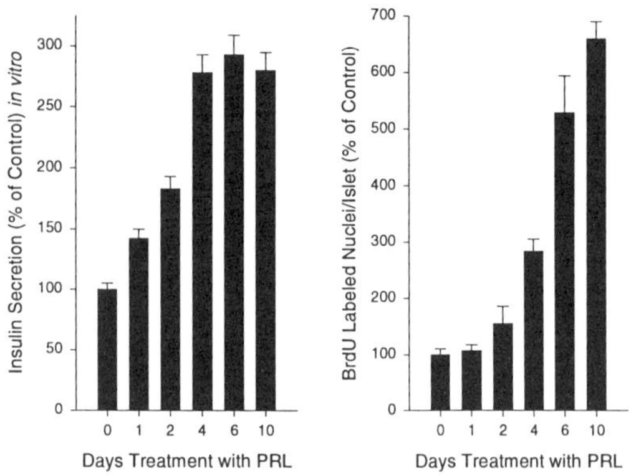

Figure 15-4. Effect of *in vitro* PRL treatment on neonatal rat islets showing the extensive period of time required for PRL to reach its maximum effect on insulin secretion and islet cell division (24, 25).

It is noteworthy that the results with human islets were similar to those in rodent islets. This is of particular interest since human PL is structurally more like human GH, whereas rodent PLs are more similar to rodent PRL. This can be explained by the observation that human PL binds much better to the human PRL receptor than to the human GH receptor, whereas rat PLs only bind to the rat PRL receptor (26, 27). This further suggests that activation of the PRL receptor whether rodent or human results in much larger effects on islets than activation of the GH receptor.

ROLE OF PRL AND PL ON ADAPTATIONS OF ISLETS OF LANGERHANS TO PREGNANCY

The most compelling evidence for lactogen activation of islets occurs during pregnancy. Pregnancy is characterized as a condition of elevated serum insulin levels, slightly lower blood glucose levels and increased peripheral insulin resistance (4, 28-30). In order to accommodate this increased demand for insulin and maintain normal glycemia, the islets must undergo long-term up-regulatory changes in function. Failure of this long-term adaptive process can lead to the development of gestational diabetes. The principal changes that occur in islets during pregnancy are an increase in islet mass and glucose-stimulated insulin secretion (11, 12, 14, 20, 24, 31).

A variety of quantitative methods have been used to investigate islet growth during pregnancy. These include quantitative immunohistochemistry of islets, determination of DNA and protein content of isolated islets, and the incorporation of tritiated thymidine or BrdU into islet DNA (20). These studies have shown that an approximate 100% increase in islet mass and 50% increase in islet cell numbers occurs during pregnancy. Therefore, both islet cell hypertrophy and hyperplasia contribute to islet growth during pregnancy. Contrary to earlier claims, there does not appear to be an increase in islet neogenesis during pregnancy (31).

These changes in islet function during pregnancy are important in the maintenance of glucose homeostasis. During pregnancy in rodents, fasting serum insulin levels are increased about 75% and glucose levels are decreased about 25% (14, 32, 33). Glucose tolerance tests are normal, but show enhanced insulin secretion (32). A similar situation is observed in humans during pregnancy (28-30, 34, 35). There have been few studies examining insulin secretion from isolated islets obtained during pregnancy. These studies showed that pregnancy results in enhanced glucose-stimulated insulin secretion from these islets (32, 36, 37). Green *et al.* also noted that there was a leftward shift in the glucose response curves for insulin secretion and insulin synthesis (36, 38). This was the first report indicating that the threshold for glucose-stimulated insulin secretion could be lowered. That is, there is an increase in the islet's sensitivity to glucose stimulation.

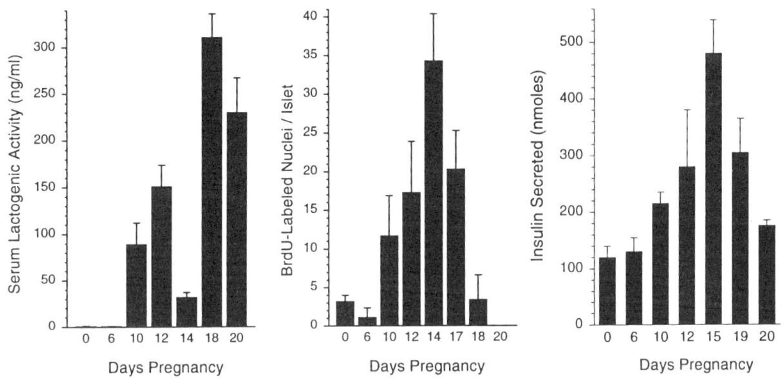

Figure 15-5. Effect of pregnancy on BrdU labeling (center panel) of islet cells and insulin secretion in rats (right panel) (12). The increase in both islet cell proliferation and insulin secretion correlates with the onset of placental lactogen secretion (left panel). The two peaks of serum lactogen activity correspond to the secretion of PL-I and PL-II during rat pregnancy.

Although these studies indicated that there was an increase in islet growth and insulin secretion during pregnancy, there were no studies examining the correlation of these parameters with changes in PL secretion throughout pregnancy. To investigate the temporal profile of changes in islets during pregnancy, we examined islet cell proliferation and insulin secretion during gestation in the rat (Figure 15-5) (12). BrdU incorporation into dividing islets cells was slightly elevated by day 10, peaked on day 14, and then returned to control levels by day 18. A similar pattern of changes in the insulin secretory profiles was observed. Differences in both the threshold of glucose-stimulated insulin secretion and the amount of insulin released above this threshold could be detected by day 10 (Figure 15-6). By day 12, the threshold was lowered from 5.7 mM glucose to 3.3 mM, remained at this level through day 15, and returned toward normal by day 20. Concomitant with the increased sensitivity of β-cells to glucose, the amount of insulin released above the threshold was increased by day 12, peaked on day 15, and returned to control levels by day 20. This lowering of the threshold for glucose-stimulated insulin secretion is an important feature of islets as they adapt to pregnancy. It is only by this maneuver that a large increase in insulin secretion can be achieved at normal blood glucose levels. The magnitude of this effect can be seen by comparing the rates of insulin secretion at 5.6 mM glucose from pregnant animals *versus* controls (Figure 15-6).

The onset of these changes in islet cell proliferation and insulin secretion correlate with the appearance of circulating PL (12). However, PL remains elevated until the end of pregnancy when islet cell proliferation and insulin secretion have returned to control levels. Since islets cultured in the presence of PRL or PL show persistent increases in islet cell division and insulin secretion, it is likely that the increase in steroids during the later stages of pregnancy are counteracting the effects of PL. This hypothesis is supported by experiments *in vitro* where islets are cultured with both PRL and progesterone (39). In this case, insulin secretion and BrdU labeling increased during the first 4 days, but subsequently returned to control levels by day 8. This temporal pattern of changes in islet function closely mimics that observed in islets during pregnancy (12). Similarly, increased plasma glucocorticoid levels during the later part of pregnancy could effectively reverse the lactogen-induced up-regulation of islet function by inhibiting insulin secretion and cell proliferation while increasing apoptosis (40).

The summary of the evidence that the adaptation of islets to pregnancy is mediated by way of lactogenic activity (*i.e.*, PRL or PL) rather than somatotogenic activity (*i.e.*, GH) is shown in Table 15-1. In addition, the magnitude of the effects observed with the homologous PRL or PL *in vitro* is comparable to those observed during pregnancy.

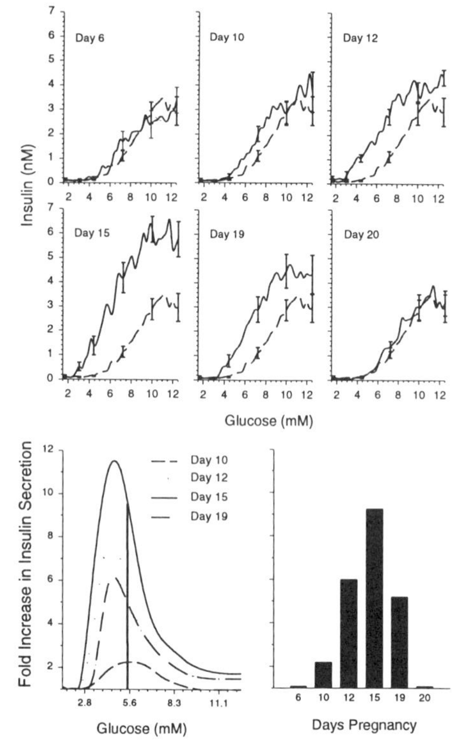

Figure 15-6 (left). The top six panels show the glucose-dependent insulin secretion from perfused pancreases during the course of pregnancy in comparison to the controls (12). The shift in the glucose-stimulation threshold and above threshold insulin secretion is first detected on day 10. The differences increase until day 15 and then return to control levels by day 20 of pregnancy. The bottom left panel shows the ratio of insulin secretion during pregnancy compared to controls for each glucose concentration. The bottom right panel shows the fold increase in insulin secretion compared to control at a normal blood glucose concentration of 5.6 mM.

Table 15-1. Comparison of the effects of pregnancy with species homologous lactogenic hormones (prolactin and or placental lactogen) and growth hormone on islets.

Islet structure and function	Pregnancy	PRL/PL	GH
β-cell proliferation	+++	+++	+
Islet volume	+++	+++	+,0
Glucose-stimulated insulin secretion	+++	+++	+,0
Lower threshold for glucose stimulation	+++	+++	0
Insulin synthesis	+++	+++	+

Table 15-1. continued

Islet structure and function	Pregnancy	PRL/PL	GH
Insulin content	+++	+++	+,0
β-cell junctional coupling	+++	+++	?
Glucose utilization	+++	+++	0
Glucose oxidation	+++	+++	0
Glucokinase activity	+++	+++	?
Gluckinase expression	+++	+++	?
Glucose transporter 2 expression	+++	+++	?
c-AMP metabolism	+++	+++	?

It should also be mentioned that the physiological relevance of PL secretion has been questioned because placentas of many species do not produce a distinct PL (for example, rabbits, pigs, dogs, and cats; (41). In addition, the occurrence of normal pregnancies in women with undetectable serum concentrations of human PL because of the deletion of their genes has been reported (42-44). Unfortunately, insulin secretion during pregnancy has not been examined in either of these conditions. An elevation in insulin secretion may still occur because other mechanisms may compensate for an absence of a distinct PL. First, the decreased capacity for glucose disposal from the induction of peripheral insulin resistance may be sufficient to stimulate an increase in islet function. Second, an increase in serum levels of PRL may compensate for the absence of a distinct PL during pregnancy. An increase in serum levels of PRL during pregnancy has been shown in dogs (45) and humans (46-48). Furthermore, the extent of the PRL receptor expression in target tissues may also be important because the binding of PRL to the mammary gland is much greater in species without a distinct PL (41). In the absence of additional information, these observations are insufficient to disprove the need for increased islet function as an essential feature in altered maternal metabolism during pregnancy in mammalian species.

MECHANISMS BY WHICH PRL/ PL AND GH UP-REGULATE ISLETS

The long form of the PRL receptor is present in islets (49-54). The intensity of immunohistochemical staining in control islets is quite variable among β-cells, with many showing a low level of PRL receptor expression. By day 14 of pregnancy in rats, this cellular heterogeneity for PRL receptors is markedly decreased with nearly all cells having a uniformly high level of staining (54). Also, an increase in PRL receptor mRNA has been reported in islets during pregnancy and after treatment with PRL *in vitro* (55). These studies suggest that lactogens induce the expression of the PRL receptor and this likely has a role in the islet's adaptation to pregnancy. The long form of the GH receptor is also present in islets (50, 52). However, in contrast to the PRL receptor, GH has no effect on the expression of the GH receptor (55).

PRL and GH receptors belong to the cytokine superfamily of receptors, which do not have intrinsic tyrosine kinase activity but interact with members of the JAK (Janus kinase) family of tyrosine kinases (56). The activated JAK kinases phosphorylate a member of the Stat (Signal Transduction and Activators of Transcription) family of transcription factors, which then dimerize and translocate to the nucleus to regulate the expression of specific genes (57). The principal components involved in PRL and GH receptor signaling are JAK-2 and Stat5, although others may be involved. In recent immunohistochemical studies, we have identified JAK2 in rat islets and INS-1 β-cell line and have observed the translocation of Stat5 to the nucleus after treatment with PRL (58). This translocation of Stat5 is a rapid event, detectable within a few minutes, and is maximal within 30 minutes.

To explore the differences in the effects of PRL and GH on islets, we examined the translocation of Stat5b from the cytoplasm to the nucleus in intact rat islets. PRL treatment for 30 minutes results in Stat5b translocation in islet β-cells but not α- or δ-cells (Figure 15-7, left panel) and corresponds to the cellular localization of PRL receptors in islets. Surprisingly, GH treatment results in an even greater translocation of Stat5b in all types of islet cells (Figure 15-7, left panel). Such a response would suggest that GH should have equal if not greater effects on islets. However observations of Stat5b translocation over longer periods of treatment with PRL or GH leads to a different conclusion (Figure 15-7, right panel). The effect of GH on Stat5b translocation is transient and then returns to levels slightly above those seen in the control. In contrast the effect of PRL on Stat5b translocation peaks early, dips at 2 hours and then returns to a maximum level at 4 hours and remains elevated. Changes in Stat5b translocation could be detected with PRL concentrations greater than 50 ng/ml and GH concentrations greater than 10 ng/ml (data not shown) which are within the physiological concentrations of these hormones. These results closely reflect the differences in the effects of PRL and GH on insulin secretion and cell division. This indicates that a prolonged activation of Stat5b is needed for changes in islet function. In addition, they suggest that the basis for the differences is focused in the early events of receptor activation and regulation.

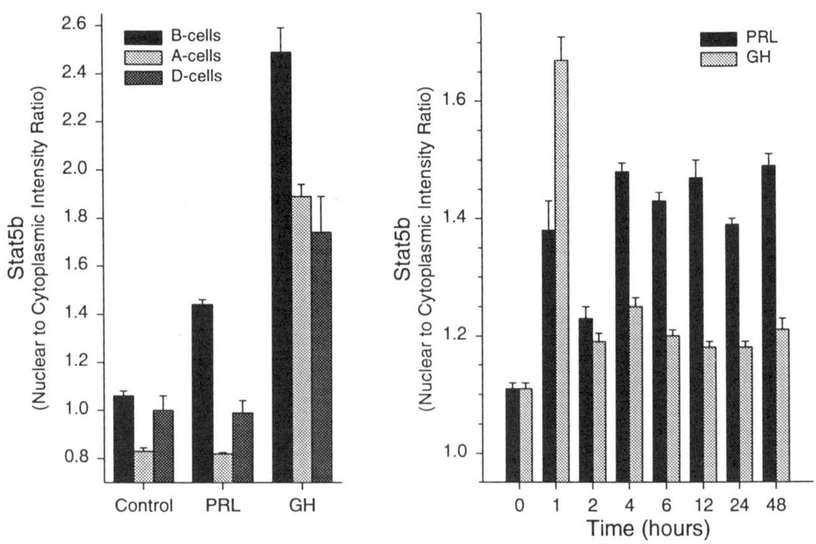

Figure 15-7. The left panel shows the effect of PRL and GH treatment for 30 minutes on Stat5b translocation to the nucleus in islet β-, α- and δ- cells of rat islets. The right panel shows the time dependent effect of PRL and GH treatment on Stat5b translocation to the nucleus.

With the demonstration that lactogens induce the changes in islets during pregnancy, it is important to identify the cellular mechanisms involved. In particular, the lowering of the threshold of glucose-stimulated insulin secretion is an essential feature of the successful adaptation of islets to pregnancy. Although conditions have been identified with an increase in the glucose-stimulation threshold, for example somatostatin, fasting and reduced glucokinase expression (59-61), few studies have examined insulin secretion when there is a lowering of the threshold (36, 62, 63). Therefore, we undertook a study to investigate whether the threshold can change *in vivo* after PRL or GH infusion, glucose loading or fasting, or *in vitro* after treatment with forskolin, glucagon, cholecystokinin, carbamylcholine or gastric inhibitory peptide (10). Unexpectedly, all treatments that enhance insulin secretion lowered the threshold of glucose stimulation into the range of 3.0-4.0 mM glucose. This suggests that the metabolism of approximately 3.0 mM glucose activates all of the pathways necessary for glucose-stimulated insulin secretion. Further increases in glucose metabolism would increase the activity of these pathways and/or stimulate additional regulatory pathways. This model of insulin secretion is supported by the observation that most secretagogues require stimulatory or near stimulatory concentrations of glucose to have an effect on insulin secretion.

Glucose metabolism is central to regulated insulin secretion and glucokinase is regarded as the critical glucose sensor. Moreover, a reduction in glucokinase activity increases the threshold of glucose-stimulated insulin secretion (59). Therefore, we examined whether changes in glucose metabolism, glucokinase activity and expression, and the expression of the glucose transporter Glut2, correlated with the changes in insulin secretion observed in islets during pregnancy or after culture with PRL *in vitro* (64). The rates of glucose oxidation and utilization in islets were elevated during pregnancy. This enhanced glucose sensitivity results in a shift of the glucose metabolism profiles to lower concentrations. A similar increase in glucose metabolism was found in islets treated with PRL *in vitro*. This increase in glucose metabolism correlates with a 50% increase in glucokinase activity observed in islets during pregnancy and after treatment with PRL *in vitro*. This change also correlates with the increase in glucokinase protein in islets during pregnancy and after treatment with PRL *in vitro*. The glucose transporter Glut2 was also increased more than 2-fold in islets during pregnancy and after treatment with PRL *in vitro*. These changes support the hypothesis that the up-regulation of key components in glucose metabolism have a primary role in the long-term adaptation of islets to pregnancy. The best candidate for having a central role in this process is glucokinase, which is the rate-limiting step in islet glucose metabolism (65). The 50% increase in glucokinase activity is close to the predicted amount (66) needed to lower the glucose-stimulation threshold to that observed during pregnancy and in islets treated with PRL *in vitro*.

Previous studies have also reported an increase in cAMP metabolism during pregnancy and proposed that it was responsible for the increase in glucose-stimulated insulin secretion (62, 63). An increase in cAMP levels has been shown to potentiate insulin secretion and lower the threshold for glucose-stimulated insulin secretion in control islets (10, 67-69). Recently, we re-examined this issue by studying cAMP metabolism in islets during pregnancy and after treatment with PRL *in vitro* (25). This study indicated that these changes are most likely a consequence of the increased glucose metabolism and not a primary event in altering the pattern of insulin secretion. Support for this idea is that under all conditions studied, whether pregnancy, PRL treated or with different glucose concentrations, the levels of c-AMP are comparable for similar rates of glucose metabolism (25).

Although the amount of insulin secretion in pregnancy or PRL treated islets is higher than that in control islets at the same glucose concentration, it is appropriate for the amount of glucose oxidation observed in control islets at a higher glucose concentration (25). This is strong evidence that the increased insulin secretion seen in pregnancy or PRL treated islets *in vitro* is primarily due to an increase in glucose oxidation because of the increase in glucokinase activity. Thus, it is expected that all pathways normally influenced by glucose metabolism in islets will be up regulated in parallel with the enhanced glucose metabolism.

Another feature shared by islets during pregnancy and PRL treatment is enhanced intercellular communication by way of gap junctions (13, 22, 70, 71). An association between enhanced junctional communication and increased insulin secretion has been suggested in several studies (72). Therefore, an increase in junctional coupling may be involved in the increased insulin secretion observed during the adaptation of islets to pregnancy. Interestingly, enhanced junctional communication was observed after only 90 minutes of treatment with PRL *in vitro* (70). This suggests that it may be an early event in PRL's effects on islets. However, we have not observed an effect of PRL on insulin secretion or β-cell division at this very early time point. Thus, it is presently unclear as to how this cellular alteration affects islet function.

Further evidence that the previously described changes are important in the upregulation of islet function is the presence of Stat5 binding sites in the promoters of glucokinase, glucose transporter II, PRL receptor and insulin. The regulation of the PRL receptor and insulin genes by Stat5 has been characterized (53, 73). The relative importance of Stat5 compared to other regulators of these genes is currently under investigation. In particular, the interaction between Stat5 and glucose regulatory factors in

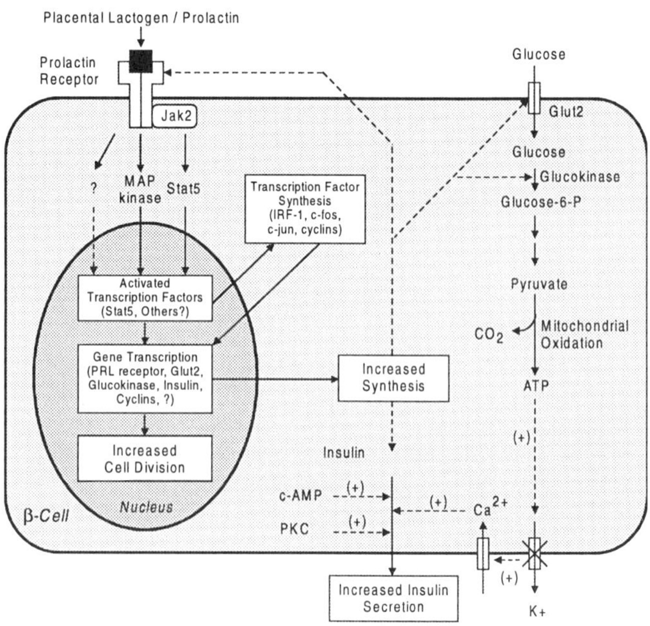

Figure 15-8. Proposed model of the mechanisms by which PRL and PL induce changes in β-cells (see text for details).

glucokinase expression needs to be examined. This information will provide additional insight into the regulation of glucose sensitivity in normal islets.

Based on the available information, we have been able to develop a model for lactogen regulation of islets (Figure 15-8). During pregnancy, islets undergo a number of changes to adapt to the increased demand for insulin. These changes occur under direct influence of lactogenic hormones and under normal blood glucose conditions. This is not an acute short-term regulatory process, but rather occurs over a long period of time. The most significant outcomes are a large increase in β-cell proliferation and enhanced insulin secretion with a lowering of the threshold of glucose-stimulated insulin secretion. These are two quite diverse cellular processes, one which is common to many cells (mitosis) and one which is highly differentiated (insulin secretion). How the intracellular signaling cascade diverges to affect these two different processes is presently unclear. However, the first steps are activation of the PRL receptor and interaction with JAK2. JAK2 phosphorylates Stat5, which then dimerizes and translocates to the nucleus. Through activated transcription factors, an increase in gene expression for PRL receptor, glucose transporter2, glucokinase, insulin and others occurs. The increased expression of the PRL receptor leads to increased β-cell sensitivity to lactogens. Increases in glucokinase and glucose transporter Glut2 would lead to enhanced glucose metabolism. This increase in glucose metabolism would also lead to enhancement of other regulatory pathways such as c-AMP and phospholipase C metabolism (10, 74, 75). All of these changes in metabolism contribute to the shift in the glucose-stimulated insulin response curve. Although, it is most likely that PRL directly stimulates β-cell division, elevated c-AMP and PLC metabolism may also contribute to enhanced β-cell proliferation. Although this model focuses on PRL/PL regulation of islets, it is presumed that GH action on islets is similar with the important difference being that its effects are limited by the rapid down-regulation of its receptor activity.

EFFECT OF PRL AND GH ON ISLET FUNCTION IN OTHER CONDITIONS

Our studies have provided new insight into the physiological consequences of the regulation of islet function by PRL, GH, and PL. The presence of the long forms of both the PRL and GH receptors on the β-cells of islets suggests each hormone could be involved in the regulation of insulin secretion. Furthermore, the occurrence of abnormalities in carbohydrate metabolism *in vivo* with both an absence and excess of these hormones supports this idea. However, it has been difficult to determine to what extent these changes due to their direct effect on islet function or as a consequence of their widespread effects on carbohydrate and lipid metabolism in many tissues. In particular, their ability to reduce the effectiveness of insulin's actions in peripheral tissues (*i.e.*, peripheral insulin resistance) may alter islet function by inducing an increase in serum glucose concentrations. With the new perspective presented in this chapter we can examine how these hormones regulate islet function under normal physiological conditions and those with an absence or excess of these hormones.

Normal islet function appears to require a basal level of stimulation by these hormones. An impairment of glucose tolerance because of a reduction in insulin secretion is observed in both humans and rodents with hypopituitarism and GH deficiency (76-81). However, the reduced level of insulin secretion is compensated for by the increased insulin sensitivity of the peripheral tissues. This effect was recognized in the studies in the 1930's by Bernado Houssay who demonstrated that hypophysectomy partially ameliorated the diabetic symptoms of pancreatectomized dogs (1, 76, 82, 83). The reduced insulin secretion with GH deficiency is not a consequence of a reduced islet mass because the total number of islets and the ratio of total islet mass to body weight in dwarf mice are comparable to that observed in controls (31). This indicates that islets require basal stimulation by pituitary hormones to maintain a normal responsiveness to glucose stimulation. The ability of either GH or PRL to restore normal insulin secretion in hypophysectomized rats suggests that either hormone can meet this requirement (79). In humans, a similar improvement in insulin secretion has been shown with GH treatment of children with GH deficiency (80, 84, 85) and with PL treatment of patients with hypopituitarism (86). This also suggests that the normal serum concentrations of PRL are insufficient to compensate for the absence of GH. This concept is further supported by the inability to demonstrate any changes in insulin secretion in adult rats chronically treated with bromocryptine to suppress endogenous PLR secretion (unpublished observations).

This dependence of normal islet function on a basal level of stimulation of GH is particularly interesting. In all mammalian species examined so far, spontaneous episodes of GH secretion occur several times over a 24-hour period (87-89). In the rat, the pituitary secretes GH in a pulsatile manner in males and in a nearly continuous fashion in females (87, 90). Plasma GH pulses in male rats are approximately one hour in duration with peak blood plasma concentration of 200 ng/ml, followed by a two to three hour period where GH is very low or undetectable (< 2 ng/ml). In contrast, circulating levels in female rats are more continuous with smaller changes in serum concentrations. *In vivo* studies in hypophysectomized rats have shown that the time interval between GH pulses is a critical factor in the ability of tissues to maintain responsiveness to subsequent GH pulses (88). The mechanism underlying this phenomenon is based on the desensitization of the response to GH with prolonged exposure. Therefore, the differences in plasma GH patterns in male and female rats lead to sex-dependent differences in its biological effects of peripheral tissues. In particular, the effect of these GH pulses on the liver of male rats has been shown to be responsible for the increased body growth of male compared to female rats and expression of sex-specific genes in the liver (90, 91). This desensitization to GH stimulation is based on changes in the activation of Stat5b by tyrosine phosphorylation (92-94). This results in levels of tyrosine phosphorylation of Stat5b that are approximately 10% of the maximal levels observed during the first one hour of GH stimulation. The importance of the pulsatile activation of Stat5b in the liver is demonstrated in male mice with a knockout of the Stat5b gene in which their body size and sex-specific gene expression in the liver is comparable to females (91). These observations suggest that pulsatile GH stimulation leads to higher average levels of Stat5b activation in males compared to the more continuous GH stimulation in females with desensitization.

Although a difference in the effects of pulsatile *versus* continuous GH stimulation on the liver has been demonstrated, it is currently unknown whether an analogous relationship occurs with islets. The presence in islets of a similar desensitization of Stat5b activation by GH is consistent with this view (Figure 15-7). However, the absence of marked differences in insulin secretion between males and females suggests that the low level of Stat5b activation with continuous GH is sufficient to maintain the normal glucose sensitivity of islets. It remains to be determined whether pulsed *versus* continuous GH has greater or unique effects on islet function. In addition, we have been unable to demonstrate that pulsatile GH is more effective than continuous GH in the stimulation of β-cell proliferation in neonatal rat islets (unpublished observations). This suggests that the effects of GH on islet function may be limited to those observed with continuous GH stimulation.

Paradoxically, an impairment of glucose tolerance and an increased incidence of diabetes are associated with an excess secretion of GH (76, 82, 83, 95-97). These abnormalities are undoubtedly the result of GH counteracting the insulin stimulation of glucose uptake in peripheral tissues. This decreased capacity for the disposal of glucose increases the demand for insulin, which if not met by the pancreatic islets will develop into diabetes. The chronic administration of GH to normal adult dogs induces glucose intolerance and hyperinsulinemia that eventually develops into diabetes (76, 82, 83, 96, 97). Initially, an increase in islet mass and insulin secretion is observed before the permanent loss of β-cells (83). However, this diabetogenic effect is both age and species dependent with adult dogs being very susceptible, while puppies and rats are resistant (97). Acromegaly is also associated with an increased incidence of diabetes in humans (95, 96). These observations led to the proposal that GH is only diabetogenic when islets cannot compensate for the increased demand for insulin by increasing the number of β-cells (97). The stimulation of β-cell proliferation by GH in immature, but not adult islets, supports this idea.

Although an excess of GH increases the need for more insulin, the progression to diabetes must be dependent on more factors than the absence of a compensating direct effect of GH on β-cells. For example, an abnormal glucose tolerance due to insulin resistance is observed in 68% of the human patients with acromegaly, but less than 15% are diabetic (98). This suggests that another mechanism may be involved in the development of diabetes induced by an excess secretion of GH. It is important to recognize that diabetes will only develop when the production of additional insulin is not sufficient to meet the increased demand for insulin. The availability of insulin can be increased by enhancing the glucose sensitivity of the islets or by an expansion of the total mass of β-cells. A deficit in either response would then be expected to result in the degranulation and exhaustion of the existing β-cells. Because the direct effects of the homologous GH on islets are so small (8, 9, 11, 24), the major stimulus for this adaptation is probably the increased glycemia produced by the decreased rate of glucose disposal. The long-term change in the total glucose flux in β-cells rather than a continuous elevation in the serum glucose concentration is likely responsible for these changes in islet function. An increase in insulin secretion and β-cell proliferation has been reported in rats infused *in vivo* with large amounts of glucose for 24 to 72 hours (99). This suggests that a difference in the β-cell's sensitivity to small changes in glycemia may be important in the age and species differences in the diabetogenic action of GH. Furthermore, it is important to consider that an elevation in PRL is also detected in 25 to 50% of the patents with acromegaly (100). This situation is similar to that in rats with a transplantable pituitary tumor (101, 102). Although often referenced as evidence for a direct effect of GH, the dramatic increase in islet mass in these rats is more likely due to the large amounts of PRL rather than GH secreted by this tumor (14, 102). These observations are consistent with the view that the diabetogenic effect of GH is the result of normal functioning β-cells having to meet an increased demand for insulin. In this regard, this situation is similar to gestational diabetes where the islets are unable to increase insulin secretion sufficiently to overcome the peripheral insulin resistance. However, the underlying defect in islets in gestational diabetes is different.

Unlike GH, a chronic elevation of PRL is not associated with an impairment of glucose tolerance or an increased incidence of diabetes (100, 103-106). This suggests that the enhancement of islet function by PRL is more than sufficient to compensate for the increased demand for insulin. Similar to pregnancy (see the previous section), the lowering of the threshold for glucose-stimulated insulin secretion by PRL will dramatically increase the amount of insulin secreted at normal serum glucose concentrations. The stimulation of β-cell proliferation may be important in maintaining a sufficient islet mass for this increased rate of insulin secretion. We have observed a 25% increase in the ratio of total islet mass to body weight in transgenic mice expressing human GH, which binds to both the mouse PRL and GH receptors (31). The above changes are consistent with the enhanced glucose-stimulated insulin secretion and slight decrease in fasting serum glucose concentrations with moderate hyperinsulinemia observed in many conditions with an excess secretion of PRL (103-106).

The differences in time required for the effects of PRL on individual tissues may be responsible for a large part of the controversy concerning whether hyperprolactinemia induces glucose intolerance. The impaired glucose tolerance reported with a short-term infusion of PRL or PL in humans (106) is most likely the result of an induction of insulin resistance in peripheral tissues. Because the effects on islet function are slower, only with an elevation of PRL or PL for several days will the glucose tolerance return to normal or even be enhanced because of the increased insulin secretion. This view is supported by the normal glucose tolerance with enhanced insulin secretion observed in women with naturally occurring hyperprolactemia (103-105), and in rats with extra pituitaries transplanted into their kidneys (107). This enhancement of islet function accounts for the inability of PRL to induce diabetes in any species unless the islet mass is severely reduced by removing most of the pancreas (108). This suggests that the enhancement of islet function by PRL is more than adequate to meet the increased insulin demand resulting from the induction of peripheral insulin resistance.

In summary, the differences in the response of islets to PRL and GH help explain the changes in carbohydrate metabolism observed in conditions with an altered secretion of these hormones. A tonic stimulation of islets by either PRL or GH is required for the normal responsiveness of islets to glucose stimulation. Although the excess secretion of either hormone can induce peripheral insulin resistance, the mechanisms by which this increased demand for insulin is met are different. The limited capacity of GH to stimulate islet function implies that islets with normal glucose sensitivity must meet this demand. Diabetes can occur when the total islet mass is too small to respond to this increased demand. In contrast, the ability of PRL to increase the glucose sensitivity, insulin secretion and β-cell proliferation in islets is more than sufficient to meet this increased demand without an elevation in serum glucose concentrations. Therefore, the difference in the response of islets to these hormones has important implications for conditions with altered secretion of these hormones.

CONCLUSION

It has been difficult to understand the physiological roles of PRL, GH, and PL in the regulation of islet function because of the questionable interpretation of published data. The use of heterologous hormones is the basis of much of this confusion. To avoid this problem, we examined the direct effect of the homologous PRL, GH, and PL on insulin secretion and β-cell proliferation in islets isolated from rats, mice, and humans. We found that homologous PRL and PL are potent stimulators of islet function in each of these species. The homologous GH was much less effective with little to no effect observed on adult islets. Many studies have erroneously concluded that GH is a potent growth factor for β-cells because they were done in rats using the heterologous human GH that actually binds to both the rat PRL and GH receptor. We have also demonstrated that PRL and PL are the most potent growth factors of differentiated β-cells that have been identified (21, 24).

These findings have intriguing implications for our views about the physiological roles of these hormones in the regulation of islet function. The presence of the long form of both the PRL and GH receptors suggests that islets may be sensitive to changes in the circulating levels of these hormones. A tonic stimulation of islets by either PRL or GH is required for the normal responsiveness of islets to glucose stimulation. The inability of the endogenous PRL to maintain normal insulin secretion in conditions with GH deficiency suggests that the normal levels of PRL are too low to compensate for the absence of GH. The rapid desensitization of the GH response explains why its effects on islet function are so small compared to those with a continuous stimulation with PRL or PL. This difference between the effects of PRL and GH on islet function may explain the increased incidence of diabetes observed with the elevated secretion of GH, but not with PRL, in many species. More importantly, it suggests that the physiological role of PRL and PL on islet function is to increase insulin production, under normal glucose concentrations, during conditions of increased demand. The best example of this is during pregnancy where elevated insulin secretion counter balances the induction of peripheral insulin resistance. Although an elevation in serum glucose concentration by itself could stimulate islet function, the resulting hyperglycemia would be detrimental to both the mother and the developing fetus.

REFERENCES

1. Houssay BA. The hypophysis and metabolism. N Engl J Med 1936;214:961-963.
2. Houssay BA, Anderson E, Bates RW. Diabetogenic action of prolactin. Endocrinology 1955;57:55-63.
3. Houssay BA, Penhos JC. Diabetogenic action of pituitary hormones on adrenalectomized hyphophysectomised dogs. Endocrinology 1956;59:637-641.
4. Jovanovic-Peterson L, Peterson CM. Pregnancy and the endocrine pancreas. In: E. S, editor. The Endocrine Pancreas. New York: Raven Press; 1991. p. 229-252.
5. Nicoll CS. Prolactin and growth hormone: Specialists on one hand and mutual mimics on the other. Perspect Biol Med 1982;25:369-381.
6. Nicoll CS, Mayer GL, Russell SM. Structural features of prolactins and growth hormones that can be related to their biological properties. Endocr Rev 1986;7:169-203.
7. Ogren L, Talamantes F. Prolactins of pregnancy and their cellular source. Int Rev Cytol 1988;112:1-65.
8. Brelje TC, Allaire P, Hegre O, Sorenson RL. Effect of prolactin versus growth hormone on islet function and the importance of using homologous mammosomatotropic hormones. Endocrinology 1989;125:2392-9.
9. Brelje TC, Scharp DW, Lacy PE, Ogren L, Talamantes F, Robertson M, *et al.* Effect of homologous placental lactogens, prolactins, and growth hormones on islet B-cell division and insulin secretion in rat, mouse, and human islets: implication for placental lactogen regulation of islet function during pregnancy. Endocrinology 1993;132:879-87.
10. Brelje TC, Sorenson RL. Nutrient and hormonal regulation of the threshold of glucose-stimulated insulin secretion in isolated rat pancreases. Endocrinology 1988;123:1582-90.
11. Brelje TC, Sorenson RL. Role of prolactin versus growth hormone on islet B-cell proliferation *in vitro*: implications for pregnancy. Endocrinology 1991;128:45-57.

12. Parsons JA, Brelje TC, Sorenson RL. Adaptation of islets of Langerhans to pregnancy: increased islet cell proliferation and insulin secretion correlates with the onset of placental lactogen secretion. Endocrinology 1992;130:1459-66.
13. Sorenson RL, Brelje TC, Hegre OD, Marshall S, Anaya P, Sheridan JD. Prolactin (*in vitro*) decreases the glucose stimulation threshold, enhances insulin secretion, and increases dye coupling among islet B cells. Endocrinology 1987;121:1447-53.
14. Sorenson RL, Parsons JA. Insulin secretion in mammosomatotropic tumor-bearing and pregnant rats. A role for lactogens. Diabetes 1985;34:337-41.
15. Billestrup N, Nielsen JH. The stimulatory effect of growth hormone, prolactin, and placental lactogen on beta-cell proliferation is not mediated by insulin-like growth factor-I. Endocrinology 1991;129:883-8.
16. Swenne I. Pancreatic beta-cell growth and diabetes mellitus. Diabetologia 1992;35:193-201.
17. Rabinovitch A, Quigley C, Rechler MM. Growth hormone stimulates islet B-cell replication in neonatal rat pancreatic monolayer cultures. Diabetes 1983;32:307-12.
18. Nielsen JH. Effects of growth hormone, prolactin, and placental lactogen on insulin content and release, and deoxyribonucleic acid synthesis in cultured pancreatic islets. Endocrinology 1982;110:660-606.
19. Whittaker PG, Taylor KW. Direct effects of rat growth hormone in rat islets of langerhans in tissues culture. Diabetologia 1980;18:323-8.
20. Sorenson RL, Brelje TC. Adaptation of islets of Langerhans to pregnancy: beta-cell growth, enhanced insulin secretion and the role of lactogenic hormones. Horm Metab Res 1997;29:301-7.
21. Brelje TC, Sorenson RL. The physiological roles of prolactin, growth hormone and placental lactogen in the regulation of islet beta cell proliferation. In: N S, editor. Pancreatic Growth and Regeneration. New York: Karger Landes Systems; 1997. p. 1-30.
22. Sorenson RL, Johnson MG, Parsons JA, Sheridan JD. Decreased glucose stimulation threshold, enhanced insulin secretion, and increased beta cell coupling in islets of prolactin-treated rats. Pancreas 1987;2:283-8.
23. Kawai M, Kishi K. *In vitro* studies of the stimulation of insulin secretion and B-cell proliferation by rat placental lactogen-II during pregnancy in rats. J Reprod Fertil 1997;109:145-152.
24. Brelje TC, Parsons JA, Sorenson RL. Regulation of islet beta-cell proliferation by prolactin in rat islets. Diabetes 1994;43:263-73.
25. Weinhaus AJ, Bhagroo NV, Brelje TC, Sorenson RL. Role of cAMP in upregulation of insulin secretion during the adaptation of islets of Langerhans to pregnancy. Diabetes 1998;47:1426-35.
26. Goffin V, Shiverick KT, Kelly PA, Martial JA. Sequence-function relationships within the expanding family of prolactin, growth hormone, placental lactogen, and related proteins in mammals. Endocr Rev 1996;17:385-410.
27. Lowman HB, Cunningham BC, Wells JA. Mutational analysis and protein engineering of receptor-bindg determinants in human placental lactogen. JBC 1991;266:10982-10988.
28. Spellacy WN, Goetz FC. Plasma insulin in normal late pregnancy. NEJM 1962;268:988-991.
29. Buchanan TA. Metabolic changes during normal and diabetic pregnancies. In: Reece EA, Coustan DR, editors. Diabetes Mellitus in Pregnancy: Churchill Livingstone; 1995. p. 59-77.
30. Hare JW, Brown FM. Pathophysiology. In: Brown FM, Hare JW, editors. Diabetes Complicating Pregnancy: The Joslin Clinic Method. New York: John Wiley & Sons; 1995. p. 3-13.
31. Parsons JA, Bartke A, Sorenson RL. Number and size of islets of Langerhans in pregnant, human growth hormone-expressing transgenic, and pituitary dwarf mice: effect of lactogenic hormones. Endocrinology 1995;136:2013-21.

32. Costrini NV, Kalkhoff RK. Relative effects of pregnancy, estradiol, and progesterone on plasma insulin and pancreatic islet content. J Clin Inv 1991;50:992-999.
33. Metzger B, Pek S, Hare J, Freinkel N. Relationships between glucose, insulin and glucagon during fasting in late gestation in the rat. Life Sci 1974;15:301-8.
34. Spellacy WN, Goetz FC, Greenberg BZ, Ells J. Plasma insulin in normal "early" pregnancy. Obst Gyne 1965;25:862-865.
35. Spellacy WN, Goetz FC, Greenberg BZ, Ells J. Plasma insulin in normal midpregnancy. Am J Obst Gyne 1965;92:11-15.
36. Green IC, Howell SL, Montague W, Taylor KW. Regulation of insulin release from isolated islets of Langerhans of the rat in pregnancy. Biochem J 1973;134:481-487.
37. Dunger A, Lucke S, Besch W, Hahn H-J. The rat pancreatic B-cell during pregnancy and after delivery. Int. Journal of Feto-Maternal Med. 1989;2:55-61.
38. Bone AJ, Taylor KW. Metabolic adaptation to pregnancy shown by increased biosynthesis of insulin in islets of Langerhans isolated from pregnant rats. Nature 1976;262:501-502.
39. Sorenson RL, Brelje TC, Roth C. Effects of steroid and lactogenic hormones on islets of Langerhans: a new hypothesis for the role of pregnancy steroids in the adaptation of islets to pregnancy. Endocrinology 1993;133:2227-34.
40. Weinhaus A, Bhagroo NV, Brelje TC, Sorenson RL. Dexamethasone counteracts the effect of prolactin on islet function: Implications for islet regulation in late pregnancy. Endocrinology 2000;141.
41. Forsyth IA. Variation among species in the endocrine control of mammary growth and function: the roles of prolactin, growth hormone, and placental lactogen. J Dairy Sci 1986;69:886-903.
42. Nielsen PV, Pedersen H, Kampmann EM. Absence of human placental lactogen in an otherwise uneventful pregnancy. Am J Obstet Gynecol 1979;135:322-6.
43. Sideri M, De Virgiliis G, Guidobono F, Borgese N, Sereni LP, Nicolini U, *et al.* Immunologically undetectable human placental lactogen in a normal pregnancy. Case report. Br J Obstet Gynaecol 1983;90:771-3.
44. Simon P, Decoster C, Brocas H, Schwers J, Vassart G. Absence of human chorionic somatomammotropin during pregnancy associated with two types of gene deletion. Hum Genet 1986;74:235-8.
45. Onclin K, Verstegen JP. Secretion patterns of plasma prolactin and progesterone in pregnant compared with nonpregnant dioestrous beagle bitches. J Reprod Fertil Suppl 1997;51:203-8.
46. Tyson JE, Hwang P, Guyda H, Friesen HG. Studies of prolactin secretion in human pregnancy. Am J Obstet Gynecol 1972;113:14-20.
47. Yuen BH, Cannon W, Woolley S, Charles E. Maternal plasma and amniotic fluid prolactin levels in normal and hypertensive pregnancy. Br J Obstet Gynaecol 1978;85:293-8.
48. Luciano AA, Varner MW. Decidual, amniotic fluid, maternal and fetal prolactin in normal and abnormal pregnancies. Obstet Gynecol 1984;63:384-8.
49. Asfari M, De W, Postel-Vinay MC, Czernichow P. Expression and regulation of growth hormone (GH) and prolactin (PRL) receptors in a rat insulin producing cell line (INS-1). Mol Cell Endocrinol 1995;107:209-14.
50. Moldrup A, Billestrup N, Nielsen JH. Rat insulinoma cells express both a 115-kDa growth hormone receptor and a 95-kDa prolactin receptor structurally related to the hepatic receptors. J Biol Chem 1990;265:8686-90.
51. Polak M, Scharfmann R, Band E, Haour F, Postel-Vinay MC, Czernichow P. Demonstration of lactogenic receptors in rat endocrine pancreases by quantitative autoradiography. Diabetes 1990;39:1045-1049.

52. Nielsen JH, Moldrup A, Billestrup N. Expression of the growth hormone receptor gene in insulin producing cells. Biomed Biochim Acta 1990;49:1151-5.
53. Galsgaard ED, Nielsen JH, Moldrup A. Regulation of prolactin receptor (PRLR) gene expression in insulin- producing cells. Prolactin and growth hormone activate one of the rat PRLR gene promoters via STAT5a and STAT5b. J Biol Chem 1999;274:18686-92.
54. Sorenson RL, Stout LE. Prolactin receptors and JAK2 in islets of Langerhans: an immunohistochemical analysis. Endocrinology 1995;136:4092-8.
55. Moldrup A, Petersen ED, Nielsen JH. Effects of sex and pregnancy hormones on growth hormone and prolactin receptor gene expression in insulin-producing cells. Endocrinology 1993;133:1165-72.
56. Bole-Feysot C, Goffin V, Edery M, Binart N, Kelly PA. Prolactin (PRL) and its receptor: actions, signal transduction pathways and phenotypes observed in PRL receptor knockout mice. Endocr Rev 1998;19:225-68.
57. Schindler C, Darnell JE, Jr. Transcriptional responses to polypeptide ligands: the JAK-STAT pathway. Annu Rev Biochem 1995;64:621-51.
58. Stout LE, Svensson AM, Sorenson RL. Prolactin regulation of islet-derived INS-1 cells: characteristics and immunocytochemical analysis of STAT5 translocation. Endocrinology 1997;138:1592-603.
59. Burch PT, Trus MD, Berner DK, Leontire A, Zawalich KC, Matchinsky FM. Adaptation of glycolytic enzymes: glucose use and insulin release in rat pancreatic islets during fasting and refeeding. Diabetes 1981;30:923-928.
60. Sorenson RL, Lindell DV, Elde RP. Glucose stimulation of somatostatin and insulin release from the isolated, perfused rat pancreas. Diabetes 1980;29:747-51.
61. Zawalich WS, Dye ES, Pagliara AS, Rognstad R, Matschinsky FM. Starvation diabetes in the rat: onset, recovery and specificity of reduced responsiveness of pancreatic B-cells. Endocrinology 1979;104:1344-1350.
62. Green IC, Perrin D, Howell SL. Insulin release in isolated islets of Langerhans of pregnant rats: Relationship between glucose metabolism and cyclic AMP. Horm Metab Res 1978;10:32-35.
63. Howell SL, Green IC, Montague W. A possible role of adenylate cyclase in the long-term dietary regulation of insulin secretion from rat islets of Langerhans. Biochem J 1973;136:343-349.
64. Weinhaus AJ, Stout LE, Sorenson RL. Glucokinase, hexokinase, glucose transporter 2, and glucose metabolism in islets during pregnancy and prolactin-treated islets *in vitro*: mechanisms for long term up-regulation of islets. Endocrinology 1996;137:1640-9.
65. Matschinsky FM. Banting Lecture 1995. A lesson in metabolic regulation inspired by the glucokinase glucose sensor paradigm. Diabetes 1996;45:223-41.
66. Davis EA, Cuesta-Munoz A, Raoul M, Buettger C, Sweet I, Moates M, *et al.* Mutants of glucokinase cause hypoglycaemia- and hyperglycaemia syndromes and their analysis illuminates fundamental quantitative concepts of glucose homeostasis. Diabetologia 1999;42:1175-86.
67. Malaisse WJ, Malaisse-Lagae F. The role of cyclic AMP in insulin release. Experientia 1984;40:1068-74.
68. Lipson LG, Sharp GW. Insulin release in pregnancy: studies on adenylate cyclase, phosphodiesterase, protein kinase, and phosphoprotein phosphatase in isolated rat islets of Langerhans. Endocrinology 1978;103:1272-80.
69. Sharp GW. The adenylate cyclase-cyclic AMP system in islets of Langerhans and its role in the control of insulin release. Diabetologia 1979;16:287-96.
70. Michaels RL, Sorenson RL, Parsons JA, Sheridan JD. Prolactin enhances cell-to-cell communication among beta-cells in pancreatic islets. Diabetes 1987;36:1098-103.

71. Sheridan JD, Anaya PA, Parsons JA, Sorenson RL. Increased dye coupling in pancreatic islets from rats in late-term pregnancy. Diabetes 1988;37:908-11.
72. Meda P, Perrelet A, Orci L. Gap junction and cell-to-cell coupling in endocrine glands. Modern Cell Biology 1984;3:131-196.
73. Galsgaard ED, Gouilleux F, Groner B, Serup P, Nielsen JH, Billestrup N. Identification of a growth hormone-responsive Stat5-binding element in the rat insulin 1 gene. Mol Endocrin 1996;10:652-660.
74. Zawalich WS, Karl RC, Ferrendelli JA, Matschinsky FM. Factors governing glucose induced elevation of cyclic 3'5' AMP levels in pancreatic islets. Diabetologia 1975;11:231-235.
75. Zawalich WS, Zawalich KC. Phosphoinositide hydrolysis and insulin release from isolated perifused rat islets: Studies with glucose. Diabetes 1988;37:1294-1300.
76. Davidson MB. Effect of growth hormone on carbohydrate and lipid metabolism. Endocrine Reviews 1987;8:115-131.
77. Martin JM, Gagliardino JJ. Effect of growth hormone on the isolated pancreatic islets of rat *in vitro*. Nature 1967;213:630-1.
78. Malaisse WJ, Malaisse-Lagae F, King S, Wright PH. Effect of growth hormone on insulin secretion. J Physiol 1968:423-428.
79. Bennett LL, Curry DL, Li CH. Enhancement of insulin secretion by human chorionic somatomammotropin and related hormones. Proc Soc Exp Biol Med 1976;152:281-3.
80. Lippe BM, Kaplan SA, Golden MP, Hendricks SA, Scott ML. Carbohydrate tolerance and insulin receptor binding in children with hypopituitarism: response after acute and chronic human growth hormone administration. J Clin Endocrinol Metab 1981;53:507-13.
81. Pierluissi J, Pierluissi R, Ashcroft SJ. Effects of hypophysectomy and growth hormone on cultured islets of Langerhans of the rat. Diabetologia 1982;22:134-7.
82. Campbell J, Davidson IWF, Lei HP. The production of permanent diabetes by highly purified growth hormones. Endocrinology 1950;46:588-590.
83. Pierluissi J, Campbell J. Metasomatotrophic diabetes and its induction: basal insulin secretion and insulin release responses to glucose, glucagon, arginine and meals. Diabetologia 1980;18:223-8.
84. Gold H, Spector S, Samaan NA, Pearson OH. Effect of growth hormone on carbohydrate metabolism in hypopituitary dwarfs. Metabolism 1968;17:74-83.
85. Walker J, Chaussain JL, Bougneres PF. Growth hormone treatment of children with short stature increases insulin secretion but does not impair glucose disposal. J Clin Endocrinol Metab 1989;69:253-8.
86. Grumbach MM, Kaplan SL, Sciarra JJ, Burr IM. Chorionic growth hormone-prolactin (CGP): secretion, disposition, biologic activity in man and postulated function as the "growth hormone" of the second half of pregnancy. Annals New York Acad Science 1968;148:501-531.
87. Eden S. Age- and sex-related differences in episodic growth hormone secretion in the rat. Endocrinology 1979;105:555-560.
88. Jansson JO, Eden S, Isaksson O. Sexual dimorphism in the control of growth hormone secretion. Endocr Rev 1985;6:128-50.
89. Bizot-Espiard JG, Double A, Guardiola-Lemaitre B, Delagrange P, Ktorza A, Penicaud L. Diurnal rhythms in plasma glucose, insulin, growth hormone and melatonin levels in fasted and hyperglycaemic rats. Diabetes Metab 1998;24:235-40.
90. Waxman DJ, Pampori NA, Ram PA, Agrawal AK, Shapiro BH. Interpulse interval in circulating growth hormone patterns regulates sexually dimorphic expression of hepatic cytochrome P450. Proc Natl Acad Sci U S A 1991;88:6868-72.
91. Udy GB, Towers RP, Snell RG, Wilkins RJ, Park SH, Ram PA, *et al.* Requirement of STAT5b for sexual dimorphism of body growth rates and liver gene expression. Proc Natl Acad Sci U S A 1997;94:7239-44.

92. Gebert CA, Park SH, Waxman DJ. Regulation of signal transducer and activator of transcription (STAT) 5b activation by the temporal pattern of growth hormone stimulation. Mol Endocrinol 1997;11:400-14.
93. Gebert CA, Park SH, Waxman DJ. Termination of growth hormone pulse-induced STAT5b signaling. Mol Endocrinol 1999;13:38-56.
94. Gebert CA, Park SH, Waxman DJ. Down-regulation of liver JAK2-STAT5b signaling by the female plasma pattern of continuous growth hormone stimulation. Mol Endocrinol 1999;13:213-27.
95. Luft R, Cerasi E, Hausberger CA. Studies on the pathogenesis of diabetes in acromegaly. Acta Endocrinol (Copenh) 1967;56:593-607.
96. Fineberg SE, Merimee TJ, Rabinowitz D, Edgar PJ. Insulin secretion in acromegaly. J Clin Endocrinol Metab 1970;30:288-92.
97. Young FG, Korner A. Growth hormone. In: Williams RH, editor. Diabetes. New York: Paul B Hoeber; 1960. p. 216-233.
98. Merimee TJ, Grant MB. Growth hormone and its disorders. In: Becker KL, editor. Principles and Practice of Endocrinology and Metabolism. Philadelphia: J B Lippincott; 1990. p. 125-134.
99. Chen C, Hosokawa H, Bumbalo LM, Leahy JL. Regulatory effects of glucose on the catalytic activity and cellular content of glucokinase in the pancreatic beta cell. Study using cultured rat islets. J Clin Invest 1994;94:1616-20.
100. Evans WS, Carlsen E, Ho KY. Prolactin and its disorders. In: Becker KL, editor. Principles and Practice of Endocrinology and Metabolism. Philadelphia: JB Lippincott; 1990. p. 134-139.
101. Martin JM, Akerblom HK, Gary G. Insulin secretion in rats with elevated circulating growth hormone due to Mtt-W15 tumor. Diabetes 1968;17:661-667.
102. Parsons JA, Hartfel MA, Hegre OD, McEvoy RC. Effect of MtTW15 mammosomatotropic tumors on pancreatic islet hormones. Diabetes 1983;32:67-74.
103. Landgraf R, Landgraf-Leurs M, M,C,, Weissmann A, Hörl R, von Werder K, Scriba PC. Prolactin: A diabetogenic hormone. Diabetologiz 1977;13:99-104.
104. Johnston DG, Alberti KG, Nattrass M, Burrin JM, Blesa-Malpica G, Hall K, *et al.* Hyperinsulinaemia in hyperprolactinaemic women. Clin Endocrinol (Oxf) 1980;13:361-8.
105. Gustafson AB, Banasiak MF, Kalkhoff RK, Hagen TC, Kim HJ. Corrleation of hyperprolactinemia with altered plasma insulin and glucagon: Similarity to effects of late human pregnancy. J Clin Endocrinol Metab 1980;51:242-246.
106. Rojdmark S, Lamminpaa K. Does acute endogenous hyperprolactinemia affect intravenous glucose tolerance in humans? Metabolism 1984;33:567-71.
107. Adler RA, Sokol HW. Glucose tolerance in rats with elevated circulating prolactin levels. Horm Metab Res 1982;14:307-9.
108. Houssay BA. Other hormones. In: Williams RH, editor. Diabetes. New York: Paul B Hoeber; 1960. p. 233-256.

Chapter 16

In Vitro Effects of Prolactin on the Lympho-Hemopoietic System

Elizabeth L. Hooghe-Peters, Zeynep Dogusan and Robert Hooghe

Free University of Brussels (V.U.B.), Brussels, Belgium and Flemish Institute for Technological Research (VITO) Mol, Belgium

Functional receptors for prolactin (PRL-R) are expressed on many cell types in various tissues, including leukocytes. The PRL-R belongs to the hemopoietin-cytokine receptor family. PRL itself is also expressed in the lympho-hemopoietic system (LHS) where it induces discrete effects. PRL thus qualifies as a cytokine. Effects on the LHS are modest in rodents and man and, though dispensable, PRL is clearly able to affect leukocyte physiology. The present contribution summarizes mainly recent *in vitro* studies on the role of PRL in the hemopoietic and in the immune systems.

INTRODUCTORY PERSPECTIVE

Following observations of clear-cut immune deficiency in hypophysectomized rats and its correction by PRL, it was proposed that PRL plays an important role in the immune system (reviewed in references 1-8). Studies in various strains of dwarf mice (lacking PRL, growth hormone and thyrotropin) and in bromocriptine-treated rats (thus with reduced serum PRL levels) further seemed to support this view (1-5, 7, 8). The establishment of the PRL-dependent Nb2 rat T-lymphoma cell line (see below) provided a model to explore the place of PRL in the immune system (9). Lymphoma cells are considered as the neoplastic equivalent of a normal lymphocyte, frozen in a given stage of differentiation. Although a normal counterpart of the Nb2 cell has not been identified, the dramatic effects of PRL in Nb2 cells strongly influenced the perception of the role of PRL in the immune system. Interest for PRL in immunology also stems from the gender differences, where estrogens are considered to play the key role, in both the normal and the diseased immune system (10). As estrogens induce PRL secretion, it is possible that some effects of estrogens, also on the LHS, are PRL-mediated. Furthermore, in murine models of lupus and in systemic lupus erythematosus in the human (a condition more prevalent in women), serum levels of PRL are elevated (11, 12). PRL lowering drugs have therefore been used -with limited success- in several autoimmune conditions (13 - 15). At the other end of the clinical spectrum, reports of stimulatory (or protective, or restoring) effects of PRL in the LHS have raised hopes that therapy with

PRL may be beneficial in conditions such as bone marrow aplasia and immune deficiency syndromes (16, 17). Finally, although there is no known equivalent of the Nb2 PRL-dependent T-cell lymphoma in man, a search for PRL-responsive hemopoietic malignancies is warranted (18, 19): PRL deprivation may be an effective therapy in neoplastic conditions that are responsive to PRL. In sharp contrast to *in vivo* and *in vitro* experiments showing effects (nearly always stimulatory) of PRL, detailed investigation of PRL- and PRL-R knock-out (KO) mice revealed a normal LHS and immune responses of normal amplitude. PRL is clearly dispensable in the development of the LHS (8, 20-22). It is not yet known whether PRL plays no role in the developing LHS or whether, in KO mice, its functions are taken over by other factors: redundancy is an important feature of the cytokine network. In view mainly of the possible clinical implications, it is important to identify targets for (physiological and pharmacological concentrations of) PRL in the LHS.

HYPOTHESES

Extreme views hold either that PRL has no specific role on the LHS or, on the opposite, that the LHS contains both PRL-producing cells and targets for PRL. In the first case, PRL is simply considered as a hormone with many targets, a growth factor for many cell types. According to the second view, some leukocyte and stromal cell populations produce PRL, that, possibly together with endocrine PRL, exerts discrete, specific effects on some cells in the LHS. We will not address in detail PRL production in the LHS and concentrate on *in vitro* molecular and cellular effects of PRL on LHS targets, irrespectively of the origin (endocrine or paracrine) of PRL, trying to answer the question: is PRL a growth and differentiation factor, with specific targets in the LHS? As said before, PRL is globally seen as an immunostimulatory molecule, in many areas of hematology and immunology. In a few instances, an immunosuppressive role has been proposed: this is the case of tolerance to the fetus and remission of some autoimmune diseases during pregnancy.

PRL IN THE LHS SYSTEM: GENERAL COMMENTS AND CAVEATS

1. The role of PRL cannot be evaluated in experiments performed in the presence of fetal calf serum as it contains lactogenic substances (9, 23). This could be one of the reasons why the role of PRL has not been noticed earlier.
2. Even in serum-free conditions, experiments are not necessarily PRL-free.
    a. Leukocytes express small amounts of PRL, which may however lead to high local concentrations (5, 7, 24).
    b. Uptake of PRL from serum has been reported, and this PRL can later be secreted. Thus, uptake in vivo or *in vitro* (during culture in the presence of serum) could lead to secretion of PRL in serum-free culture conditions (25).
3. In the past, pituitary hormone preparations were contaminated, and as GH and PRL have similar MW, cross-contamination could hardly be avoided. Nowadays, recombinant PRL (r-PRL) is produced, not only in bacteria, but also in yeast or murine cell expression systems (17). Different batches of PRL may differ in biological activity. Bioactivity of natural or recombinant PRL used in biological experiments should therefore be evaluated in the Nb2 assay (and/or the pigeon crop sac assay).
4. In the case of r-PRL, attention should be given to contamination by endotoxin. This should be a constant worry as PRL also signals through p38 MAP-kinase like endotoxin (26 and see below). In addition, LPS can induce cytokines that signal

through Stat 5 as does PRL (27). Removal or inhibition of LPS can prove to be difficult.

5. Without resorting to mutational analysis of cellular proteins, the use of PRL and GH of different origins (both natural and recombinant, and from different species) will provide the most convincing evidence for a role of PRL in a given system (28). There is a wide and well-known cross-reactivity of PRL in the various species. In the best case, analogs (both active and inactive) can be used. For assessing the role of PRL, neutralizing antibody is often used. Controls with pre-immune serum or isotype matched antibodies should be included.
6. The hormones most likely to interfere with the evaluation of the role of PRL are GH (which, in primates, binds the PRL-R in addition to its own R) and estrogens (which are among the most potent inducers of PRL) (5, 10). In many normal and pathological conditions, it is not clear whether estrogen or PRL plays the major role (10). Such questions are best evaluated *in vitro*.

THE NB2 PARADIGM

PRL clearly is a growth factor for the Nb2 cell line, which was established 20 years ago from neoplastic lymph nodes taken from an estrogen-treated male rat. The cells in culture are dependent on lactogenic substances or interleukin (IL)-2 for growth (9, 23, 29). The facts that the lymphoma arose in an hyperprolactinemic (estrogen-treated) rat and was first propagated in estrogen-conditioned rats suggests that PRL-dependency was already present *in vivo*. This cell line is still used for the bioassay of lactogenic hormones and has been of invaluable help to study the mode of action of PRL, not only in leukocytes but also in other targets for PRL. Detailed studies on PRL-R, signal transduction mechanisms (mainly through the Jak-Stat pathway) and gene induction by PRL have been performed in Nb2 cells (6, 30, 31).

The Nb2 cell has an immature T cell phenotype (CD4+, CD8+). The expression of the T-cell receptor is up regulated by lactogen withdrawal. PRL represses the expression of α and β chains but stimulates the expression of the γ chain of the T-cell receptor (32) and induces the phosphorylation of several components of the T-cell receptor (33). In Nb2 cells, PRL signals through an aberrant receptor (called intermediate form, i.e. the long form, with a deletion, see below), induces numerous growth-related and other genes, has anti-apoptotic effects and stimulates proliferation (34-40). The Jak 2 - Stat 5 - GAS pathway has been extensively analyzed in Nb2 cells, but other signaling pathways are also activated: by phosphorylating the PRL-R, Jak 2 enables the association of Shc with the receptor, which in turn activates the Shc/Grb2/Vav/Sos/Ras/Rac/Raf/MEK/MAPK and S6 kinase pathways (41–43). Vav and Sos are also activated by Fyn, a src-family kinase that, together with Jak 2 is activated by PRL (44). PRL-activation of Fyn may also lead to the activation of PI3-kinase (45). So, PRL induces the expression or activation of many transcription factors of which IRF-1 has received most attention (6, 31, 46). Much effort has been devoted to the analysis (following substractive hybridization) of genes induced by PRL in Nb2 cells (6, 46, 47).

The Nb2 cell thus proved to be an ideal tool for molecular studies of PRL action. The next question is to which extent this wealth of information can be extrapolated to normal cells of the LHS. The activation of Stat 5, the subsequent induction of IRF-1 (to quote only two events) are key steps in lymphocyte activation. How often is PRL involved in these phenomena-that can also be induced by several other factors - in leukocyte physiology or pathology? Clearly, the PRL KO do not have the immune defects present in the IRF-1 KO (20, 48)!

PRL AND LACTOGENIC SUBSTANCES

PRL IN THE LHS:

The biochemistry of PRL and PRL-related molecules is discussed elsewhere in this book. The present chapter will hardly deal with the production of PRL by leukocytes. It should however be remembered that PRL production in the pituitary and in extra-pituitary sites is regulated by different factors (5, 24, 49-51). PRL is found in bone marrow (BM) cells from Pit-1 deficient, Snell dw/dw dwarf mice where pituitary cells do not express PRL (R. Kooijman, unpublished results). As in rodents, PRL expression in human leukocytes is low and can be more conveniently studied in several IM-9 B-lymphoblastoid cell clones where expression is high (52).

In man, PRL expression in extra-pituitary sites is under the control of a distal promoter, that is not responsive to thyrotropin-releasing hormone nor to dopamine (in contrast to the pituitary promotor; however, see below). As in the pituitary, expression in extra-pituitary tissues is decreased by glucocorticoids. Cyclic AMP and the T-cell activator phytohaemagglutinin (PHA) alone or in combination with phorbol myristate acetate (PMA) increased reporter gene expression in Jurkat lymphoma T-cells transfected with the extra-pituitary (upstream) promoter, and cAMP acted in synergy with PMA or PHA to increase it further (51). The regulation of PRL expression in extra-pituitary tissues is largely unknown. The protein expressed by IM-9 cells is indistinguishable by MW and antigenicity from pituitary PRL: the mRNA is longer than in the pituitary but the coding region is identical.

Also, little is known about extra-pituitary tissues and cell types expressing PRL. The major source are stromal cells from the decidualized endometrium (24). Among all lympho-hemopoietic tissues tested from rat or mouse, only rat thymus gave a clear signal in PCR (53). However, Pellegrini *et al.* demonstrated PRL biosynthesis in mouse thymocytes and splenocytes and in human peripheral blood mononuclear cells (PBMC) (54). Also, Delhase *et al.* documented the expression of PRL in rat BM, spleen and thymus by PCR and *in situ* hybridization (55). The cells expressing PRL mRNA also contain immunoreactive PRL (4). PCR analysis indicated that human T-cells (thymocytes or T cells from peripheral blood) express more PRL than B cells, where Southern analysis was required to generate a signal. Monocytes were negative. PRL expression in the human LHS was also confirmed by *in situ* hybridization in lymphocytes, epithelial cells, and vascular endothelial cells (4, 56).

In contrast to the data showing no PRL expression in monocytes from PBMC, it was recently concluded from experiments using the very sensitive suppression substractive hybridization technique that PRL expression is a marker of monocytic differentiation. Indeed, both human CD14+ mature monocytes from peripheral blood and U937 cells differentiated with vitamin D3 + transforming growth factor β expressed PRL whereas CD34+ stem cells isolated from bone marrow and the undifferentiated U937 cells did not (57).

In conclusion, in the LHS, the expression of PRL is best documented in T cells. It is increased by T-cell stimulation and by cAMP and reduced by glucocorticoids.

Many immuno-reactive PRL forms have been identified on Western blot but characterization of the proteins was often incomplete so that it is usually not known whether they are indeed products of the PRL (or a PRL-related) gene (5, 7). In addition, it is usually not known whether they are actually synthesized by leukocytes (rather than secreted after being taken up from serum).

Macroprolactin represents a special challenge for the endocrinologist. A large proportion of patients with hyperprolactinemia have mainly high molecular weight PRL (150-250 kDa) (58-62). The exact nature of this substance is still under investigation. (a) In certain cases, it is simply polymeric PRL, (b) in a large proportion of cases, both PRL and immunoglobulin (Ig) are found in the complex (^{125}I-PRL can form complexes with patients' serum and is displaced by an excess of cold PRL) (58, 60) and (c) in rare cases, PRL is covalently linked to Ig, probably in plasmocytes (61, 62 and A. Walker, personal communication).

The biological activity of macroprolactin is usually low *in vivo*, so that patients have only mild symptoms despite high serum PRL values. *In vitro*, however, most samples have the expected bioactivity in the Nb2 assay, except in case (c) where it is low. The discrepancy between bioactivity *in vitro* and *in vivo* is best explained by the fact that macroprolactin poorly crosses the capillary barrier and thus cannot reach target cells. *In vitro*, PRL binds to and activates the PRL-R. In case (c), macroprolactin (PRL covalently bound to Ig), stimulates the proliferation of leukemic cells whereas neither PRL nor Ig had any effect. Only by engaging both the PRL-R and the Fc-R was stimulation achieved (62).

TOOLS FOR RESEARCH

Most of the experiments discussed below rely on the use of exogenous PRL, either natural or r-PRL. Among PRL variants, only a phosphorylated PRL has been investigated in some detail (63). S179PRLD is a mimic of phosphorylated PRL that acts as a PRL-R antagonist (64).

Natural glycosylated PRL was consistently reported as less bioactive (65). This is not the case with r-PRL (66).

The contribution of paracrine (or autocrine) PRL is mainly explored through the use of specific antibodies: *in vitro*, an inhibitory action of anti-PRL serum and restoration of the effect by PRL strongly suggests a role for PRL.

PRL-R

Only one gene coding for a receptor for PRL has been identified in rodents and man. Hormones that bind to this receptor are called lactogenic (PRL, placental lactogens, primate GH) (22, 67). The PRL-R and signaling mechanisms are described in depth in other chapters. An important feature of PRL-R that may be relevant also for the LHS is the bell-shaped curve of physiological responses: the amplitude of the response is dose-dependent, until a certain point, where higher concentrations of PRL generate smaller effects. At higher receptor occupancy, dimerization and signaling cannot occur.

Adult lympho-hemopoietic tissues from rodents and man express PRL-R. There is still considerable uncertainty about the level of expression.

In the **rat**, alternative splicing results in the expression of a long form of the PRL-R, responsible for Jak-Stat activation and other forms of signaling and a short form, unable to signal through Jak-Stat. This short form does, however, interfere with signaling through the long form (22, 67). Most rat tissues, including those of the LHS, express both long and short forms (53, 68-70). By *in situ* hybridization, in spleen and thymus, signals were stronger for the long form of the PRL-R than for the short form. In the spleen, grains were more concentrated over some cells of the red pulp. In the thymus, the labeling was higher over epithelial cells than over lymphocytes (69). By PCR analysis, expression was found in all lympho-hemopoietic tissues (see below). By single color flow-cytometry, using the T6 monoclonal antibody (Mab), 50 % (BM) to 80 % (thymus) of the cells were found positive in various lympho-hemopoietic tissues. The fluorescence intensity was low, except in 15% of the spleen cells, where it was high. After immunization, there was a clear shift towards higher intensity, peaking at day 6 after immunization (71). Using the U5 MAb, labeling was highest in BM cells (95%) and lowest in lymph node cells (67 %); 80 % of thymocytes were positive (72).

Similar observations were made in the mouse: positive PCR signals were found in all lympho-hemopoietic tissues, with the highest signals in lymph nodes and weakest in BM. Using the U5 Mab, the percentage of positive cells closely parallelled values in the rat, somewhat lower only for lymph node cells (50 % positive). Labeling intensity is also higher after *in vitro* stimulation with Con A. Using three-color flow cytometry, it was shown that PRL-R expression is high on CD8+ and on double negative thymocytes and very low on double positives and on CD4+ cells. In murine splenocytes, labeling was more intense on Mac1+ cells (macrophages) than on B220+ cells (B-lymphocytes) (72). Using the same U5 Mab, Gala and Schevach found that 80% of B cells (B220+) express the PRL-R. 85 % of thioglycolate-induced peritoneal macrophages were positive. PRL-R expression was low in lymph node cells (only 5% of the CD4+ and 20 % of the CD8+), but these figures were higher in ovariectomized mice. In both control and ovariectomized mice, *in vivo* stimulation with Con A (in the footpad) increased the percentage of positive cells (73).

In man, in addition to the classical receptor (long form, homologous to the long form of the rat PRL-R), a second form has been cloned and characterized (74). It is called intermediate form as it shares characteristics with the PRL-R present only in rat Nb2 cells. This receptor is able to signal through Jak 2 (as does the long form) and through Fyn (but less efficiently than the long form). Dot-blot analysis of mRNA expression indicated that both the long and the intermediate forms are expressed in most tissues investigated. In the LHS, expression of PRL-R was very low in the thymus. The intermediate form predominated in BM and the long form in peripheral blood leukocytes. Both forms were expressed in lymph nodes.

Flow cytometry, after labeling with the U5 MAb identified PRL-R on a large percentage (above 80%) of cells in all leukocyte populations. Recently, however, these data were reconsidered and it was proposed, on the basis of experiments with biotinylated PRL, that no more than 22-34% of PBMC express PRL-R. As no double labeling was performed, the cells expressing receptors have not been identified (75). The earlier studies indicated that, as in the mouse, human T cells expressed less receptors than macrophages and B cells. In human PBMC, labeling was strongest on B-lymphocytes (76). There are no detailed reports describing the expression of PRL-R on human granulocytes. As others and we have observed effects of PRL on these cells, we assume that the mRNA for the PRL-R has a very slow turnover in BM cells and an even lower one in granulocytes.

The number of receptors detected on leukocytes is relatively low: about 350 receptors per cell, as calculated independently by several workers. However, it is not very different from the amount of receptors found on the epithelial cells in the mammary gland, the classical target for PRL. This number of receptors should therefore be sufficient to induce effects. Higher numbers of receptors have been found on human large granular lymphocytes (77).

In conclusion, the expression of the PRL-R is highest in NK cells, B-cells and monocytes-macrophages, lowest in resting T-cells. Immune activation increases PRL-R expression in T cells. Expression of PRL-R on granulocytes has not been documented.

RECENT *IN VIVO* DATA

In vivo effects of PRL on the LHS have been reviewed in detail (1, 4, 5). Unfortunately, there are no recent studies of the LHS in hypophysectomized rats. Such studies are needed, in part because older reconstitution studies were done with PRL still contaminated with GH and possibly other hormones. The reported immunological defects following hypophysectomy do not compromise survival. In one study, administration of anti PRL antibodies to hypophysectomized rats was lethal for all animals within 7 weeks. Hematological failure was suspected but was not properly documented (78).

Another open question is the mode of action of bromocriptine. This dopaminergic agent is a potent inhibitor of PRL secretion and is also immunosuppressive *in vivo*. The effects of bromocriptine on immunological parameters are reversed by administration of PRL. This was taken as evidence for an immunostimulatory role of PRL. However, more experiments are needed as bromocriptine acts directly on many tissues. In particular, it reduced B and T cell activation *in vitro* and PRL is probably not involved in these effects (79, 80 ; see however below: synovial cells).

Transgenic mice overexpressing PRL show little pathology, and none has been reported in the LHS (81). More important, the detailed description of PRL-/- and PRL-R-/- mice has been published. Neither mutation is responsible for any immunological or hematological phenotype, even after immunization with various antigens. No real defects could be evidenced in those animals where immunization was the sole challenge. It cannot be excluded that different approaches (e.g. idiotypic studies, study of γδ-T cells) would show some alterations. It is also hypothesized that major injuries or stresses provided e.g. by hemorrhage, X-irradiation, glucocorticoid treatment, ovariectomy would reveal differences in the LHS between KO and wild type animals. PRL could play a protective or homeostatic role, which is not apparent in KO mice, as a result of redundancy in the cytokine network. Also, as compensation is more likely to occur when a protein is absent throughout ontogeny, conditional KO would be more likely to show defects reminiscent of those seen after hypophysectomy in rats. Based on the observation that neither Stat 5a nor Stat 5b KO mice exhibit clear-cut defects in the LHS but that the double KO (Stat 5a + 5b) have major deficits (peripheral T cells are profoundly deficient in proliferation and there are no NK cells (82)), double KO (lacking both PRL and GH genes) are being generated. Indeed, PRL signals predominantly through Stat 5a and GH through Stat 5a and 5b. Our feeling is that mice lacking another cytokine (or its receptor or another signaling molecule) would be more likely to suffer from the lack of PRL. In particular, PRL is expected to compensate for cytokines signaling through Stat 5, such as IL-2, thrombopoietin or stem cell factor (if the target cell also expresses receptors for PRL, see below The erythroid lineage). Compensation by PRL or other factors is obviously incomplete in e.g. Epo, IL-2, IL-7, IL-7-R KO mice, as these show clear-cut deficits..

In the rat, L. Yang *et al.* have recently observed strong effects on pups after delivering PRL to the mother during pregnancy via minipumps to reach constant levels of 50 ng/ml. Thymus was small, lymphopoiesis was inhibited, erythropoiesis was stimulated. The latter effect is in agreement with *in vitro* effects described below. A PRL-R antagonist (S179D PRL, a molecular mimic of phosphorylated PRL) (64), administered in the same way, markedly increased granulocyte numbers in peripheral blood and lead to massive apoptosis in the thymus. γδ-T cells developed normally but then underwent apoptosis; as a result, the number of γδ-T cell homing to the skin (and presumably in other sites) was reduced (L. Yang and A. Walker, personal communication). These observations do not contradict data from KO mice, where only the αβ-T cell receptor was studied (19). So, maybe there is not so much difference between the situation in a PRL-KO mouse and a rat exposed *in utero* to a PRL-R antagonist!

In adult rats carrying a pituitary graft (secreting only PRL) or given ovine PRL, neutrophils from a carragenin-induced pleural exsudate produced more NO and TNF-α than those from control rats whereas neutrophils from bromocriptine-treated rats produced less NO and TNF-a than controls (83). These observations are in agreement with earlier data from hypophysectomized rats and with *in vitro* experiments (see below).

In man, the situation is somewhat more complex as GH from primates is also lactogenic (i.e. binds the PRL-R) and could to a certain extent compensate for the lack of PRL. As is the case for PRL, small amounts of GH are produced by leukocytes (5, 23). Neither hypoprolactinemia, whether present from birth or not, nor hyperprolactinemia has been correlated with important changes in the human LHS. There is a contrast between the situation in rats, where hypophysectomy is well tolerated but accompanied by immunological defects and that in man, where hypophysectomy is lethal unless adrenal steroid or ACTH replacement therapy is provided. It can be speculated that in hypophysectomized rat a higher level of glucocorticoid is maintained (responsible for immunological defects) whereas in man the parallel reduction in PRL and ACTH alters less the homeostasis in the immune system.

Few data analyzing correlations between PRL levels and hematological or immune parameters are available. Patients undergoing hemodialysis with hyperprolactinemia have higher numbers of circulating erythroid precursors than normoprolactinemic patients. Although the number of cases was small, the results of this study agree with other *in vitro* and *in vivo* data and suggest that PRL may compensate for the low erythropoietin (Epo) activity in patients with kidney failure (84). Another interesting correlation was reported in lactating women (with higher serum PRL concentrations than control, bottle-feeding, mothers). They also have lower CD19+ B-cell percentages in their blood. More frequent or extended breast-feeding sessions were correlated with lower CD19+ percentages, reduced serum, and higher serum IgA and IgM concentrations (85, 86). In chronic hyperprolactinemia, as seen in prolactinoma, there are very few changes in blood parameters. An increased CD4/CD8 ratio in the hyperprolactinemic patients could be detected as compared with healthy controls but as this ratio remained high after treatment with dopamine agonists, it is hard to draw any conclusion (87).

For many years, the role for PRL in autoimmune disease has been debated (4, 5, 13-15; the contribution of paracrine PRL to autoimmune disease is briefly discussed below: see synovial cells). Patients with hyperprolactinemia, mostly without clinical signs of autoimmune disease, have more frequently autoantibodies (e.g. anti-endothelial cell antibodies or anti-thyroid antibodies) than normoprolactinemic subjects (88, 89). An increased incidence of autoimmune thyroid disorders was reported in hyperprolactinemic women (89). Also, in patients with systemic lupus erythematosus, serum PRL levels are increased and flares correlate with surges in PRL levels (13-15). It is however still not clear whether PRL plays a pathogenic role. Actually, a raise in PRL secretion also accompanies normal immune responses (90, 91). High PRL levels in the serum could thus simply be a marker for the disease. Some reports indicate that inflammatory cytokines stimulate the release of PRL from the pituitary (4, 5) but clinical observations and animal data rather suggest a causative role for PRL in disease exacerbations (4, 5, 11-13).

An important question is whether autoantibodies in hyperprolactinemic patients without autoimmune disease are an early sign of autoimmune disease.

Taken together, these and earlier *in vivo* studies show effects of PRL on natural immunity (4, 5, 8, 16). (Restorative or protecting) effects on specific immunity were mainly seen in animals that underwent hormonal or immunosuppressive treatment. In rodents and man, hyperprolactinemia is a frequent feature in auto-immunity. *In vivo* studies are mainly concerned with effects of endocrine PRL. Extra-pituitary PRL is still present after hypophysectomy and in Pit-1 deficient dwarf mice (and presumably also in Pit-1 deficient patients), and most probably after administration of bromocriptine. Even so, it is very difficult to identify effects of extra-pituitary paracrine or autocrine PRL *in vivo*: *in vitro* experiments make it easier to trace effects of the paracrine whisper in the absence of the endocrine shout!

BIOLOGICAL *IN VITRO* EFFECTS OF PRL

PRL stimulates proliferation or differentiation or functional activity in most cell types of the LHS. Some of these data, in particular recent ones, are reviewed here. Whenever experiments are performed with mixed cell populations, it should be kept in mind that the effect seen in one cell type might be mediated by another cell type present in the culture.

CD34+ PROGENITOR CELLS

The group of Matera has shown that PRL (50 ng/ml) in combination with IL-3, granulocyte macrophage-colony stimulating factor (GM-CSF) and Epo slightly stimulated purified CD34+ progenitor cells from human cord blood. The effect was on both burst forming erythroid (BFU-e) colonies and on colony forming units-granulocyte (CFU-g). Limiting dilution analysis showed that PRL acts directly on the hemopoietic progenitor cells (92). There is also evidence that PRL is produced by cultured BM stroma cells in response to platelet-activating factor (PAF). As the development of BFU–e colonies from CD34+ progenitor cells cultured on BM stroma cells was significantly reduced in the presence of anti-PRL antibody, it is proposed that locally produced PRL contributes to the growth of hemopoietic progenitors (93).

As the recently described prolactin-like protein (PLP)- E does not signal through the PRL-R but through a member of the gp130 family, it will not be discussed here (94). This placental hormone acts on myeloid differentiation, in particular megakaryocytopoiesis.

THE ERYTHROID LINEAGE

In contrast to above-mentioned *in vivo* and *in vitro* data, the group of Lodish has evidence that PRL has no effect on the development of erythroid cells from CFU-e or BFU-e progenitors. They are mainly interested in signaling through the Epo-R and have performed experiments in both normal and Epo-R KO mice (95). In a series of experiments, they have shown that the PRL-R (or hybrid receptors) can substitute for the Epo-R. However, in their experiments, cells with transfected PRL-R are always used.

Judging from the control experiments with cells not transfected with PRL-R, there is no functional PRL-R on CFU-e or BFU-e cells. Indeed, PRL has no effect in cells that have not been transfected with PRL-R. Unfortunately, their results with PRL in non-transfected cells have not been reported in detail. The experimental conditions can account for the differences between the data from Matera's and from Lodish's groups. Indeed, Matera *et al.* are working with human progenitor cells whereas Lodish *et al.* are using murine cells already committed to the erythroid lineage. It can only be concluded that, in the mouse, committed erythroid progenitor cells have no functional PRL-R.

THE MYELOID LINEAGE

As mentioned above (A), PRL enhances the effects of GM-CSF and IL-3 on the proliferation and differentiation of myeloid progenitors (93).

In addition, a few studies report effects of PRL on the function of macrophages or granulocytes. Supraphysiological concentrations of PRL or GH primed human monocytes for enhanced H_2O_2 production induced by PMA (96). Human GH also primed human neutrophils for superoxide anion secretion in response to PMA. The effect of GH was mediated by the PRL-R as clearly shown by the use of hormone analogs (28). It is therefore possible that other reported effects of GH on monocytes, macrophages or granulocytes are mediated by the PRL-R and could thus equally well be induced by PRL.

DENDRITIC CELLS

Matera *et al.* have recently shown that PRL synergized with GM-CSF to promote the functional maturation (e.g. increase in MHC class II) of dendritic cells from monocytic precursors. The effects of physiological concentrations (10-20 ng/ml) of PRL were comparable to those of IL-4. In the absence of GM-CSF, the action of PRL was inhibitory (at physiological concentrations) or stimulatory (at 80 ng/ml). Physiological concentrations of PRL stimulated the expression of the GM-CSF-R and this could account for the effect of PRL on the maturation of monocyte-derived dendritic cells (97). Similar data have recently been obtained by S. Richards (17). Also, physiological concentrations of PRL (12 ng/ml) were shown to cooperate with IL-2 to induce IFN-γ production by dendritic cells (98).

B-LYMPHOCYTES

PRL promoted a modest expansion of B-lineage cells in short term cultures of murine bone-marrow cells grown on mitomycin-treated S17 stromal cells (that support B cell development). This effect could thus be mediated by the feeder cells. In addition to data on normal cells, information was also obtained from pre-B BaF-3 cells transfected with the long form of the rat PRL-R. Upon exposure to PRL, they exhibited some changes characteristic of B-cell differerentiation: expression of the IL-2-R, of the lambda 5 pseudo-light chain and of the Bcl-2 protein (99). The conclusion is that signals provided by the transfected PRL-R favor B-cell differentiation.

PRL enhanced DNA synthesis in human B cells stimulated with submitogenic concentrations of *S. aureus* Cowan (100). r-hPRL also stimulated proliferation of murine hybridomas and exhibited an additive effect with IL-4, IL-5 and IL-6. It also reversed TGF-β proliferative inhibition (101). PRL synergized with IL-2 to enhance IL-2-R expression on stimulated B cells (17). Ig production was increased in murine hybridomas exposed to r-hPRL (101). Finally, PRL also increased IgM and IgG secretion by human B cells stimulated by IL-2 in combination with anti-IgM (102) and r-hPRL increased Ig production by human B cells stimulated with anti-CD40 (17).

T-LYMPHOCYTES

Following observations in Nb2 cells, Witorsch and colleagues demonstrated that PRL could inhibit the dexamethasone-induced apoptosis of murine thymocytes (40).

A series of experiments has been performed on the murine T-helper lymphocyte clone L2, which is not transformed and remains dependent on repeated stimulation by allogeneic (irradiated) cells. In this cell line, exogenous PRL is required for IL-2 driven proliferation (103). Stimulation with IL-2 resulted in up-regulation of PRL-R and (PI-3K-dependent) nuclear translocation of PRL. Nuclear translocation of PRL was also documented in splenocytes stimulated with Con A and in Nb2 cells stimulated with PRL (104).

In human thymocytes and PBMC, PRL induced rapid phosphorylation of multiple, TCR/CD3 complex proteins, including CD3ε and ZAP-70 tyrosine kinase, indicating that the PRL-R can directly affect T-lymphocyte activation by means of signaling cross-talk with the TCR/CD3 complex (33).

In rat splenocytes, PRL (at the pharmacological concentration of 1 μg/ml) was nearly as efficient as Con A for inducing IL-2 production (105). In further studies, stimulation of thymidine incorporation was seen only in splenocytes and lymph node cells from ovariectomized and diestrous female rats, but not in cells from males or estrous females. After incubation of splenocytes and thymocytes with PRL, bioactive IL-2 was only detected in culture medium from cells of ovariectomized female rats. Flow cytometric analysis of PRL-stimulated splenocytes revealed increases in percentages (after 18 h) of IL-2-R+, (after 4 days) of CD8+ and, to a lesser extent, CD4+ (106). The IL-2-R being a known target of Stat 5, its increased expression after treatment with PRL is thus not unexpected (107).

Using human PBMC, Cesario *et al.* found that PRL in concentrations of 10^{-8} M (230 ng/ml) or greater, significantly enhances the production of IFN-γ. The effect was dose related and was observed in cultures stimulated with lectins (PHA and Con A), but not when anti CD3 antibodies, ionophore, or IL-2 were used to induce the cells. The effect was not seen when cells were treated with the tyrosine kinase inhibitor genistein (108). We have extended these observations and measured the production of a number of cytokines in PBMC or in tonsillar cells stimulated with PHA. At doses up to 100 ng/ml, there was no significant effect of PRL on the production of IFN-γ, TNF-α or IL-1β. There was a minor, but significant decrease of IL-5 production in PBMC grown in the presence of PRL (10 ng/ml). The latter effect was also observed with GH. Both hormones could act through the PRL-R (109 and our unpublished data). Recently, Matera and Mori (98) have shown that PRL cooperates in a dose-dependent manner with IL-12 to induce IFN-γ release from T-cells.

NK AND LAK CELLS

The presence of PRL-R on human NK cells has been established in 1988 by Matera *et al.* (77). Actually, NK cells probably have more receptors per cell (about 660) than other lymphocytes. The same group has investigated the effects of PRL on NK cells. At 12 - 50 ng/ml, PRL stimulated thymidine incorporation in NK cells. PRL and IL-2 had synergic effects. At 200 ng/ml, PRL had inhibitory effects (100). Also, PRL alone or in synergy with IL-2 favors the generation of lymphokine-activated killer (LAK) cells (110). The inhibition of NK activity by high levels of PRL is best explained by properties of the PRL-R: at higher receptor occupancy, dimerization and signaling cannot occur.

It has been proposed that PRL favors the establishment of tolerance to the fetus. Although the decidua is an important site of extra-pituitary PRL production, the hypothesis that PRL silences potential rejection mechanisms has not been explored. However, PLP-A, a member of the PRL family, expressed in trophoblast cells, was shown to specifically interact with a NK lymphocytes of the decidua from pregnant and pseudo-pregnant rats (111). This PRL-related molecule may indeed contribute to the immunological protection of the fetus.

In addition to its effects on effector cytotoxic cells, PRL also affects the target cells: the susceptiblity of leukemic cell lines and of freshly isolated leukemic blasts to NK and LAK cells was increased by pre-treatment with PRL, possibly as a result of modulation of cell surface antigens (112).

SYNOVIAL CELLS

To test the possible contribution of PRL to the pathogenesis of rheumatoid arthritis (RA), Nagafuchi *et al.* have isolated primary synovial cells (CD14+, 35%; CD4+, 26%; CD8+, 20%; fibrobast-like, 19%) from patients with RA. Unseparated cells express PRL, TNF-α, IL-6 and matrix metalloproteinases (MMP). Upon incubation in the presence of bromocriptine, there was a dose-dependent decrease in expression of all these factors. PRL (100 or 1000 ng/ml) restored IL-6 secretion to control levels. The fibroblast-like cells were grown in culture and found to express PRL-R. In those cells, PRL (500 ng/ml) induced the translocation of Stat 5 into the nucleus. Also, in cells incubated in the presence of PRL, a dose-dependent increase in proliferation and in the production of MMP-3 and a decrease of tissue inhibitor of metalloproteinase-1 were induced (113). Taken together, these data suggest an involvement of PRL in the tissue destruction caused by RA: synovial cells (both T cells and fibroblasts) produce PRL, which stimulates in an autocrine or paracrine way the proliferation of fibroblasts, collagenase activity and the expression of pro-inflammatory cytokines. Bromocriptine inhibits the whole sequence of events. The inhibition of PRL expression in extra-pituitary sites by bromocriptine has not been investigated in detail and certainly requires confirmation.

STROMAL CELLS

BM and thymic stromal cells express PRL and PRL-R. In the past, there has been much interest for the thymic stroma but no recent data implicate PRL in its function. Primary BM stromal cells show a dose-dependent increase in PRL-R expression following treatment with adipogenic agonists, which at first sight does not particularly point to a role for PRL in hemopoiesis or immunity (114). It is likely that, as shown in NIH-3T3 cells, PRL enhances C/EBPβ and PPARγ mRNA expression and so augments adipogenic conversion (115). What may be relevant to the immune system is that PRL stimulates the expression of PPARγ, which is the PPAR isoform most expressed in the immune system (116).

Only few studies have focused on the effects of hormones on BM stroma. Recently it was shown that markers for osteogenic potential were more expressed in rat BM stroma cell cultures from hypophysectomized mice than in stroma from normal rats. The increased cell proliferation and bone protein expression was attributed to a higher response to factors present in fetal calf serum, after *in vivo* deprivation. The authors focus on GH and IGF. Stimulation by GH or other lactogenic substances through the PRL-R should certainly be considered.

The role of hormones in bone formation and osteoporosis is now a major field of investigation and estrogens receive much more attention than PRL. After menopause, there is a drop in both hormones. Recently, osteopontin, a protein present in bone and in the extra-cellular matrix, was recognized as a cytokine (Eta-1), important for type 1 T-cell responses (117). During the involution that follows weaning, a decrease in osteopontin expression in the mammary gland was observed and attributed to the drop in PRL (118). Whether PRL is indeed responsible for osteopontin gene expression, in the breast and in the LHS is presently not known. A positive answer would point to a possible link between PRL and type 1 T-cell responses.

BIOCHEMICAL AND MOLECULAR STUDIES IN HEMOPOIETIC CELLS EXPOSED TO PRL *IN VITRO*

What are the mechanisms involved in PRL action in the LHS? To which extent is the Nb2 model representative of normal leukocytes? Providing an answer to those questions has been an arduous task: so far, only part of the data obtained with Nb2 could be confirmed in normal leukocytes. This could be due in part to the low number of receptors per cell and to the lack of some key reagents (such as potent anti PRL-R antibodies). As far as could be judged, our recent studies in rat and man rather support the idea that the Nb2 model applies to a large extent also to PRL-R signal transduction in lymphocytes.

A functional PRL-R is present on rat leukocytes since upon stimulation with a physiological concentration of PRL (10 ng/ml), the Jak 2/ Stat 5b pathway was activated and the target IRF-1 gene was induced (119).

The PRL-R is indeed expressed in rat lympho-hemopoietic tissues. As judged by PCR analysis, Western blotting, immunocytochemistry and metabolic labeling, receptor expression was higher in spleen than in BM cells. In BM cells, strong signals were obtained in Western blot and in immunocytochemistry and low signals with PCR and with metabolic labeling (119). Such discrepancy (also reported in the mouse) between PRL-R mRNA and protein levels is best explained by assuming a low turnover of PRL-R in BM cells.

In the human, PRL-R expression was demonstrated by PCR on PBMC, on BM cells and on tonsillar cells (120 and manuscript in preparation). We were unable to document the expression of PRL-R on polymorphonuclear cells. Yet, PRL has effects on these cells and antibodies to PRL abrogated these effects. PRL-R signaling occurred via Stat 5 and MAP kinase p38 in PBMC. Although PRL signaling through MAP kinase has been documented in other systems (135), this is the first report documenting activation of p38 mitogen-activated protein kinase in PRL signal transduction. It was however recently shown that GH also signals through p38 (121). As in the rat, a physiological concentration of PRL led to the induction of various target genes. In particular, in PBMC PRL induced suppressor of cytokine signaling (SOCS)-3 and iNOS expression within 15 minutes. In BM cells, physiological concentrations of PRL resulted in the induction of the IRF-1, cytokine-induced-src homology domain 2 containing protein (CIS) and SOCS-2. In polymorphonuclear cells, IRF-1, CIS and SOCS-2 (but not SOCS-3) were induced by PRL. In tonsillar cells from adult patients, the expression of CIS was increased and that of SOCS-2 and SOCS-7 was induced by PRL (120). In addition, Merchav *et al.* (manuscript in preparation) recently showed that PRL, like IL-3, induced CIS and SOCS-2 in CD34+ progenitor cells from cord blood.

Several members of the recently described family of SOCS factors are involved in cytokine signaling: CIS binds to receptors and SOCS-1 and SOCS-3 bind to Jak (122-128). These 3 factors act as inhibitors of signal transduction and contribute to termination of signaling. In addition to inhibiting homologous signal transduction, they can also interfere with signal transduction by other cytokines. Such interference has been clearly demonstrated for IL-4 signal transduction that is inhibited (presumably through SOCS factors) by pretreatment with IL-10 or IFN-γ (129, 130). The function of SOCS-2 is not an inhibitory one: SOCS-2 was shown to relieve PRL signal transduction from inhibition by SOCS-1 (128). So far, there are no published data about the function of SOCS-7.

It is thus of particular interest that PRL, as other cytokines, can induce the expression of various SOCS in leukocytes. As a result, PRL could modulate signal transduction originating from receptors for other cytokines. We have evidence that PRL interferes with IL-6-induced Stat 3 phosphorylation in human tonsillar cells. This effect could be mediated by SOCS factors (manuscript in preparation).

Taken together, biochemical and molecular studies indicate that functional PRL-R are expressed in the normal LHS. PRL signals through Jak 2 and Stat 5 and MAP kinase. It induces genes that are important for leukocyte survival, proliferation and function. Most likely, the multiple signaling pathways induced by PRL regulate distinct classes of genes. Thus, both similarities and differences have been observed between normal leukocytes and Nb2 cells.

CONCLUSIONS

Effects of PRL on the LHS in rodents and man are barely detectable in baseline conditions. Neither lack nor excess of PRL, when present from embryonic stages, strongly affects hematological or immunological parameters nor responses to simple antigenic challenges. Adult hypophysectomy (or treatment with bromocriptine) leads to well-defined immunological deficits (in rat but not in man) and these are corrected by PRL. Possibly, the better-preserved corticosteroid secretion in rats is responsible for the immune deficiency. In man, the effects of a drop in PRL secretion may be neutralized by a simultaneous drop in glucocorticoid secretion. Stimulatory/reparing effects of PRL have also been reported following hemorrhage, ovariectomy or treatment with glucocorticoid or azathioprine (16, 106, 131). The most obvious mechanism of action for PRL is by activating the signaling pathways first identified in Nb2 cells (Jak – Stat- GAS, MAP-kinase, etc). Dorshkind and Horseman have proposed that PRL counteracts effects of stress in the immune system, possibly by interfering with the glucocorticoid-R (GR) (8). An interaction between the GR and Stat 5 has indeed been demonstrated and results in increased signaling by Stat 5 and reduced signaling by the GR (132). Another possible basis for interaction with other hormone- and growth factor signaling mechanisms is through the induction of SOCS genes. Several SOCS genes can be induced by PRL, in particular in leukocytes (120, 128). SOCS factors limit signaling through various receptors for cytokines: this inhibition can be homologous (leading to termination of signaling) or heterologous (leading to interference). Interestingly, PRL could thus either synergize with several cytokines (which signal through the same tyrosine kinases, Stats or MAP kinases) or antagonize/buffer the effects of cytokines by inducing the expression of SOCS factors.

PERSPECTIVES

Data supporting the hypothesis that PRL modulates survival, proliferation, differentiation and cell function in the LHS are abundant in rodents and man. There is no contradiction with the proposal that PRL and PRL-R KO mice and PRL-deficient humans compensate their deficit totally (as far as can be judged today). A deficit might become obvious only when another cytokine or hormone (or receptor or other signaling molecule) is missing at the same time or when additional challenges are present as is the case in antigenic competition, after hemorrhage, irradiation, glucocorticoid treatment, cancer, severe burns or other major illness. Still, there is a broad gap between the over-enthusiastic reports including plans for clinical trials and a detailed understanding of the role of PRL in the LHS. The complexity of the hormone and cytokine networks makes analysis *in vivo* very difficult. Epidemiological studies combining clinical and genetic data (133) may provide valuable information. *In vitro* analysis, in contrast, allowing a step-by-step dissection of the problem, may be more fruitful. Even so, the low expression of PRL-R and the lack of basic tools such as potent anti-PRL-R antibodies hamper progress. The existence of a second receptor for PRL should be excluded. New analogs of PRL are becoming available for targeting the PRL-PRL-R axis (28, 64, 134). The serial analysis of gene expression (sage) has already been applied to compare gene expression in cells from control and PRL-R KO mice (67) and in the very near future, the DNA array technique should establish which subpopulations of leukocytes respond to PRL and, in each of them, which gene clusters are activated/repressed by PRL. The same approach will also tell which effects are mediated by PRL or by other hormones (GH, estrogens). The data obtained will delineate the contribution of PRL in selected hematological and autoimmune diseases and may provide a better rationale for pursuing (or not) clinical trials with PRL in e.g. BM aplasia and immune deficiency syndromes.

ACKNOWLEDGEMENTS

By lack of space, many important papers were not quoted. We have often quoted recent papers, where prior references can be found. We thank C. Clevenger, K. Dorshkind, N. Horseman, R. Kooijman, L. Matera, R. McMurray, S. Merchav, S. Richards, A. Walker, S. Walker and L. Yang for discussion and communication of unpublished data, M. Socolovsky and L.-y. Yu-Lee for discussion. The authors' research is supported by the Gemeenschappelijke Onderzoeksacties (GOA), the Fonds voor Wetenschappelijk Onderzoek-VL, the Ministry of Research of the Brussels-Capital Region and the Free University of Brussels (Onderzoeksraad VUB). PRL and antibodies were generously provided by Dr A.F. Parlow (NIDDK's National Hormone & Pituitary Program).

REFERENCES

1. Gala RR. Prolactin and growth hormone in the regulation of the immune system. Proc Soc Exp Biol Med 1991;198:513-27.
2. Arkins S, Dantzer R, Kelley KW. Somatolactogens, somatomedins, and immunity. J Dairy Sci 1993;76:2437-50.
3. Hooghe R, Delhase M, Vergani P, Malur A, Hooghe-Peters EL. Growth hormone and prolactin are paracrine growth and differentiation factors in the haemopoietic system. Immunol Today 1993;14:212-4.
4. Hooghe-Peters EL and Hooghe R Growth hormone, prolactin and IGF-I as lymphohematopoietic cytokines. RG Landes, Austin/Springer-Verlag, New York, 1995
5. Kooijman R, Hooghe-Peters EL, Hooghe R. Prolactin, growth hormone, and insulin-like growth factor-I in the immune system. Adv Immunol 1996;63:377-454
6. Yu-Lee L-y, Luo G, Moutoussamy S, Finidori J. Prolactin and growth hormone signal transduction in lymphohaemopoietic cells. Cell Mol Life Sci 1998;54: 1067-75.
7. Matera L Action of pituitary and lymphocyte prolactin. Neuro- immunomodulation 1997; 4:171-80.

8. Dorshkind K, Horseman, N. The roles of prolactin (PRL), growth hormone (GH), insulin-like growth factor-I (IGF-I), and thyroid hormones in lymphocyte development and function: insights from genetic models of hormone and hormone receptor deficiency. Endocrine Rev. 2000; 21:292-312.
9. Gout PW, Beer CT, Noble RL 1980 Prolactin-stimulated growth of cell cultures established from malignant Nb2 rat lymphomas. Cancer Res 40:2433-36.
10. Kincade, PW, Medina KL, Smithson G 1995 Sex hormones as negative regulators of lymphopoiesis. Immunol Rev 137:119-34.
11. McMurray RW, Keisler D, Izui S, Walker SE. Effects of parturition, suckling and pseudopregnancy on variables of disease activity in the B/W mouse model of systemic lupus erythematosus. J Rheumatol 1993;20:1143-51.
12. Blanco-Favela F, Quintal-Alvarez G, Leanos-Miranda A. Association between prolactin and disease activity in systemic lupus erythematosus. Influence of statistical power. J Rheumatol 1999;26:55-9.
13. McMurray RW, Weidensaul D, Allen SH, Walker SE. Efficacy of bromocriptine in an open label therapeutic trial for systemic lupus erythematosus. J Rheumatol 1995;22:2084-91.
14. Alvarez-Nemegyei J, Cobarrubias-Cobos A, Escalante-Triay F, Sosa-Munoz J, Miranda JM, Jara LJ. Bromocriptine in systemic lupus erythematosus: a double-blind, randomized, placebo-controlled study. Lupus 1998;7:414-9.
15. Funauchi M, Ikoma S, Enomoto H, Sugiyama M, Ohno M, Hamada K, Kanamaru A. Prolactin modulates the disease activity of systemic lupus erythematosus accompanied by prolactinoma. Clin Exp Rheumatol 1998;16:479-82.
16. MA, Welniak LA, Sun R, Tian Z-G, Henry M, Richards S, Raziuddin A, Longo DL, Murphy WJ. Prolactin exerts hematopoietic growth-promoting effects *in vivo* and partially counteracts myelosuprression by azidothymidine. Exp Hematol 1999;27:811-16.
17. Richards SM, Murphy WJ. Use of human prolactin as a therapeutic protein to potentiate immunohematopoietic function. J. Neuroimmunol 2000, in press.
18. Matera L, Cutufia M, Geuna M, Contarini M, Buttiglieri S, Galin S, Fazzari A, Cavaliere C. Prolactin is an autocrine growth factor for the Jurkat human T-leukemic cell line. J Neuroimmunol 1997;79:12-21.
19. Matera L, Geuna M, Pastore C, Buttiglieri S, Gaidano G, Savarino A, Marengo S, Vonderhaar BK. Expression of prolactin and prolactin receptors by non-Hodgkin's lymphoma cells. Int J Cancer 2000;85:124-30.
20. Horseman ND, Zhao W, Montecino-Rodriguez E, Tanaka M, Nakashima K, Engle SJ, Smith F, Markoff, Dorshkind K. Defective mammopoiesis, but normal hematopoiesis, in mice with a targeted disruption of the prolactin gene. EMBO J 1997;16:6926-35.
21. Bouchard B, Ormandy CJ, Di Santo JP, Kelly PA Immune system development and function in prolactin receptor deficient mice. J Immunol 1999;163:576-82.
22. Bole-Feysot C, Goffin V, Edery M, Binart N, Kelly PA 1998 Prolactin (PRL) and its receptor: actions, signal transduction pathways and phenotypes observed in PRL receptor knockout mice. Endocr Rev 19:225-68.
23. Shiu RPC, Elsholtz HP, Tanake T, Friesen FG, Gout PW, Geeer CT, Noble RL Receptor-mediated mitogen action of prolactin in a rat lymphoma cell line. Endocrinology 1983,113:159-65.
24. Ben-Jonathan N, Mershon JL, Allen DL, Steinmetz RW. Extrapituitary prolactin: distribution, regulation, functions, and clinical aspects. Endocr Rev 1996;17:639-69.
25. Clevenger CV, Russell DH, Appasamy PM, Prystowsky MB. Regulation of interleukin 2-driven T-lymphocyte proliferation by prolactin. Proc Natl Acad Sci U S A 1990;87:6460-4.

26. Nick JA, Young SK, Brown KK, Avdi NJ, Arndt PG, Suratt BT, Janes MS, Henson PM, Worthen GS. Role of p38 mitogen-activated protein kinase in a murine model of pulmonary inflammation. J Immunol 2000;164:2151-9.
27. Yamaoka K, Otsuka T, Niiro H, Arinobu Y, Niho Y, Hamasaki N, Izuhara K. Activation of STAT5 by lipopolysaccharide through granulocyte-macrophage colony-stimulating factor production in human monocytes. J Immunol 1998;160:838-45.
28. Fu YK, Arkins S, Fuh G, Cunningham BC, Wells JA, Fong S, Cronin MJ, Dantzer R, Kelley KW. Growth hormone augments superoxide anion secretion of human neutrophils by binding to the prolactin receptor. J Clin Invest 1992; 89:451-7.
29. Rayhel EJ, Fields TJ, Albright JW, Diamantstein T, Hughes JP. Interleukin 2 and a lactogen regulate proliferation and protein phosphorylation in Nb2 cells. Biochem J 1988;249:333-8.
30. Clevenger CV, Freier DO, Kline JB 1998 Prolactin receptor signal transduction in cells of the immune system. J Endocrinol 157:187-97.
31. Horseman ND, Yu-Lee LY. Transcriptional regulation by the helix bundle peptide hormones: growth hormone, prolactin, and hematopoietic cytokines. Endocr Rev 1994;15:627-49.
32. Hosokawa Y, Yang M, Kaneko S, Tanaka M, Nakashima K. Prolactin induces switching of T-cell receptor gene expression from alpha to gamma in rat Nb2 pre-T lymphoma cells. Biochem Biophys Res Commun 1996;220:958-62.
33. Montgomery DW, Krumenacker JS, Buckley AR. Prolactin stimulates phosphorylation of the human T-cell antigen receptor complex and ZAP-70 tyrosine kinase: a potential mechanism for its immunomodulation. Endocrinology 1998;139:811-4.
34. Krumenacker JS, Montgomery DW, Buckley DJ, Gout PW, Buckley AR. Prolactin receptor signaling: shared components with the T-cell antigen receptor in Nb2 lymphoma cells. Endocrine 1998;9:313-20.
35. Ali S, Edery M, Pellegrini I, Lesueur L, Paly J, Djiane J, Kelly PA. The Nb2 form of prolactin receptor is able to activate a milk protein gene promoter. Mol Endocrinol 1992;6:1242-8.
36. Krumenacker J, Buckley D, Leff M, McCormack J, de Jong G, Gout P, Reed J, Miyashita T, Magnuson N, Buckley A 1998 Prolactin-regulated apoptosis of Nb2 lymphoma cells: pim-1, bcl-2, and bax expression. Endocrine 9:163-70.
37. Leff MA, Buckley DJ, Krumenacker JS, Reed JC, Miyashita T, Buckley AR. Rapid modulation of the apoptosis regulatory genes, bcl-2 and bax by prolactin in rat Nb2 lymphoma cells. Endocrinology 1996;137:5456-62.
38. Borg KE, Zhang M, Hegge D, Stephen RL, Buckley DJ, Magnuson NS, Buckley AR. Prolactin regulation of pim-1 expression: positive and negative promoter elements. Endocrinology 1999;140:5659-68.
39. Weimann E, Baixeras E, Zamzami N, Kelly PA. Prolactin blocks glucocorticoid induced cell death by inhibiting the disruption of the mitochondrial membrane. Leuk Res 1999;23:751-62.
40. Witorsch RJ, Day EB, Lavoie HA, Hashemi N, Taylor JK. Comparison of glucocorticoid-induced effects in prolactin-dependent and autonomous rat Nb2 lymphoma cells. Proc Soc Exp Biol Med 1993;203:454-60.
41. Camarillo IG, Linebaugh BE, Rillema JA. Differential tyrosyl-phosphorylation of multiple mitogen-activated protein kinase isoforms in response to prolactin in Nb2 lymphoma cells. Proc Soc Exp Biol Med 1997;215:198-202.
42. Yu TX, Rillema JA. The MEK inhibitor PD 098059 inhibits prolactin-induced Nb2 cell mitogenesis but not milk product synthesis in cultured mouse mammary tissues. Biochim Biophys Acta 1998;1448:126-34.

43. Erwin RA, Kirken RA, Malabarba MG, Farrar WL, Rui H. Prolactin activates Ras via signaling proteins SHC, growth factor receptor bound 2, and son of sevenless. Endocrinology 1995;136:3512-8.
44. Clevenger CV, Ngo W, Sokol DL, Luger SM, Gewirtz AM. Vav is necessary for prolactin-stimulated proliferation and is translocated into the nucleus of a T-cell line. J Biol Chem 1995;270:13246-53.
45. Hunter S, Burton EA, Wu SC, Anderson SM. Fyn associates with Cbl and phosphorylates tyrosine 731 in Cbl, a binding site for phosphatidylinositol 3-kinase. J Biol Chem 1999;274:2097-106.
46. Too CK, Knee R, Pinette AL, Li AW, Murphy PR. Prolactin induces expression of FGF-2 and a novel FGF-responsive NonO/p54nrb-related mRNA in rat lymphoma cells. Mol Cell Endocrinol 1998;137:187-95.
47. Nemeth E, Bole-Feysot C, Tashima LS. Suppression subtractive hybridization (SSH) identifies prolactin stimulation of p38 MAP kinase gene expression in Nb2 T lymphoma cells: molecular cloning of rat p38 MAP kinase. J Mol Endocrinol 1998;20:151-6.
48. Duncan GS, Mittrucker HW, Kagi D, Matsuyama T, Mak TW. The trans-cription factor interferon regulatory factor-1 is essential for natural killer cell function in vivo. J Exp Med. 1996;184:2043-8.
49. Berwaer M, Martial JA, Davis JR. Characterization of an up-stream promoter directing extrapituitary expression of the human prolactin gene. Mol Endocrinol 1994;8:635-42.
50. Gellersen B, Kempf R, Telgmann R, DiMattia GE 1994 Nonpituitary human prolactin gene transcription is independent of Pit-1 and differentially controlled in lymphocytes and in endomedtrial stroma. Mol Endocrinol 8:356-73.
51. Reem GH, Ray DW, Davis JR. The human prolactin gene upstream promoter is regulated in lymphoid cells by activators of T-cells and by cAMP. J Mol Endocrinol 1999;22:285-92.
52. DiMattia GE, Gellersen B, Bohnet HG, Friesen HG. A human B-lympho-blastoid cell line produces prolactin. Endocrinology 1988;122:2508-17.
53. Touraine P, Leite de Moraes MC, Dardenne M, Kelly PA. Expression of short and long forms of prolactin receptor in murine lymphoid tissues. Mol Cell Endocrinol 1994;104:183-90.
54. Pellegrini I, Lebrun JJ, Ali S, Kelly PA 1992 Expression of prolactin and its receptor in human lymphoid cells. Mol Endocrinology 6:1023-1031
55. Delhase M, Vergani P, Malur A, Hooghe-Peters EL, Hooghe RJ. The transcription factor Pit-1/GHF-1 is expressed in hemopoietic and lymphoid tissues. Eur J Immunol. 1993;23:951-5.
56. Wu H, Devi R, Malarkey WB. Expression and localization of prolactin messenger ribonucleic acid in the human immune system. Endocrinology 1996;137:349-53.
57. Gingras MC, Margolin JF.Differential expression of multiple unexpected genes during U937 cell and macrophage differentiation detected by suppressive subtractive hybridization. Exp Hematol 2000;28:65-76.
58. Cavaco B, Leite V, Santos MA, Arranhado E, Sobrinho LG. Some forms of big big prolactin behave as a complex of monomeric prolactin with an immunoglobulin G in patients with macroprolactinemia or prolactinoma. J Clin Endocrinol Metab 1995;80:2342-6.
59. Olukoga AO, Kane JW. Macroprolactinaemia: validation and application of the polyethylene glycol precipitation test and clinical characterization of the condition. Clin Endocrinol (Oxf) 1999;51:119-26.
60. Hattori N, Inagaki C. Immunological aspects of human growth hormone and prolactin. Domest Anim Endocrinol 1998;15:371-5.

61. Bonhoff A, Vuille JC, Gomez F, Gellersen B. Identification of macroprolactin in a patient with asymptomatic hyperprolactinemia as a stable PRL-IgG complex. Exp Clin Endocrinol 1995;103:252-5.
62. Walker AM, Montgomery DW, Saraiya S, Ho TW, Garewal HS, Wilson J, Lorand L. Prolactin-immunoglobulin G complexes from human serum act as costimulatory ligands causing proliferation of malignant B lymphocytes. Proc Natl Acad Sci U S A 1995;92:3278-82.
63. Wang YF, Liu JW, Mamidi M, Walker AM. Identification of the major site of rat prolactin phosphorylation as serine 177. J Biol Chem. 1996;271:2462-9.
64. Chen TJ, Kuo CB, Tsai KF, Liu JW, Chen DY, Walker AM. Development of recombinant human prolactin receptor antagonists by molecular mimicry of the phosphorylated hormone. Endócrinology. 1998;139:609-16.
65. Ellis LA, Picciano MF. Bioactive and immunoreactive prolactin variants in human milk. Endocrinology 1995;136:2711-20
66. Jurcovicova J, Krueger KS, Rokyta R. Effect of glycosylated and non-glycosylated prolactin on the proliferation of normal human lymphocytes. Physiol Res 1997;46:199-201.
67. Goffin V, Binart N, Clement-Lacroix P, Bouchard B, Bole-Feysot C, Edery M, Lucas BK, Touraine P, Pezet A, Maaskant R, Pichard C, Helloco C, Baran N, Favre H, Bernichtein S, Allamando A, Ormandy C, Kelly PA. From the molecular biology of prolactin and its receptor to the lessons learned from knockout mice models. Genet Anal 1999;15:189-201.
68. Royster M, Driscoll P, Kelly PA, Freemark M. The prolactin receptor in the fetal rat: cellular localization of messenger ribonucleic acid, immunoreactive protein, and ligand-binding activity and induction of expression in late gestation. Endocrinology 1995;136:3892-900
69. Ouhtit A, Morel G, Kelly PA. Visualization of gene expression of short and long forms of prolactin receptor in the rat. Endocrinology 1993;133:135-44.
70. McAveney KM, Gimble JM, Yu-Lee L. Prolactin receptor expression during adipocyte differentiation of bone marrow stroma. Endocrinology 1996;137:5723-6.
71. Koh CY, Phillips JT. Prolactin receptor expression by lymphoid tissues in normal and immunized rats. Mol Cell Endocrinol. 1993;92:R21-5.
72. Gagnerault MC, Touraine P, Savino W, Kelly PA, Dardenne M. Expression of prolactin receptors in murine lymphoid cells in normal and autoimmune situations. J Immunol 1993;150:5673-81.
73. Gala RR, Shevach EM. Identification by analytical flow cytometry of prolactin receptors on immunocompetent cell populations in the mouse. Endocrinology 1993; 133:1617-23.
74. Kline JB, Roehrs H, Clevenger CV. Functional characterization of the intermediate isoform of the human prolactin receptor. J Biol Chem 1999;274:35461-8.
75. Bresson J-L, Sébastien J, Gagnerault M-C, Kayser C, Beressi N, Wu Z, Kinet S, Dardenne M, Postel-Vinay, M-C 1999 Growth hormone (GH) and prolactin receptors in human peripheral blood mononulcear cells: relation with age and GH-binding protein. Endocrinology 140:3203-3209
76. Dardenne M, de Moraes M do C, Kelly PA, Gagnerault MC. Prolactin receptor expression in human hematopoietic tissues analyzed by flow cytofluorometry. Endocrinology. 1994;134:2108-14.
77. Matera L, Muccioli G, Cesano A, Bellussi A, Gennazzani E. Prolactin receptors on large granular lymphocytes: dual recognition by cyclosporin A. Brain Behav Immunol 1988;2:1-10.
78. Nagy E, Berczi I. Hypophysectomized rats depend on residual prolactin for survival. Endocrinology. 1991;128:2776-84.

79. Morikawa K, Oseko F, Morikawa S 1993 Immunosuppressive properties of bromocriptine on human B lymphocyte function *in vitro*. Clin Exp Immunol 93:200-5.
80. Morikawa K, Oseko F, Morikawa S 1994 Immunosupressive activity of bromocriptine on human T lymphocyte function *in vitro*. Clin Exp Immunol 95:514-18.
81. Wennbo H, Kindblom J, Isaksson OG, Tornell J. Transgenic mice overexpressing the prolactin gene develop dramatic enlargement of the prostate gland. Endocrinology 1997;138:4410-5.
82. Moriggl R, Topham DJ, Teglund S, Sexl V, McKay C, Wang D, Hoffmeyer A, van Deursen J, Sangster MY, Bunting KD, Grosveld GC, Ihle JN. Stat5 is required for IL-2-induced cell cycle progression of peripheral T cells. Immunity 1999;10:249-59.
83. Meli R, Raso GM, Gualillo O, Pacilio M, Di Carlo R. Prolactin modulation of nitric oxide and TNF-alpha production by peripheral neutrophils in rats. Life Sci 1997;61(14):1395-403.
84. Bellone G, Rollino C, Borsa S, Ferrero I, Martina G, Carbone A, Mareschi K, Quarello F, Piccoli G, Emanuelli G, Matera L. Association Between Elevated Prolactin Levels and Circulating Erythroid Precursors in Dialyzed Patients. Proc Soc Exp Biol Med 2000;223:367-71.
85. Zimmer JP, Garza C, Heller ME, Butte N, Goldman AS. Relationship between serum prolactin, lactation and changes in maternal blood B-cell (CD19+) percents during the first 8 months post-partum. J Reprod Immunol 1996;30:81-95.
86. Zimmer JP, Garza C, Butte NF, Goldman AS. Maternal blood B-cell (CD19+) percentages and serum immunoglobulin concentrations correlate with breast-feeding behavior and serum prolactin concentration. Am J Reprod Immunol 1998 ;40:57-62.
87. Köller M, Kotzmann H, Clodi M, Riedl M, Luger A Effect of elevated serum prolactin concentrations on the immunophenotype of human lymphocytes, mitogen-induced proliferation and phagocytic activity of polymorphonuclear cells. Eur J Clin Invest 1997;27:662-6.
88. Krause I, Blumenfeld Z, Malchinsky M, Cohen S, Blank M, Eldor A, Weksler B, Schweitzer K, Shoenfeld Y. Anti-endothelial cell antibodies in the sera of hyperprolactinemic women. Lupus 1998;7:377-82.
89. Ferrari C, Boghen M, Paracchi A, Rampini P, Raiteri F, Benco R, Romussi M, Codecasa F, Mucci M, Bianco M. Thyroid autoimmunity in hyperprolactinaemic disorders. Acta Endocrinol 1983;104:35-41.
90. Neidhart M. Bromocriptine microcapsules inhibit ornithine decarboxylase activity induced by Freund's complete adjuvant in lymphoid tissues of male rats. Endocrinology 1989;125:2846-52.
91. Perez Castro C, Penalva R, Paez Pereda M, Renner U, Reul JMHM, Stalla GK, Holsboer F, Arzt E 1999 Early activation of thyrotropin-releasing hormone and prolactin plays a critical role during a T cell-dependent immune response. Endocrinology 140:690-7.
92. Bellone G, Geuna M, Carbone A, Silvestri S, Foa R, Emanuelli G, Matera L. Regulatory action of prolactin on the *in vitro* growth of CD34+ve human hemopoietic progenitor cells. J Cell Physiol 1995;163:221-31.
93. Bellone G, Astarita P, Artusio E, Silvestri S, Mareschi K, Turletti A, Buttiglieri S, Emanuelli G, Matera L. Bone marrow stroma-derived prolactin is involved in basal and platelet-activating factor-stimulated *in vitro* erythropoiesis. Blood 1997;90:21-7.
94. Lin J, Linzer DI. Induction of megakaryocyte differentiation by a novel pregnancy-specific hormone. J Biol Chem. 1999;274:21485-9.
95. Socolovsky M, Fallon AE, Lodish HF. The prolactin receptor rescues EpoR-/- erythroid progenitors and replaces EpoR in a synergistic interaction with c-kit. Blood 1998;92:1491-6.

96. Warwick-Davies J, Lowrie DB, Cole PJ. Growth hormone is a human macrophage activating factor. Priming of human monocytes for enhanced release of H2O2. J Immunol. 1995;154:1909-18.
97. Matera L, Galetto A, Geuna M, Vekemans K, Ricotti E, Contarini F, Moro F, Basso G. Individual and combined effect of GM-CSF and prolactin on maturation of dendritic cells from blood monocytes under serum-free conditions. J Neuroimmunol 2000, in press.
98. Matera L, Mori, M. Different cooperation of prolactin with IL-2 and IL-12 on the production of IFN-γ by NK and T cells. Ann. NY Acad Sc. 2000, in press.
99. Morales P, Carretero MV, Geronimo H, Copin SG, Gaspar ML, Marcos MA, Martin-Perez J. Influence of prolactin on the differentiation of mouse B-lymphoid precursors. Cell Growth Differ 1999;10:583-90.
100. Matera L, Cesano A, Bellone G, Oberholtzer E. Modulatory effect of prolactin on the resting and mitogen-induced activity of T, B, and NK lymphocytes. Brain Behav Immun 1992;6:409-17.
101. Richards SM, Garman RD, Keyes L, Kavanagh B, McPherson JM. Prolactin is an antagonist of TGF-beta activity and promotes proliferation of murine B cell hybridomas. Cell Immunol 1998;184:85-91.
102. Lahat N, Miller A, Shtiller R, Touby E. Differential effects of prolactin upon activation and differentiation of human B lymphocytes. J Neuroimmunol 1993;47:35-40.
103. Clevenger CV, Sillman AL, Hanley-Hyde J, Prystowsky MB. Requirement for prolactin during cell cycle regulated gene expression in cloned T-lymphocytes. Endocrinology. 1992;130:3216-22.
104. Belkowski SM, Levine JE, Prystowsky MB. Requirement of PI3-kinase activity for the nuclear transport of prolactin in cloned murine T lymphocytes. J Neuroimmunol 1999;94:40-7.
105. Mukherjee P, Mastro AM, Hymer WC Prolactin induction of interleukin-2 receptors on rat splenic lymphocytes. Endocrinology 1990;126:88-94.
106. Viselli SM, Stanek EM, Mukherjee P, Hymer WC, Mastro AM Prolactin-induced mitogenesis of lymphocytes from ovariectomized rats. Endocrinology 1991;129:983-90.
107. John S, Robbins CM, Leonard WJ. An IL-2 response element in the human IL-2 receptor alpha chain promoter is a composite element that binds Stat5, Elf-1, HMG-I(Y) and a GATA family protein. EMBO J 1996;15:5627-35.
108. Cesario TC, Yousefi S, Carandang G, Sadati N, Le J, Vaziri N. Enhanced yields of gamma interferon in prolactin treated human peripheral blood mononuclear cells. Proc Soc Exp Biol Med 1994;205:89-95.
109. Thellin O, Coumans B, Devos S, Hooghe-Peters EL, Zorzi W, Igout A, Heinen E, Hooghe R. Limited effects of placental and pituitary growth hormone on cytokine expression *in vitro*. Eur.Cytokine Network 2000; 11: in press.
110. Matera L, Bellone G, Lebrun JJ, Kelly PA, Hooghe-Peters EL, Di Celle PF, Foa R, Contarini M, Avanzi G, Asnaghi V. Role of prolactin in the *in vitro* development of interleukin-2-driven anti-tumoural lymphokine-activated killer cells. Immunology 1996;89: 619-26.
111. Müller H, Liu B, Croy BA, Head JR, Hunt JS, Dai G, Soares MJ. Uterine natural killer cells are targets for a trophoblast cell-specific cytokine, prolactin-like protein A. Endocrinology 1999;140:2711-20.
112. Oberholtzer E, Contarini M, Veglia F, Cossarizza A, Franceschi C, Geuna M, Provinciali M, Di Stefano G, Sissom J, Brizzi MF, Pegoraro L, Matera L. Prolactin increases the susceptibility of primary leukemia cells to NK and LAK effectors. Adv Neuroimmunol 1996;6:233-47.
113. Nagafuchi H, Suzuki N, Kaneko A, Asai T, Sakane T. Prolactin locally produced by synovium infiltrating T lymphocytes induces excessive synovial cell functions in patients with rheumatoid arthritis. J Rheumatol 1999;26:1890-900.

114. McAveney KM, Gimble JM, Yu-Lee L 1996 Prolactin receptor expression during adipocyte differentiation of bone marrow stroma. Endocrinology 137:5723-26.
115. Nanbu-Wakao R, Fujitani Y, Masuho Y, Muramatu M, Wakao H. Prolactin enhances CCAAT enhancer-binding protein-beta (C/EBP beta) and peroxisome proliferator-activated receptor gamma (PPAR gamma)messenger RNA expression and stimulates adipogenic conversion of NIH-3T3 cells. Mol Endocrinol 2000;14:307-16.
116. Clark RB, Bishop-Bailey D, Estrada-Hernandez T, Hla T, Puddington L, Padula SJ. The nuclear receptor PPAR gamma and immunoregulation: PPAR gamma mediates inhibition of helper T cell responses. J Immunol. 2000;164:1364-71.
117. Ashkar S, Weber GF, Panoutsakopoulou V, Sanchirico ME, Jansson M, Zawaideh S, Rittling SR, Denhardt DT, Glimcher MJ, Cantor H. Eta-1 (osteopontin): an early component of type-1 (cell-mediated) immunity. Science 2000;287:860-4.
118. Baik MG, Lee MJ, Choi YJ. Gene expression during involution of mammary gland (review). Int J Mol Med. 1998;2:39-44.
119. Dogusan Z, Book ML, Verdood P, Yu-Lee L-y, Hooghe-Peters EL. Prolactin activates interferon regulatory factor-1 expression in normal lympho-hemopoietic cells. Eur. Cytokine Network, 2000,11:in press.
120. Dogusan Z, Hooghe-Peters EL, Berus D, Velkeniers B, Hooghe R. Expression of SOCS genes in normal and leukemic human leukocytes stimulated by prolactin, growth hormone and cytokines. J Neuroimmunol 2000, in press.
121. Zhu T, Lobie PE. Janus kinase 2-dependent activation of p38 mitogen-activated protein kinase by growth hormone. Resultant transcriptional activation of ATF-2 and CHOP, cytoskeletal re-organization and mitogenesis. J Biol Chem 2000;275:2103-14.
122. Yoshimura A, Ohkubo T, Kiguchi T, Jenkins NA, Gilbert DJ, Copeland NG, Hara T, Miyajima A. 1995 A novel cytokine-inducible gene CIS encodes an SH2-containing protein that binds to tyrosine-phosphorylated interleukin 3 and erythropoietin receptors. EMBO J. 14: 2816-26.
123. Matsumoto A, Masuhara M, Mitsui K, Yokouchi M, Ohtsubo M, Misawa H, Miyajima A, Yoshimura 1997 CIS, a cytokine inducible SH2 protein, is a target of the JAK-STAT5 pathway and modulates STAT5 activation. Blood 89: 3148-54.
124. Hilton DJ 1999 Negative regulators of cytokine signal transduction. Cell Mol Life Sci 1999 55: 1568-77.
125. Verdier F, Rabionet R, Gouilleux F, Beisenherz-Huss C, Varlet P, Muller O, Mayeux P, Lacombe C, Gisselbrecht S, Chretien S 1998 A sequence of the CIS gene promoter interacts preferentially with two associated STAT5A dimers: a distinct biochemical difference between STAT5A and STAT5B Mol Cell Biol 18: 5852-60.
126. Matuoka K, Miki H, Takahashi K, Takenawa T 1997 A novel ligand for an SH3 domain of the adaptor protein Nck bears an SH2 domain and nuclear signaling motifs. Biochem Biophys Res Commun 239: 488-92.
127. Helman D, Sandowski Y, Cohen Y, Matsumoto A, Yoshimura A, Merchav S, Gertler A. 1998 Cytokine-inducible SH2 protein (CIS3) and JAK2 binding protein (JAB) abolish prolactin receptor-mediated STAT5 signaling. FEBS Lett 441: 287-91.
128. Pezet A, Favre H, Kelly PA, Edery M. Inhibition and restoration of prolactin signal transduction by suppressors of cytokine signaling. J Biol Chem 1999;274:24497-502.
129. Dickensheets HL, Venkataraman C, Schindler U, Donnelly RP. 1999. Interferons inhibit activation of STAT6 by interleukin 4 in human monocytes by inducing SOCS-1 gene expression. Proc Natl Acad Sci U S A 96: 10800-5.
130. Ito S, Ansari P, Sakatsume M, Dickensheets H, Vazquez N, Donnelly RP, Larner AC, Finbloom DS 1999 Interleukin-10 inhibits expression of both interferon alpha- and interferon gamma- induced genes by suppressing tyrosine phosphorylation of STAT1. Blood 93: 1456-63.

131. Zellweger R, Zhu XH, Wichmann MW, Ayala A, DeMaso CM, Chaudry IH. Prolactin administration following hemorrhagic shock improves macrophage cytokine release capacity and decreases mortality from subsequent sepsis. J Immunol. 1996;157:5748-54.
132. Stoecklin E, Wissler M, Schaetzle D, Pfitzner E, Groner B. Interactions in the transcriptional regulation exerted by Stat5 and by members of the steroid hormone receptor family. J Steroid Biochem Mol Biol. 1999;69:195-204.
133. Stevens FR, Hajeer A, John S, Thomson W, Worthington J, Davis JR, Ollier WE. The Bg/II polymorphism of the human prolactin gene lies within intron C and can be detected by PCR/RFLP. Eur J Immunogenet 1999;26:261-3.
134. Fuh G, Wells JA. Prolactin receptor antagonists that inhibit the growth of breast cancer cell lines. J Biol Chem. 1995;270:13133-7.
135. Goupille O, Barnier JV, Guibert B, Paly J, Djiane J. Effect of PRL on MAPK activation: negative regulatory role of the C-terminal part of the PRL receptor. Mol Cell Endocrinol. 2000;159:133-46.

Chapter 17

Prolactin Receptors

**Marc Edery, Nadine Binart,
Brigitte Bouchard, Vincent Goffin, Paul A. Kelly**

*INSERM U344, Faculte de Medecine Necker,
Paris FRANCE*

The initial step in the mechanism of action of prolactin (PRL) is the binding to a cell surface receptor. Binding of PRL to its receptor occurs with a high affinity (association constants range from 10^7 to 10^{10} M^{-1}) and a structural specificity. PRL receptors (PRLR) have been identified in a number of cell types and tissues in several species including non-mammalian vertebrates. PRLR belongs to the class I cytokine receptor superfamily. Signal transduction by the PRLR is mainly mediated by two families of signaling molecules; the Janus tyrosine kinases and the signal transducers and activators of transcription (STATs). Disruption of the PRLR gene has provided a new mouse model with which to identify functions directly associated with PRL. Several phenotypes have been analyzed which are described in this review. Applied to the SAGE technology, the PRLR knock-out model allows the quantitative and qualitative evaluation of the expression patterns of hepatic genes in two physiological situations : transcriptomes corresponding to livers from both wild type and PRLR KO mice are being compared, and following statistical analyses, candidate genes presenting a differential profile will be further characterized.

INTRODUCTION

The initial step in the action of PRL is the binding to a specific membrane receptor, the PRL receptor. As is true for the ligand, the PRLR has been shown to be a member of the same family as the the GH receptor, and as well part of the larger class of receptors, known as the class-1 cytokine receptor superfamily.

In this chapter, we will discuss many of the molecular steps by which PRL exerts its various functions via the PRLR.

The PRLR was identified more than two decades ago as a specific, high affinity, membrane-anchored protein. PRLRs are found in the plasma membrane and also in endosomes, Golgi fractions and enriched lysosomes (1,2) . In the lactating mammary gland, the majority of receptors are localized in intracellular membranes. This probably reflects the rapid receptor synthesis and degradation that occurs, even though the receptors are not occupied by PRL(3). In addition , PRLR have ben identified in a cytosolic fraction of rabbit mammary gland (4) and in milk (5). Conflicting results have been reported concerning the potential localization of PRLR in the nucleus (6,7). PRLR have been identified in a number of cells and tissues in embryonic and adult mammals. The expression of short and long isoform of the receptor has been shown to vary as a function of the stage of the estrous cycle, pregnancy and lactation (8). The hormonal regulation of the PRLR differs depending on the organ considered. In the mammary gland the number of PRLR increases gradually during pregnancy, but rises markedly early in lactation. The high circulating concentration of progesterone during pregnancy is probably responsible for limiting the increase in PRLR in the mammary gland (9). PRL is able to up- and down-regulate the level of its own receptor, depending on the concentration and duration of the exposure to the hormone. Down-regulation precedes up-regulation, is rapidly reversible and is observed in the presence of high concentrations of hormone(10). This down-regulation is presumably a consequence of the accelerated internalization of hormone receptor complexes and their degradation by lysosomes. Up-regulation occurs after several days of exposure to moderate levels of PRL (11) or GH (12).

Treatment of rabbit mammary explants with bivalent but not monovalent antibodies, specific to PRLR, mimics the effect of PRL by increasing casein and DNA synthesis. This suggested that receptor aggregation is necessary for signal transduction which was confirmed later with the availability of PRLR cDNAs. However, both bivalent and monovalent antibodies induce the internalization and this results in down-regulation of receptors (13). Numerous studies have been conducted on the binding of PRL to various tissues in mammals. These studies are based on the association of radioactive iodinated ligand with membrane preparations. By modifying the hormone-receptor complex with bifunctional reagents, it is possible to generate a covalently linked complex. By this means it has been possible to determine the size of the receptor by polyacrylamide gel electrophoresis. Prior to cloning the cDNA, estimation of the molecular weight of the PRLR varied from 30,000 to 300,000. In 1988, the cDNA encoding the rat PRLR was isolated in our laboratory (14) and as is true for the closely related receptor for GH (GHR), is a single-pass transmembrane chain (Figure 17-1). In the early 1990's, sequence comparisons with newly-identified membrane receptors led to the identification of a new family of receptors including both PRLR and GHR (15). Termed class-1 cytokine receptors, this superfamily includes receptors for several interleukins, granulocyte-colony stimulating factor (G-CSF), granulocyte-macrophage-colony stimulating factor (GM-CSF), leukemia inhibitory factor (LIF), oncostatin M (OM), erythropoietin (EPO), thrombopoietin (TPO), gp130 and the obesity factor, leptin (16) (Figure 17-2). These membrane chains are apparently phylogenetically related, they contain stretches of highly conserved amino acids, mostly in the extracellular domain. The gene encoding human PRLR is located on chromosome 5 (p13 $\rightarrow$14) and contains at least ten exons for an overall length > 100 kb (17). Multiple isoforms of membrane-bound PRLR resulting from alternative splicing of the primary transcript have been identified in several species (8). These different PRLR isoforms differ in the length and amino acids composition of their cytoplasmic tail and are referred to as short, intermediate or long form of the PRLR with respect to their size (Figure 17-1).

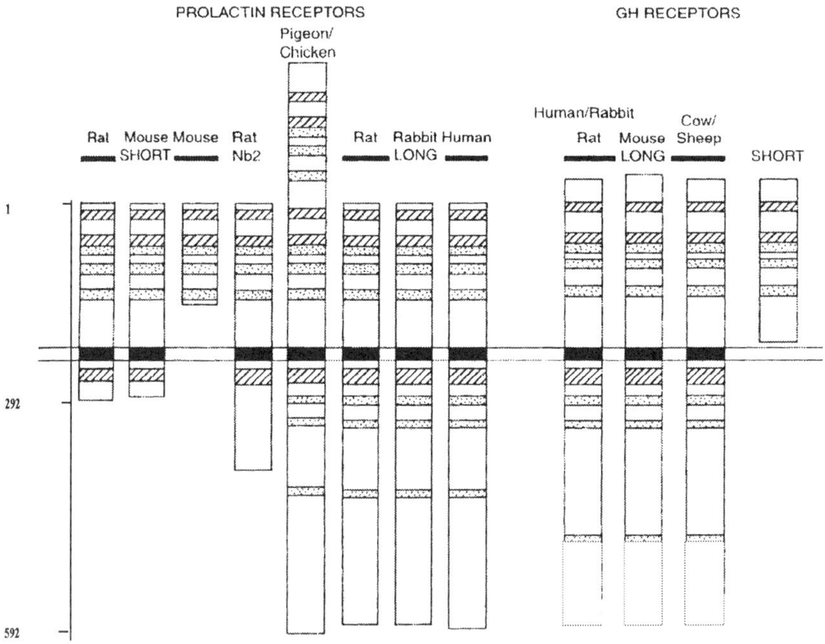

Figure 17-1. Schematic representation of various forms of prolactin and growth hormone receptors. The first and last amino acids of the mature protein are indicated. The transmembrane domain is represented by a black box. Regions of high (~70%) amino acid identity are cross-hatched and those of moderate identity (40-60%) are stippled.

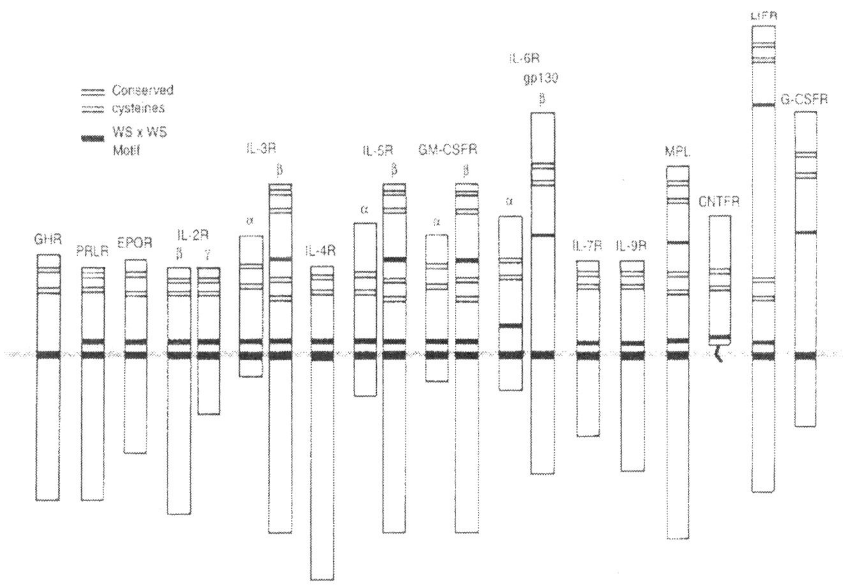

Figure 17-2. Schematic representation of the GH/PRL Cytokine receptor family. The plasma membrane is indicated by a stippled rectangle.

In mice, one long and three short isoforms have been identified, the short forms only differing by a few amino acids in the C-terminal part of their cytoplasmic tail (18). In addition to the membrane-anchored PRLR, soluble isoforms have also been identified but whether they result from alternative splicing of the primary mRNA or proteolytic cleavage of membrane-bound PRLR (or both) is uncertain. In all cases, however, the extracellular, ligand-binding domain is identical, whatever the isoform.

Typically, a cytokine receptor extracellular domain (ECD) is composed of a module of 200 aa, referred to as the cytokine receptor homology region (16). This region can be divided into two subdomains of ~100aa, each showing homology with the fibronectin type III module (8).Two highly conserved features are found in the cytokine receptor ECDs : the first is two pairs of disulfide-linked cysteines in the N-terminal subdomain, and the second is a pentapeptide termed « WS motif » (Trp-Ser-any amino acid-Trp-Ser) found in tthe membrane proximal region of the C-terminal subdomain. The cytoplasmic domain of cytokine receptors display more restricted sequence similarity than the extra-cellular domain. Two regions, called Box1 and Box2 are relatively conserved (8). Box1 is a membrane-proximal region composed of amino acids highly rich in prolines and hydrophobic residues (aa 243-250 in PRLR). The 3D structure of genetically engineered hPRLRECD has been determined by crystallographic analysis (19). It folds in two domains of 7β-strands forming a sandwich of two anti-parallel β-sheets. As anticipated from sequence comparison, this folding pattern is likely to be shared by several, if not all, cytokine receptors (8). Determining the structure of PRLR cytoplasmic domains is one of the objectives of our laboratory.

TISSUE DISTRIBUTION OF PRLR

Prolactin receptors have been detected in a number of cells and tissues of embryonic and adult mammals as well as non-mammals. The expression of short and long isoforms of receptor has been shown to vary as a function of the stage of the estrous cycle, pregnancy and lactation (8). PRL receptors are widely distributed throughout vertebrate tissues. We have determined the cellular distribution and developmental expression of the PRLR in the fetal rat tissues by in situ hybridization, immunocytochemistry and radioligand binding (20). Sense and antisense strand probes were prepared that encoded the long and short isoforms of the rat PRLR and hybridized to various fetal tissues obtained at the end of pregnancy (days 17.5-20.5). These studies indicated that the mRNA encoding the short and long isoforms were widely expressed in tissues from all three germ layers and that, in addition to the classical target organs of PRL, tissues not known previously to contain PRLR, such as olfactory neuronal epithelium and bulb, fetal adrenal cortex, gastrointestinal and bronchial mucosae, renal tubular epithelia, choroid plexus, thymus, liver pancreas and epidermis, also express the PRLR mRNA. The wide distribution of the PRLR is obviously correlated with the extremely large spectrum of activities of its ligand(s) (8).

INTERACTION BETWEEN PRL AND ITS RECEPTOR

Although stochiometric analysis of the interaction between different PRLR ECDs and lactogenic hormones resulted in a 1 :1 (21) or 1 :2 (22) complex depending on the species involved, dimerization of the PRLR upon ligand binding has now been clearly demonstrated. We have shown that at least two regions of hPRL are involved in the binding of the hormone to the PRLR. The first, referred to as binding site 1, encompasses several residues belonging to helices 1 and 4, while the second, termed binding site 2, involves helices 1 and 3. Elucidation of the events occuring upon PRL-induced activation of membrane-bound PRLR has resulted from the analysis of the shape of experimental curves obtained with point-mutated hPRL analogs in PRL-responsive bio-assays performed over a wide range of hormone concentrations (23). As first described for the closely related GHR, activation of the PRLR involves ligand-induced sequential receptor dimerization. In a first step, interaction of PRL binding site 1 with one PRLR occurs and leads to the formation of an inactive $H_1:R_1$ (1 hormone, 1 receptor) complex. Formation of this complex appears to be a prerequisite for PRL binding site 2 to interact with another PRLR, then achieving an active trimeric complex ($H_1:R_2$), composed of one hormone and one receptor homodimer. Based on this model of receptor activation, we and others have developed PRL, PL and GH analogs able to block the PRLR in an inactive conformation and therefore behaving as PRL antagonists (24,25). Most often, such analogs carry a mutation in binding site 2 and are inactive since they are unable to induce PRLR dimerization, but display antagonistic properties due to their ability to still bind to the receptor through their binding site 1 and to form an inactive $H_1:R_1$ complex (23). Such PRL antagonists may be potentially interesting in PRL-related pathologies.

Signal transduction pathways will not be presented here since they will be analysed in details in another chapter in this book.

PRLR KNOCKOUT (KO) MICE.

PRL was originally isolated as a pituitary factor capable of inducing milk secretion in rabbits and thereafter to stimulate the growth of the crop sac in pigeons. PRL was also shown to be luteotrophic, that is to promote the formation and function of the corpus luteum. Subsequently, a number of additional activities have been associated with this hormone in various vertebrate species. In the now classical review by Nicoll and Bern, 85 different biological functions were attributed to PRL (26). We have recently expanded the list of PRL activities in vertebrates to include more than 300 separate functions involving this multifaceted hormone (8). Classically, two experimental models have been primarly used for the examination of the role of PRL and its receptor *in vivo*. The first is pituitary ablation, or the administration of dopamine D2 receptor agonists to reduce pituitary PRL secretion, and the second is based on the use of mice strains lacking pituitary somatotroph and lactotroph cells. The possibility of incomplete PRL suppression and/or the possible suppression of other pituitary and non pituitary hormones has limited the validity of these models. To circumvent these technical limitations, we have generated PRLR deficient mice by gene targeting (27), PRL KO mice have been recently developed by others (28). The use of the PRLR knockout model offers the possibility of directly examining the effects of PRLR-mediated signaling *in vivo*. In the following section, we will briefly describe the generation of these PRLR KO mice. The technique of gene targeting in mice has been used to develop the first experimental model in which the effects of the complete absence of any lactogen or PRL- mediated effects can be studied. To generate mice carrying a germ null mutation of the PRLR gene, a targeting vector was constructed with 7.5 kb of overall homology in which a 1.5 kb fragment containing exon 5 of the PRLR gene was replaced with the similarly sized Tk-NEO cassette, which resulted in a mutation creating an in-frame stop codon (the encoded protein contains only the 44 N-terminal residues of the PRLR). After electroporation into embryonic stem cells and neomycin selection, selected clones were injected into blasto-cysts and were able to generate germline chimeras. F1 intercrosses allowed the generation of $PRLR^{+/-}$ and $PRLR^{-/-}$ animals. No PRLR mRNA containing exon 5 was transcribed and no PRLR protein could be detected in the liver of $PRLR^{-/-}$ animals.

REPRODUCTION AND LACTATION PHENOTYPES OF PRLR KO MICE

Some aspects of these phenotypes will be described in detail in another chapter; mainly the PRLR$^{+/-}$ females were unable to lactate their first litter indicating that two functional alleles of the PRLR are required for efficient lactation. The second important phenotype was the sterility of homozygous females : the PRLR$^{-/-}$ females never became pregnant and showed a number of reproductive defects ; an alteration of estrous cyclicity, and an absence of pseudopregnancy. Fewer eggs were fertilized, oocytes at the germinal vesicle stage were released from the ovary and fragmented embryos were found. Single cells fertilized eggs were recovered, suggesting that, for most oocytes, an arrest of development occured immediately after fertilization. Only a few blastocysts were recovered in the uterus of PRLR$^{-/-}$ females. The lack of blastocyst implantation leading to the complete sterility in PRLR$^{-/-}$ mice may be due to a deficient environment in both oviduct and uterus. The failure of a significant proportion of eggs to undergo germinal vesicle breakdown within maturing PRLR$^{-/-}$ follicles directly demonstrates the important influence of PRL and its receptor on oocyte maturation or atresia. Divergent effects of PRL on the rate of implantation development of mouse embryos have been reported (8). A number of factors in the oviduct which influence preimplantation development may be affected by the PRLR mutation. Estrogen and progesterone can influence the rate of ovum transport and preimplantation development (29). Uterine preparation for embryo implantation is dependent upon continuous estrogen and progesterone secretion by the corpus luteum of the ovary, which is supported by the pituitary in rodents during the first half of pregnancy. PRL has been shown to stimulate progesterone synthesis by dispersed ovarian cells from mid-pregnant mice (30), demonstrating that lactogenic hormones can directly stimulate ovarian progesterone secretion. Furthermore a nidatory ovarian estrogen surge is required to allow embryon implantation (31). Thus PRLR$^{-/-}$ females probably cannot support the implantation of blastocysts because the corpus luteum does not receive pituitary PRL support, and thus progesterone (Pg)- and estrogen-dependent signals for implantation cannot occur. Pg administration rescued infertility in PRLR$^{-/-}$ mice from the periimplantation period to midgestation. This result shows that the luteal Pg production via ovarian PRLR signaling is required for implantation and early pregnancy (32). In conclusion, the phenotypes of animals lacking functional genes encoding PRLR or PRL confirm the essential role of these molecules in the sinaling pathways leading to mammary gland developmend and reproductive functions, whereas estradiol and progesterone receptors are probably involved to a lesser extent in mammary gland development.

In order to more fully understand the molecular mechanisms involved in PRL-mediated events in the ovary, oviduct and uterus, we are using approaches such as differential display to identify genes that are either induced or repressed in the PRLR deficient mice and characterize new gene targets of PRL regulation. In-depth molecular characterization should help to elucidate the actual control mechanisms as well as PRL-dependent signaling pathways which are involved in the early stages of pregnancy.

IMMUNE PHENOTYPE OF PRLR KO MICE

PRL has also been reported to play a role in immunomodulation. Since initial observations showing that hypophysectomized rats present abnormalities of the immune system, including an increased thymic atrophy and lymphopenia, a number of studies have focused on the potential immunomodulatory roles of PRL. In a recent study, we have characterized the development and function of the immune system in PRLR-deficient mice. Compared to wild-type control mice, PRLR$^{-/-}$ mice did not demonsrate any alteration in thymic or splenic structure cellularity. There was no difference in the composition of lymphocytes subsets present in primary (bone marrow and thymus) or secondary (spleen and lymph nodes) lymphoid organs. All lymphocyte populations tested from PRLR$^{-/-}$ mice were functional *in vitro* : they could proliferate normally to mitogens, cytokines and allogeneic cells. PRLR$^{-/-}$ splenocytes displayed normal NK-mediated cytotoxicity. Additionally, *in vivo* studies were performed and revealed that PRLR$^{-/-}$ mice could generate normal steady-state Ig levels, and mount a normal specific Ig response following immunization with T-dependent antigens. PRLR$^{-/-}$ animals were able to control the growth of tumor cells and infection by *Listeria monocytogenes.* Taken together, these results show that the immune system development and function can proceed normally in the absence of PRL-mediated signaling and suggest that PRLR pathways are not essential for immunomodulation *in vivo.*

BONE PHENOTYPE OF PRLR KO MICE

Bone development is a multistep process that includes patterning of skeletal elements, commitment of hematopoietic and/or mesenchymal cells to chondrogenic and osteogenic lineages, and further differentiation into three specialized cell types : chondrocytes in cartilage and osteoblasts and osteoclasts in bone. Bone formation occurs mostly during embryonic development and post-natal growth, but is also important in adults for subsequent bone remodeling to maintain calcium homeostasis and for adaptation to physical forces. Bone formation requires recruitment, proliferation and differentiation of osteoprogenitor cells (32,33). Little was known about the putative role of PRL, with respect to bone cells and bone metabolism. Lactation in humans and mammals is associated with a vitamin D-induced maternal bone responsive state, and it has also been suggested that PRL might induce the release of a bone resorptive agent. Many clinical studies have reported a possible role of PRL, during pregnancy, lactation or hyperprolactinemia, but other hormonal parameters could also be implicated and no definitive results have been published (34-37). A direct demonstration of the PRLR protein expression in bone cells has been lacking and the putative role of PRL in bone remains discussed. Recently, studies on human, mouse and rat fetal tissues have shown expression of PRLR mRNA and the appearance, during early fetal development of intense PRLR immunoreactivity in skeletal tissues, including long bones, vertebrae and ribs (20,38). Also, PRLR transcripts were detected in two human osteoblastic lines (MG63 and Saos-2) and their regulation by vitamin D3 and dexamethasone has been demonsrated (39). The approach developed in our laboratory has been to identify potential PRLR expression in murine bone cells as a first step in assessing a direct effect of PRL on bone. The expression of PRLR mRNA was measured by reverse transcription-polymerase chain reaction (RT-PCR) combined with Southern analysis, using total RNA from primary cultures of normal murine osteoblasts and osteoclast-like cells. To avoid any problems of contamination, osteoclast-like cells were prepared by co-culturing primary murine osteoblasts from PRLR knockout mice with bone marrow cells from wild-type mice in the presence of 1,25 (OH) 2D3. Long and all but one of the short forms of the PRLR were detected in normal murine osteoblasts. Clearly, more long form mRNA was observed than that of the two other short forms. No expression of any PRLR mRNA was seen in murine osteoclast-like cells. Finally, comparative western blot analysis of PRLR protein immunoprecipitated from primary cultures of osteoblasts from PRLR$^{+/+}$ or PRLR$^{-/-}$ pups using an anti-PRLR monoclonal antibody revealed that only the long form could be detected in wild-type osteoblasts, and this band was absent in cells from PRLR$^{-/-}$ animals. This demonsration of PRLR gene expression in murine osteoblasts supports the hypothesis of a direct action of PRL on bone.

In order to evaluate the potential role of PRL in bone development, we studied this phenotype in PRLR knockout mice (40). In calvariae of PRLR$^{-/-}$ embryos (18.5 days), bone formation appeared to be delayed compared to calvariae of PRLR$^{+/+}$ and PRLR$^{+/-}$ mice. Double calcein labeling in post-pubescent PRLR$^{+/+}$ and PRLR$^{+/-}$ mice resulted in two distinct fluorescent bands representing normal mineral appositional rates (MAR). In contrast, in PRLR$^{-/-}$ mice, the distance between the two labels was markedly decreased, suggesting a possible defect in bone formation. Analysis of serum from PRLR$^{+/+}$ and PRLR$^{-/-}$ indicated that knockout mice had an increase in serum calcium with normal phosphate and magnesium levels. Our approaches have been the first to show, using a knockout model, that PRL may play an important role in bone remodeling and that PRL may play a direct role in bone metabolism. Some systemic problems observed in knockout mice, including hypercalcemia, led us to investigate their global hormonal status. We studied the level of several hormones implicated in bone metabolism and/or calcium homeostasis such as estradiol (E2), Pg, parathyroid hormone (PTH), intact PTH-related protein (PTH-rP) and PRL. Plasma and/or serum of 3-month old knockout versus wild–type males and females were collected and used for specific radioimmunoassays. Briefly, our results demonstrate that the presence of PRLR is important for the overall hormonal status of mice, because Pg, E2 and PTH are essential factors with regard to the bone alterations observed in PRLR deficient mice. This suggests that an effect of PRL on osteoblasts could be required for normal bone formation and maintenance of bone mass. Thus, the PRLR knockout mouse model provides a new tool to investigate the involvment of PRL in bone metabolism. The consequences of null mutation of the PRLR gene on the mouse hormonal regulation must be considered as an indirect factor to explain the bone alterations observed in these mice. However, the presence of PRLR on murine osteoblasts, as previously demonstrated, suggests direct effects of PRL and its receptor on bone cells. A similar approach has been recently used in our laboratory to evaluate the bone phenotype of mice deficient for the GHR-GHRBP as well as other transgenic models. Thus we hope to investigate the interaction and the role of PRL and GH on bone metabolism during embryogenesis and on growth.

MATERNAL BEHAVIOR PHENOTYPE OF PRLR KO MICE

With the evolution of live birth synchronized with lactation, maternal behavior toward young is essential for species survival in mammals. In rodents, such behaviors include nest building, pup retrieval and grouping, anogenital licking and crouching over the pups in the nest to provide warmth and nourishment. It has also been shown that these behaviors can be induced in virgin females with repeated exposure to pups. There has been increasing correlative evidence of hormonal control of the onset and maintenance of nurturing behavior (41,42). PRL, the levels of which increase markedly in the blood just before parturition, is often implicated as playing a key role in the regulation of maternal behavior. The use of PRLR knockout model offers the possibility of directly examining the effects of PRLR-mediated signaling on maternal behavior in the mouse. We studied pup-directed maternal behavior in PRLR KO mice (43). Each female tested was exposed to three 1-3 day old foster pups, which were placed in the opposite corner of the cage from the nest. For the succeeding 30 min, the test female was continually observed and the following endpoints were recorded : latency to retrieve each pup and place it into the nest, total number of pups retrieved, and latency to crouch over all three pups for more than 5 min at a time. Since PRLR knockout mice are infertile, it was not possible to study behavior in PRLR$^{-/-}$ pregnant or lactating animals. However, primiparous heterozygous females were studied, a profound deficit was observed in maternal care when challenged with foster pups.

The typical behavior of wild-type females is to immediately begin retrieving the pups into the nest and crouching over them, remaining with the pups in the nest for the remainder of the test. PRLR heterozygous mutant mice took notice, made immediate contact with and sniffed the pups as did the wild-type animals. PRLR$^{+/-}$ then typically returned to the nest, returning once or twice to sniff the pups before retrieving one, or at most two pups over the next several minutes before settling into the nest and ignoring the unretrieved pups for the remainder of the test. Wild-type females typically displayed full maternal behavior within several minutes, while none of the PRLR$^{+/-}$ mice exhibited such behavior within the testing period. PRLR$^{-/-}$ and PRLR$^{+/-}$ nulliparous females showed a clear deficiency in pup-induced maternal behavior. Full maternal behavior, defined as retrieving all three pups and crouching over them in the nest, was displayed by wild-type virgin females with a latency of 1 or 2 days, while PRLR$^{+/-}$ virgin females showed an average latency of 4 days before displaying this behavior, and PRLR$^{-/-}$ virgins did not exhibit such behavior at all during the 6 day testing period. Morris maze studies revealed normal configural learning in the heterozygous and homozygous animals. In addition, eating, locomoter activity, sexual behavior and exploration, all processes regulated by the hypothalamus, were normal in PRLR mutant mice, suggesting specificity in the response. Olfactory function was also tested in an aversive conditioning paradigm. The results indicated that heterozygous and homozygous PRLR mutant mice were not anosmic, ruling out the lack of smell as the explanation of the effect on behavior. Although the direct target of PRL and its receptor in the brain involved in the mediation of maternal behavior is not known, the use of the PRLR knockout model clearly establishes the PRL and its receptor as regulators of maternal behavior in the mouse, and provides a new tool to understand this process.

In order to systematically study behavior, we have decided to develop a battery of tests to evaluate behavioral phenotyping, based on a series of standard methods proposed (44). In general, before beginning the actual tests, mice should be evaluated for general health, gross neurological functions, locomotor activity, measurements of motor abilities, simple evaluations of sensory functions and tests of higher brain function. As a function of these preliminary tests, a series of behavioral tests will be performed and those will be limited to a number of six general categories of behavior among the long list that is available; learning and memory, feeding, pain-analgesia, agression, reproduction and anxiety. We plan to apply the above tests to PRLR KO mice and mice over-expressing the PRL gene.

IDENTIFICATION OF NEW TARGET GENES OF PRL

As mentioned above, PRL is implicated in a number of physiological processes, with perhaps a large number still yet to be determined. Although the effects of PRL and the genes specifically induced by PRL are best known during lactation and pregnancy, it is also well established that other extrapituitary sites synthesize and secrete PRL, and that the PRLR is expressed at various levels in many organs. In most cases, the functions of PRL in these different target-tissues, such as the liver, the prostate and the brain are not clear. Through different studies, PRL has been shown to be involved in proliferation and differenciation mechanisms, in the immune response or in apoptosis. Thus the

PRL-PRLR system can be viewed as playing a role in the development of normal cellular functions as well as in the progression towards some pathological processes ; using various approaches, we have initiated several projects with the goal of identifying the genes induced in response to PRL.

With respect to the proliferative effects of PRL, the PRL-dependent Nb2 rat lymphoma cell line was used as a model to identify genes whose expression is modulated during PRL-induced cell cycle. Five new differential screening techniques were applied at different stages of the Nb2 cell cycle to isolate cDNAs corresponding to differential transcripts. Three of these approaches, namely differential display, representational difference analysis (RDA) and suppressive substractive hybridization (SSH), consist of a selective and/or suppressive PCR of cDNA obtained from the cell populations to be compared (unstimulated vs. PRL stimulated Nb2 cells). These PCR-based techniques render them more sensitive than any other classical substraction libraries. The two other approaches used were the screening of an organized cDNA library and analysis of the expression of low expressed candidate genes. These methods are based on the hybridization of identical high density nylon filters with complex radiolabeled cDNA probes obtained from the mRNAs of the different cell populations to be compared. Some of these new genes are being currently characterized : two of them are presumably transcription factors involved in the control of cellular proliferation and one is likely to be involved in cytotoxicity. Another approach, the recently developed serial analysis of gene expression (SAGE) allows the definition of the cellular transcriptome, representing the entire set of transcripts in a given tissue, as well as their relative level of expression. The model that we have developed in our laboratory is designed to identify genes expressed in the liver, an organ in which the level of expression of the PRLR is particularly high. As discussed earlier, long and short PRLR isoforms exist in rodents, and both these forms are present in the liver, although their respective roles remain to be determined. The principle of the SAGE technique is based on the fact that a sequence as short as 14 bp is sufficient to identify a given transcript, if the particular position of this 14 bp sequence is known in the transcript of interest. The concatenation of these minimal sequences or tags allows a direct and rapid sequencing of serial transcripts. To date, the elaboration of the genome projects has resulted in the accumulation of a large amount of data, but does not give any indication with respect to the potential function of the genes identified. Other technological advances, such as the possibility of generating EST (expressed sequences tags) have greatly increased strategies to identify genes that are regulated in a specific manner. The model of the PRLR KO mouse is unique as it allows the analysis of the consequences following complete ablation of the signaling pathway induced by PRL, at the level of the entire organism. The use of this model, coupled with the SAGE technique, will allow the qualitative as well as quantitative evaluation of the expression pattern of hepatic genes in two physiological situations. The transcriptomes corresponding to livers from both a wild type and a PRLR KO mouse are being compared, and following statistical analyses, candidate genes presenting a differential profile will be further characterized. Initial experiments currently underway include Northern blots to confirm the expression patterns. The normal liver transcriptome will represent a unique databank, independent of any hormonal status. On the other hand comparison with the KO profiles will orient our future projects of research on PRL.

CONCLUSION

Because of the wild range of biological actions associated with PRL and the existence of various forms of PRL receptors, it is doubtful that one unifying mechanism of action will be found for this hormone. The development and application of multiple functional tests and models both *in vivo* and in vitro should permit precise identification of signaling pathways involved in the various functions of the hormone.

REFERENCES

1. Bergeron JM, Searle N, Khan MN, Posner BI. Differential and analytical subfractionation of rat liver components internalizing insulin and prolactin. Biochemistry 1986 ;25 :1756-64.
2. Bergeron JM, Posner BI, Josefszberg Z, Sikstrom R. Intracellular polypeptide hormone receptors. The demonstration of specific binding sites for insulin and human growth hormone in Golgi fractions isolated from the liver of female rats. J Biol Chem 1978 ;253 :4058-66
3. Djiane J, Delouis C, Kelly PA. Prolactin receptor turnover in explants of pseudopregnant rabbit mammary gland. Mol Cell Endocrinol 1982 ;25 ;163-70
4. Ymer SI, Kelly PA, Herington AC, Djiane J. Immunological relationship and binding capacity of prolactin receptors in cytosolic and membrane fractions of rabbit mammary gland. Mol Cell Endocrinol 1987 ;53 ;67-73
5. Postel-Vinay MC, Belair L, Kaiser C, Kelly PA, Djiane J. Identification of prolactin and growth hormone binding proteins in milk. Proc Natl Acad Sci USA 1991 ;88 ;6687-90
6. Buckley AR, Crowe PD, Russel DH. Rapid activation of a protein kinase C in isolated rat liver nuclei by prolactin, a known hepatic mitogen. Proc Natl Acad Sci USA 1988 ;85 ;8649-53
7. Perrot-Applanat M, Gualillo O, Buteau H, Edery M, Kelly PA. Internalization of prolactin receptor and prolactin in transfected cells does not involve nuclear translocation. J Cell Sci 1997 ;110 ;123-132
8. Bole-Feysot C, Goffin V, Edery M, Binart N, Kelly PA. Prolactin (PRL) and its receptor : Actions, signal transduction pathways and phenotypes observed in PRL receptor knockout mice. Endocr Rev 1998 ;19 ;225-68
9. Djiane J, Durand P, Kelly PA. Evolution of prolactin receptors in rabbit mammary gland during pregnancy and lactation. Endocrinology 1977 ;100 ;1348-56
10. Djiane J, Clauser H, Kelly PA. Rapid down-regulation of prolactin receptors in mammary gland and liver. Biochem Biophys Res Commun 1979 ;90 ;1371-78
11. Posner BI, Kelly PA, Friesen HG. Prolactin receptors in rat liver : possible induction by prolactin. Science 1975 ;188 ;57-9
12. Norstedt G, Husman B, Mode A, Eneroth P, Lewis UJ, Gustafsson JA. Induction of prolactin receptors in the liver is more closely related to the growth-promoting than to the lactogenic potency of peptides. Acta Endocrinol (Copenh) 1987 ;114 ;350-6
13. Djiane J, Kelly PA, Katoh M, Dusanter-Fourt I. Prolactin receptor : identification of the binding unit by affinity labelling and characterization of poly and monoclonal antibodies. Horm Res 1985 ;22 ;179-88
14. Boutin JM, Jolicoeur C, Okamura H, Gagnon J, Edery M, Shirota M, Banville D, Dusanter-Fourt I, Djiane J, Kelly PA. Cloning and expression of the rat prolactin receptor, a member of the growth hormone/ prolactin receptor gene family. Cell 1988 ;53 ;69-77
15. Bazan F. A novel family of growth factor receptors : a commonn binding domain in the growth hormone, prolactin, erythropoietin and IL-6 receptors, and p75 IL-2 receptor b-chain. Biochem Biophys Res Commun 1989 ;164 ;788-95

16. Wells JA, De Vos AM. Hematopoietic receptor complexes. Annu Rev Biochem 1996 ;65 ;609-34
17. Arden KC, Boutin J-M, Djiane J, Kelly PA, Cavenee WK. The receptors for prolactin and growth hormone are localized in the same region of human chromosome 5. Cytogenet Cell Genet 1990 ;53 ;161-5
18. Davis JA, Linzer DJH. Expression of multiple forms of the prolactin receptor. Mol Endocrinol 1989 ;3 ;674-80
19. Somers W, Ultsch M, De Vos AM, Kossiakoff AA. The X-ray structure of the growth hormone-prolactin receptor complex. Nature 1994 ;372 ;478-81
20. Royster M, Driscoll P, Kelly PA, Freemark M. The prolactin receptor in the fetal rat : cellular localization of messenger RNA, immunoreactive protein, and ligand binding activity and induction of expression in late gestation. Endocrinology 1995 ;136 ;3892-900
21. Gertler A, Petridou B, Kriwi GG, Djiane J. Interaction of lactogenic hormones with purified recombinant extracellular domain of rabbit prolactin receptor expressed in insect cells. FEBS Lett 1993 ;319 ;277-81
22. Hooper KP, Padmanabhan R, Ebner KE. Expression of the extracellular domain of the rat liver prolactin receptor and its interaction with ovine prolactin. 1993 J Biol Chem ;268 ;22347-52
23. Goffin V, Shivericck KT,, Kelly PA, Martial JA. Sequence-function relationships within the expanding family of prolactin, growth hormone, placental lactogen and related proteins in mammals. Endocr Rev 1996 ;17 ;385-410
24. Goffin V, Kinet S, Ferrag F, Binart N, Martial JA, Kelly PA. Antagonistic properties of human prolactin analogs that show paradoxical agonistic activity in the Nb2 bioassay. J Biol Chem 1996 ;271 ;16573-9
25. Fuh G, Wells JA. Prolactin receptor antagonists that inhibit the growth of breast cancer cell lines. J Biol Chem 1995 ;270 ;13133-7
26. Nicoll CS, Bern H. On the actions of PRL among the vertebrates : is there a common denominator ? In : Wolstenholme G, Knight J editors. Lactogenic Hormone. London : Churchill Livingstone 1972 ;299-327
27. Ormandy CJ, Camus A, Barra J, Damotte D, Lucas BK, Buteau H, Edery M, Brousse N, Babinet C, Binart N, Kelly PA. Null mutation of the prolactin receptor gene produces multiple reproductive defects in the mouse. Genes Dev 1997; 11;167-78
28. Horseman ND, Zhao W, Montecino-Rodriguez E, Tanaka M, Nakashima K, Engle SJ, Smith F, Markoff E, Dorshkind K. Defective mammopoiesis, but normal hematopoiesis, in mice with atargeted disruption of the prolactin gene. EMBO J 1997 ;16 ;6926-35
29. Roblero L,Garavagno A. Effect of estradiol 17-beta and progesterone on oviductal transport and early development of mouse embryos. J Reprod Fertil 1979 ;57 ;91-5
30. Galosy S, Talamantes F. Luteotropic action of placental lactogens at mid pregnancy in the mouse. Endocrinology 1995 ;136 ;3993-4003
31. O'Neill C, Quinn P. Interaction of uterine flushings with mouse blastocysts *in vitro* as assessed by the incorporation of [3H] uridine. J Reprod Fertil 1981 ;62 ;257-62
32. Reese J, Binart N, Brown N, Ma WG, Paria BC, Das SK, Kelly PA, Dey SK. Implantation and decidualisation defects in prolactin receptor (PRLR)-deficient mice are mediated by ovarian but not uterine PRLR. Endocrinology 2000 ;141 ;1872-81
33. Hogan BL. Bone morphogenetic protein : multifunctional regulators of vertebrate development. Genes Dev 1996 ;10 ;1580-94
34. Cross NA, Hillman LS, Allen SH, Krause GF. Changes in bone mineral density and markers of bone remodeling during lactation and post weaning in women consuming high amounts of calcium. J Bone Miner Res 1995 ;10 ;1312-20
35. Schlechte JA. Clinical impact of hyperprolactinemia. Baillere Clin Endocrinol Metab 1995 ;9 ;359-66

36. Sowers MF, Hollis BW, Shapiro B, Randolph J, Janney CA, Zhang D, Schork A, Crutchfield M, Syanczyk F, Russel-Aulet M. Elevated levels of parathyroid hormone-related peptide associated with lactation and bone density loss. JAMA 1996 ;276 ;549-54
37. Lopez JM, Gonzalez G, Reyes V, Campino C, Diaz S. Bone turnover and density in healthy women during breastfeeding and after weaning. Osteoporos Int 1997 ;6 ;153-9
38. Freemark M, Driscoll P, Maaskant R, Petryk A, Kelly PA. Ontogenesis of prolactin receptors in the human fetus in early gestation. Implications for tissue differntiation and development. J Clin Invest 1997 ;99 ;1107-17
39. Bataille-Simoneau N, Gerland K, Chappard D, Basle MF,Mercier L. Expression of prolactin receptors in human osteosarcoma cells. Biochem Biophys Res Commun 1996 ;229 ;323-8
40. Clement-Lacroix P, Ormandy C, Lepescheux L, Ammann P, Damotte D, Goffin V, Bouchard B, Amling M, Gaillard-Kelly M, Binart N, Baron R, Kelly PA. Osteoblasts are a new target for prolactin/analysis of bone formation in prolactin receptor knockout mice. Endocrinology 1999 ;140 ;96-105
41. Bridges RS, Numan M, Ronsheim PM, Mann PE, Lupini CE. Central prolactin infusions stimulate maternal behavior in steroid-treated,nulliparous female rats. Proc Natl Acad Sci USA 1990 ;87 ;8003-7
42. Bridges RS Dibiase R, Loundes DD, Doherty PC. Prolactin stimulation of maternal behavior in female rats. Science 1985 ;227 ;782-4
43. Lucas BK, Ormandy C,Binart N, Bridges RS, Kelly PA. Null mutation of the prolactin receptor gene produces a defect in maternal behavior. Endocrinology 1998 ;139 ;4102-7
44. Crawley JN, Paylor R. Aproposed test battery and constellations of specific behavioral paradigms to investigate the behavioral phenotypes of transgenic and knockout mice. Horm Behav 1997 ;31 ;197-211

Chapter 18

Prolactin Receptor Signal Transduction

Charles V. Clevenger, Michael A. Rycyzyn, Farhat Syed, and J. Bradford Kline

University of Pennsylvania Medical Center, Philadelphia, PA

INTRODUCTION

The pleiotropic actions of prolactin (PRL) are mediated by ligand-induced dimerization of its receptor, the PRLR. Structural analysis of the PRLR has revealed significant homology to other members of the cytokine receptor family, including the receptors for growth hormone (GH), erythropoietin, granulocyte-macrophage colony stimulating factor (GM-CSF) and interleukins 2-7 (1,2). While the amino acid residues found within the extracellular domain of the PRLR are critical for its engagement of ligand, the structure of the PRLR intracellular domain also coordinates the interaction of several signaling networks, that orchestrate PRL induced gene expression (recent related overviews on this subject can be found in (3-6). It is the purpose of this chapter to examine the association and activation of these multimeric PRLR-associated signaling complexes at a molecular level. In doing so, several recently identified kinases, phosphatases, exchange factors, chaperones, isomerases, GTP-binding and adaptor/docking proteins will be overviewed. Finally, how these transduction networks act to mediate the specific signal initiated by PRL at both the genomic and non-genomic level will be discussed.

INSIGHTS FROM THE HUMAN PRLR ISOFORMS – STRUCTURAL AND KINETIC CONSIDERATIONS MODULATE SIGNALING

PRL exerts its effects at the molecular level by inducing the dimerization of its type 1 transmembrane receptor, the PRLR (7). This event initiates a cascade of signaling events that induces the proliferation and differentiation of lobular units (8,9) in mammary tissues as a prelude to lactation. While the diversity of PRL function in other organs is mediated by the alternative utilization of signaling cascades unique to each tissue, differences in tissue-specific PRL action may also be attributed to the variety of PRLR isoforms observed in nature. Several isoforms of the PRLR have been identified in both mammals (10-16) and birds (17-19). The most well characterized isoforms are those found in the rat: the short form (45 kDa) (20), long form (80-85 kDa) (21), and a mutant intermediate form found on the PRL-dependent rat T cell lymphoma line Nb2 (65 kDa) (21,22). Until recently, the only human PRLR isoform characterized thus far was the

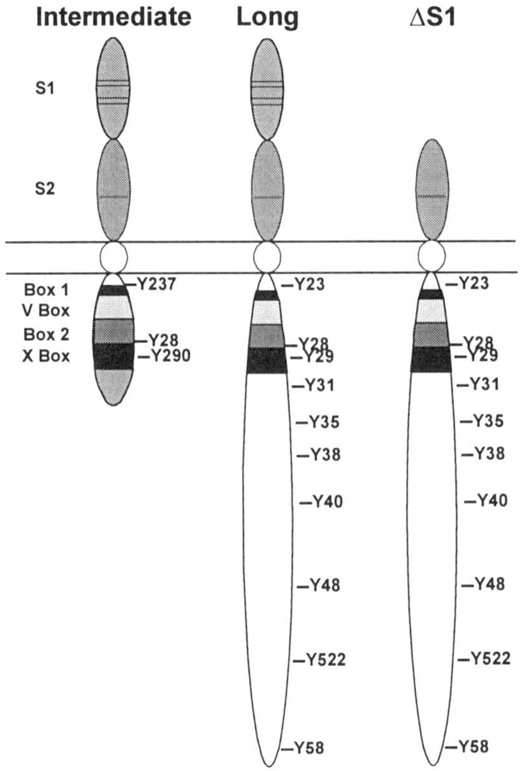

Figure 18-1. Human PRLR Isoform Structure. The three human isoforms of the PRLR found in normal and malignant human breast tissues consist of the initially cloned long isoform and two novel isoforms recently identified and fully sequenced by our laboratory. The intermediate isoform mRNA results from an alternative splice within exon 10 of the PRLR locus, resulting in a frameshift. This mRNA encodes for a unique 13 amino acid tail (indicated by darkest box), terminated by a consensus stop codon. The ΔS1 isoform is also generated from alternative splice, resulting in the deletion of exons 4 and 5, producing an isoform that completely lacks its S1 domain. The removal of the S1 domain removes three of five ligand interaction motifs; however two ligand interaction motifs are retained, as well as the receptor-receptor interaction motif remain. Tyrosine residues in the intracellular domain are denoted with a "Y".

long form cloned from the liver (23). Previous studies have, however, provided evidence that other human PRLR isoforms may be expressed in human tissues (24,25). This was confirmed in our lab, as two new human PRLR isoforms have been identified by RT-PCR, namely the human intermediate PRLR (26) and the human ΔS1 PRLR (Kline and Clevenger, manuscript in preparation).

Unlike its rat counterpart, the human intermediate PRLR results from an RNA processing event occurring at a consensus splice junction. This isoform was termed the intermediate PRLR because of the similarities it shares with the rat intermediate PRLR in both the proximity of the splice junctions and the extent of the nucleotide deletions observed. Unlike the rat intermediate PRLR, the splice also induces a frameshift after residue 312, causing the addition of 13 heterologous amino acids and a premature stop codon, resulting in a glycosylated transmembrane receptor of approximately 50 kDa. As a member of the cytokine receptor superfamily, the intracytoplasmic domain of the long PRLR contains the membrane proximal Box 1 and Box 2 motifs. Box 1 is a hydrophobic, proline-rich region that resembles an SH3 binding domain (27-29). The Box 2 domain is hydrophobic and acidic, and its signaling function is largely uncharacterized. Like the long form, the human intermediate PRLR contains the Box 1 and Box 2 motifs. However, due to the frame shift, C-terminal regions of the long PRLR are absent in the intermediate PRLR. This may contribute to differences observed in proximal signaling capabilities of the two isoforms. Ba/F3 cells transfected with the intermediate isoform exhibited no proliferation at physiological concentrations of PRL and only modest proliferation at pharmacologic concentrations of PRL. Ba/F3 cells transfected with the long PRLR, in comparison, demonstrated robust proliferation to physiologic concentrations of PRL (26). In contrast, the enhancement of cell survival over a 72-hour period was comparable between the intermediate and long PRLR. Taken together, these results suggest that fundamental signaling differences exist between these two isoforms. These results were found to include differences in proximal signaling. Addition of PRL to intermediate PRLR-expressing transfectants resulted in the activation of Jak2, while the activation of Fyn was severely diminished. This was in contrast to the human long PRLR, which induced the activation of both Jak2 and Fyn (26). These proximal signaling differences may help explain the physiological characteristics attributable to cells expressing the intermediate PRLR. Previous mutational analysis indicated that residues in the C-terminus may be necessary for maximal signaling and proliferation due to their associations with Stat5 and Fyn (26,30-33). This is significant in that the human intermediate PRLR is truncated by a frameshift upstream of these tyrosine residues.

In contrast to the intermediate PRLR, the intracellular domain of the ΔS1 PRLR isoform is homologous to the long PRLR. However, due to alternative RNA splicing, the ΔS1 isoform lacks exons 4 and 5 that encode the first fibronectin-like motif within the extracellular domain. Splicing of the extracellular domain is not new, as previous studies have observed this phenomenon in birds (18,19,34). As expected, this splicing has significant consequences on ligand binding, as approximately 66% of PRL contact residues are deleted in this isoform. Real-time kinetic analysis using surface plasmon resonance bears this out, as the overall affinity of the ΔS1 isoform for PRL is 36-fold less than that of the long PRLR. More specifically, the k_{on} time for ligand association is longer, while the k_{off} time is shorter. Counterintuitively, the altered kinetics of association with ligand observed in the ΔS1 isoform resulted in a more effective transducer. This effect has been observed in our laboratory in PRL-stimulated COS-1 cells transiently transfected with either the ΔS1 or long PRLR in conjunction with Jak2 or Fyn. While transfectants expressing the long PRLR showed maximal activation of Jak2 and Fyn after 10 minutes, transfectants expressing the ΔS1 PRLR exhibited maximal Jak2 and Fyn activation after 5 minutes. Thus, the ΔS1 isoform with weaker ligand binding exhibited faster signaling capabilities. This fits well with the latest models of PRL-PRLR interactions, which suggest the relative binding contributions of each receptor of the homodimerized PRL/PRLR complex may be more crucial than the overall affinity of PRL for both receptors (35).

As most of the studies of the PRLR isoforms illustrated here have been performed *in vitro*, it is important to understand the significance of PRLR isoforms in a physiological sense. Examination of mRNA levels of all three human PRLR isoforms across a wide array of tissues revealed variable levels of PRLR isoform expression, suggesting different possible roles for the isoforms within different organs. Studies of PRLR isoform expression in other species have shown they are not equally represented in all tissues and may regulate the effects of PRL through both hetero- and homodimerization. In the mouse, for example, examination of the transcripts of both short and long isoforms showed variation of expression in a tissue-specific manner, depending on the stage of estrous cycle, pregnancy, and lactation (36,37). Thus, the observed variability in expression of the long, intermediate, and ΔS1 PRLR isoforms may represent yet another physiologic mechanism through which the tissue-specific actions of the PRL-PRLR complex are regulated in human tissues.

JAK FAMILY OF TYROSINE KINASES

First identified through the use of RT-PCR screening techniques, the Jak family of protein tyrosine kinases consists of four members, Jak1, Jak2, Jak3, and Tyk2 (38,39). All members of the cytokine receptor superfamily appear to activate one or another of these kinases shortly after stimulation with ligand. The activation of Jak2 by the receptors for erythropoietin (40), growth hormone (41), and PRL (42,43) occurs rapidly after ligand binding. The critical, non-redundant role of Jak2 has been made evident by studies mutating the *Jak2* locus (44). Such genetic disruption resulted in embryonic lethality in Jak2 deficient mice, and the inability of such embryonic tissues to respond to erythropoietin, thrombopoietin, interleukin-3, granulocyte-colony stimulating factor, and γ-interferon. As discussed in greater detail below, loss of Jak2-signaling also precludes PRL-induced gene expression. Thus, the Jak family of kinases are important early mediators of PRLR and cytokine receptor signaling.

Structurally, the Jak kinases demonstrate molecular masses of 120-140 kiloDaltons and identity at the amino acid level of 36-47%. Most of the homology among Jak family members resides in one of seven Jak homology (JH) domains (39). JH1, the most C-terminal, contains the tyrosine kinase domain, while JH2 represents a pseudokinase domain. The function of JH3-7 remain poorly understood, however, mutagenesis studies have revealed that JH5-7 are necessary for the interaction of Jak2 with the growth hormone receptor (45). Unlike members of the Src or Tec family of tyrosine kinases, Jak kinases lack other well-recognized interaction motifs, such as Src-homology-2 (SH2), SH3, or pleckstrin homology (PH) domains. Jak kinases contain several tyrosine residues. These residues undergo both phosphorylation and dephosphorylation during Jak activation, and are thought to modulate the overall kinase activity and the interactions of the activated Jak. However, the functional significance of tyrosines on a per residue basis within the Jak kinase family (with the exception of Y^{1007} in Jak2, as discussed below) remains to be established.

Both Jak1 and Jak2 interact with the PRLR (43,46,47). The transfection of mutant cell lines lacking either Jak1 or Jak2 with a β-lactoglobulin promoter-CAT construct has revealed that Jak2, but not Jak1, is necessary for PRL-induced gene expression from this reporter (48). Unlike the growth hormone receptor, where the association between the receptor and Jak2 is induced by ligand binding (41), Jak2 appears to be constitutively associated with the PRLR (43,46,47). Thus, association per se with the PRLR is not sufficient for Jak2 activation, and would argue that other structural alterations (i.e. receptor dimerization) or protein modifications (i.e. phosphorylation/dephosphorylation) are necessary to activate PRLR-associated Jak2. Several lines of evidence indicate that two Jak2 proteins need to associate with the ligand-engaged PRLR complex for effective functional activation. First, studies with heterodimerizing cytokine receptors, such as those for interferon α and γ, and IL2, that bind and activate heterologous pairs of Jak kinases (i.e. IFNα - Jak1/Tyk2, IFNγ - Jak1/2, IL2 - Jak2/3), have revealed that loss of either member of the kinase pair results in functional inactivation of the receptor (49,50). Use of chimeras between the extracellular domain of the granulocyte-macrophage colony-stimulating factor receptor and the intracellular domain of the PRLR, have also revealed that two intact Jak2-association domains are required for effective PRL-driven signaling and proliferation (51).

The association of Jak2, and its subsequent activation by ligand-induced PRLR dimerization, remains under study. Jak2 activation occurs rapidly after PRLR dimerization, with detectable autophosphorylation occurring within 0.5-1.0 minute, reaching maximal levels after 5-10 minutes (43,46,47). It is known that the membrane-proximal, proline-rich Box 1 motif in the PRLR is necessary for Jak2 association (52). This motif, containing the amino acid residues -IFPPVPGP-, is highly conserved across members of the cytokine receptor superfamily. Mutational analysis of this domain, affected by introducing replacement alanines into each of the amino acid residues within this domain in the PRLR has revealed that the C-terminal proline is particularly critical for Jak2 association (53). Whether the Box 1 motif is sufficient for Jak2 activation, however, has been debated. Overexpression of both rat PRLR mutants and Jak2 in the 293LA line has revealed a requirement for only the Box 1 motif during PRL-induced Jak2 activation (54). The high levels of Jak2 found in this transfected line however somewhat cloud the obtained results, as Jak2 is known to autoactivate at high concentrations. Other studies have suggested that other elements in the PRLR C-terminal to the Box 1, such as the Variable Box ("V-box") and Box 2 motifs also contribute to Jak2 activation (33,51,55). These studies have utilized cell lines that do not overexpress Jak2, suggesting that cellular Jak2 concentrations can influence the association of this kinase with the PRLR. The V-Box represents a region of low homology between cytokine receptors immediately C-terminal to the Box 1 motif. The Box 2 motif resides C-terminal to the V-box, is highly conserved between cytokine receptors, and consists of hydrophobic and acidic residues. Like the PRLR, structural mutagenesis of the growth hormone receptor indicates the importance of both the Box 1 and Box 2 motifs to Jak2 engagement and activation (56,57). C-terminal tyrosine residues known to be phosphorylated during PRL stimulation may further enhance Jak2 activation. When these tyrosines were replaced with phenylalanine residues in a PRLR chimera, dramatically reduced levels of Jak2 activation were noted in Ba/F3 transfectants (33). The precise mechanisms through which Jak2 associates with the PRLR, i.e. which residues within Jak2 are necessary for PRLR interaction, however, remains open to additional study. How PRLR dimerization serves to activate the juxtaposed, PRLR-associated Jak2 proteins is also poorly understood, although both autophosphorylation (58) and dephosphorylation (see the section on SHP-2 phosphatase below) contribute to such activation.

The precise role of Jak2 in PRLR action remains open to interpretation. While all data clearly indicate that Jak2 activity, as induced by PRLR dimerization, is necessary for PRL action (48), whether Jak2 activation is sufficient for all other PRLR-associated signaling is highly controversial. Some groups have argued that Jak2 activation is both necessary and sufficient for the mediation of signals originating from associated cytokine receptors (44). However, the majority of studies involving the PRLR indicate that other proximally-activated signaling cascades, not directly influenced by Jak2 activity, also contribute to PRL-induced gene expression (26,59,60). As discussed below, signaling cascades activated by PRLR-associated Src family members appear to be independent of Jak2 activity. Correspondingly, PRLR mutants lacking the C-terminus of their intracellular domain fail to demonstrate PRL induced gene-expression (59). While this may in part be due to the loss of phosphorylated tyrosine residues required for Stat5 engagement, the loss of interaction with other signaling proteins such as Tec, Vav, and Fyn may also contribute to this phenotype (26).

Three major substrates for the catalytic activity of Jak2 are the PRLR, Stat5, and Jak2 itself. As discussed above, Jak2 rapidly autophosphorylates after receptor dimerization, resulting in upregulated activity. This results in additional phosphorylation of the C-terminal tyrosine residues in the PRLR and Stat5. The phosphorylation of the PRLR C-terminal tyrosines is necessary for the association and subsequent tyrosine phosphorylation/activation of the transcription factor Stat5 (the structure and function of Stat5 are discussed in detail in Chapter 19). These tyrosine "docking" sites within the PRLR demonstrate little homology, save for the relative conservation of a hydrophobic amino acid in the +1 position (61). However, other studies (62) have shown that PRLR mutants truncated in their C-terminus were perfectly capable of Stat5 association and phosphorylation. Indeed, one study has suggested that the C-terminal tyrosines are not required for Stat5 phosphorylation, but instead contribute to the nuclear translocation of Stat5 through an unspecified mechanism (63). These findings would indicate that Stat5 may also associate with the PRLR/Jak2 complex via interaction with phosphorylated tyrosine residues in Jak2. Taken together, these data would suggest that the formation of the PRLR/Jak2/Stat5 complex is cell type dependent, and may rely on mutual interaction between each of these proteins, perhaps resulting in an overall increase in the affinity of the individual members of this complex for each other. As a closing note, it should be mentioned that Jak2 is a very "promiscuous" kinase demonstrating little specificity for the residues surrounding the tyrosines it phosphorylates. Indeed, many other PRLR-associated signaling proteins, such as SHP2 (64) and Tec (65), may serve as downstream substrates.

SHP-2 PHOSPHATASE

Tyrosine phosphorylation can have profound up- and down-regulatory effects on the kinase activity of members of the Jak, Src, and Tec tyrosine kinase families. An important regulator of tyrosine dephosphorylation for both Jak1 and 2 is the SH2 containing phosphatase SHP-2 (also known as Syp, PTP1D, SHPTP-2, SHPTP-3, and PTP2C). SHP-2 is 72 kiloDaltons in size, containing two SH2 domains in its N-terminus, and a consensus phosphatase in its C-terminus (66). SHP-2 is a member of SH2-containing tyrosine phosphatases, including SHP-1, and the *Drosophila* homologue Csw, which demonstrate 55-75% amino acid identity in their conserved domains.

SHP-2 has been implicated in the upregulation of signaling from several members of the cytokine receptor family, through its cleavage of inhibitory phosphotyrosine residues (67). In terms of PRLR-mediated signaling, SHP-2 was found to enhance PRLR-associated Jak2 mediated signaling in 293 and Cos transfectant model systems (68,69). Indeed, the phosphatase activity of SHP-2 is necessary for the tyrosine phosphorylation and functional activation of the Stat5 transcription factor (69). SHP-2 is itself phosphorylated on tyrosine residues during PRLR signaling by Jak2 (68). This phosphorylation occurs on tyrosines Y304 and Y327 within the phosphatase domain of SHP-2 (64) and increases the phosphatase activity of SHP-2. Inactivation, via point mutagenesis, of the phosphatase domain within SHP-2, blocks the upregulation of Jak2 kinase. While the SH2 of SHP-2 may modulate the binding of Jak2 to SHP-2 (68), the interaction between these proteins appears to be mediated by the N-terminal JH domains of Jak2 and a forty amino acid region in the N-terminal region of the phosphatase domain of SHP-2 (64).

Several questions remain regarding the association, activation, and role of SHP-2 in PRLR signaling. The question of how SHP-2 itself is activated remains something of a conundrum, i.e. if SHP-2 is required for full Jak2 activity and is both phosphorylated by Jak2, how is it that SHP-2 itself is activated? As discussed below, other members of the Src family of tyrosine kinases could conceivably contribute to such activation. Alternatively, low levels of spontaneous Jak2 activity, induced by the juxtaposition of two Jak2 molecules as a consequence of PRL-induced receptor dimerization, could perhaps account for the activation of SHP-2. Another remaining question relates to the function of the two SH2 domains in SHP-2 – if they are not required for Jak interaction, what is their function? Some recent evidence suggests that SHP-2 may utilize these SH2 motifs to induce multimerization of signaling complexes. Indeed SHP-2, in addition to Jak1 and Jak2, has been noted to bind Grb2 (64) and the transmembrane glycoprotein SHPS-1 (for SHP-substrate-1, also known as SIRPα) (70). The binding of SHPS-1 to SHP-2 is regulated by tyrosine phosphorylation; this phosphorylation, in turn is regulated by integrin-receptor signaling (71,72). Thus, a larger complex between SHPS-1/SHP-2/PRLR could serve to integrate signals between the PRLR complex and signals from the extracellular matrix.

SOCS FAMILY OF SUPPRESSORS

The SOCS (for Suppressors of cytokine signaling) family represents a unique family of modulators of cytokine signaling. Ranging in size from 42 to 65 kiloDaltons, members of this family (CIS and SOCS 1-7) are related by structural homologies (73,74). In resting, non-cytokine stimulated cells, expression of the mRNA and corresponding proteins are very low; within one hour of cytokine stimulation rapid induction of mRNA and protein can be seen. In PRL-responsive cells, SOCS1 and 3 are transiently induced, while the mRNA encoding for SOCS2 and CIS remain elevated up to 24 hours after PRL stimulation (75). Each member contains a central consensus SH2 and a C-terminal region of homology, termed the SOCS box. Functionally, the SOCS differ in their effects on PRLR signaling; in general most studies have demonstrated that SOCS 1 and 3 inhibit PRLR signaling, by blocking Jak and/or Stat phosphorylation, while SOCS 2 and CIS block the inhibitory effects of SOCS 1 and 3 (75-77). Both SOCS1 and 3 markedly inhibit PRL-induced Jak2 kinase activity in 293 transfectants (75). It has been shown that SOCS 1 associates with PRLR-associated Jak2 (75), and appears to do so through binding of its SH2 motif to phospho-Y^{1007} within Jak2 (73). In contrast, SOCS2 appears to bind the PRLR directly, although the motifs in SOCS2 required for this interaction are incompletely characterized, as are the PRLR elements involved in engaging SOCS 2 and CIS. Recent studies with the growth hormone receptor, however, would indicate that SOCS3 binds to membrane proximal growth hormone receptor phosphotyrosines, while SOCS2 and CIS bind more distal phosphotyrosines (78). These data would indicate that Jak inhibition by SOCS proteins occurs either as a binary complex between JAK-SOCS or as a ternary complex constituted of JAK-SOCS-receptor.

How the SOCS proteins inhibit Jak kinase activity remain an area of active investigation, however, recent data would indicate that this occurs through multiple interactions between the Jak kinase activation loop, specifically phospho-Y^{1007}, the SH2 motif of SOCS proteins, as well as the 25 amino acids N-terminal to the SOCS SH2 motif (79). These data would suggest that SOCS interaction with the activation loop blocks access of either substrates and/or ATP to the catalytic pocket of Jak2. The conserved C-terminal "SOCS-box" motif does not appear necessary for inhibition of Jak2 activity; instead this element may serve to protect the SOCS protein from proteolytic degradation (73).

SRC FAMILY OF TYROSINE KINASES

The observation of PRLR utilization of the Src family of protein tyrosine kinases was made concurrently with the identification of the PRLR/Jak2 association (80). The Src family of protein tyrosine kinases consist of the tyrosine kinases Src, Lyn, Lck, Fyn, Yes, Hck, Fgr, and Yrk, ranging in size from 53 to 62 kiloDaltons (81,82). Structurally, the proteins share in common three prototypical Src homology (SH) domains that consist of the SH1 tyrosine kinase domain, the SH2 phosphotyrosine binding domain, and the SH3 poly-proline binding domain. These SH domains demonstrate 60-75% amino acid identity between Src family members. The N-terminal of individual Src family members is

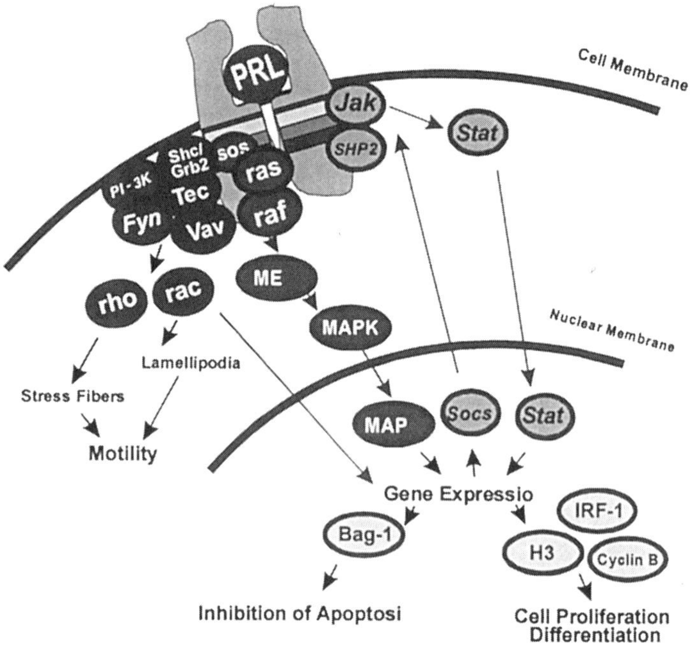

Figure 18-2. PRLR Signal Transduction. PRLR signal transduction activates multiple signaling pathways. Ligand induced dimerization of the PRLR complex induces the association and activation of several signaling networks including the JAK/Stat/Socs, Bag, Fyn, Shc/Grb2/Sos/Ras/Raf/MEK/MAPK, Fyn and PI3K/Tec/Vav/Rac/Rho pathways.

unique, but each is myristolated at a glycine residue in the second position. The combined effect of myristolation and the interaction of SH2/3 motifs, result in the localization of Src family kinases to the cell membrane and close proximity to transmembrane receptors. Phosphorylation of the C-terminal tyrosine (Y527) of Src family members (by tyrosine kinases such as Csk) induces the binding of the phosphotyrosine by the internal SH2 motif, resulting in kinase inactivation (i.e. the kinase folds on itself). Conversely, autophosphorylation (after dephosphorylation of Y527) of internal tyrosines (i.e. Y416) upregulates tyrosine kinase activity of Src. Although numerous, the substrates of Src family kinase activity (i.e. ZAP70, phosphatidylinositol-3-kinase, phospholipase C-γ, etc.) are of a more restricted nature than those observed for the Jak family of kinases. Interestingly, recent data has also indicated that Src is necessary for ErbB receptor-associated Stat5 activation, and may serve as an activator of the Jak kinases (83).

The first Src-family member noted in association with the PRLR was the $p59^{Fyn}$ tyrosine kinase in the PRL-dependent Nb2 T cell line. The kinetics of activation of Fyn by PRL parallel those of Jak2 activation with a maxima of autophosphorylation achieved within 10-20 minutes of ligand stimulation. In the Nb2 cell, Fyn is constitutively associated with the PRLR, an observation that would argue against an interactive role for the Fyn SH2 motif. Subsequent studies have also revealed that the $p60^{Src}$ interacts with the PRLR in liver (84) and in breast tissues (Clevenger, unpublished observations). The molecular basis for Src-family association with the PRLR and its subsequent activation remains to be detailed, however, recent data in chicken embryo fibroblast and in Cos-1 transfectants would indicate that PRLR-associated Src-family activation is independent of the association/activation of Jak2 with the PRLR (60,85). While the C-terminus of the PRLR appears to be necessary for activation of Fyn (26,33), the role of this domain in the association of Fyn with the PRLR is currently under investigation by our laboratory. It should also be mentioned that the PRLR/Fyn complex is most likely multimeric in nature, as Src-family members have been reported to associate with other PRLR-associated signaling proteins such as Vav, PI3K, and Tec family members (86-89). Indeed, even Stat5B has been reported to associate with and undergo phosphorylation and nuclear translocation in response to Src activation (85).

TEC FAMILY OF TYROSINE KINASES

The 70 kDa protein tyrosine kinase Tec belongs to an emerging subfamily of signaling molecules encompassing Tec, Btk, Itk/Tsk/Emt, Rlk/Txk, and Bmx (90). The structural features of Tec kinases include a pleckstrin homology (PH) domain (91), a Tec homology domain (92), as well as a kinase domain and C-terminal SH2 and SH3 motifs. Tec is activated through a variety of cytokine receptor superfamily members including those for IL-2 (93,94), IL-6 (95), GM-CSF (65), erythropoietin (94), and G-CSF (96). It has also been shown to be activated via lymphocyte receptors such as the T cell receptor and CD28 (97). Tec has been found to contribute with Jak2 to the activation of several downstream transduction / transcription factors including Rho (98) and c-fos (65).

Recent studies in our lab have indicated that Tec contributes to proximal PRLR signaling, as stimulation of responsive cell lines with PRL resulted in the phosphorylation of Tec kinase within five minutes of ligand binding. As signaling cascades are complex and may involve crosstalk between many differing pathways, it comes as little surprise that the association between the PRLR and Tec may constitute part of a larger signal protein complex. Studies have previously revealed the activation and association of the guanine nucleotide exchange factor (GEF) $p95^{Vav}$ with the PRLR (99), while others have observed an interaction between Tec and Vav (94-96). Expanding upon these observations, our lab has recently observed the formation of a ternary PRLR/Tec/Vav complex upon ligand stimulation (Kline and Clevenger, submitted). In COS-1 cells transfected with cDNAs encoding all three proteins, Tec and Vav form a constitutive association. When PRL is added to the cells the Vav/Tec complex associates with the PRLR, and Tec is activated. The structural basis for the interaction of the PRLR with Tec and Vav was further examined through the use of GST-chimeras containing portions of the intracellular domain of the PRLR. These studies revealed that both Tec and Vav interact with a central region (amino acids 323 to 527) within the human PRLR C-terminal to the Box1/Variable Box/Box2 motifs. Further mapping revealed a bipartite binding site for both Vav and Tec in each half of this domain (amino acids 323 to 425 and 425 to 527). Analysis of this domain has revealed a conserved motif in each of these subdomains (W(L/P)LPQ), a structural element also found in the receptors for erythropoietin, platelet-derived growth factor, and serotonin.

Our knowledge of the interactions of the Vav/Tec complex with other PRLR-associated signaling proteins continues to evolve. Members of both the Jak and Src kinase families are known to bind, phosphorylate, and upregulate the kinase activity of Tec family members (65,93,100). Additionally, negative regulation of Tec is in part mediated by the SOCS proteins, as SOCS1 binds to Tec and down regulates its kinase activity (101), an action that may contribute to the SOCS1 mediated inhibition of PRL/PRLR action (102). Indeed, the proximal role of Tec in PRLR-mediated signal transduction is increasingly multifaceted, and further studies will be required to determine how the interactions between Tec-associated proteins may modulate receptor associated signaling.

VAV FAMILY OF GUANINE NUCLEOTIDE EXCHANGE FACTORS

The Vav family consists of three members, Vav1, Vav2, and Vav3. As outlined below, this family contains multiple protein and phosphoinositide interaction motifs, and mediates the exchange of guanine nucleotides on small G-proteins. Our initial observations of the exchange of GDP for GTP on the small G-proteins Rac and Ras led us to question the involvement of Vav family members during PRLR signaling. Subsequent analysis by our laboratory revealed that most of the PRL-induced guanine nucleotide exchange factor (GEF) activity within the Nb2 T-cell model system was within anti-Vav1 immunoprecipitates. The activation of Vav1 occurred very rapidly (within 5 minutes), returning to baseline by 10-20 minutes. Concomitant with this PRL-induced activation three other phenomenon were observed: 1) a five-fold increase in Vav1 tyrosine phosphorylation, 2) a PRL-induced association between Vav and the PRLR, and 3) after 10-20 minutes of PRL stimulation a translocation of Vav into the nucleus occurred. The expression of Vav1 was shown to be necessary for PRL-driven proliferation, as the ablation of endogenous Vav1 expression through the use of antisense oligonucleotides inhibited PRL-driven proliferation in the PRL-dependent Nb2 T cell line (103). Thus, Vav1 represents a critical proximate component in PRLR transduction in T cells.

Analysis of Vav family structure provides insight into its role during PRLR-associated signaling, and these structures are highly conserved between Vav 1-3. Vav family members contain multiple functional domains with high levels of homology between family members. In its amino terminal, two domains with homology to a helix-loop-helix-like motif and a leucine zipper domain have been identified; these domains may serve to facilitate the interaction of Vav1 with other proteins and suppress an intrinsic transforming capability of this protein. Overlapping this region is a calponin homology domain, which may facilitate the interaction of Vav1 with filamentous actin. Centrally located within Vav are two nuclear localization motifs, a cysteine-rich zinc finger-like site, a Dbl-GEF enzymatic site, and a pleckstrin homology (PH) domain. The Dbl-GEF site serves to facilitate the exchange of GDP off for GTP on the small G-proteins RhoA, Rac-1 and Cdc-42, thereby activating these downstream effectors (104). The PH domain binds phosphoinositide metabolites; in doing so it upregulates Vav GEF activity and concentrates Vav1 to regions of high lipid concentration (i.e. the plasma membrane). Finally, in the C-terminus of Vav1 reside one SH2 and two SH3 domains, with striking overall homology to the signal transduction adaptor protein Grb-2. These domains, which respectively bind phosphotyrosine residues and proline rich regions, enable many of the reported interactions of Vav1 with signaling factors such as SLP76, Lck, Fyn, Nck, Zap-70, Cbl, and Syk (87,105-109). In terms of PRLR signaling, both an interaction of Vav1 with the tyrosine kinases Fyn (33) and Tec have been observed by our laboratory following PRL stimulation.

The presence of several structural domains within Vav1 clearly indicates that Vav1 is involved in the integration of multiple signaling inputs. Vav1 is widely expressed in the mature lymphoid, myeloid, and erythroid lineages and selectively in their differentiating precursors (110-114). The phosphorylation of Vav1 has been observed after stimulation of the T-cell receptor (115), membrane immunoglobulin (116), Fc receptor (117), interferon α receptor (118), c-*kit* (119), PRLR (103) and the receptors for interleukins (IL)-1 (120) and -2 (121). Knockout mice lacking the Vav1 locus are immunosuppressed (122) and demonstrate defective T cell signaling, proliferation, and development. Taken together, these data indicate that Vav1 expression contributes to immune system function, mediating critical growth and differentiation signals from ligand-stimulated hematopoietic receptors.

The expression of Vav1 is largely limited to cells of the immune system. This limited expression, despite widespread expression of Rho family members throughout the body, has led several groups to identify related Rho-GEF proteins expressed in other tissues. Two new homologous family members have been identified Vav2 and Vav3 (123,124). Both Vav2 and Vav 3 are widely expressed in many, but not all, epithelial tissues and demonstrate 50-55% identity to Vav1 at the amino acid level. Vav2 and 3 are functionally distinct from Vav1, in that they serve as GEFs for RhoA and RhoG, but not Rac (in contrast, Vav1 activates both Rac and RhoA). Thus, alternative Vav family members probably participate in PRLR signaling in non-hematopoietic tissues.

Activation of Vav1 may involve tyrosine phosphorylation at position Y^{174} (109,125), presumably by members of the Src family of tyrosine kinase. The functional effect of phosphorylation at this site however has been a matter of some debate, as others have found that tyrosine phosphorylation at this site failed to stimulate GEF activity, but instead potentiated the oncogenic activity of Vav (109,125,126). Vav activity is also potentiated by the binding of the phosphoinositide metabolite 3,4,5-PIP_3 to its PH domain (127,128). The generation of phosphoinositides at the plasma membrane results from the activation of phosphoinositide 3-kinase (PI3K) by the allosteric actions of the G-protein ras. Thus, the regulation of Vav activity requires the integration of signaling from both Src family members (such as PRLR associated Fyn) and Ras/PI3K (129,130). How other receptor- and/or Vav-associated kinases, either tyrosine or serine threonine, contribute to and/or regulate Vav activity, however, is currently unknown.

The substrates of Vav1/2/3 activity are members of the G-protein Rho family (131,132). As described above Vav enhances the off rate for GDP bound to Rho family members, enabling the rapid binding of GTP (given the 10:1 GTP:GDP within the cytosol). GTP induces an allosteric shift in Rac/Rho structure enabling these proteins to interact with their downstream mediators. The mediators of activated Rac and Rho include a number of proteins involved in the regulation of the filamentous cytoskeleton including the kinases PAK and citron, and the motility protein IQGAP (133). Indeed associations between Vav and the actin membrane-anchoring proteins vinculin and talin have also been noted (134).

What are the physiologic consequences of Vav activation? The Vav knockout mouse has provided some clues. Aside from the T and B-cell defects noted in the vav^-/vav^- mice noted above, a profound deficiency in these animals in their ability to cap engaged T-cell receptors has been reported (135). This event is driven by the actin cytoskeleton and serves to cluster receptor complexes, an event that may modulate receptor signaling. This event also serves as a necessary prelude to receptor endocytosis, which functions to re-cycle or degrade receptor and internalize bound ligand. Manipulation of the downstream effectors of Vav family activity has also provided insights into Vav function. Indeed, both Rac and Rho are central regulators of the cytoskeleton, cellular morphology, and motility. Overexpression or enhanced activity of Rac induces the generation of cellular lamellipodia, while the overexpression of Rho leads to the generation of focal adhesions and stress fibers (131,136-138). In addition, the enhanced activity of either Vav or Rho family members is transforming and the increased expression of one Rho-family member (Tiam1) leads to an enhancement of the metastatic phenotype (139). Thus, the Vav family and its effectors are central transduction proteins integrating receptor-associated inputs, resulting in the modulation of the cytoskeleton, cellular motility and morphology, and receptor localization. Indeed, the association of Vav family members with the PRLR, led our laboratory to hypothesize and demonstrate that PRL stimulation could alter the cytoskeleton and motility of human breast cancer cells (140).

BAG-1 FAMILY OF PROTEIN CHAPERONES

Bag-1 (Bcl-2-associated gene product-1) encodes a 219 amino acid protein (141). Three forms of Bag-1 exist as a consequence of alternative RNA translation start sites, and include Bag-1, Bag-1M (46 kDa), and Bag-1L (58 kDa) (142,143). The larger Bag family (Bag-1 through Bag-5) shares homology in a C-terminal 47-amino acid Bag domain, a motif of unknown function. In addition, the Bag-1 isoforms contain a central ubiquitin-like motif, while the Bag-1L isoform contains a N-terminal nuclear localization motif. Bag-1 was identified originally by library screens for Bcl-2 interacting proteins (141). Indeed, Bag-1 does interact with Bcl-2 and potentiate its anti-apoptotic activity (144). With respect to the cytokine receptors, Bag-1 levels are intimately related to PRLR or IL3 stimulation, in that withdrawal of either ligand respectively from Nb2 or Ba/F3 cells results in a profound loss of Bag-1 protein well before the onset of apoptosis (145). Consistent with the hypothesis that Bag-1 contributes to PRL-mediated cell survival, is the observation that the PRL-independent, dexamethasone resistant, Nb2 subline SFJCD1 demonstrated no reduction in Bag-1 levels during PRL withdrawal or dexamethasone treatment. Overexpression of Bag-1 in Ba/F3 cells, mediated by transfection of a constitutive expression construct, rendered Ba/F3 growth factor independent, enabling both survival and proliferation in IL3 deficient medium. How Bag-1 mediates such survival signals in cytokine-dependent cell lines and in human tissues remains an intense area of study. Bag-1 is known to associate with Bcl-2 (141), Raf-1(144), the PRLR (Clevenger, unpublished data), the receptors for hepatocyte growth factor and platelet-derived growth factor (146), and the heat shock protein (hsp)-70 (Kanelakis *et al.* 1999). It is this association with hsp70 and glucocorticoid receptor folding mediated by the hsp90-based chaperone machinery that has led to the recent hypothesis that Bag-1 directly contributes to protein folding and recycling. Indeed, recent data have revealed that Bag-1 associates with the 26 S proteosome in HeLa cells via its ubiquitin-like domain and promotes binding of Hsc70 and Hsp70 to the proteolytic complex (147). Given that PRLR endocytosis, degradation, and recycling significantly modulate the activity of this receptor complex (148), Bag-1 may serve as an important link between PRLR folding and degradation.

OTHER PRL-ASSOCIATED SIGNALING PATHWAYS

ZAP-70. The 70 kiloDalton tyrosine kinase ζ-chain associated protein (ZAP-70) has been found to be inducibly associated and phosphorylated by the PRLR complex (149). The PRLR/Zap70 complex is multimeric as the exchange factor Vav1 can also be found in such complexes. As its name implies, Zap-70 was originally identified in association with the ζ-chain of the T-cell receptor (TCR), an interaction mediated by the phosphotyrosine residues in ITAM motifs of the TCR. Zap-70 is a member of the Syk family of tyrosine kinases and contains two N-terminal SH2 motifs and a C-terminal tyrosine kinase motif, whose activity is regulated by tyrosine phosphorylation, in a manner similar to Src-family kinases. Both Fyn and Zap-70 contribute to Vav activation (87,150,151). Thus, the cross-talk mediated by Zap-70 between the PRLR and the TCR may serve to regulate Vav guanine nucleotide exchange and the activity of its downstream substrates Rac and Rho (152).

Shc/Grb2/Sos/Rac/Raf/MEK/MAPK Pathway. Several members of this signaling cascade have been identified to associate inducibly with the PRLR complex in response to ligand stimulation (153-155). Consecutively consisting of two adaptor proteins, one guanine nucleotide exchange factor, a small G-protein, a serine-threonine kinase, and two dual-specificity kinases, activation of this cascade ultimately results in the nuclear translocation of MAPK and the consequent phosphorylation and activation of several transcription factors including Myc, Jun, and TCF (156). While signaling from this pathway invariably contributes to cell cycle progression, the signaling specificity of this pathway is limited in that virtually every integral cell membrane receptor can activate this cascade. Like Jak2 and Fyn, elements in the C-terminus of the PRLR may contribute to the activation of this cascade (154).

Phosphatidylinositol-3-kinase (PI3K). The generation of phosphoinositide metabolites that bind to pleckstrin homology domains in a variety of signaling molecules [i.e. see Vav above; (157,158)], occurs as a consequence of the activation of the PI3K. PI3K exists as a dimeric molecule, consisting of a 110 kiloDalton catalytic unit, and a 85 kiloDalton regulatory unit. Engagement of the p85 subunit by tyrosine kinase receptors and growth factor receptors can enhance the overall activity of PI3K. Binding of the p85 subunit is mediated by its intrinsic SH3 and two SH2 domains. The p85 PI3K subunit associates with the PRLR complex following PRL stimulation (88,159), resulting in the enhanced generation of the metabolites PI3,4,5P3 and PI3,4P2. PRLR-associated Fyn and Ras may contribute to the PRL-induced activation of PI3K (88,130). The association of PI3K with the PRLR may be enhanced by its interactions with the adaptor proteins Cbl (89) and insulin receptor substrate-1 (IRS-1; (159)). PRLR-associated PI3K activity may contribute to PRL-induced cell motility via Vav and its internal PH domain (140).

Others. Several other signaling proteins have been found to be associated and/or activated by the PRLR complex, including: protein kinase C (PKC; (160,161)), focal adhesion kinase (FAK; (162)), intracellular calcium (163,164), Jnk kinase (165), and the insulin receptor substrate-1 (IRS-1; (159)). The molecular basis for association and/or activation of these signaling proteins, however, remains to be determined.

PROLACTIN AS AN INTRANUCLEAR MEDIATOR OF ITS OWN SIGNAL – ROLE OF CYCLOPHILINS

The aforementioned signaling pathways are not uniquely associated with the PRLR and can be found in association with other members of the cytokine receptor superfamily, the protein tyrosine kinase receptor superfamily, and the seven-transmembrane spanning receptor superfamily. Thus, while the PRLR and its associated signaling cascades are indisputably necessary for PRL action, the pathways identified to date, cannot explain the pleiotropism of PRL action in tissues of the mammary gland and immune system (166-171). Therefore, how PRL (and indeed all cytokines) confer a specific signal while sharing signaling cascades has become a central question.

It is known that PRL is internalized within 30 minutes of its addition to cells expressing PRLR and can be found within the endosomes and multivesicular bodies (172,173). Thereafter, up to 10% of exogenously applied PRL can be found within the cell nucleus. To examine the biological significance of this finding, the signaling sequences within PRL were manipulated to target this hormone into the extracellular space, the cytosol, or the nucleus. This was accomplished by the transfection of PRL mutants lacking their ER leader sequence with or without an SV40 nuclear localization sequence. In the presence of IL2 and a neutralizing anti-PRL antiserum, only T-cells transfected with the PRL construct targeted into the nucleus demonstrated significant proliferation (174). These findings revealed that the intranuclear PRL retrotransport functionally contributed to proliferation, and was not an epi-phenomenon of cell growth. Translocation of PRL to the nucleus has been observed in several PRL-responsive tissues including the breast, T-lymphocytes, liver (160), adrenal (175), and ovary (176). Numerous other peptide hormones such as EGF, NGF, and PDGF (177), FGF (178), IL-5 (179), and insulin (180) have also been observed within the nucleus after their introduction into the extracellular medium. Thus, the nuclear retrotranslocation of peptide hormones is a widespread phenomenon that could regulate numerous physiologic processes.

How could extracellular hormones, such as PRL, undergo retrotranslocation into the nucleus and function to stimulate cell proliferation? Since PRL has neither an intrinsic enzymatic activity nor a recognizable nuclear localization signal sequence (NLS), it was reasoned that the nuclear retrotranslocation and function of PRL must be mediated by the interaction of PRL with other proteins. One such PRL-binding protein identified by yeast two-hybrid analysis, using PRL as "bait", was cyclophilin B (CypB). CypB is a 22kDa member of the greater immunophilin family comprised of CypA-D, Cyp40, FK506 binding protein and Tor (181). These immunophilins are a family of peptidyl-prolyl isomerases (PPI) that serve as protein chaperones and mediate the effects of cyclosporin A (CsA), FK506, and rapamycin respectively. In human tissues and cells, CypB is localized to the ER, nucleus, and extracellular space, and it is also found in breast milk and serum at concentrations of 150ng/ml (182).

Subsequent to the identification of a PRL-CypB interaction, the existence of this direct complex both *in vitro* and *in vivo* by biochemical and functional analysis was confirmed (183). Using purified recombinant proteins a direct interaction between CypB and PRL was demonstrated, a finding also confirmed *in vivo* analysis of human serum. Functional testing of CypB on a PRL-dependent cell line revealed a significant potentiation of ligand-driven growth. The exogenous addition of CypB at physiological concentrations into the defined medium of responsive cell lines potentiated PRL-driven proliferation up to eighteen-fold, while it was found not to contribute to proliferation driven by related members of the cytokine superfamily, such as IL-2 or IL-3. Thus, CypB could significantly contribute to the *in vivo* actions of lactogenic hormones.

How does CypB potentiate PRL action? Several hypothetical mechanisms are feasible, including the possibility that the CypB/PRL interaction facilitates the interaction of PRL with its receptor, or modulates the signal output from the PRLR. However, radioligand-binding studies followed by Scatchard analysis revealed that CypB did not alter the affinity of binding between PRL and its receptor, nor did CypB alter the activation of receptor-associated Jak2/STAT5 pathway. Collectively, these findings indicate that the potentiation of PRL action secondary to its interaction with CypB is most likely not occurring as the result of an alteration in PRLR structure/function.

An alternative mechanism of CypB-mediated potentiation is that CypB might facilitate the retrotransport of PRL into the nucleus. Given the previous data indicating that a nuclear retrotranslocation of PRL occurred during the IL-2-driven proliferation of T cells and the EGF-driven proliferation of breast cancer cells, the effects of CypB and PRL on internalization were tested. When co-stimulated with CypB, PRL internalization appeared to be quantitatively increased, resulting in a 5-10 fold increase in the number of cell nuclei demonstrating internalized PRL (i.e. virtually 100% of the cell nuclei revealed intense anti-PRL indirect immunofluorescence).

The mechanism through which CypB facilitated the nuclear retrotransport of PRL became more evident following a careful examination of the primary sequence of CypB. This analysis revealed a putative nuclear localization sequence (NLS) very similar to the known NLS of SV40 large T-antigen in the N-terminus of CypB. To test whether such a motif in CypB could directly contribute to the retrotranslocation of PRL, a CypB mutant, termed CypB-NT, was generated that lacked this NLS and its function tested in both human breast cancer cells (T47D) and rat T cells. This mutant exhibited the same PRL binding properties and enzymatic activity as wild-type CypB, however, the inclusion of CypB-NT into the culture medium resulted in the absence of detectable anti-PRL immunofluorescence within the nuclei of T47D cells with a corresponding loss of enhanced PRL-driven growth. In fact, inclusion of CypB-NT with CypB at equimolar concentrations completely abrogated the enhanced PRL-driven growth exhibited by CypB alone. This confirmed a role for the N-terminus of CypB in the nuclear retrotransport of PRL and the growth enhancing properties of nuclear PRL.

Given the preceding evidence that demonstrates CypB does facilitate the retrotransport of PRL to the nucleus, the obvious questions arise as to how this transport occurs and what role the PRL/CypB complex plays in the nucleus. While this exact mechanism remains a mystery, certain lines of evidence point towards a retrotranslocation similar to that of Shiga toxin (184-186). Studies have shown that upon endocytosis by clathrin-coated pits, Shiga toxin can retrotransport through the *trans*-golgi network (TGN) into and then out of the ER, even while lacking an ER retrieval signal (i.e. KDEL)(185) Prior studies indicate colloidal gold-labeled PRL can clearly be visualized by electron microscopy within the ER/Golgi apparatus 0.5-2 hours after its internalization via a vesicular pathway (173), suggesting its internalization may follow pathways parallel to Shiga toxin. Once in the ER, it is speculated that the ER transporter complex Sec61 may assist in the translocation of the CypB/PRL complex into the cytosol. A 22kDa protein that exhibits PPI activity (CypB) has been shown to associate with the Sec61 complex along with nascent pre-PRL (187). Upon Sec61 facilitating the transport of the PRL/CypB complex into the cytosol, this complex could target to the nucleus via the nuclear localization sequence within CypB through known nuclear importers such as the importin family.

So, what is the role of the PRL/CypB complex in the nucleus? While it has been demonstrated that cyclophilins can directly associate with and modulate the activity of transcription factors (188), the action of the PRL/CypB complex within the nucleus may not be exclusively CypB mediated. Yeast-two hybrid analysis has also indicated an interaction between PRL and known transcription factors, a finding confirmed by co-immunoprecipitation studies. Therefore, it appears that CypB, via its ability to transport peptide hormones into the nucleus, may enable peptide hormones to act directly upon gene transcription in a manner analogous to that of steroid hormones. It has been known for some time that upon binding their respective receptors in the cytoplasm, the steroid hormone/receptor complex translocates to the nucleus where it serves as either a transcriptional co-activator or repressor. It appears that the somatolactogens may parallel steroid hormone transduction pathways (189) mediating both rapid signaling from the cell surface while providing a secondary signal within the cell nucleus. Thus peptide hormone retrotransport and intranuclear action may provide a mechanism for peptide hormone specificity, while using signaling machinery common to multiple pathways.

SUMMARY

It has been gratifying to follow the explosion of information regarding PRLR signal transduction, all of which has occurred since the early 1990's. Collectively, these studies indicate that networks of PRLR-associated proteins mediate the signal triggered by ligand-induced PRLR dimerization, resulting in PRL-induced gene expression and action. These findings also present several possible mechanisms for crosstalk between the PRLR and other cell surface receptors including those for other growth factors and extracellular matrix proteins.

Despite these insights, however, several salient questions regarding PRLR signaling remain incompletely resolved. Precisely how does receptor dimerization trigger the activation of the proximal kinases such as Jak2 and Fyn? How does altering the structure and affinity of the ligand-binding extracellular domain modulate this activity? How can megaDaltons worth of signaling complexes associate with the correspondingly meager 40 kiloDalton intracellular domain of the PRLR – what are the spatial and temporal mechanisms for such association? Finally, how do the non-genomic (i.e. PRLR-associated signaling cascades) and genomic (i.e. intranuclear PRL) actions of the PRL/PRLR complex summate to induce specific patterns of PRL-induced gene expression and action? The answers to these question no doubt will provoke further rounds of inquiry, but in doing so provide novel insights into the mechanisms of action of the PRLR and the cytokine receptor superfamily to which it belongs.

ACKNOWLEDGEMENTS

The authors are grateful from funding from the National Institutes of Health R01CA69294, R01DK50771 (to CVC), F32DK09727 (to JBK), F32DK10043 (to MAR) and from the American Cancer Society RPG00307TBE (to CVC) for their research into the topic of this chapter.

REFERENCES

1. Bazan JF 1990 Structural design and molecular evolution of a cytokine receptor superfamily. Proc Natl Acad Sci USA 87:6934-6938
2. Bazan JF 1990 Haematopoietic receptors and helical cytokines. Immunol Today 11:350-354
3. Bole-Feysot C, Goffin V, Edery M, Binart N, Kelly PA 1998 Prolactin (PRL) and tis receptor: Actions, signal transduction pathways and phenotypes observed in PRL receptor knockout mice. Endocrine Rev 19:225-268
4. Clevenger CV, Plank TL 1997 Prolactin as an autocrine/paracrine growth factor in breast tissue. J Mammary Gland Biol Neoplasia 2:59-68
5. Clevenger CV, Freier DO, Kline JB 1998 Prolactin receptor signal transduction in cells of the immune system. Journal of Endocrinology 157:187-197
6. Yu-Lee L-Y, Luo G, Book ML, Morris SM 1998 Lactogenic hormone signal transduction. Biol Reprod 58:302-311
7. Gertler A, Grosclaude J, Strasburger CJ, Nir S, Djiane J 1996 Real-time measurements of the interactions between lactogenic hormones and prolactin-receptor extracellular domains from several species support the model hormone-induced transient receptor dimerization. J Biol Chem 271:24482-24491
8. Horseman ND, Zhao W, Montecino-Rodriguiez E, Tanaka M, Nakashima K, Engle SJ, Smith F, Markoff E, Dorshkind K 1997 Defective mammopoiesis, but normal hematopoiesis, in mice with a targeted disruption of the prolactin gene. EMBO J 16:6926-6935
9. Riddle O, Bates RW, Dykshorn SW 1933 The preparation, identification and assay of prolactin - a hormone of the anterior pituitary. Am J Physiol 105:191-216
10. Kelly PA, Djiane J, Postel-Vinay MC, Edery M 1991 The Prolactin/Growth Hormone Receptor Family. Endo Rev 12(3):235-251
11. Goffin V, Kelly PA 1996 The prolactin / growth hormone receptor family: structure / function relationships. J of Mammary Gland Biol Neopl 2:7-17
12. Postel-Vinay M-C 1996 Groth hormone- and prolactin-binding proteins: souluble forms of receptors. Horm Res 45:178-181

13. Anthony RV, Smith GW, Duong A, Pratt SL, Smith MF 1995 Two forms of the prolactin receptor messenger ribonucleic acid are present in ovine fetal liver and adult ovary. Endocrine 3:291-295
14. Davis JA, Linzer DIH 1989 Expression of multiple forms of the prolactin receptor in mouse liver. Mol Endocrinol 3:674-680
15. Nagano M, Chastre E, Choquet A, Bara J, Gespach C, Kelly PA 1995 Expression of prolactin and growth hormone receptor genes and their isoforms in the gastrointestinal tract. Am J Physiol 268:G431-G442
16. Schuler LA, Nagel RJ, Gao J, Horseman ND, Kharbanda S 1997 Prolactin receptor heterogeneity in bovine fetal and maternal tissues. Endocrinology 138:3187-3194
17. Chen X, Horseman ND 1994 Cloning, expression, and mutational analysis of th pigeon prolactin receptor. Endocrinology 135:269-176
18. Tanaka M, Maeda K, Okubo T, Nakashima K 1992 Double antenna structure of chicken prolactin receptor deduced from the cDNA sequence. Biochem Biophys Res Commun 188:490-496
19. Mai JNC, Burnside J, Li L, Tang J, Davolos C, Cogburn LA 1999 Characterization of unique truncated prolactin receptor transcripts, corresponding to the intracellular domain, in the testis of the sexually mature chicken. Endocrinology 140:1165-1174
20. Shirota M, Banville D, Ali S, Jolicouer C, Boutin JM, Edery M, Djiane J, Kelly PA 1990 Expression of two forms of prolactin receptor in rat ovary and liver. Mol Endocrinol 4:1136-1143
21. Boutin J-M, Jolicoeur C, Okamura H, Gagnon J, Edery M, Shirota M, Banville D, Dusanter-Fourt I, Djiane J, Kelly PA 1988 Cloning and Expression of the Rat Prolactin Receptor, a Member of the Growth Hormone / Prolactin Receptor Gene Family. Cell 53:69-77
22. Ali S, Pellegrini I, Kelly PA 1991 A prolactin-dependent immune cell line (Nb2) expresses a mutant form of prolactin receptor. J Biol Chem 266:20110-20117
23. Boutin JM, Edery M, Shirota M, Jolicoeur C, Lesueur L, Ali S, Gould D, Djiane J, Kelly PA 1989 Identification of a cDNA encoding a long form of prolactin receptor in human hepatoma and breast cancer cells. Mol Endocrinol 3:1455-1461
24. Nagano M, Chastre E, Choquet A, Bara J, Gespach C, Kelly PA 1995 Expression of prolactin and growth hormone receptor genes and their isoforms in the gastrointestinal tract. Amer J Phys 268:G431-G442
25. Clevenger CV, Chang W-P, Ngo W, Pasha TLM, Montone KT, Tomaszewski JE 1995 Expression of prolactin and prolactin receptor in human breast carcinoma: Evidence for an autocrine/paracrine loop. Am J Pathol 146:1-11
26. Kline JB, Roehrs H, Clevenger CV 1999 Functional characterization of the intermediate isoform of the human prolactin receptor. J Biol Chem 274:35461-35468
27. Goujon L, Allevato G, Simonin G, Paquereau L, Le Cam A, Clark J, Nielsen JH, Djiane J, Postel-Vinay M-C, Edery M, Kelly PA 1994 Cytoplasmic sequences of the growth hormone receptor necessary for signal transduction. Proc Natl Acad Sci USA 91:957-961
28. Sato N, Sakamaki K, Terada N, Arai K-I, Miyajima A 1993 Signal transduction by the high-affinity GM-CSF receptor: two distinct cytoplasmic regions of the common beta subunit responsible for different signaling. The EMBo Journal 12:4181-4189
29. Venkitaraman AR, Cowling RJ 1992 Interleukin 7 receptor functions by recruiting tyrosine kinase p59fyn through a segment of its cytoplasmic tail. Proc Natl Acad Sci USA 89:12083-12087
30. Gilmour KC, Reich NC 1994 Receptor to nucleus signaling by prolactin and interleukin 2 via activation of latent DNA-binding factors. Proc Natl Acad Sci USA 91:6850-6854
31. Groner B, Altoik S, Meier V 1994 Hormonal regulation of transcription factor activity in mammary epithelial cells. Mol Cell Endocrinol 100:109-114

32. Buteau H, Pezet A, Ferrag F, Perrot-Applanat M, Kelly PA, Edery M 1998 N-glycosylation of the prolactin receptor is not required for activation of gene transcription but is crucial for its cell surface targeting. Mol Endocrinol 12:544-555
33. Chang W-P, Ye Y, Clevenger CV 1998 Stoichiometric structure/function analysis of the prolactin receptor signaling domains by receptor chimeras. Mol Cell Biol 18:896-905
34. Chen X, Horseman ND 1994 Cloning, expression, and mutational analysis of the pigeon prolactin receptor. Endocrinology 135:269-276
35. Kinet S, Bernichtein S, Kelly PA, Martial JA, Goffin V 1999 Biological properties of human prolactin analogs depend not only on global hormone affinity, but also on the relative affinities of both receptor binding sites. J Biol Chem 274:26033-26043
36. Nagano M, Kelly PA 1994 Tissue Distribution and Regulation of Rat Prolactin Receptor Gene Expression. J Biol Chem 269:13337-13345
37. Clarke DL, Linzer DLH 1993 Changes in prolactin receptor expression during pregnancy in the mouse ovary. Endocrinology 133:224-232
38. Ihle JN 1995 Cytokine receptor signaling. Nature 377:591-594
39. Ihle JN, Kerr IM 1995 Jaks and Stats in signaling by the cytokine receptor superfamily. Trends Genet 11:69-74
40. Witthuhn BA, Quelle FW, Silvennoinen O, Yi T, Tang B, Miura O, Ihle JN 1993 JAK2 associates with the erythropoietin receptor and is tyrosine phosphorylated and activated following stimulation with erythropoietin. Cell 74:227-236
41. Argetsinger LS, Campbell GS, Yang X, Witthuhn BA, Silvennoinen O, Ihle JN, Carter-Su C 1993 Identification of JAK2 as a growth hormone receptor-associated tyrosine kinase. Cell 74:237-244
42. Campbell GS, Argetsinger LS, Ihle JN, Kelly PA, Rillema JA, Carter-Su C 1994 Activation of JAK2 tyrosine kinase by prolactin receptors in Nb2 cells and mouse mammary gland explants. Proc Natl Acad Sci USA 91:5232-5236
43. Rui H, Kirken RA, Farrar WL 1994 Activation of receptor-associated tyrosine kinase JAK2 by prolactin. J Biol Chem 269:5364-5368
44. Parganas E, Wang D, Stravopodis D, Topham DJ, Marine J-C, Teglund S, Vanin EF, Bodner S, Calamonici OR, van Deursen JM, Grosveld G, Ihle JN 1998 Jak2 is essential for signaling through a variety of cytokine receptors. Cell 93:385-395
45. Zhao Y, Wagner F, Frank SJ, Kraft AS 1995 The amino-terminal portion of the JAK2 protein kinase is necessary for binding and phosphorylation of the granulocyte-macrophage colony-stimulated factor receptor b_c chain. J Biol Chem 270:13814-13818
46. Dusanter-Fourt I, Muller O, Ziemicki A, Mayeux P, Drucker B, Dijane J, Wilks A, Harpur AG, Fischer S, Gisselbrecht S 1994 Identification of JAK protein tyrosine kinases as signaling molecules for prolactin. Functional analysis of prolactin receptor and prolactin-erythropoietin receptor chimera expressed in lymphod cells. EMBO J 13:2583-2591
47. Lebrun JJ, Ali S, Sofer L, Ullrich A, Kelly PA 1994 Prolactin-induced proliferation Nb2 cells involves tyrosine phosphorylation of the prolactin receptor and its associated tyrosine kinase JAK2. J Biol Chem 269:14021-14026
48. Han Y, Watling D, Rogers NC, Stark GR 1997 JAK2 and STAT5, but not JAK1 and STAT1, are required for prolactin-induced b-lactoglobulin transcription. Mol Endocrinol 11:1180-1188
49. Kawahara A, Minami Y, Miyazaki T, Ihle JN, Taniguchi T 1995 Critical role of the interleukin 2 (IL2) receptor gamma-chain associated Jak3 in the IL-2 induced c-*fos* and c-*myc*, but not *bcl-2* gene, induction. Proc Natl Acad Sci USA 92:8724-8728
50. Muller M, Briscoe J, Laxton C, Guschin D, Ziemiecki A, Silvennoinen O, Harpur AG, Barbieri G, Witthuhn BA, Schindler C 1993 The protein tyrosine kinase Jak1 complements defects in interferon-alpha/beta and -gamma signaling. Nature 366:129-135

51. Chang W-P, Clevenger CV 1996 Modulation of growth factor receptor function by isoform heterodimerization. Proc Natl Acad Sci USA 93:5947-5952
52. Lebrun J-J, Ali S, Ullrich A, Kelly PA 1995 Proline-rich sequence-mediated Jak2 association to the prolactin receptor is required but not sufficient for signal transduction. J Biol Chem 270:10664-10670
53. Pezet A, Buteau H, Kelly PA, Edery M 1997 The last proline of box 1 is essential for association with JAK2 and functional activation of the prolactin receptor. Mol Cell Endocrinol 129:199-208
54. Lebrun J-J, Ali S, Goffin V, Ullrich A, Kelly PA 1995 A single phosphotyrosine residue of the prolactin receptor is responsible for activation of gene transcription. Proc Natl Acad Sci USA 92:4031-4035
55. Goupille O, Daniel N, Bignon C, Jolivet J, Djiane J 1997 Prolactin signal transduction to milk protein genes: carboxy-terminal part of the prolactin receptor and its tyrosine phosphorylation are not obligatory for JAK2 and STAT5 activation. Mol Endocrinol 127:155-169
56. Hackett RH, Wang Y-D, Larner AC 1995 Mapping of the cytoplasmic domain of the human growth hormone receptor required for the activation of Jak2 and Stat proteins. J Biol Chem 270:21326-21330
57. Frank SJ, Gilliland G, Kraft AS, Arnold CS 1994 Interaction of the growth hormone receptor cytoplasmic domain with the JAK2 tyrosine kinase. Endocrinology 135:2228-2239
58. Feng J, Witthuhn BA, Matsuda T, Kohlhuber F, Kerr IM, Ihle JN 1997 Activation of Jak2 catalytic activity requires phosphorylation of Y^{1007} in the kinase activation loop. Mol Cell Biol 17:2497-2501
59. Bignon C, Daniel N, Belair L, Djiane J 1999 *In vitro* expression of long and short ovine prolactin receptors: activation of Jak2/STAT5 pathway is not sufficient to account for prolactin signal transduction to the ovine b-lactoglobulin promoter. J Mol Endocrinol 23:125-136
60. Fresno Vara JA, Carretero MV, Geronimo H, Ballmer-Hofer K, Martin-Perez J 2000 Stimulation of c-Src by prolactin is independent of Jak2. Biochem J 345:17-24
61. Pezet A, Ferrag F, Kelly PA, Edery M 1997 Tyrosine docking sites of the rat prolactin receptor required for association and activation Stat5. J Biol Chem 272:25043-25050
62. DaSilva L, Rui H, Erwin RA, Howard OMZ, Kirken RA, Malabarba MG, Hackett RH, Larner AC, Farrar WL 1996 Prolactin recruits STAT1, STAT3 and STAT5 independent of conserved receptor tyrosines TYR402, TYR479, TYR515, and TYR580. Mol Cell Endocrinol 117:131-140
63. Ali S 1998 Prolactin receptor regulates Stat5 tyrosine phosphorylation and nuclear translocation by two separate pathways. J Biol Chem 273:7709-7716
64. Yin T, Shen R, Feng G-S, Yang Y-C 1997 Molecular characterization of specific interactions between SHP-2 phosphatase and Jak tyrosine kinases. J Biol Chem 272:1032-1037
65. Yamashita Y, Watanabe S, Miyazato A, Ohya K, Ikeda U, Shimada K, Komatsu N, Hatake K, Miura K, Ozawa K, Mano H 1998 Tec and Jak2 kinases cooperate to mediate cytokine-driven activation of c-fos transcription. Blood 91:1496-1507
66. Stein-Gerlach M, Wallasch C, Ullrich A 1998 SHP-2, SH2-containing protein tyrosine phosphatase-2. Internanational Journal of Biochemistry and Cell Biology 30:559-566
67. Huyer G, Alexander DR 1999 Immune signalling: SHP-2 docks at multiple ports. Current Biol 9:R129-R132
68. Ali S, Chen Z, Lebrun J-J, Vogel W, Karitonenkov A, Kelly PA, Ullrich A 1996 PTP1D is a positive regulator of the prolactin signal leading to b-casein promoter activation. EMBO J 15:135-142
69. Berchtold S, Volarevic S, Moriggi R, Mercep M, Groner B 1998 Dominant negative variants of the SHP-2 tyrosine phosphatase inhibit prolactin activation of Jak2 and induction of Stat5-dependent transcription. Mol Endocrinol 12:556-567

70. Fujioka Y, Matozaki T, Noguchi T, Iwamatsu A, Yamao T, Takahashi N, Tsuda M, Takada T, Kasuga M 1996 A novel membrane glycoprotein, SHPS-1, that binds the SH2-domain containing protein tyrosine phosphatase SHP-2 in response to mitogens and cell adhesion. Mol Cell Biol 16:6887-6899
71. Timms JF, Swanson KD, Marie-Cardine A, Raab M, Rudd CE, Schraven B, Neel BG 1999 SHPS-1 is a scaffold for assembling distinct adhesion-regulated multi-protein complexes in macrophages. Current Biol 9:927-939
72. Stofega MR, Wang H, Ulrich A, Carter-Su C 1998 Growth hormone regulation of SIRP and SHP-2 tyrosyl phosphorylation and association. J Biol Chem 273:7112-7117
73. Naka T, Fujimoto M, Kishimoto T 1999 Negative regulation of cytokine signaling: STAT-induced Stat inhibitor. TIBS 24:394-398
74. Starr R, Hilton DJ 1999 Negative regulation of the JAK/STAT pathway. Bioessays 21:47-52
75. Pezet A, Favre H, Kelly PA, Edery M 1999 Inhibition and restoration of prolactin signal transduction by suppressors of cytokine signaling. J Biol Chem 274:24497-24502
76. Starr R, Hilton DJ 1998 SOCS: suppressors of cytokine signaling. International Journal of Biochemistry and Cell Biology 30:1081-1085
77. Helmanm D, Sandowski Y, Cohen Y, Matsumoto A, Yoshimura A, Merchav S, Gertler A 1998 Cytokine-inducible SH2 protein (CIS3) and JAK2 binding protein (JAB) abolish prolactin receptor-mediated STAT5 signaling. FEBS Letters 441:287-291
78. Ram PA, Waxman DJ 1999 SOCS/CIS protein inhibition of growth hormone-stimulated STAT5 signaling by multiple mechanisms. J Biol Chem 274:35553-35561
79. Yasukawa H, Misawa H, Sakamoto H, Masahuara M, Sasaki A, Wakioka T, Ohtsuka S, Imaizumi T, Matsuda T, Ihle JN, Yoshimura A 1999 The JAK-binding protein JAB inhibits Janus tyrosine kinase activity through binding in the activation loop. EMBO J 18:1309-1320
80. Clevenger CV, Medaglia MV 1994 The protein tyrosine kinase $p59^{fyn}$ is associated with prolactin receptor and is activated by prolactin stimulation of T-lymphocytes. Mol Endocrinol 8:674-681
81. Kefalas P, Brown TP, Brickell PM 1995 Signalling by the $pp60^{c\text{-}src}$ family of protein-tyrosine kinases. International Journal of Biochemistry and Cell Biology 27:551-563
82. Courtneidge SA 1994 Protein tyrosine kinases, with emphasis on the Src family. Cancer Biology 5:239-246
83. Olayioye MA, Beuvink I, Horsch K, Daly JM, Hynes NE 1999 ErbB receptor-induced activation of stat transcription factors is mediated by Src tyrosine kinases. J Biol Chem 274:17209-17218
84. Berlanga JJ, Fresno Vara JA, Martin-Perez J, Garcia-Ruiz JP 1995 Prolactin receptor is associated with c-Src in rat liver. Mol Endocrinol 9:1461-1467
85. Kazansky AB, Kabotyanski EB, Wyszomierski SL, Mancini MA, Rosen JM 1999 Differential effects of prolactin and src/abl kinases on the nuclear translocation of STAT5B and STAT5A. J Biol Chem 274:22484-22492
86. Heyeck SD, Wilcox HM, Bunnel SC, Berg LJ 1997 Lck phosphorylates the activaiton loop tyrosine of the Itk kinase domain and activates Itk kinase activity. J Biol Chem 272:25401-25408
87. Michel F, Grimaud L, Tuosto L, Acuto O 1998 Fyn and ZAP-70 are required for Vav phosphorylation in T cells stimulated by antigen-presenting cells. J Biol Chem 273:31932-31938
88. Al-Sakkaf KA, Dobson PRM, Brown BL 1997 Prolactin induced tyrosine phosphorylation of p59fyn may mediate phosphatidylinositol 3-kinase activation in Nb2 cells. J Mol Endocrinol 19:347-350

89. Hunter S, Burton EA, Wu SC, Anderson SM 1999 Fyn associates with Cbl and phosphorylates tyrosine 731 in Cbl, a binding site for phosphatidylinositol 3-kinase. J Biol Chem 274:2097-2106
90. Bolen JB 1995 Protein tyrosine kinases in the initiation of antigen receptor signaling. Curr Opin Immunol 7:306-311
91. Musacchio A, Gibson T, Rice P, Thompson J, Saraste M 1993 The PH domain: a common piece in the structural patchwork of signalling proteins. TIBS 18:
92. Vihinen M, Nilsson L, Smith CI 1994 Tec homology (TH) adjacent to the PH domain. FEBS Lett 350:263-265
93. Mano H, Sato K, Yazuki Y, Hirai H 1994 Tec protein kinase directly associates with Lyn protein tyrosine kinase through its N-terminal unique domain. Oncogene 9:3205-3211
94. Machide M, Mano H, Todokoro K 1995 Interleukin 3 and Erythropoietin Induce Association of Vav with Tec Kinase through Tec Homology Domain. Oncogene 11:625
95. Takahashi M, Hibi M, Fujitani M, Fukada T, Yamaguchi T, Hirano T 1997 Tec tyrosine kinase links the cytokine receptors to PI-3 kinase probably through Jak. Oncogene 14:2273-2282
96. Miyazato A, Yamashita Y, Hatake K, Miura Y, Ozawa K, Mano H 1996 Tec protein kinase is involved in the signaling mechanism of granulocyte colony-stimulating factor receptor. Cell Growth and Differentiation 7:1135-1139
97. Da Silva AJ, Yamamoto M, Zalvan CH, Rudd CE 1992 Engagement of the Tcr/CD3 complex stimulates p59fyn activity: Detection of associated proteins at 72 and 120-130 KD. Mol Immunol 29:1417-1425
98. Mao J, Xie W, Yuan H, Simon M, Mano H, Wu D 1998 Tec/Bmx non-receptor tyrosine kinases are involved in regulation of Rho and serum response factor by Ga12/13. EMBO J 17:5638-5646
99. Clevenger CV, Ngo W, Sokol DL, Luger SM, Gewirtz AM 1995 Vav Is Necessary for Prolactin-Stimulated Proliferation and Is Translocated into the Nucleus of a T-cell Line. J Biol Chem 270(22):13246-13253
100. Rawlings DJ, Scharenberg AM, Park H, Wahl MI, Lin S, Kato RM, Fluckiger A-C, Witte ON, Kinet J-P 1996 Activation of BTK by a phosphorylation mechanism initiated by SRC family kinases. Science 271:822-825
101. Ohya K-I, Kajigaya S, Yamashita Y, Miyazato A, Hatake K, Miura Y, Ikeda U, Shimada K, Ozawa K, Mano H 1997 SOCS-1/JAB/SSI-1 can bind to and suppress Tec protein-tyrosine kinase. J Biol Chem 272:27178-27182
102. Pezet A, Favre H, Kelly PA, Edery M 1999 Inhibition and restoration of prolactin signal transduction by suppressors of cytokine signaling. J Biol Chem 274:24497-24502
103. Clevenger CV, Ngo W, Luger SM, Gewirtz AM 1995 Vav is necessary for prolactin-stimulated proliferation and is translocated into the nucleus of a T-cell line. J Biol Chem 270:13246-13253
104. Crespo P, Schuebel KE, Ostrom AA, Gutkind JS, Bustelo XR 1997 Phosphotyrosine-dependent activation of Rac-1 GDP/GTP exchange by the *vav* proto-oncogene product. Nature 385:169-172
105. Deckert M, Tartare-Deckert S, Couture C, Mustelin T, Altman A 1996 Functional and physical interactions of syk family kinases with the vav proto-oncogene product. Immunity 5:591-604
106. Fernandez JA, Keshvara LM, Peters JD, Furlong MT, Harrison ML, Geahlen RL 1999 Phosphorylation- and activation-independent association of the tyrosine kinase Syk and the tyrosine kinase substrates Cbl and Vav with tubulin in B-cells. J Biol Chem 274:1401-1406
107. Miranti CK, Leng L, Maschberger P, Brugge JS, Shattil SJ 1998 Identification of a novel integrin signaling pathway involving the kinase Syk and the guanine nucleotide exchange factor Vav1. Current Biol 8:1289-1299

108. Arni S, Senaldi G, Poincelet M, Hoessli DC 1993 Selective associateion of the p59^{fyn} tyrosine kinase with murine T lymphoma membrane phophoproteins. Oncogene 8:2485-2491
109. Han J, Das B, Wei W, Van Aelst L, Mosteller RD, Khosravi-Far R, Westwick JK, Der CJ, Broek D 1997 Lck regulates Vav activation of members of the Rho family of GTPases. Mol Cell Biol 17:1346-1353
110. Bustelo XR, Rubin SD, Suen K-L, Carrasco D, Barbacid M 1993 Developmental expression of the *vav* protooncogene. Cell Growth and Differentiation 4:297-308
111. Bustelo XR, Ledbetter JA, Barbacid M 1992 Product of *vav* proto-oncogene defines a new class of tyrosine protein kinase substrates. Nature 356:68-71
112. Coppola J, Bryant S, Koda T, Conway D, Barbacid M 1991 Mechanism of activation of the *vav* protooncogene. Cell Growth and Differentiation 2:95-105
113. Katzav S 1993 Single point mutation in the SH2 domain impair the transforming potential of *vav* and fail to activate proto-*vav*. Oncogene 8:1757-1763
114. Katzav S, Martin-Zanca D, Barbacid M 1989 *vav,* a novel human oncogene derived from a locus ubiquitously expressed in hematopoietic cells. EMBO J 8:2283-2290
115. Gulbins E, Coggeshall KM, Baier G, Katzav S, Burn P, Altman A 1993 Tyrosine-kinase stimulated guanine nucleotide exchange activity of vav in T cell activation. Science 260:822-825
116. Gulbins E, Langlet C, Baier G, Bonnefoy-Berard N, Herbert E, Altman A, Coggeshall EM 1994 Tyrosine phosphorylation and activation of vav GTP/GDP exchange activity in Antigen receptor-triggered B cells. J Immunol 152:2123-2129
117. Margolis B, Hu P, Katzav S, Li W, Oliver JM, Ullrich A, Weiss A, Schlessinger J 1992 Tyrosine phosphorylation of *vav* proto-oncogene product containing SH2 domain and transcription factor motifs. Nature 356:71-74
118. Platanias LC, Sweet MC 1994 Interferon α induces rapid tyrosine phosphorylation of the *vav* proto-oncogene product in hematopoietic cells. J Biol Chem 269:3143-3146
119. Alai M, Mui ALF, Cutler RL, Bustelo XR, Barbacid M, Krystal G 1992 Steel factor stimulates the tyrosine phosphorylation of the proto-oncogene product, p95^{vav}, in human hematopoietic cells. J Biol Chem 267:18021-18025
120. Gulbins E, Coggeshall KM, Baier G, Telford D, Langlet C, Baier-Bitterlich G, Bonnefoy-Berard N, Burn P, Wittinghofer A, Altman A 1994 Direct stimulation of vav guanine nucleotide exchange activity for ras by phorbol esters and diglycerides. Mol Cell Biol 14:4749-4758
121. Evans GA, Howard OZM, Erwin R, Farrar WL 1993 Interleukin-2 induces tyrosine phophorylation of the *vav* proto-oncogene product in human T cells: lack of requirement for the tyrosine kinase lck. Biochem J 294:339-342
122. Zhang R, Alt FW, Davidson L, Orkin SH, Swat W 1995 Defective signalling through the T- and B- Cell antigen receptors in lymphoid cells lacking the *vav* proto-oncogene. Nature 374:470-473
123. Movilla N, Bustelo XR 1999 Biological and regulatory properties of Vav-3, a new member of the Vav family of oncoproteins. Mol Cell Biol 19:7870-7885
124. Schuebel KE, Movilla N, Rosa JL, Bustelo XR 1998 Phosphorylation-dependent and constitutive activation of Rho proteins by wild-type and oncogenic Vav-2. EMBO J 17:6608-6621
125. Lopez-Lago M, Lee H, Cruz C, Movilla N, Bustelo XR 2000 Tyrosine phosphorylation mediates both activation and downmodulation of the biological activity of Vav. Mol Cell Biol 20:1678-1691
126. Abe K, Whitehead IP, O'Bryan JP, Der CJ 1999 Involvement of NH_2-terminal sequences in the negative regulation of Vav signaling and transforming activity. J Biol Chem 274:30410-30418

127. Han J, Luby-Phelps K, Das B, Shiu X, Mosteller RD, Krishna UM, Falck JR, White MA, Broek D 1998 Role of substrates and products of PI 3-kinase in regulating activation of rac-related guanosine triphosphatases by Vav. Science 279:558-560
128. Nimnual AS, Yatsula BA, Bar-Sagi D 1998 Coupling of ras and rac guanosine triphosphatases through the ras exchanger Sos. Science 279:560-563
129. Downward J 1998 Role of phosphoinositide-3-OH kinase in ras signaling. Signal Transduction in Health and Disease 31:1-10
130. Rodriguez-Viciana P, Warne PH, Dhand R, Vanhaesbroek B, Gout I, Fry MJ, Waterfield MD, Downward J 1994 Phosphotidylinositol-3-OH kinase as a direct target of Ras. Nature 370:527-532
131. Hall A 1998 Rho GTPases and the actin cytoskeleton. Science 279:509-514
132. Henning SW, Cantrell DA 1998 GTPases in antigen receptor signalling. Current Opinion in Immunology 10:322-329
133. Aspenstrom P 1999 Effectors for the Rho GTPases. Curr Op Cell Biol 11:95-102
134. Fischer K-D, Kong YY, Nishina H, Tedford K, Marengere LEM, Kozieradzki I, Sasaki T, Starr M, Chan G, Gardener S, Ngheim MP, Bouchard D, Barbacid M, Bernstein A, Penninger JM 1998 Vav is a regulator of cytoskeletal reorganization mediated by the T-cell receptor. Current Biol 8:554-562
135. Holsinger LJ, Graef A, Swat W, Chi T, Bautista DM, Davidson L, Lewis RS, Alt FW, Crabtree GR 1998 Defects in actin-cap formation in Vav-defecient mice implicate an actin requirement for lymphocyte signal transduction. Current Biol 8:563-572
136. Allen WE, Jones GE, Pollard JW, Ridley AJ 1997 Rho, rac, and cdc42 regulate actin organization and cell adhesion in macrophages. J Cell Sci 110:707-720
137. Chrzanowska-Wodnicki M, Burridge K 1996 Rho-stimulated contractility drives the formation of stress fibers and focal adhesions. J Cell Biol 133:1403-1415
138. Zigmond SH 1996 Signal transduction and actin filament organization. Current Biol 8:66-73
139. Keely P, Parise L, Juliano R 1998 Integrins and GTPases in tumour cell growth, motility and invasion. Trends in Cell Biology 8:101-106
140. Maus MV, Reilly SC, Clevenger CV 1999 Prolactin as a chemoattractant for human breast carcinoma. Endocrinology 140:5447-5450
141. Takayama S, Sato T, Krajewski S, Kochei K, Irie S, Millan JA, Reed JC 1995 Cloning and functional analysis of BAG-1: A novel Bcl-2-binding protein with anti-cell death activity. Cell 80:279-284
142. Froesch BA, Takayama S, Reed JC 1998 Bag1L protein enhances androgen receptor function. J Biol Chem 273:11660-11666
143. Takayama S, Xie Z, Reed JC 1999 An evolutionarily conserved family of Hsp70/Hsc70 molecular chaperone regulators. J Biol Chem 274:781-786
144. Wang H-G, Takayama S, Rapp UR, Reed JC 1996 Bcl-2 interacting protein, Bag-1, binds to and activates the kinase Raf-1. Proc Natl Acad Sci USA 93:7063-7068
145. Clevenger CV, Thickman K, Ngo W, Chang W-P, Takayama S, Reed JC 1997 Role of Bag-1 in the survival and proliferation of the cytokine-dependent lymphocyte lines, Ba/F3 and Nb2. Mol Endocrinol 11:608-618
146. Bardelli A, Longati P, Albero D, Goruppi S, Schneider C, Ponzetto C, Comoglio PM 1996 HGF receptor associates with the anti-apoptotic protein Bag-1 and prevents cell death. EMBO J 15:6205-6212
147. Luders J, Demand J, Hohfeld J 2000 The ubiquitin-related Bag-1 provides a link between the molecular chaperones Hsc70/Hsp70 and the proteasome. J Biol Chem 275:4613-4617
148. Vincent V, Goffin V, Rozakis-Adcock M, Mornon J-P, Kelly PA 1997 Identification of cytoplasmic motifs required for short prolactin receptor internalization. J Biol Chem 272:7062-7068

149. Krumenacker JS, Montgomery DW, Buckley DJ, Gout PW, Buckley AR 1998 Prolactin receptor signaling: Shared components with the T-cell antigen receptor in Nb2 lymphoma cells. Endocrine 9:313-320
150. Raab M, Da Silva AJ, Findell PR, Rudd CE 1997 Regulation of Vav-SLP binding by ZAP-70 and its relevance to TCRz /CD3 induction of interleukin-2. Immunity 6:155-164
151. Salojin KV, Zhang J, Delovitch TL 1999 TCR and CD28 are coupled via ZAP-70 to the activation fo the Vav/Rac-1/PAK-1/p38 MAPK signaling pathway. J Immunol 163:844-853
152. Montgomery DW, Krumenacker JS, Buckley AR 1999 Prolactin stimulates phosphorylation of the human T-cell antigen receptor complex and ZAP-70 tyrosine kinase: A potential for immunomodulation. Endocrinology 139:811-814
153. Clevenger CV, Torigoe T, Reed JC 1994 Prolactin induces rapid phosphorylation and activation of prolactin receptor associated Raf-1 kinase in a T-cell line. J Biol Chem 269:5559-5565
154. Goupille O, Barnier J-V, Guibert B, Paly J, Djiane J 2000 Effect of PRL on MAPK activation: negative regulatory role of the C-terminal part of the PRL receptor. Mol Cell Endocrinol 159:133-146
155. Das R, Vonderhaar BK 1996 Involvement of SHC, GRB2, SOS, and RAS in prolactin signal transduction in mammary epithelial cells. Oncogene 13:1139-1145
156. Seth A, Gonzalez FA, Gupta S, Raden DL, Davis RJ 1992 Signal transduction within the nucleus by mitogen-activated protein kinase. J Biol Chem 267:24796-24804
157. Chen R-H, Corbalan-Garcia S, Bar-Sagi D 1997 The role of the PH domain in the signal-dependent membrane targeting of Sos. EMBO J 16:1351-1359
158. Lemmon MA, Ferguson KM, O'brien R, Sigler PB, Schlessinger J 1995 Specific and high-affinity binding of inositol phosphates to an isolated pleckstrin homology domain. Proc Natl Acad Sci USA 92:10472
159. Berlanga JJ, Gualillo O, Buteau H, Applanat M, Kelly PA, Edery M 1997 Prolactin activates tyrosyl phosphorylation of insulin receptor substrate 1 and phosphatidylinositol-3-OH kinase. J Biol Chem 272:2050-2052
160. Buckley AR, Crowe PD, Russell DH 1988 Rapid activation of protein kinase C in isolated rat liver nuclei by prolactin, a known hepatic mitogen. Proc Natl Acad Sci USA 85:8649-8653
161. Crowe PD, Buckley AR, Zorn NE, Rui H 1991 Prolactin activates protein kinase C and stimulates growth-related expression in rat liver. Mol Cell Endocrinol 79:29-35
162. Canbay E, Norman M, Kilic E, Goffin V, Zachary I 1997 Prolactin stimulates the JAK2 and focal adhesion kinase pathways in human breast carcinoma T47-D cells. Biochem J 324:231-236
163. Buckley AR, Montgomery DW, Kibler R, Putnam CW, Zukoski CF, Gout PW, Beer CT, Russell DH 1986 Prolactin stimulation of ornithine decarboxylase and mitogensis in NB2 node lymphoma cells: The role of protein kinase C and calcium mobilization. Immunopharmacology 12:37-51
164. Murphy PR, DeMattia GE, Friesen HG 1988 Role of calcium in prolactin-stimulated c-myc gene expression and mitogenesis in NB2 lymphoma cells. Endocrinology 122:2476-2485
165. Olazabal I, Munoz J, Ogueta S, Obregon E, Garcia-Ruiz JP 2000 Prolactin (PRL)-PRL receptor system increases cell proliferation involving JNK and AP-1 activation: Inhibition by glucocorticoids. Mol En 14:564-575
166. Gala RR 1991 Prolactin and growth hormone in the regulation of the immune system. Proc Soc Exptl Biol Med 198:513-527
167. Yu-Lee L-Y 1997 Molecular actions of prolactin in the immune system. Proc Soc Exptl Biol Med 215:35-52
168. Kooijman R, Hooghe-Peters EL, Hooghe R 1996 Prolactin, growth hormone, and insulin-like growth factor-1 in the immune system. Adv Immunol 63:377-454
169. Prystowsky MB, Clevenger CV 1994 Prolactin as a second messenger for interleukin 2. Immunomethods 5:49-55

170. Kelly PA, Ali S, Rozakis M, Goujon L, Nagano M, Pellegrini I, Gould D, Djiane J, Edery M, Finidori J, Postel-Vinay MC 1993 The growth hormone/prolactin receptor family. Rec Prog Horm Res 48:123-164
171. Shiu RPC, Murphy LC, Tsuyuki D, Myal Y, Lee-Wing M, Iwasiow B 1987 Biological actions of prolactin in human breast carcinoma. Rec Prog Horm Res 43:277-299
172. Clevenger CV, Russell DH, Appasamy PM, Prystowsky MB 1990 Regulation of IL2-driven T-lymphocyte proliferation by prolactin. Proc Natl Acad Sci USA 87:6460-6464
173. Clevenger CV, Sillman AL, Prystowsky MB 1990 Interleukin-2 driven nuclear translocation of prolactin in cloned T-lymphocytes. Endocrinology 127:3151-3159
174. Clevenger CV, Altmann SW, Prystowsky MB 1991 Requirement of nuclear prolactin for interleukin-2-stimulated proliferation of T lymphocytes. Science 253:77-79
175. Nolin JM 1978 Intracellular prolactin in rat corpus luteum and adrenal cortex. Endocrinology 102:402-406
176. Nolin JM 1980 Incorporation of regulatory peptide hormones by individual cells of the adrenal cortex: Prolactin-adrenocorticotropin differences. Peptides 1:249-255
177. Rakowicz-Szulczynska EM, Rodeck U, Herlyn M, Koprowski H 1986 Chromatin binding of epidermal growth factor, nerve growth factor, and platlet-derived growth factor in cells bearing the appropriate surface receptors. Proc Natl Acad Sci USA 83:3728-3732
178. Smith RM, Jarett L 1987 Ultrastructural evidence for the accumulation of insulin in nuclei of intact 3T3-L1 adipocytes by an insulin-receptor mediated process. Proc Natl Acad Sci USA 84:459-463
179. Baldin V, Roman AM, Bosc-Bierne I, Amalric F, Bouche G 1990 Translocation of bFGF to the nucleus is G1 phase cell cycle specific in bovine aortic endothelial cells. EMBO J 9:1511-1517
180. Jans DA, Briggs LJ, Gustin SE, Jans P, Ford S, Young IG 1997 The cytokine interleukin-5 (IL-5) effects cotransport of its receptor subunits to the nucleus *in vitro*. FEBS Letters 410:368-372
181. Schreiber SL 1991 Chemistry and biology of the immunophilins and their immunosuppressive ligands. Science 251:283-287
182. Allain F, Boutillon C, Mariller C, Spik G 1995 Selective assay for CyPA and CyPB in human blood using highly specific anti-peptide antibodies. J Immunol Meth 178:113-120
183. Rycyzyn MA, Reilly SC, O'Malley K, Clevenger CV 2000 Role of cyclophilin B in PRL signal transduction and nuclear retrotranslocation. Mol Endocrinol In Press
184. Hazes B, Read RJ 1997 Accumulating evidence suggests that several AB toxins subvert the endoplasmic reticulum-associated protein degradation pathway to enter target cells. Biochemistry 36:11051-11054
185. Johannes L, Goud B 1998 Surfing on a retrograde wave: How does Shiga toxin reach the endoplasmic reticulum. Trends in Cell Biology 8:158-162
186. Wiertz EJHJ, Tortorella D, Bogyo M, Yu J, Mothes W, Jones TR, Rapaport TA, Ploegh HL 1997 Sec61-mediated transfer of a membrane protein from the endoplasmic reticulum to the proteasome for destruction. Nature 384:432-438
187. Klappa R, Freedman RB, Zimmerman R 1995 Protein disulphide isomerase and a lumenal cyclophilin-type peptidyl prolyl *cis-trans* isomerase are in transient contact with secretory proteins during late stages of translocation. Eur J Biochem 232:755-764
188. Leverson JD, Ness SA 1998 Point mutations in v-Myb disrupt a cyclophilin-catalyzed negative regulatory mechanism. Molecular Cell 1:203-211
189. Razandi M, Peram A, Greene GL, Levin ER 1999 Cell membrane and nuclear estrogen receptors (ERs) originate from a single transcript: Studies of ERa and ERb expressed in Chinese Hamster Ovary cells. Mol Endocrinol 13:307-319

Chapter 19

Transcription Factors, Cofactors and Target Genes Mediating Prolactin Signals

Carrie S. Shemanko and Bernd Groner

Georg-Speyer-Haus, Institute for Biomedical Research, Paul Ehrlich Str. 42-44, D-60596 Frankfurt am Main, Germany

THE IMPORTANCE OF BEING PROLACTIN

More than 300 different biological functions have been ascribed to prolactin (PRL) in vertebrates. They comprise six areas of biological regulation: water and electrolyte balance, growth and development, endocrinology and metabolism, brain and behavior, reproduction, immunoregulation and protection (for review see [1]). PRL influences these processes via the regulation of gene expression in various tissues. The multitude of processes regulated in different tissues suggests differential modes of action in individual target cells, but many of these mechanisms remain undefined. PRL may affect gene expression directly, for example through transcriptional control, or may do so by indirect mechanisms involving other cellular processes such as regulation of mRNA stability, protein synthesis or secondary modifications of proteins.

The direct regulation of PRL-dependent gene promoter elements has been studied in detail. PRL-activated transcription factors and PRL-responsive elements in target gene promoters have been defined in the areas of endocrine regulation of metabolism (liver bile transport and PRL-receptor gene regulation in insulin producing cells), immunoregulation and protection (T lymphocyte proliferation), growth and development (adipocyte differentiation) and reproduction (development of the mammary gland and function of the ovarian corpus luteum) (Table 19-1).

THE SPECIFICITY OF RESPONSE

The specificity of response that is generated in the PRL signaling system is a combination of sequential component protein-protein interactions. These interactions include the tissue specific recognition of PRL by the extracellular domain of the prolactin receptor (PRL-R). The receptor couples to a non-covalently associated cytoplasmic tyrosine kinase of the JAK (Janus kinase) family, which when activated, leads to specific transcription factor activation and to target gene transcription or repression. PRL may under certain circumstances activate different pathways within the cell.

Table 19-1. Genes Regulated by PRL

FUNCTION OF PRL AND TARGET GENE INDUCED	REGULATORY FACTORS IDENTIFIED
Endocrinology and Metabolism	
Hepatic sodium-dependent bile acid cotransporter gene	Stat5
PRL-R gene in insulin-producing cells	Stat5
Immunoregulation and protection	
IRF-1 in Nb2 cells	Stat1, CBP and Stat5
Growth and Differentiation	
Adipocyte differentiation, aP2 gene	Stat5
Reproduction	
Pigeon crop sac gene, Annexin Icp35	Stat1-like
Milk protein genes:	
β-casein	Stat5, GR, YY1, and PTP1D
β-lactoglobulin, α-lactalbumin and whey acidic protein	Stat5
Ovary specific genes 20αHSD, p27, α2-macroglobulin and 3β-HSD	Stat5

The production of PRL occurs primarily in, but is not restricted to, the anterior pituitary gland. It can also be produced in other tissues and act in an autocrine or paracrine fashion (for review see [2]). This leads to PRL responses that are not under the strict control of circulating hormone. Autocrine and paracrine PRL production may contribute to the pathological growth of breast tumor cells [3-5] (for review see [6]) or fibromuscular myometrial tumors [7].

The PRL-R is a single transmembrane receptor that is expressed as two isoforms, a short and a long form. These isoforms are expressed in a variety of tissues, and at different developmental stages (for review see [1] and references therein). Through its receptor, PRL is capable of activating a variety of signaling pathways including the JAK/Stat, (signal transducers and activators of transcription) pathway (for review see [8]), the mitogen activated protein kinase (MAPK) pathway [9-13] including Shc/Sos/Grb2/Ras/Raf [14-16], as well as Src, Fyn, and phosphatidylinositol 3-kinase [17-19], and the focal adhesion kinase pathway [20]. Each PRL-activated signal cascade activates specific transcription factors, which in the case of the Stats are present in a latent form in the cytoplasm and are activated without further requirement of gene expression. In the case of MAPK, their activation requires protein synthesis.

The transcription factors relay the specificity of response by binding to specific DNA response elements or combinations of elements in the promoter regions of a restricted subset of target genes. Specific complexes of transcription factors, coactivators or corepressors and the genes that they regulate, might ultimately be responsible for the pleiotropic tissue-specific and cell-differentiation-state specific actions of PRL.

TRANSGENIC MICE WITH DISRUPTED PRL OR PRL-R GENES

Many different physiological effects of PRL have been described, though it is unclear as to which are directly influenced. The advent of homologous gene recombination in embryonic stem cells and the inactivation of the PRL gene or the PRL-R gene provided the opportunity to distinguish between effects absolutely dependent upon PRL signaling and effects in which PRL plays an accessory or subordinate role. These studies complement earlier attempts in which PRL ablation was the starting point, achieved either through chemical inhibition of PRL secretion or by the use of mutant dwarf mice strains. The effectual ablation of tissue-specific hormone production in these studies was difficult to measure. Complications with the use of transgenic mice also arose due to the transfer of maternal PRL during nursing in the PRL null mice, and there was a wider effect of disrupted lactogenic hormone signaling in the PRL-R null mice. There are different forms of the PRL-R expressed in cells, which may be of functional significance. It is generally thought that the phenotypes described for the PRL-R null mice are due to the loss of the long form of the receptor, thus limiting the capacity to study the effects of the other receptor isoforms. While transgenic systems are not without their drawbacks, they do provide an opportunity to study the direct effects of the lack of PRL signaling.

The phenotype of the mice carrying disrupted PRL or PRL-R genes demonstrated that PRL plays a significant role in mammary development [21-23], fertility [22-24], male neuroendocrinology and reproduction [24] maternal behavior [25], and bone formation [26], but not in hematopoiesis [22]. The lack of such a phenotype in transgenic mice does not rule out a function for PRL in the hematopoietic system, but it may imply an accessory or redundant role.

Mice with disrupted PRL or PRL-R genes were severely impeded in mammary development and fertility [21-23]. These mice were not able to lactate due to a block in the development of the mammary gland [22,23]. Females were sterile due to lack of embryonic implantation in the uterine wall [23]. Adult PRL knockout mice lacked terminal or lateral lobulation of the mammary gland ductal system [22], and corroborating these observations, PRL-R knockout mice lacked alveoli [23]. Transplants of mammary epithelia from PRL-R null into the fat pads of wild type mice demonstrated that the PRL-R was critical for lobuloalveolar development during pregnancy, and indirectly required for ductal growth and side branching [21]. This process also requires the progesterone receptor [21,27]. Both PRL and progesterone are required for lobuloalveolar development during pregnancy and both hormones are known to cooperate in the activation of Stat5 responsive genes. PRL and progesterone signals have been reported to synergise on the β-casein promoter [28,29]. Stat5 deficient mice have phenotypic alterations in the mammary gland and reproductive tissues [30-33] similar to the PRL and the PRL-R null mice. Together this illustrates an integrated picture which emphasizes the central role of the JAK/Stat pathway in PRL signaling, but one where PRL also functions in concert with other hormones or growth factors to result in complete mammary organogenesis and reproductive function.

Mice with targeted disruptions of other PRL signaling components are also available for study, including the receptor associated tyrosine kinase JAK2 [34,35], the Stat molecules, (for review see [36]), coactivators [37,38] and target genes such as interferon regulatory factor-1 (IRF-1) (for review see [39]). The comparison of the phenotypes observed in these mice will help to construct a comprehensive picture of the *in vivo* actions of PRL.

MEDIATORS OF PRL RESPONSE – TRANSCRIPTION FACTORS

Although PRL influences many cellular actions, few transcription factors have been identified as direct mediators of PRL action. Using cell culture systems and transgenic mice techniques, it has been determined that the most important transcription factors are the Stats. Others, however, might also be involved. In Nb2 cells the transcription factor Sp1 (specificity protein 1) is newly synthesized and activated by PRL [40]. This may play a role in the induction of cyclin D3 [41,42], but the mechanisms of upregulation and activation are not fully defined and are dependent upon protein synthesis. The PRL-R also activates the mitogen activated protein kinase (MAPK) pathway, but it does not play a role, for example, in the induction β-casein, a known PRL-regulated gene [43]. The Stats are directly activated by a PRL-dependent pathway.

SIGNAL TRANSDUCERS AND ACTIVATORS OF TRANSCRIPTION (STAT)

There are 7 different mammalian genes encoding members of the Stat family (STAT1, STAT2, STAT3, STAT4, STAT5A, STAT5B, STAT6) (for review see [44,45]). PRL can only activate a subset of these proteins in cell culture, including Stat1, Stat3, Stat5a and Stat5b [46-50].

The Stat proteins share a common domain structure (Figure 19-1a), representing distinct functional properties. The amino-terminal region is important for protein-protein interactions, especially tetramerization [51,52], as well as nuclear translocation and deactivation [53]. The DNA binding domains of Stat1 and Stat3b, a carboxyl-truncated variant of Stat3, have beeHn recently analyzed by crystallography [54,55]. Structural analysis indicated that the specificity of binding is determined by dimers of Stat molecules, which might be influenced by tetramerisation with a second Stat dimer or by other DNA-binding proteins recognising adjacent response elements.

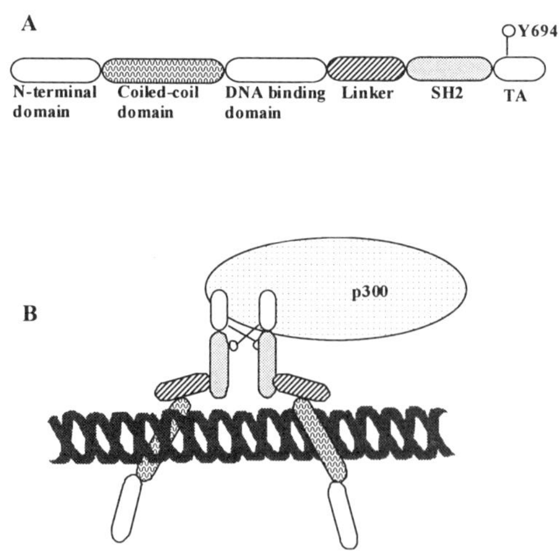

Figure 19-1. A. Domain Structure of Stats. Stat proteins share a common domain structure. Each Stat contains a tyrosine amino acid residue important for Stat activation. Shown is a representation of Stat5a with a phosphoryl group on the tyrosine at position 694. **B.** An activated Stat dimer positioned on its specific DNA consensus site. Tyrosine phosphoryl groups of each Stat molecule are interacting with the SH2 domain of the dimerization partner. P300, a coactivator, interacts with Stat5 to promote transcription. N (amino), TA (transactivation domain).

The Src-homology 2 (SH2) domain is a multi-functional domain involved in the interaction of Stats with the cytokine receptor [56], JAK (for review see [57]) and other Stat proteins [58] by the interaction with phosphotyrosine residues.

The carboxyl terminus is the most variable region when different Stat molecules are compared. It contains the essential tyrosine residue which must be phosphorylated by a JAK kinase and is required for Stat DNA binding (Figure 19-1b) [59], and various serine residues which when phosphorylated may contribute towards full activation [60-62] or stable dimer formation [63]. Serine phosphorylation may have no effect upon the stability of DNA binding of Stat1 or Stat3 [64]. The carboxyl terminus of Stat5 may also contain sequences responsible for proteasome-dependent deactivation [65]. The transactivation domain is located in the most carboxyl region of the protein, and its deletion results in a molecule that acts as a dominant negative of transcription. This region also makes protein-protein contacts.

Stats are activated by phosphorylation on their critical tyrosine residue by the receptor-associated JAK after receptor/ligand interaction. The activated Stat molecules dimerise via their SH2 domains, and are transported to the nucleus by an as of yet poorly defined mechanism. Stats bind specifically to palindromic promoter elements in the DNA, gamma interferon-activated sequences (GAS), TTCNNNGAA. The palindromic sequence in the Stat6 recognition site is separated by four nucleotides (reviewed in [45,66,67]). Stats also bind to IFNα-stimulated sites where the consensus is AGTTTCNNTTTCNC/T [68]. Stats can bind to DNA as dimers or tetramers, and tetrameric Stat5 appears to be essential for the activation of certain genes [52,69].

STAT1

Stat1 is mainly responsive to IFNα (in conjunction with Stat2) [68,70,71] and IFNγ (for review see [36,45,68]). There are two splice variants of the same gene, Stat1a and Stat1b. Stat1a is the full length variant with 750 amino acids, and the naturally occurring splice site variant, Stat1b, lacks the last 38 amino acids and is unable to mediate the IFNγ response [72]. Stat1 homodimers help mediate PRL stimulation of the IRF-1 gene in Nb2 cells [73-76]. Studies using Stat1 null mice have confirmed that Stat1 plays a major role in responding to IFN-dependent signals [77,78].

STAT3

Stat3 has been implicated in cell growth, suppression of apoptosis and cell mobility (for review see [36]). While it has been shown to be activated by PRL [47,48], it has not yet been correlated to PRL-dependent gene transcription. Stat3 is a potential oncogene [79]. It is essential for embryonic development [80]; it plays a role in the interleukin-6 mediated growth arrest and differentiation of myeloid cells [81,82], and plays an essential role in mammary gland involution [83]. Stat 3 [84] can be alternatively spliced to yield three forms, a long and two short forms [85,86]. The long form and one of the short forms are transactivating [85], and the other short form functions in a dominant negative fashion [86]. For the most part, the role of these Stat3 isoforms in PRL signaling remains to be fully elucidated.

STAT5

Stat5a was originally identified as mammary gland factor (MGF), mediating PRL signals in mammary epithelial cells, regulating the β-casein gene promoter [46,59] (for review see [8,87]). Stat5a and Stat5b are the products of two separate genes [88] with 96% amino acid sequence similarity. A truncated Stat5a molecule designed to consist of only the first 750 of a total of 794 amino acids acts as a dominant negative variant of the wild type molecule in the induction of transcription [89], even though it contains the essential tyrosine residue (Y694) responsible for Stat5 activation. The truncated molecule is activated by cytokine-induced tyrosine phosphorylation and is translocated to the nucleus where it is competent to bind DNA but is believed to inhibit transcription because it lacks a transactivation domain. It is also misregulated with regard to deactivation by phosphatases. It has a much longer activated life span than the wild-type molecules [89]. This truncated protein is similar to variants found naturally [48,90,91]. The origin of these short forms lies either in alternative splicing or proteolysis [92-94].

The transactivation domain also contains multiple sites of serine phosphorylation, which have been shown to be regulated by PRL in the case of Stat5b and to be constitutive for Stat5a [95]. The role of these phosphorylation events has yet to be determined. Transgenic animal models have shown that both Stat5 genes to differing degrees are critical for mediating effects of PRL in the mammary gland and ovary. They play different biological roles in the immune system and differ in their responses to growth hormone (for review see [36]).

STAT5 TRANSGENIC MICE

Targeted gene disruptions of Stat5a [32,33], Stat5b [30] and both Stat5a/5b [31] in mice have confirmed that these molecules are the major transducers of PRL signals. The phenotypes observed in these mice are quite similar to those found in mice in which the PRL or the PRL-R genes have been inactivated and corroborate the important role for PRL and the PRL-R in mammopoiesis and reproduction [21-23].

Stat5 deficient mice demonstrated that Stat5a is essential for the development of the mammary gland development and lactogenesis. Although Stat5b has a similar pattern of expression in the mammary gland [88], it is not able to fully compensate for the absence of Stat5a [32,33]. Stat5a null mice suffer from an inability to lactate due to lack of terminal differentiation of the mammary epithelial cells. Stat5b null mice have relatively normal alveolar development [31], but have reproductive abnormalities [30,31]. Female Stat5a/5b null mice are infertile [31].

Analysis of milk protein expression shows a decrease in α-lactalbumin in both Stat5a null and Stat5b null mice. Whey acidic protein (WAP) is severely reduced in Stat5a null mice, while expression of WAP in Stat5b null mice is initially less than wild-type it eventually increases due to relatively unimpaired development of the mammary gland. The expression of β-casein is slightly reduced in mice lacking either Stat5a or Stat5b [31]. Stat5a and Stat5b play redundant roles in milk protein expression.

Stat5a and Stat5b are transcription factors that are not limited to PRL-signal mediation. Stat5a/5b null mice also had defects in their responses to interleukin-2 (Stat5a/5b null) [96], granulocyte-macrophage colony stimulating factor (GM-CSF) (Stat5a and Stat5a/5b nulls) [31,97] interleukin-3 (Stat5a/5b null) [31], disrupted growth hormone responses [30,98,99], and stimuli for natural killer cell activation (Stat5b null) [100].

TRANSCRIPTIONAL COACTIVATORS

CBP/P300

Coactivators are proteins that act by bridging sequence specific binding factors to the transcription preinitiation complex. They are involved in modifying the chromatin by histone acetylation to make the promoter more accessible. CBP (CREB (cAMP response element binding protein) binding protein) and p300 and are two functionally homologous proteins that possess histone acetylase activity [101,102], and positively interact with adenovirus E1A [103], CREB [104-106], several other transcription factors (for review [107,108]) and nuclear receptors, including the glucocorticoid receptor (GR) [109,110].

CBP/p300 also act as coactivators of PRL-activated Stat5a and Stat5b. The amino-terminus of CBP/p300 requires the tyrosine-phosphorylated Stat transactivation domain for coactivation [111]. Many Stats use CBP/p300 as a coactivator in response to non-PRL signals, including Stat6 [112], Stat2 [113,114], Stat 1 [115,116], and Stat3 [114]. The interaction of Stats and CBP/p300 appears to be a consistent mechanism of transcriptional activation regardless of the activating signal.

THE GLUCOCORTICOID RECEPTOR

The GR is an important partner for Stat5 transactivation on the β-casein promoter in response to lactogenic signaling (Figure 19-2) [117-120]. DNA binding of the GR in this synergistic transcriptional activation with Stat5 seems not to be required [119,120], and their interaction is detectable *in vivo* in mammary epithelial cells [121]. The interaction of Stat5 and the GR results in enhanced transcription of the β-casein promoter, when compared to the effect of Stat5 alone. Stat5 activation has a negative influence on promoters carrying glucocorticoid response elements, such as the mouse mammary tumor virus LTR [29,119]. Increases in CBP/p300 levels were shown to positively influence GR action on glucocorticoid responsive as well as Stat5 responsive gene promoters. Stat5-mediated repression of GR action, however, was not mediated through limiting levels of CBP/p300 [111]. The GR pathway cross talks with the PRL signaling pathway through the direct interaction of two ligand activated transcription factors providing the potential for a greater variety of response.

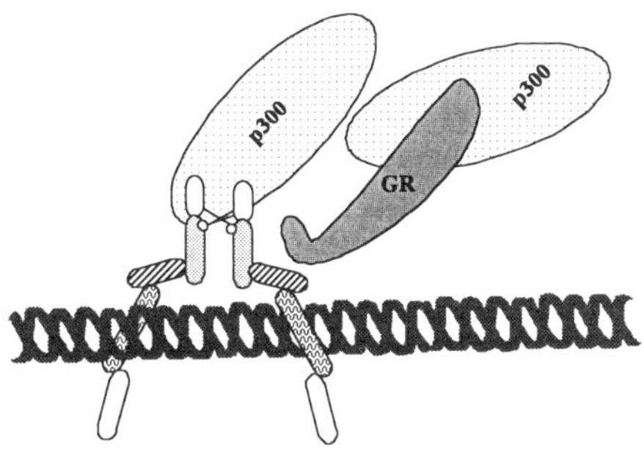

Figure 19-2. The Glucocorticoid receptor and Stat5 synergise on responsive promoters. The transcription activity of Stat5 and GR is synergistic on Stat5 responsive promoters. P300, a coactivator with histone acetylase activity, interacts with Stat5 and also GR. We propose a model where Stat5 recruits the GR to a promoter region and both Stat5 and the GR, in turn, recruit p300. The local concentration of p300 is increased and might be the ultimate regulator of the quantitative level of the response.

CBP/p300 interacts with the GR to inhibit AP-1 activity [109,110]. This is achieved by competition for limiting amounts of CBP/p300 in the cells, and this effect is seen also with Stat1 inhibition of AP1 signaling in response to IFNγ [115]. This mechanism is also used in the Stat5-induced inhibition of the Stat1-induced IRF-1 promoter. It had been demonstrated that Stat5 inhibited IRF-1 promoter activation by competing for a nuclear factor [122], which was later identified as CBP/p300 [123]. Therefore, it seems that CBP/p300 can produce negative effects on transcription, not as a direct result of its activity, but due to its limiting cellular levels.

TARGET GENES DIRECTLY REGULATED BY PROLACTIN

ENDOCRINOLOGY AND METABOLISMHEPATIC SODIUM-DEPENDENT BILE ACID COTRANSPORTER GENE

The production and regulation of bile flow is an essential function of the liver, and it had been shown that PRL upregulated hepatic bile salt transporter function during the post-partum period [124]. Specifically, the sodium-dependent cotransport of taurocholate was stimulated during a period which correlated with increases in levels of serum PRL and Stat5a and Stat5b activation [125]. The promoter region of the hepatic sodium-taurocholate cotransporting polypeptide (ntcp) was analyzed, and it was found to contain two GAS-like elements. Stat5 binds these elements in response to PRL and activation by the long form of the PRL-R [125]. In this manner PRL/Stat5 regulates bile flow in the liver.

PRL-R GENE REGULATION IN INSULIN-PRODUCING CELLS

The PRL-R has been shown to be upregulated in response to PRL, with different splice-site variants produced from different initiation sites. In pancreatic cells, the PRL-R is elevated during pregnancy and lactation [126] and an increase in mRNA for the long form of the PRL-R, comprising sequences encoding exon 1A, was found to be due to PRL-activated Stat5a and Stat5b [127]. PRL was also found to increase the mRNA which coded for the long form of the PRL-R which included exon 1C. This transcript was found to be regulated independently of Stat activation [127]. Splicing of the different exons is believed to be tissue-specific. Positive feedback by PRL on the expression of its own receptor would result in signal amplification.

IMMUNOREGULATION AND PROTECTION

INTERFERON REGULATORY FACTOR-1 – ACTIVATION BY STAT1 AND CBP, REPRESSION BY STAT5B

Interferon regulatory factor-1 (IRF-1) is a transcription factor with multiple roles in various cells. It is believed to act as a tumor-suppressor gene, and plays roles in differentiation, apoptosis, and proliferation (for review see [128]). The disruption of its gene in mice resulted in a lack of natural killer activity [129], and its activity is critical for T and B cell differentiation and macrophage function (for review see [128]). IRF-1 was discovered to be an immediate-early target gene of PRL [130-132], likely playing a role in PRL-induced cell proliferation in Nb2 rat lymphoma cells. Even though several Stat proteins have been found to be activated in Nb2 cells in response to PRL [48], it was discovered that Stat1a was the major Stat factor responsible for the induction of the IRF-1 promoter [75,76]. Stat5a was found to be a minor component [75], and both of these Stats contributed to the biphasic expression of this gene [75,76]. Reporter assays determined that while Stat1 was responsible for the induction of the promoter, Stat5a and Stat5b inhibited this induction in a manner independent of DNA binding, implying that Stat5 acted by competing for a putative DNA-binding protein or a coactivator protein [122]. This was confirmed and the factor identified as CBP/p300 [123].

IRF-1 is a multifunctional transcription factor which is also induced in nonpregnant human endometrium in response to PRL during the secretory phase of the menstrual cycle [133], likely by the activation of JAK2 and Stat1 and Stat5 [134]. Although the target genes of IRF-1 are unknown in this tissue, the temporal expression of IRF-1 points to a role in the regulation of differentiation [133]. This demonstrates that PRL can induce the same gene via a Stat factor in two different tissues for two different purposes, i.e. mitogenic and non-mitogenic functions.

Other growth related genes were also found to be induced by PRL in Nb2 cells as the result of modulation of preexisting factors. These include *c-myc*, ornithine decarboxylase, heat shock protein 70 homologue Nb29 and β-actin [135]. Genes that may be involved in the dependence of Nb2 cells upon PRL for growth were identified by differential display include elongation factor-2, α4-phosphoprotein and a Cdc5-like protein [136]. Several other genes have been postulated to be regulated by PRL in Nb2 cells: genes for fibroblast growth factor (FGF)-2, a novel FGF-responsive NonO/p54nrb-related mRNA [137], luteinizing hormone-releasing hormone (LHRH) and the LHRH-receptor [138], T cell receptor-α (TCRα) and TCRγ [139], clone 15 which is similar to the nuclear movement protein NUDC [140,141], cyclin D2 [142], cyclin D3 [41], cyclin E, cdk2, cdk5, E2F-1 [143], and the apoptotic regulatory genes bcl-2, bax [144] and pim-1 [145-147] have also been shown to be responsive to PRL signaling in Nb2 cells. This indicates, as it does in other cellular systems, that PRL potentially regulates many target genes with variable effects.

GROWTH AND DIFFERENTIATION

ADIPOCYTE DIFFERENTIATION

PRL may enhance adipogenic conversion in NIH-3T3 cells [148]. In an *in vitro* system it was discovered that PRL enhanced the mRNA expression of PPARγ (peroxisome proliferator-activated receptor γ) and C/EBPβ (CCAAT enhancer-binding protein), two transcription factors playing central roles in adipocyte differentiation. PRL was directly responsible for induction of aP2, an adipocyte-specific gene [148]. It was determined that the PRL-induced expression of aP2 was due in part by the activation of Stat5. A role for Stat5 in adipogenic differentiation was also suggested by the phenotype of the Stat5a/5b null mice, which had a significant decrease in the size of the mammary gland fat pad [31].

C/EBPβ, while not recognized as a target gene of PRL in mammary tissue, is developmentally regulated in the mammary gland [149]. C/EBPβ has been recognized for many functions where PRL is important, and is essential for ductal and lobuloalveolar development in the mammary gland [150,151], important for regulation of the β-casein gene promoter [152] and plays an essential role in ovarian granulosa cell differentiation in response to luteinizing hormone [153]. With regard to adipocytes, disruption of C/EBPα, -β or –δ, or combinations of these factors, results in defects in adipogenesis and adipocyte differentiation [154,155]. PPARγ is a ligand-activated nuclear hormone receptor which plays an important role in adipocyte differentiation, recently confirmed by observations in transgenic mice [156-158], (for review see [159]). The role for PRL in the differentiation of mammalian adipocytes is not well defined, though its receptor is also upregulated during adipocyte differentiation [160].

REPRODUCTION

PIGEON CROP SAC GENES

The crop sac in birds is a food storage organ found before the stomach, and is also known to produce a substance termed crop milk to feed the young [161]. Its epithelium proliferates and differentiates in response to PRL, and several genes in the pigeon crop are known to be PRL regulated [162]. Annexin Icp35, also known as lipocortin I and calpactin II, is regulated by PRL [163]. Analysis of the promoter region identified GAS-like elements which bound a Stat1-like protein [164]. The role of lipocortin I in humans is not well defined, though it may play a role in reproduction [165].

Milk Protein Genes

The best-described PRL/Stat5 target genes in the mammary gland are β-casein [46,166], β-lactoglobulin [167], and WAP [168] (for review see [8,169]). Induction of these genes is maximal in the presence of lactogenic hormones, and does not depend upon PRL alone. It appears that induction of these milk protein genes requires the long form of the PRL-R, or the intermediate Nb2 form [170]. The short form of the PRL-R inhibits activation of the β-casein gene promoter [171].

Various mechanisms of cross talk may also occur, as noted on the β-casein gene promoter with the involvement of PRL-induced protein kinase C alpha [172], and the possible role of Ras in Stat5-mediated β-casein expression in T cells [173]. The involvement of the MAPK pathway, while a factor in PRL signaling, is not apparent in the PRL/Stat5 induction of the β-casein gene in mammary epithelial cells [43]. There are reports of extracellular signal-regulated kinase (ERK) interaction with Stat5a [174], and a possible PRL-independent modulation of β-casein by ERK2 in Chinese hamster ovary cells [175]. Expression of the β-casein gene can be regarded as the result of interplay between different signal transduction pathways.

The milk protein gene promoters can also be regulated by a variety of factors unrelated to Stats. Studies of the β-casein gene promoter illustrate this point. A complex variety of factors regulate its expression and are shared across rodents, ruminents and humans [176]. There are also both positive [177] and negative regulatory factors [178-181] involved in the regulation of β-casein gene, for example Yin-Yang-1 (YY1) and protein tyrosine phosphatase 1D (PTP1D), also known as SHP-2.

SHP-2 is a cytoplasmic protein tyrosine phosphatase that associates with the PRL-R/JAK2 complex [177]. It is tyrosine phosphorylated upon signaling by PRL and plays an essential positive role in transcription of the β-casein promoter in Nb2 cells. It appears to interact with JAK2 independently of PRL stimulation, and relies upon JAK2 kinase activity for its PRL-induced activation. Its substrate is unknown, though its phosphatase activity and SH2 domains are required for β-casein promoter activity [177].

There are DNA-binding factors contributing to negative regulation on the β-casein promoter. The nuclear factor YY1 is a member of the GLI-Krueppel family of zinc-finger containing proteins, and is ubiquitously expressed. It has been suggested that YY1 interacts with either coactivators (histone acetyltransferases) or repressors (histone deacetylases) to either activate or repress transcription [182]. YY1 interacts with an unidentified DNA-binding protein to repress the β-casein promoter in the mammary epithelial cell line HC11. While YY1 DNA-binding itself is not regulated by hormones, the presence of Stat5a was able to decrease YY1 binding to the DNA and relieve repression [179,181]. By decreasing YY1 binding to a nearby site on the DNA, Stat5a regulates transcriptional control of the β-casein promoter by a mechanism independent of its own transactivation. In this manner it appears that the β-casein promoter is regulated by a relief of repression that is mediated by Stat5.

Milk protein gene promoters had initially been analyzed in cultured cell systems to identify the presence and activity of PRL responsive elements and Stat binding sites, to show that they are crucial for the observed gene regulation. The use of transgenic mice has extended these observations to *in vivo* situations. What is also clear from transgenic mouse models is that milk protein gene regulation is not the only essential function of PRL in the mammary gland, and that PRL plays a significant role in the maturation and differentiation of the ductal and lobuloalveolar system .

Currently there is a lack of understanding as to which PRL regulated genes are responsible for the mammary gland organogenesis that is disrupted in PRL-, PRL-R- and Stat5-knock-out mice. One candidate may be the cyclin D1 gene. Its disruption resulted in developmental defects in the mammary gland [183] that resembled the ones observed in PRL signaling deficient mice. Cyclin D1 has been identified as a Stat5 target gene in hematopoietic cells [184] and was shown to be responsive to PRL in a human breast cancer cell line T-47D [185]. One can speculate that a disrupted PRL/JAK2/Stat5 pathway results in the failure to properly induce cyclin D1 in the mouse mammary gland.

GENES OF THE OVARY

The PRL-JAK/Stat pathway also plays a significant role in the ovaries, though current data present a controversial view of many of PRL's functions there. Activated Stat5 has been detected in the ovaries in response to PRL [186], and there appears to be a preference for the use of Stat5b in the corpus luteum [187] and for Stat5a in the mammary gland. The phenotypes of the Stat5 knockout mice indicated that Stat5 plays a significant role in reproduction. Stat5b null mice required exogenous PRL to maintain pregnancy, presumably due to poor corpus luteum function [30] and Stat5a/b null mice seemed to lack corpora lutea [31]. These phenotypes corroborate the idea that Stat5 activation is essential for ovarian cell function.

While Stat5b null mice experience ovary-associated reproductive defects, only a loss of both Stat5a and Stat5b lead to complete infertility due to defects in ovarian function. These defects were primarily a result of loss of the corpus luteum and misregulation of ovarian genes such as 20a-hydroxysteroid dehydrogenase (20αHSD) and p27 [31]. p27, a negative regulator of G1 cyclin-dependent kinases, is important for ovarian function, possibly for the granulosa to luteal cell differentiation [188-190]. It was absent in the corpus luteum of Stat5a/5b null mice. PRL negatively regulates the gene encoding 20αHSD [191]. Its protein normally functions in the metabolism of progesterone to an inactive metabolite, and negative regulation of this gene would result in the metabolism of progesterone which is required for maintenance of pregnancy. Expression of 20αHSD was increased in the corpus luteum of Stat5a/5b null mice in comparison to wild-type mice. It would be expected to result in a shift to the inactive metabolite of progesterone. Taken together this indicates that PRL regulates ovarian differentiation and hormone biosynthesis by the regulation of different ovarian genes.

While it has been demonstrated that PRL is required for progesterone biosynthesis in the corpus luteum, which is important for luteotrophism and the maintenance of pregnancy, there is some evidence that PRL may also regulate luteolysis. This is accomplished through the down regulation of other enzymes involved in ovarian steroidogenesis. In the rat ovary Stat5 was shown to down regulate the type II 3b-hydroxysteroid dehydrogenase/delta5-delta4 isomerase (3β-HSD) gene [192,193], which codes for an enzyme involved in the final enzymatic step in progesterone biosynthesis. In contrast to this data, reporter assays demonstrated that PRL mediated the induction of the human promoter of 3β-HSD, also through the activation of Stat5 [194]. An interesting aspect of the studies of PRL regulation of the corpus luteum is the demonstration that Stat5 can play two apparently opposing roles in progesterone biosynthesis, which is a major role of the corpus luteum. In the rat, PRL/Stat5 down-regulates expression of 3β-HSD to produce a luteolytic effect, and in the human up-regulates the same gene to result in a luteotrophic action of PRL. These opposing results may be a reflection of the different roles of the corpus luteum in these two mammals, or possibly differences in experimental design.

Stat5 may in fact play a much wider role in the ovaries. Stat5 is preferentially activated in differentiating granulosa/luteal cells of the rat ovary [187], which corresponds to the time of α2-macroglobulin (α2M) expression [187,195-197]. α2M is a protease inhibitor with roles in cellular defense and is a regulator of cytokine activity, (for review see [198]). Stat5b and to some extent Stat5a, appears to play a major role in the activation of the α2M promoter, by binding the interleukin-6 response element [187,197]. The interleukin-6 response element carries two GAS-like sequences which are capable of binding both Stat3 and Stat5. Though Stat3 was present and active in ovarian tissue [187,195], it was not responsive to PRL induction [187,196]. At the time when α2M was expressed in the ovary [187], there was a high short-/long- PRL-R isoform expression ratio. This may indicate that the short form of the receptor does not play an inhibitory role with Stat5b in luteal cells as it does with Stat5a and the milk protein genes in myoepithelial cells [171,199]. The role of α2M in the ovary is not well defined.

CONCLUSION

PRL can produce a variety of tissue-specific functional effects that depend in part upon gene regulation and protein expression. Specificity of the genetic response is generated at multiple levels including that of the source of PRL production, PRL-R isoform expression, the choice of the signaling pathway and finally the distinct transcription complex that is built up on sequence-specific promoters. Members of the Stat family are the most well-defined PRL-signal transducing factors thus far identified that act as a result of direct activation and are not dependent upon protein synthesis. Stat members are generally thought to regulate promoters in a positive manner, but it has also been shown that they can negatively regulate genes. The induction of gene promoters may rely upon coactivators such as CBP/p300, or other transcription factors for cross talk such as GR. It is possible that Stat negatively regulates promoters (such as IRF-1 or 20αHSD) through interaction with a corepressor complex (for review see [200]).

It also is important to recognize the synergistic actions of PRL with other factors such as estrogen, growth hormone, glucocorticoids, insulin or progesterone, which cooperate to achieve mitogenic or differentiative endpoints. All of these factors contribute towards the pleiotropic effects of PRL.

The family of hormones which was initially comprised of PRL, growth hormone and placental lactogens is growing. PRL-related peptides are being described [201], which possibly are responsible for some of the actions previously attributed to PRL. These peptides signal through a receptor independent of the PRL-R, and may or may not utilize separate signaling pathways to achieve their biological purpose through defined target genes.

While the role of PRL in the induction and maintenance of mammary tumors has been well studied in model systems, its role in human breast cancer has been less clear. Overexpression of the PRL-R alone was sufficient for the induction of mammary tumors in mice [202]. PRL has a proliferative effect on breast cancer cells, and also is known to control angiogenesis, which makes it a potential therapeutic target (for review see [203]). Breast cancer cell lines and human breast tumors, in addition to normal mammary tissue, produce PRL which may regulate growth in an autocrine or paracrine fashion. It is also possible that PRL plays a cooperative role with other hormones required for mammogenesis, including that of progesterone for the progression of breast cancer [204].

Progesterone is known to increase the level of PRL-R in the mammary gland [205], and acts synergistically with PRL to activate gene targets. Progesterone, together with PRL, induces cell growth and its possible to see a pathological role if these actions were to become misregulated. Cross-talk with Stats and MAPK pathways and the dependence of breast cell lines on progesterone to respond to epidermal growth factor and PRL, has led to the suggestion that progesterone may sensitize breast cells to the subsequent and possibly synergistic actions of growth factors or cytokines. This may prime breast cells with respect to the progression of breast cancer [204]. The role of Stats in hematopoietic malignancy implies that these mediators of PRL could play a role in the loss of growth control or cellular transformation (for review see [206]).

A role of PRL in balancing the ratio of survival versus apoptotic mediators has been suggested for hematopoietic cells. PRL might have a dual role in the cell, to promote survival as well as to promote apoptosis. PRL promotes cell survival by inducing factors involved in growth and differentiation and possibly even anti-apoptotic factors such as Bcl-2 [144,147,207] and pim-1 [145-147]. Its role in the induction of apoptosis may depend on its ability to affect the expression of pro-apoptotic factors such as bax [144]. Bcl-X is a potent anti-apoptotic regulator. It is possible that cells which are terminally differentiated and highly dependent upon Stat-mediated Bcl-X transcription die upon hormone withdrawal for lack of apoptosis protection. The delicate balance of factors which promote survival versus apoptosis plays a crucial role in determining cell viability. Misregulation of the balance of PRL-regulated signal transduction pathways may contribute to the progression of cancer.

REFERENCES

1. Bole-Feysot C, Goffin V, Edery M, Binart N, Kelly PA. Prolactin (PRL) and its receptor: actions, signal transduction pathways and phenotypes observed in PRL receptor knockout mice. Endocr Rev. 1998;19:225-268.
2. Ben-Jonathan N, Mershon JL, Allen DL, Steinmetz RW. Extrapituitary prolactin: distribution, regulation, functions, and clinical aspects. Endocr Rev. 1996;17:639-669.
3. Shaw-Bruha CM, Pirrucello SJ, Shull JD. Expression of the prolactin gene in normal and neoplastic human breast tissues and human mammary cell lines: promoter usage and alternative mRNA splicing. Breast Cancer Res Treat. 1997;44:243-253.
4. Ginsburg E, Vonderhaar BK. Prolactin synthesis and secretion by human breast cancer cells. Cancer Res. 1995;55:2591-2595.
5. Clevenger CV, Chang WP, Ngo W, Pasha TL, Montone KT, Tomaszewski JE. Expression of prolactin and prolactin receptor in human breast carcinoma. Evidence for an autocrine/paracrine loop. Am J Pathol. 1995;146:695-705.
6. Vonderhaar BK. Prolactin involvement in breast cancer. Endocr Relat Cancer. 1999;6:389-404.
7. Nowak RA, Rein MS, Heffner LJ, Friedman AJ, Tashjian AH, Jr. Production of prolactin by smooth muscle cells cultured from human uterine fibroid tumors. J Clin Endocrinol Metab. 1993;76:1308-1313.
8. Groner B, Gouilleux F. Prolactin-mediated gene activation in mammary epithelial cells. Curr Opin Genet Dev. 1995;5:587-594.
9. Nohara A, Ohmichi M, Koike K, Jikihara H, Kimura A, Masuhara K, Ikegami H, Inoue M, Miyake A, Murata Y. Prolactin stimulates mitogen-activated protein kinase in human leiomyoma cells. Biochem Biophys Res Commun. 1997;238:473-477.
10. Camarillo IG, Linebaugh BE, Rillema JA. Differential tyrosyl-phosphorylation of multiple mitogen-activated protein kinase isoforms in response to prolactin in Nb2 lymphoma cells. Proc Soc Exp Biol Med. 1997;215:198-202.
11. Mitev V, Bayat-Sarmadi M, Lemnaouar M, Puissant C, Houdebine LM. The effect of prolactin on casein kinase II, MAP kinase and PKC in rabbit mammary cells and Nb2 rat lymphoid cells. Biochem Pharmacol. 1996;52:1719-1727.

12. Carey GB, Liberti JP. Stimulation of receptor-associated kinase, tyrosine kinase, and MAP kinase is required for prolactin-mediated macromolecular biosynthesis and mitogenesis in Nb2 lymphoma. Arch Biochem Biophys. 1995;316:179-189.
13. Buckley AR, Rao YP, Buckley DJ, Gout PW. Prolactin-induced phosphorylation and nuclear translocation of MAP kinase in Nb2 lymphoma cells. Biochem Biophys Res Commun. 1994;204:1158-1164.
14. Piccoletti R, Bendinelli P, Maroni P. Signal transduction pathway of prolactin in rat liver. Mol Cell Endocrinol. 1997;135:169-177.
15. Das R, Vonderhaar BK. Involvement of SHC, GRB2, SOS and RAS in prolactin signal transduction in mammary epithelial cells. Oncogene. 1996;13:1139-1145.
16. Das R, Vonderhaar BK. Activation of raf-1, MEK, and MAP kinase in prolactin responsive mammary cells. Breast Cancer Res Treat. 1996;40:141-149.
17. al-Sakkaf KA, Dobson PR, Brown BL. Prolactin induced tyrosine phosphorylation of p59fyn may mediate phosphatidylinositol 3-kinase activation in Nb2 cells. J Mol Endocrinol. 1997;19:347-350.
18. Berlanga JJ, Vara JAF, Martin-Perez J, Garcia-Ruiz JP. Prolactin receptor is associated with c-src kinase in rat liver. Molecular Endocrinology. 1995;9:1461-1467.
19. Clevenger CV, Medaglia MV. The protein tyrosine kinase P59fyn is associated with prolactin (PRL) receptor and is activated by PRL stimulation of T-lymphocytes. Mol Endocrinol. 1994;8:674-681.
20. Canbay E, Norman M, Kilic E, Goffin V, Zachary I. Prolactin stimulates the JAK2 and focal adhesion kinase pathways in human breast carcinoma T47-D cells. Biochem J. 1997;324:231-236.
21. Brisken C, Kaur S, Chavarria TE, Binart N, Sutherland RL, Weinberg RA, Kelly PA, Ormandy CJ. Prolactin controls mammary gland development via direct and indirect mechanisms. Dev Biol. 1999;210:96-106.
22. Horseman ND, Zhao W, Montecino-Rodriguez E, Tanaka M, Nakashima K, Engle SJ, Smith F, Markoff E, Dorshkind K. Defective mammopoiesis, but normal hematopoiesis, in mice with a targeted disruption of the prolactin gene. Embo J. 1997;16:6926-6935.
23. Ormandy CJ, Camus A, Barra J, Damotte D, Lucas B, Buteau H, Edery M, Brousse N, Babinet C, Binart N, Kelly PA. Null mutation of the prolactin receptor gene produces multiple reproductive defects in the mouse. Genes Dev. 1997;11:167-178.
24. Steger RW, Chandrashekar V, Zhao W, Bartke A, Horseman ND. Neuroendocrine and reproductive functions in male mice with targeted disruption of the prolactin gene. Endocrinology. 1998;139:3691-3695.
25. Lucas BK, Ormandy CJ, Binart N, Bridges RS, Kelly PA. Null mutation of the prolactin receptor gene produces a defect in maternal behavior. Endocrinology. 1998;139:4102-4107.
26. Clement-Lacroix P, Ormandy C, Lepescheux L, Ammann P, Damotte D, Goffin V, Bouchard B, Amling M, Gaillard-Kelly M, Binart N, Baron R, Kelly PA. Osteoblasts are a new target for prolactin: analysis of bone formation in prolactin receptor knockout mice. Endocrinology. 1999;140:96-105.
27. Lydon JP, DeMayo FJ, Funk CR, Mani SK, Hughes AR, Montgomery CA, Jr., Shyamala G, Conneely OM, O'Malley BW. Mice lacking progesterone receptor exhibit pleiotropic reproductive abnormalities. Genes Dev. 1995;9:2266-2278.
28. Richer JK, Lange CA, Manning NG, Owen G, Powell R, Horwitz KB. Convergence of progesterone with growth factor and cytokine signaling in breast cancer. Progesterone receptors regulate signal transducers and activators of transcription expression and activity. J Biol Chem. 1998;273:31317-31326.
29. Stoecklin E, Wissler M, Schaetzle D, Pfitzner E, Groner B. Interactions in the transcriptional regulation exerted by Stat5 and by members of the steroid hormone receptor family. J Steroid Biochem Mol Biol. 1999;69:195-204.

30. Udy GB, Towers RP, Snell RG, Wilkins RJ, Park SH, Ram PA, Waxman DJ, Davey HW. Requirement of STAT5b for sexual dimorphism of body growth rates and liver gene expression. Proc Natl Acad Sci U S A. 1997;94:7239-7244.
31. Teglund S, McKay C, Schuetz E, van Deursen JM, Stravopodis D, Wang D, Brown M, Bodner S, Grosveld G, Ihle JN. Stat5a and Stat5b proteins have essential and nonessential, or redundant, roles in cytokine responses. Cell. 1998;93:841-850.
32. Liu X, Robinson GW, Wagner KU, Garrett L, Wynshaw-Boris A, Hennighausen L. Stat5a is mandatory for adult mammary gland development and lactogenesis. Genes Dev. 1997;11:179-186.
33. Liu X, Gallego MI, Smith GH, Robinson GW, Hennighausen L. Functional rescue of Stat5a-null mammary tissue through the activation of compensating signals including Stat5b. Cell Growth Differ. 1998;9:795-803.
34. Neubauer H, Cumano A, Muller M, Wu H, Huffstadt U, Pfeffer K. Jak2 deficiency defines an essential developmental checkpoint in definitive hematopoiesis. Cell. 1998;93:397-409.
35. Parganas E, Wang D, Stravopodis D, Topham DJ, Marine JC, Teglund S, Vanin EF, Bodner S, Colamonici OR, van Deursen JM, Grosveld G, Ihle JN. Jak2 is essential for signaling through a variety of cytokine receptors. Cell. 1998;93:385-395.
36. Akira S. Functional roles of STAT family proteins: lessons from knockout mice. Stem Cells. 1999;17:138-146.
37. Oike Y, Takakura N, Hata A, Kaname T, Akizuki M, Yamaguchi Y, Yasue H, Araki K, Yamamura K, Suda T. Mice homozygous for a truncated form of CREB-binding protein exhibit defects in hematopoiesis and vasculo-angiogenesis. Blood. 1999;93:2771-2779.
38. Yao TP, Oh SP, Fuchs M, Zhou ND, Ch'ng LE, Newsome D, Bronson RT, Li E, Livingston DM, Eckner R. Gene dosage-dependent embryonic development and proliferation defects in mice lacking the transcriptional integrator p300. Cell. 1998;93:361-372.
39. Nguyen H, Hiscott J, Pitha PM. The growing family of interferon regulatory factors. Cytokine Growth Factor Rev. 1997;8:293-312.
40. Too CK. Induction of Sp1 activity by prolactin and interleukin-2 in Nb2 T- cells: differential association of Sp1-DNA complexes with Stats. Mol Cell Endocrinol. 1997;129:7-16.
41. Yang M, Hosokawa Y, Kaneko S, Tanaka M, Nakashima K. Structure and characterization of rat cyclin D3 promoter. Gene. 1996;181:153-159.
42. Yang M, Nomura H, Hu Y, Kaneko S, Kaneko H, Tanaka M, Nakashima K. Prolactin-induced expression of TATA-less cyclin D3 gene is mediated by Sp1 and AP2. Biochem Mol Biol Int. 1998;44:51-58.
43. Wartmann M, Cella N, Hofer P, Groner B, Liu X, Hennighausen L, Hynes NE. Lactogenic hormone activation of Stat5 and transcription of the beta- casein gene in mammary epithelial cells is independent of p42 ERK2 mitogen-activated protein kinase activity. J Biol Chem. 1996;271:31863-31868.
44. Pellegrini S, Dusanter-Fourt I. The structure, regulation and function of the Janus kinases (JAKs) and the signal transducers and activators of transcription (STATs). Eur J Biochem. 1997;248:615-633.
45. Darnell JE, Jr. STATs and gene regulation. Science. 1997;277:1630-1635.
46. Wakao H, Gouilleux F, Groner B. Mammary gland factor (MGF) is a novel member of the cytokine regulated transcription factor gene family and confers the prolactin response. Embo J. 1994;13:2182-2191.
47. DaSilva L, Rui H, Erwin RA, Zack Howard OM, Kirken RA, Malabarba MG, Hackett RH, Larner AC, Farrar WL. Prolactin recruits STAT1, STAT3 and STAT5 independent of conserved receptor tyrosines TYR402, TYR479, TYR515 and TYR580. Mol Cell Endocrinol. 1996;117:131-140.

48. Kirken RA, Malabarba MG, Xu J, Liu X, Farrar WL, Hennighausen L, Larner AC, Grimley PM, Rui H. Prolactin stimulates serine/tyrosine phosphorylation and formation of heterocomplexes of multiple Stat5 isoforms in Nb2 lymphocytes. J Biol Chem. 1997;272:14098-14103.
49. Mayr S, Welte T, Windegger M, Lechner J, May P, Heinrich PC, Horn F, Doppler W. Selective coupling of STAT factors to the mouse prolactin receptor. Eur J Biochem. 1998;258:784-793.
50. Schaber JD, Fang H, Xu J, Grimley PM, Rui H. Prolactin activates Stat1 but does not antagonize Stat1 activation and growth inhibition by type I interferons in human breast cancer cells. Cancer Res. 1998;58:1914-1919.
51. Vinkemeier U, Cohen SL, Moarefi I, Chait BT, Kuriyan J, Darnell JE, Jr. DNA binding of *in vitro* activated Stat1 alpha, Stat1 beta and truncated Stat1: interaction between NH2-terminal domains stabilizes binding of two dimers to tandem DNA sites. Embo J. 1996;15:5616-5626.
52. John S, Vinkemeier U, Soldaini E, Darnell JE, Jr., Leonard WJ. The significance of tetramerization in promoter recruitment by Stat5. Mol Cell Biol. 1999;19:1910-1918.
53. Strehlow I, Schindler C. Amino-terminal signal transducer and activator of transcription (STAT) domains regulate nuclear translocation and STAT deactivation. J Biol Chem. 1998;273:28049-28056.
54. Becker S, Groner B, Muller CW. Three-dimensional structure of the Stat3beta homodimer bound to DNA. Nature. 1998;394:145-151.
55. Chen X, Vinkemeier U, Zhao Y, Jeruzalmi D, Darnell JE, Jr., Kuriyan J. Crystal structure of a tyrosine phosphorylated STAT-1 dimer bound to DNA. Cell. 1998;93:827-839.
56. Pezet A, Ferrag F, Kelly PA, Edery M. Tyrosine docking sites of the rat prolactin receptor required for association and activation of stat5. J Biol Chem. 1997;272:25043-25050.
57. Stark GR, Kerr IM, Williams BR, Silverman RH, Schreiber RD. How cells respond to interferons. Annu Rev Biochem. 1998;67:227-264.
58. Shuai K, Horvath CM, Huang LH, Qureshi SA, Cowburn D, Darnell JE, Jr. Interferon activation of the transcription factor Stat91 involves dimerization through SH2-phosphotyrosyl peptide interactions. Cell. 1994;76:821-828.
59. Gouilleux F, Wakao H, Mundt M, Groner B. Prolactin induces phosphorylation of Tyr694 of Stat5 (MGF), a prerequisite for DNA binding and induction of transcription. Embo J. 1994;13:4361-4369.
60. Ng J, Cantrell D. STAT3 is a serine kinase target in T lymphocytes. Interleukin 2 and T cell antigen receptor signals converge upon serine 727. J Biol Chem. 1997;272:24542-24549.
61. Horvath CM, Darnell JE, Jr. The antiviral state induced by alpha interferon and gamma interferon requires transcriptionally active Stat1 protein. J Virol. 1996;70:647-650.
62. Bromberg JF, Horvath CM, Wen Z, Schreiber RD, Darnell JE, Jr. Transcriptionally active Stat1 is required for the antiproliferative effects of both interferon alpha and interferon gamma. Proc Natl Acad Sci U S A. 1996;93:7673-7678.
63. Zhang X, Blenis J, Li HC, Schindler C, Chen-Kiang S. Requirement of serine phosphorylation for formation of STAT-promoter complexes. Science. 1995;267:1990-1994.
64. Wen Z, Darnell JE, Jr. Mapping of Stat3 serine phosphorylation to a single residue (727) and evidence that serine phosphorylation has no influence on DNA binding of Stat1 and Stat3. Nucleic Acids Res. 1997;25:2062-2067.
65. Wang D, Moriggl R, Stravopodis D, Carpino N, Marine JC, Teglund S, Feng J, Ihle JN. A small amphipathic alpha-helical region is required for transcriptional activities and proteasome-dependent turnover of the tyrosine-phosphorylated stat5. Embo J. 2000;19:392-399.
66. Decker T, Kovarik P, Meinke A. GAS elements: a few nucleotides with a major impact on cytokine-induced gene expression. J Interferon Cytokine Res. 1997;17:121-134.

67. Ihle JN. STATs: signal transducers and activators of transcription. Cell. 1996;84:331-334.
68. Darnell JE, Jr., Kerr IM, Stark GR. Jak-STAT pathways and transcriptional activation in response to IFNs and other extracellular signaling proteins. Science. 1994;264:1415-1421.
69. Verdier F, Rabionet R, Gouilleux F, Beisenherz-Huss C, Varlet P, Muller O, Mayeux P, Lacombe C, Gisselbrecht S, Chretien S. A sequence of the CIS gene promoter interacts preferentially with two associated STAT5A dimers: a distinct biochemical difference between STAT5A and STAT5B. Mol Cell Biol. 1998;18:5852-5860.
70. Leaman DW, Leung S, Li X, Stark GR. Regulation of STAT-dependent pathways by growth factors and cytokines. Faseb J. 1996;10:1578-1588.
71. Schindler C, Darnell JE, Jr. Transcriptional responses to polypeptide ligands: the JAK-STAT pathway. Annu Rev Biochem. 1995;64:621-651.
72. Müller M, Laxton C, Briscoe J, Schindler C, Improta T, Darnell JE, Jr., Stark GR, Kerr IM. Complementation of a mutant cell line: central role of the 91 kDa polypeptide of ISGF3 in the interferon-alpha and -gamma signal transduction pathways. Embo J. 1993;12:4221-4228.
73. Yu-Lee LY, Luo G, Book ML, Morris SM. Lactogenic hormone signal transduction. Biol Reprod. 1998;58:295-301.
74. Wang Y, O'Neal KD, Yu-Lee L. Multiple prolactin (PRL) receptor cytoplasmic residues and Stat1 mediate PRL signaling to the interferon regulatory factor-1 promoter. Mol Endocrinol. 1997;11:1353-1364.
75. Wang YF, Yu-Lee LY. Multiple stat complexes interact at the interferon regulatory factor-1 interferon-gamma activation sequence in prolactin-stimulated Nb2 T cells. Mol Cell Endocrinol. 1996;121:19-28.
76. Stevens AM, Wang YF, Sieger KA, Lu HF, Yu-Lee LY. Biphasic transcriptional regulation of the interferon regulatory factor- 1 gene by prolactin: involvement of gamma-interferon-activated sequence and Stat-related proteins. Mol Endocrinol. 1995;9:513-525.
77. Durbin JE, Hackenmiller R, Simon MC, Levy DE. Targeted disruption of the mouse Stat1 gene results in compromised innate immunity to viral disease. Cell. 1996;84:443-450.
78. Meraz MA, White JM, Sheehan KC, Bach EA, Rodig SJ, Dighe AS, Kaplan DH, Riley JK, Greenlund AC, Campbell D, Carver-Moore K, DuBois RN, Clark R, Aguet M, Schreiber RD. Targeted disruption of the Stat1 gene in mice reveals unexpected physiologic specificity in the JAK-STAT signaling pathway. Cell. 1996;84:431-442.
79. Bromberg JF, Wrzeszczynska MH, Devgan G, Zhao Y, Pestell RG, Albanese C, Darnell JE, Jr. Stat3 as an oncogene. Cell. 1999;98:295-303.
80. Takeda K, Noguchi K, Shi W, Tanaka T, Matsumoto M, Yoshida N, Kishimoto T, Akira S. Targeted disruption of the mouse Stat3 gene leads to early embryonic lethality. Proc Natl Acad Sci U S A. 1997;94:3801-3804.
81. Nakajima K, Yamanaka Y, Nakae K, Kojima H, Ichiba M, Kiuchi N, Kitaoka T, Fukada T, Hibi M, Hirano T. A central role for Stat3 in IL-6-induced regulation of growth and differentiation in M1 leukemia cells. Embo J. 1996;15:3651-3658.
82. Minami M, Inoue M, Wei S, Takeda K, Matsumoto M, Kishimoto T, Akira S. STAT3 activation is a critical step in gp130-mediated terminal differentiation and growth arrest of a myeloid cell line. Proc Natl Acad Sci U S A. 1996;93:3963-3966.
83. Chapman RS, Lourenco PC, Tonner E, Flint DJ, Selbert S, Takeda K, Akira S, Clarke AR, Watson CJ. Suppression of epithelial apoptosis and delayed mammary gland involution in mice with a conditional knockout of Stat3. Genes Dev. 1999;13:2604-2616.
84. Zhong Z, Wen Z, Darnell JE, Jr. Stat3: a STAT family member activated by tyrosine phosphorylation in response to epidermal growth factor and interleukin-6. Science. 1994;264:95-98.
85. Schaefer TS, Sanders LK, Nathans D. Cooperative transcriptional activity of Jun and Stat3 beta, a short form of Stat3. Proc Natl Acad Sci U S A. 1995;92:9097-9101.

86. Caldenhoven E, van Dijk TB, Solari R, Armstrong J, Raaijmakers JAM, Lammers JWJ, Koenderman L, de Groot RP. STAT3beta, a splice variant of transcription factor STAT3, is a dominant negative regulator of transcription. J Biol Chem. 1996;271:13221-13227.
87. Groner B, Fritsche M, Stoecklin E, Berchtold S, Merkle C, Moriggl R, Pfitzner E. The transactivation potential of Stat5 is regulated through its DNA binding activity and interactions with heterologous transcription factors. Proceedings of the International Conference on Growth Hormone Action, Boston, 1999. 2000;in press.
88. Liu X, Robinson GW, Gouilleux F, Groner B, Hennighausen L. Cloning and expression of Stat5 and an additional homologue (Stat5b) involved in prolactin signal transduction in mouse mammary tissue. Proc Natl Acad Sci U S A. 1995;92:8831-8835.
89. Moriggl R, Gouilleux-Gruart V, Jahne R, Berchtold S, Gartmann C, Liu X, Hennighausen L, Sotiropoulos A, Groner B, Gouilleux F. Deletion of the carboxyl-terminal transactivation domain of MGF-Stat5 results in sustained DNA binding and a dominant negative phenotype. Mol Cell Biol. 1996;16:5691-5700.
90. Wang D, Stravopodis D, Teglund S, Kitazawa J, Ihle JN. Naturally occurring dominant negative variants of Stat5. Mol Cell Biol. 1996;16:6141-6148.
91. Bovolenta C, Testolin L, Benussi L, Lievens PM, Liboi E. Positive selection of apoptosis-resistant cells correlates with activation of dominant-negative STAT5. J Biol Chem. 1998;273:20779-20784.
92. Garimorth K, Welte T, Doppler W. Generation of carboxy-terminally deleted forms of STAT5 during preparation of cell extracts. Exp Cell Res. 1999;246:148-151.
93. Azam M, Lee C, Strehlow I, Schindler C. Functionally distinct isoforms of STAT5 are generated by protein processing. Immunity. 1997;6:691-701.
94. Lee C, Piazza F, Brutsaert S, Valens J, Strehlow I, Jarošinski M, Saris C, Schindler C. Characterization of the Stat5 protease. J Biol Chem. 1999;274:26767-26775.
95. Yamashita H, Xu J, Erwin RA, Farrar WL, Kirken RA, Rui H. Differential control of the phosphorylation state of proline-juxtaposed serine residues Ser725 of Stat5a and Ser730 of Stat5b in prolactin- sensitive cells. J Biol Chem. 1998;273:30218-30224.
96. Moriggl R, Topham DJ, Teglund S, Sexl V, McKay C, Wang D, Hoffmeyer A, van Deursen J, Sangster MY, Bunting KD, Grosveld GC, Ihle JN. Stat5 is required for IL-2-induced cell cycle progression of peripheral T cells. Immunity. 1999;10:249-259.
97. Feldman GM, Rosenthal LA, Liu X, Hayes MP, Wynshaw-Boris A, Leonard WJ, Hennighausen L, Finbloom DS. STAT5A-deficient mice demonstrate a defect in granulocyte-macrophage colony-stimulating factor-induced proliferation and gene expression. Blood. 1997;90:1768-1776.
98. Park SH, Liu X, Hennighausen L, Davey HW, Waxman DJ. Distinctive roles of STAT5a and STAT5b in sexual dimorphism of hepatic P450 gene expression. Impact of STAT5a gene disruption. J Biol Chem. 1999;274:7421-7430.
99. Davey HW, Park SH, Grattan DR, McLachlan MJ, Waxman DJ. STAT5b-deficient mice are growth hormone pulse-resistant. Role of STAT5b in sex-specific liver p450 expression. J Biol Chem. 1999;274:35331-35336.
100. Imada K, Bloom ET, Nakajima H, Horvath-Arcidiacono JA, Udy GB, Davey HW, Leonard WJ. Stat5b is essential for natural killer cell-mediated proliferation and cytolytic activity. J Exp Med. 1998;188:2067-2074.
101. Ogryzko VV, Schiltz RL, Russanova V, Howard BH, Nakatani Y. The transcriptional coactivators p300 and CBP are histone acetyltransferases. Cell. 1996;87:953-959.
102. Bannister AJ, Kouzarides T. The CBP co-activator is a histone acetyltransferase. Nature. 1996;384:641-643.
103. Eckner R, Ewen ME, Newsome D, Gerdes M, DeCaprio JA, Lawrence JB, Livingston DM. Molecular cloning and functional analysis of the adenovirus E1A- associated 300-kD protein (p300) reveals a protein with properties of a transcriptional adaptor. Genes Dev. 1994;8:869-884.

104. Arias J, Alberts AS, Brindle P, Claret FX, Smeal T, Karin M, Feramisco J, Montminy M. Activation of cAMP and mitogen responsive genes relies on a common nuclear factor. Nature. 1994;370:226-229.
105. Chrivia JC, Kwok RP, Lamb N, Hagiwara M, Montminy MR, Goodman RH. Phosphorylated CREB binds specifically to the nuclear protein CBP. Nature. 1993;365:855-859.
106. Kwok RP, Lundblad JR, Chrivia JC, Richards JP, Bachinger HP, Brennan RG, Roberts SG, Green MR, Goodman RH. Nuclear protein CBP is a coactivator for the transcription factor CREB. Nature. 1994;370:223-226.
107. Eckner R. p300 and CBP as transcriptional regulators and targets of oncogenic events. Biol Chem. 1996;377:685-688.
108. Janknecht R, Hunter T. Transcription. A growing coactivator network. Nature. 1996;383:22-23.
109. Chakravarti D, LaMorte VJ, Nelson MC, Nakajima T, Schulman IG, Juguilon H, Montminy M, Evans RM. Role of CBP/P300 in nuclear receptor signalling. Nature. 1996;383:99-103.
110. Kamei Y, Xu L, Heinzel T, Torchia J, Kurokawa R, Gloss B, Lin SC, Heyman RA, Rose DW, Glass CK, Rosenfeld MG. A CBP integrator complex mediates transcriptional activation and AP-1 inhibition by nuclear receptors. Cell. 1996;85:403-414.
111. Pfitzner E, Jahne R, Wissler M, Stoecklin E, Groner B. p300/CREB-binding protein enhances the prolactin-mediated transcriptional induction through direct interaction with the transactivation domain of Stat5, but does not participate in the Stat5- mediated suppression of the glucocorticoid response. Mol Endocrinol. 1998;12:1582-1593.
112. Gingras S, Simard J, Groner B, Pfitzner E. p300/CBP is required for transcriptional induction by interleukin-4 and interacts with Stat6. Nucleic Acids Res. 1999;27:2722-2729.
113. Bhattacharya S, Eckner R, Grossman S, Oldread E, Arany Z, D'Andrea A, Livingston DM. Cooperation of Stat2 and p300/CBP in signalling induced by interferon- alpha. Nature. 1996;383:344-347.
114. Paulson M, Pisharody S, Pan L, Guadagno S, Mui AL, Levy DE. Stat protein transactivation domains recruit p300/CBP through widely divergent sequences. J Biol Chem. 1999;274:25343-25349.
115. Horvai AE, Xu L, Korzus E, Brard G, Kalafus D, Mullen TM, Rose DW, Rosenfeld MG, Glass CK. Nuclear integration of JAK/STAT and Ras/AP-1 signaling by CBP and p300. Proc Natl Acad Sci U S A. 1997;94:1074-1079.
116. Zhang JJ, Vinkemeier U, Gu W, Chakravarti D, Horvath CM, Darnell JE, Jr. Two contact regions between Stat1 and CBP/p300 in interferon gamma signaling. Proc Natl Acad Sci U S A. 1996;93:15092-15096.
117. Lechner J, Welte T, Doppler W. Mechanism of interaction between the glucocorticoid receptor and Stat5: role of DNA-binding. Immunobiology. 1997;198:112-123.
118. Lechner J, Welte T, Tomasi JK, Bruno P, Cairns C, Gustafsson J, Doppler W. Promoter-dependent synergy between glucocorticoid receptor and Stat5 in the activation of beta-casein gene transcription. J Biol Chem. 1997;272:20954-20960.
119. Stoecklin E, Wissler M, Gouilleux F, Groner B. Functional interactions between Stat5 and the glucocorticoid receptor. Nature. 1996;383:726-728.
120. Stoecklin E, Wissler M, Moriggl R, Groner B. Specific DNA binding of Stat5, but not of glucocorticoid receptor, is required for their functional cooperation in the regulation of gene transcription. Mol Cell Biol. 1997;17:6708-6716.
121. Cella N, Groner B, Hynes NE. Characterization of Stat5a and Stat5b homodimers and heterodimers and their association with the glucocortiocoid receptor in mammary cells. Mol Cell Biol. 1998;18:1783-1792.
122. Luo G, Yu-Lee L. Transcriptional inhibition by Stat5. Differential activities at growth-related versus differentiation-specific promoters. J Biol Chem. 1997;272:26841-26849.

123. Luo G, Yu-Lee L. Stat5b inhibits NFkappaB-mediated signaling. Mol Endocrinol. 2000;14:114-123.
124. Ganguly TC, Liu Y, Hyde JF, Hagenbuch B, Meier PJ, Vore M. Prolactin increases Na+/ taurocholate cotransport in isolated hepatocytes from postpartum rats and ovariectomized rats. J Pharmacol Exp Ther. 1993;267:82-87.
125. Ganguly TC, O'Brien ML, Karpen SJ, Hyde JF, Suchy FJ, Vore M. Regulation of the rat liver sodium-dependent bile acid cotransporter gene by prolactin. Mediation of transcriptional activation by Stat5. J Clin Invest. 1997;99:2906-2914.
126. Møldrup A, Petersen ED, Nielsen JH. Effects of sex and pregnancy hormones on growth hormone and prolactin receptor gene expression in insulin-producing cells. Endocrinology. 1993;133:1165-1172.
127. Galsgaard ED, Nielsen JH, Moldrup A. Regulation of prolactin receptor (PRLR) gene expression in insulin- producing cells. Prolactin and growth hormone activate one of the rat PRL-R gene promoters via STAT5a and STAT5b. J Biol Chem. 1999;274:18686-18692.
128. Yu-Lee LY. Molecular actions of prolactin in the immune system. Proc Soc Exp Biol Med. 1997;215:35-52.
129. Duncan GS, Mittrucker HW, Kägi D, Matsuyama T, Mak TW. The transcription factor interferon regulatory factor-1 is essential for natural killer cell function *in vivo*. J Exp Med. 1996;184:2043-2048.
130. Yu-Lee LY, Hrachovy JA, Stevens AM, Schwarz LA. Interferon-regulatory factor 1 is an immediate-early gene under transcriptional regulation by prolactin in Nb2 T cells. Mol Cell Biol. 1990;10:3087-3094.
131. Clevenger CV, Sillman AL, Hanley-Hyde J, Prystowsky MB. Requirement for prolactin during cell cycle regulated gene expression in cloned T-lymphocytes. Endocrinology. 1992;130:3216-3222.
132. Stevens AM, Yu-Lee LY. The transcription factor interferon regulatory factor-1 is expressed during both early G1 and the G1/S transition in the prolactin-induced lymphocyte cell cycle. Mol Endocrinol. 1992;6:2236-2243.
133. Jabbour HN, Critchley HOD, Yu-Lee LY, Boddy SC. Localization of interferon regulatory factor-1 (IRF-1) in nonpregnant human endometrium: expression of IRF-1 is upregulated by prolactin during the secretory phase of the menstrual cycle. J Clin Endocrinol Metab. 1999;84:4260-4265.
134. Jabbour HN, Critchley HOD, Boddy SC. Expression of functional prolactin receptors in nonpregnant human endometrium: janus kinase-2, signal transducer and activator of transcription-1 (STAT1), and STAT5 proteins are phosphorylated after stimulation with prolactin. J Clin Endocrinol Metab. 1998;83:2545-2553.
135. Yu-Lee LY. Prolactin stimulates transcription of growth-related genes in Nb2 T lymphoma cells. Mol Cell Endocrinol. 1990;68:21-28.
136. Too CK. Differential expression of elongation factor-2, alpha4 phosphoprotein and Cdc5-like protein in prolactin-dependent/independent rat lymphoid cells. Mol Cell Endocrinol. 1997;131:221-232.
137. Too CK, Knee R, Pinette AL, Li AW, Murphy PR. Prolactin induces expression of FGF-2 and a novel FGF-responsive NonO/p54nrb-related mRNA in rat lymphoma cells. Mol Cell Endocrinol. 1998;137:187-195.
138. Wilson TM, Yu-Lee LY, Kelley MR. Coordinate gene expression of luteinizing hormone-releasing hormone (LHRH) and the LHRH-receptor after prolactin stimulation in the rat Nb2 T-cell line: implications for a role in immunomodulation and cell cycle gene expression. Mol Endocrinol. 1995;9:44-53.
139. Hosokawa Y, Yang M, Kaneko S, Tanaka M, Nakashima K. Prolactin induces switching of T-cell receptor gene expression from alpha to gamma in rat Nb2 pre-T lymphoma cells(1). Biochem Biophys Res Commun. 1996;220:958-962.

140. Axtell SM, Truong TM, O'Neal KD, Yu-Lee LY. Characterization of a prolactin-inducible gene, clone 15, in T cells. Mol Endocrinol. 1995;9:312-318.
141. Morris SM, Anaya P, Xiang X, Morris NR, May GS, Yu-Lee L. A prolactin-inducible T cell gene product is structurally similar to the Aspergillus nidulans nuclear movement protein NUDC. Mol Endocrinol. 1997;11:229-236.
142. Yang M, Hosokawa Y, Hu Y, Kaneko S, Kaneko H, Tanaka M, Nakashima K. Cloning and functional analysis of rat cyclin D2 promoter: multiple prolactin-responsive elements. Biochem Mol Biol Int. 1997;43:749-754.
143. Hosokawa Y, Yang M, Kaneko S, Tanaka M, Nakashima K. Synergistic gene expressions of cyclin E, cdk2, cdk5 and E2F-1 during the prolactin-induced G1/S transition in rat Nb2 pre-T lymphoma cells. Biochem Mol Biol Int. 1995;37:393-399.
144. Leff MA, Buckley DJ, Krumenacker JS, Reed JC, Miyashita T, Buckley AR. Rapid modulation of the apoptosis regulatory genes, bcl-2 and bax by prolactin in rat Nb2 lymphoma cells. Endocrinology. 1996;137:5456-5462.
145. Borg KE, Zhang M, Hegge D, Stephen RL, Buckley DJ, Magnuson NS, Buckley AR. Prolactin regulation of pim-1 expression: positive and negative promoter elements. Endocrinology. 1999;140:5659-5668.
146. Buckley AR, Buckley DJ, Leff MA, Hoover DS, Magnuson NS. Rapid induction of pim-1 expression by prolactin and interleukin-2 in rat Nb2 lymphoma cells. Endocrinology. 1995;136:5252-5259.
147. Krumenacker JS, Buckley DJ, Leff MA, McCormack JT, de Jong G, Gout PW, Reed JC, Miyashita T, Magnuson NS, Buckley AR. Prolactin-regulated apoptosis of Nb2 lymphoma cells: pim-1, bcl-2, and bax expression. Endocrine. 1998;9:163-170.
148. Nanbu-Wakao R, Fujitani Y, Masuho Y, Muramatu M, Wakao H. Prolactin enhances CCAAT enhancer-binding protein-beta (C/EBP beta) and peroxisome proliferator-activated receptor gamma (PPAR gamma) messenger RNA expression and stimulates adipogenic conversion of NIH-3T3 cells. Mol Endocrinol. 2000;14:307-316.
149. Raught B, Liao WS, Rosen JM. Developmentally and hormonally regulated CCAAT/enhancer-binding protein isoforms influence beta-casein gene expression. Mol Endocrinol. 1995;9:1223-1232.
150. Seagroves TN, Krnacik S, Raught B, Gay J, Burgess-Beusse B, Darlington GJ, Rosen JM. C/EBPbeta, but not C/EBPalpha, is essential for ductal morphogenesis, lobuloalveolar proliferation, and functional differentiation in the mouse mammary gland. Genes Dev. 1998;12:1917-1928.
151. Robinson GW, Johnson PF, Hennighausen L, Sterneck E. The C/EBPbeta transcription factor regulates epithelial cell proliferation and differentiation in the mammary gland. Genes Dev. 1998;12:1907-1916.
152. Doppler W, Welte T, Philipp S. CCAAT/enhancer-binding protein isoforms beta and delta are expressed in mammary epithelial cells and bind to multiple sites in the beta-casein gene promoter. J Biol Chem. 1995;270:17962-17969.
153. Sterneck E, Tessarollo L, Johnson PF. An essential role for C/EBPbeta in female reproduction. Genes Dev. 1997;11:2153-2162.
154. Tanaka T, Yoshida N, Kishimoto T, Akira S. Defective adipocyte differentiation in mice lacking the C/EBPbeta and/or C/EBPdelta gene. Embo J. 1997;16:7432-7443.
155. Wang ND, Finegold MJ, Bradley A, Ou CN, Abdelsayed SV, Wilde MD, Taylor LR, Wilson DR, Darlington GJ. Impaired energy homeostasis in C/EBP alpha knockout mice. Science. 1995;269:1108-1112.
156. Barak Y, Nelson MC, Ong ES, Jones YZ, Ruiz-Lozano P, Chien KR, Koder A, Evans RM. PPAR gamma is required for placental, cardiac, and adipose tissue development. Mol Cell. 1999;4:585-595.

157. Kubota N, Terauchi Y, Miki H, Tamemoto H, Yamauchi T, Komeda K, Satoh S, Nakano R, Ishii C, Sugiyama T, Eto K, Tsubamoto Y, Okuno A, Murakami K, Sekihara H, Hasegawa G, Naito M, Toyoshima Y, Tanaka S, Shiota K, Kitamura T, Fujita T, Ezaki O, Aizawa S, Nagai R, Tobe K, Kimura S, Kadowaki T. PPAR gamma mediates high-fat diet-induced adipocyte hypertrophy and insulin resistance. Mol Cell. 1999;4:597-609.
158. Rosen ED, Sarraf P, Troy AE, Bradwin G, Moore K, Milstone DS, Spiegelman BM, Mortensen RM. PPAR gamma is required for the differentiation of adipose tissue *in vivo* and *in vitro*. Mol Cell. 1999;4:611-617.
159. Lowell BB. PPARgamma: an essential regulator of adipogenesis and modulator of fat cell function. Cell. 1999;99:239-242.
160. McAveney KM, Gimble JM, Yu-Lee L. Prolactin receptor expression during adipocyte differentiation of bone marrow stroma. Endocrinology. 1996;137:5723-5726.
161. Horseman ND, Buntin JD. Regulation of pigeon cropmilk secretion and parental behaviors by prolactin. Annu Rev Nutr. 1995;15:213-238.
162. Pukac LA, Horseman ND. Regulation of cloned prolactin-inducible genes in pigeon crop. Mol Endocrinol. 1987;1:188-194.
163. Xu YH, Horseman ND. Nuclear proteins and prolactin-induced annexin Icp35 gene transcription. Mol Endocrinol. 1992;6:375-383.
164. Sidis Y, Horseman ND. Prolactin induces rapid p95/p70 tyrosine phosphorylation, and protein binding to GAS-like sites in the anx Icp35 and c-fos genes. Endocrinology. 1994;134:1979-1985.
165. Sun M, Liu Y, Gibb W. Distribution of annexin I and II in term human fetal membranes, decidua and placenta. Placenta. 1996;17:181-184.
166. Groner B, Altiok S, Meier V. Hormonal regulation of transcription factor activity in mammary epithelial cells. Mol Cell Endocrinol. 1994;100:109-114.
167. Burdon TG, Maitland KA, Clark AJ, Wallace R, Watson CJ. Regulation of the sheep beta-lactoglobulin gene by lactogenic hormones is mediated by a transcription factor that binds an interferon-gamma activation site-related element. Mol Endocrinol. 1994;8:1528-1536.
168. Li S, Rosen JM. Nuclear factor I and mammary gland factor (STAT5) play a critical role in regulating rat whey acidic protein gene expression in transgenic mice. Mol Cell Biol. 1995;15:2063-2070.
169. Djiane J, Daniel N, Bignon C, Paly J, Waters M, Vacher P, Dufy B. Prolactin receptor and signal transduction to milk protein genes. Proc Soc Exp Biol Med. 1994;206:299-303.
170. Ali S, Edery M, Pellegrini I, Lesueur L, Paly J, Djiane J, Kelly PA. The Nb2 form of prolactin receptor is able to activate a milk protein gene promoter. Mol Endocrinol. 1992;6:1242-1248.
171. Berlanga JJ, Garcia-Ruiz JP, Perrot-Applanat M, Kelly PA, Edery M. The short form of the prolactin (PRL) receptor silences PRL induction of the beta-casein gene promoter. Mol Endocrinol. 1997;11:1449-1457.
172. Marte BM, Meyer T, Stabel S, Standke GJ, Jaken S, Fabbro D, Hynes NE. Protein kinase C and mammary cell differentiation: involvement of protein kinase C alpha in the induction of beta-casein expression. Cell Growth Differ. 1994;5:239-247.
173. Chida D, Wakao H, Yoshimura A, Miyajima A. Transcriptional regulation of the beta-casein gene by cytokines: cross- talk between STAT5 and other signaling molecules. Mol Endocrinol. 1998;12:1792-1806.
174. Pircher TJ, Petersen H, Gustafsson JA, Haldosen LA. Extracellular signal-regulated kinase (ERK) interacts with signal transducer and activator of transcription (STAT) 5a. Mol Endocrinol. 1999;13:555-565.
175. Gao J, Horseman ND. Prolactin-independent modulation of the beta-casein response element by Erk2 MAP kinase. Cell Signal. 1999;11:205-210.

176. Winklehner-Jennewein P, Geymayer S, Lechner J, Welte T, Hansson L, Geley S, Doppler W. A distal enhancer region in the human beta-casein gene mediates the response to prolactin and glucocorticoid hormones. Gene. 1998;217:127-139.
177. Ali S, Chen Z, Lebrun JJ, Vogel W, Kharitonenkov A, Kelly PA, Ullrich A. PTP1D is a positive regulator of the prolactin signal leading to beta- casein promoter activation. Embo J. 1996;15:135-142.
178. Altiok S, Groner B. beta-Casein mRNA sequesters a single-stranded nucleic acid-binding protein which negatively regulates the beta-casein gene promoter. Mol Cell Biol. 1994;14:6004-6012.
179. Raught B, Khursheed B, Kazansky A, Rosen J. YY1 represses beta-casein gene expression by preventing the formation of a lactation-associated complex. Mol Cell Biol. 1994;14:1752-1763.
180. Schmitt-Ney M, Doppler W, Ball RK, Groner B. Beta-casein gene promoter activity is regulated by the hormone-mediated relief of transcriptional repression and a mammary-gland-specific nuclear factor. Mol Cell Biol. 1991;11:3745-3755.
181. Meier VS, Groner B. The nuclear factor YY1 participates in repression of the beta-casein gene promoter in mammary epithelial cells and is counteracted by mammary gland factor during lactogenic hormone induction. Mol Cell Biol. 1994;14:128-137.
182. Thomas MJ, Seto E. Unlocking the mechanisms of transcription factor YY1: are chromatin modifying enzymes the key? Gene. 1999;236:197-208.
183. Sicinski P, Donaher JL, Parker SB, Li T, Fazeli A, Gardner H, Haslam SZ, Bronson RT, Elledge SJ, Weinberg RA. Cyclin D1 provides a link between development and oncogenesis in the retina and breast. Cell. 1995;82:621-630.
184. Matsumura I, Kitamura T, Wakao H, Tanaka H, Hashimoto K, Albanese C, Downward J, Pestell RG, Kanakura Y. Transcriptional regulation of the cyclin D1 promoter by STAT5: its involvement in cytokine-dependent growth of hematopoietic cells. Embo J. 1999;18:1367-1377.
185. Kaneko H, Fujikawa T, Alam KS, Kaneko S, Tanaka M, Hibasami H, Nakashima K. Cooperative and differential effects of estrogen, prolactin, 22K and 20K human growth hormones on cyclin D1/PRAD1 gene expression in T-47D human breast cancer cells. Biochem Mol Biol Int. 1998;46:411-414.
186. Ruff SJ, Leers-Sucheta S, Melner MH, Cohen S. Induction and activation of Stat 5 in the ovaries of pseudopregnant rats. Endocrinology. 1996;137:4095-4099.
187. Russell DL, Richards JS. Differentiation-dependent prolactin responsiveness and stat (signal transducers and activators of transcription) signaling in rat ovarian cells. Mol Endocrinol. 1999;13:2049-2064.
188. Fero ML, Rivkin M, Tasch M, Porter P, Carow CE, Firpo E, Polyak K, Tsai LH, Broudy V, Perlmutter RM, Kaushansky K, Roberts JM. A syndrome of multiorgan hyperplasia with features of gigantism, tumorigenesis, and female sterility in p27(Kip1)-deficient mice. Cell. 1996;85:733-744.
189. Kiyokawa H, Kineman RD, Manova-Todorova KO, Soares VC, Hoffman ES, Ono M, Khanam D, Hayday AC, Frohman LA, Koff A. Enhanced growth of mice lacking the cyclin-dependent kinase inhibitor function of p27(Kip1). Cell. 1996;85:721-732.
190. Nakayama K, Ishida N, Shirane M, Inomata A, Inoue T, Shishido N, Horii I, Loh DY, Nakayama K. Mice lacking p27(Kip1) display increased body size, multiple organ hyperplasia, retinal dysplasia, and pituitary tumors. Cell. 1996;85:707-720.
191. Albarracin CT, Parmer TG, Duan WR, Nelson SE, Gibori G. Identification of a major prolactin-regulated protein as 20 alpha- hydroxysteroid dehydrogenase: coordinate regulation of its activity, protein content, and messenger ribonucleic acid expression. Endocrinology. 1994;134:2453-2460.

192. Martel C, Labrie C, Dupont E, Couet J, Trudel C, Rheaume E, Simard J, Luu-The V, Pelletier G, Labrie F. Regulation of 3 beta-hydroxysteroid dehydrogenase/delta 5-delta 4 isomerase expression and activity in the hypophysectomized rat ovary: interactions between the stimulatory effect of human chorionic gonadotropin and the luteolytic effect of prolactin. Endocrinology. 1990;127:2726-2737.
193. Martel C, Gagné D, Couet J, Labrie Y, Simard J, Labrie F. Rapid modulation of ovarian 3 beta-hydroxysteroid dehydrogenase/delta 5- delta 4 isomerase gene expression by prolactin and human chorionic gonadotropin in the hypophysectomized rat. Mol Cell Endocrinol. 1994;99:63-71.
194. Feltus FA, Groner B, Melner MH. Stat5-mediated regulation of the human type II 3beta-hydroxysteroid dehydrogenase/delta5-delta4 isomerase gene: activation by prolactin. Mol Endocrinol. 1999;13:1084-1093.
195. Russell DL, Norman RL, Dajee M, Liu X, Henninghausen L, Richards JS. Prolactin-induced activation and binding of stat proteins to the IL-6RE of the alpha 2-macroglobulin (alpha 2M) promoter: relation to the expression of alpha 2M in the rat ovary. Biol Reprod. 1996;55:1029-1038.
196. Dajee M, Kazansky AV, Raught B, Hocke GM, Fey GH, Richards JS. Prolactin induction of the alpha 2-Macroglobulin gene in rat ovarian granulosa cells: stat 5 activation and binding to the interleukin-6 response element. Mol Endocrinol. 1996;10:171-184.
197. Dajee M, Fey GH, Richards JS. Stat 5b and the orphan nuclear receptors regulate expression of the alpha2-macroglobulin (alpha2M) gene in rat ovarian granulosa cells. Mol Endocrinol. 1998;12:1393-1409.
198. Borth W. Alpha 2-macroglobulin, a multifunctional binding protein with targeting characteristics. Faseb J. 1992;6:3345-3353.
199. Perrot-Applanat M, Gualillo O, Pezet A, Vincent V, Edery M, Kelly PA. Dominant negative and cooperative effects of mutant forms of prolactin receptor. Mol Endocrinol. 1997;11:1020-1032.
200. Knoepfler PS, Eisenman RN. Sin meets NuRD and other tails of repression. Cell. 1999;99:447-450.
201. Linzer DI, Fisher SJ. The placenta and the prolactin family of hormones: regulation of the physiology of pregnancy. Mol Endocrinol. 1999;13:837-840.
202. Wennbo H, Gebre-Medhin M, Gritli-Linde A, Ohlsson C, Isaksson OG, Tornell J. Activation of the prolactin receptor but not the growth hormone receptor is important for induction of mammary tumors in transgenic mice. J Clin Invest. 1997;100:2744-2751.
203. Goffin V, Touraine P, Pichard C, Bernichtein S, Kelly PA. Should prolactin be reconsidered as a therapeutic target in human breast cancer? Mol Cell Endocrinol. 1999;151:79-87.
204. Lange CA, Richer JK, Horwitz KB. Hypothesis: Progesterone primes breast cancer cells for cross-talk with proliferative or antiproliferative signals. Mol Endocrinol. 1999;13:829-836.
205. Nagasawa H, Miura K, Niki K, Namiki H. Interrelationship between prolactin and progesterone in normal mammary gland growth in SHN virgin mice. Exp Clin Endocrinol. 1985;86:357-360.
206. Ward AC, Touw I, Yoshimura A. The Jak-Stat pathway in normal and perturbed hematopoiesis. Blood. 2000;95:19-29.
207. Morales P, Carretero MV, Geronimo H, Copin SG, Gaspar ML, Marcos MA, Martin-Perez J. Influence of prolactin on the differentiation of mouse B-lymphoid precursors. Cell Growth Differ. 1999;10:583-590.

Index

D

E

F

G

H

I

N

O

P

R